Green Synthesis and Emerging Applications of Frontier Nanomaterials

Edited by

Martin F. Desimone[1,2]
Rajshree B. Jotania[3]
Ratiram G. Chaudhary[4]

[1]Cátedra de Química Analítica Instrumental, IQUIMEFA-CONICET, Facultad de Farmacia y Bioquímica, Universidad de Buenos Aires, (1113) Junin 956 Piso 3. Buenos Aires. Argentina

[2]Instituto de Ciências Biológicas (ICB), Universidade Federal do Rio Grande

– FURG, Rio Grande, RS, Brazil

[3]Department of Physics, Electronics and Space science, University school of sciences, Gujarat University, Ahmedabad 380 009, India

[4]Post Graduate Department of Chemistry, Seth Kesarimal Porwal College of Arts, Science and Commerce, Kamptee, RTM Nagpur University, Nagpur, Maharashtra-441001, India

Published by **Materials Research Forum LLC**
Millersville, PA 17551, USA

Published as part of the book series
Materials Research Foundations
Volume 169 (2024)
ISSN 2471-8890 (Print)
ISSN 2471-8904 (Online)

Print ISBN 978-1-64490-326-1
eBook ISBN 978-1-64490-327-8

Distributed worldwide by

Materials Research Forum LLC
105 Springdale Lane
Millersville, PA 17551
USA
https://www.mrforum.com

Manufactured in the United States of America
10 9 8 7 6 5 4 3 2 1

Table of Contents

About the Editors

Preface

In the burgeoning field of nanotechnology, the green synthesis of nanoparticles has emerged as a sustainable and environmentally friendly alternative to traditional chemical methods. This approach utilizes natural resources, such as plant extracts, microorganisms, and biomolecules, to reduce the toxicity and environmental impact associated with conventional synthesis techniques.

This book titled "Green synthesis and emerging applications of frontier nanomaterials" offers a comprehensive overview of the latest advancements in this field. The chapters delve into the eco-friendly synthesis and applications of various nanoparticles, including silica, silver, iron, zinc, copper, nickel, ceria, carbon nanotubes, zirconium, rhodium, quantum dots, titanium, bismuth, and gold. Each chapter provides detailed information on the synthesis methods, characterization techniques, and potential applications of these nanomaterials.

Chapter 1 by Gabriel O. Ostapchuk *et al.*, illustrates the green synthesis of silica nanoparticles and their applications in drug delivery, catalysis, and biosensing. Given their exceptional properties and wide-ranging applications, silica nanoparticles have become invaluable materials in various fields. This chapter delves into the green synthesis methods for producing these nanoparticles, offering a sustainable and environmentally friendly alternative to traditional approaches.

Green synthesis, using plants, bacteria, or fungi, offers a sustainable alternative to traditional methods for producing silver nanoparticles (AgNPs). These natural agents reduce silver salts and stabilize the nanoparticles. The choice of synthesis method influences the size and shape of AgNPs. In Chapter 2, Fátima Ibarra *et al.*, provide a comprehensive overview of the current state of knowledge on green synthesis of AgNPs.

Chapter 3, led by Rajshree B. Jotania, explores the green synthesis and applications of iron-based nanomaterials. These materials have diverse uses in fields such as environmental remediation, food packaging, biosensing, and biomedical applications. This chapter highlights the benefits of green synthesis and the potential of iron-based nanomaterials for a sustainable future.

Chapter 4 by P. R. Bhilkar *et al.*, delves into the bioinspired synthesis of zinc-based nanomaterials and explores their applications in biomedical, environmental, and agricultural fields. This chapter provides a comprehensive overview of various biological sources used to synthesize zinc-based nanomaterials and discusses their potential applications. Additionally, it explores future perspectives and advancements in this area.

Chapter 5 by M. Deshmukh *et al.*, describes the bioinspired fabrication of copper nanoparticles and their applications. This chapter focuses on the biogenic synthesis of copper/copper oxide nanoparticles and their applications in antimicrobial activity, photocatalytic degradation, and plant growth responses.

Chapter 6 by Chiranjibi Dhakal *et al.*, explores the synthesis of nickel-based nanomaterials. This chapter discusses the synthesis methods for nickel nanoparticles (Ni-NPs) and nickel-based nanomaterials, focusing on both bottom-up and top-down approaches. The study emphasizes green synthesis as a sustainable approach for producing Ni-NPs. Biosynthesized Ni-NPs have diverse applications, including antimicrobial, antileishmanial, anticancer, anti-diabetic activities, drug delivery, battery electrodes, wastewater management, and biosensors.

Chapter 7 by N. B. Singh team, highlights the applications of cerium oxide nanoparticles (CeO_2 NPs) synthesized through green methods. Green synthesis offers a sustainable and environmentally friendly alternative to traditional chemical methods discussed in the chapter (precipitation, hydrothermal, solvothermal, and sol-gel methods). Following the synthesis methods, the chapter delves into the diverse biomedical applications of green-synthesized CeO_2 NPs. These include their antibacterial activity, antioxidant properties, potential for drug and gene delivery, and use in treating various diseases. Additionally, the chapter explores their use in cerium oxide-based biosensors and enzyme mimetic applications.

Chapter 8 by A.A. Abdala and collaborators explores the synthesis, environmental, and biomedical applications of carbon nanotubes (CNTs). The focus is on various synthesis methods of CNTs, including electrolysis, chemical vapor deposition, mechanical thermal, laser ablation, flame synthesis, arc discharge, and green synthesis. It also discusses the potential applications of CNTs in environmental remediation (e.g., removing organic solvents, oils, dyes, heavy metal ions, and pollutants) and biomedical fields.

Md. Ahad Ali and Md. Abu Bin Hasan Susan, in Chapter 9 present a comprehensive overview of the green synthesis and emerging applications of zirconium (Zr) and rhodium (Rh)-based nanomaterials. The chapter focuses on Zr-based nanomaterials, including metallic Zr NPs, ZrO_2 NPs, Zr-based metal-organic frameworks (MOFs), and other Zr-based nanostructures. It discusses their green synthesis methods and applications in environmental remediation (adsorption and catalysis), sensing, solar cells, and other fields. Additionally, the chapter delves into the green synthesis and applications of Rh-based nanomaterials. It highlights their potential in energy, biomedical applications, biosensing, and discusses future prospects and challenges in this area.

Chapter 10 by Yogita Sahu *et al.*, provides a comprehensive overview of the green synthesis of quantum dots (QDs). The chapter explores both traditional chemical methods (colloidal synthesis, thermal decomposition, hydrothermal/solvothermal synthesis) and green approaches (microbial synthesis, biological synthesis, plant extract-mediated synthesis, and microwave-assisted synthesis). It also discusses the characterization of QDs and their potential applications.

Chapter 11 by the Ajaya Kumar Singh group explores the synthesis and applications of titanium-based nanomaterials, focusing on titanium dioxide (TiO_2) nanoparticles. The chapter discusses various synthesis methods, including chemical methods (sol-gel, solvothermal, hydrothermal, micelle/inverse micelle, microwave, template, vapor deposition, electrochemical, oxidation, and sono-chemical) and green methods (using plant extracts, bacterial/fungal extracts, or other biological sources). It delves into the properties of TiO_2 NPs (crystal, optical, electrochemical, and thermal properties) and their diverse applications. These include nanomedicine (anticancer therapy), antimicrobial activity, agriculture, food preservation, fuel cells, paints and coatings, and wastewater treatment.

Chapter 12 by T. L. Lambat and colleagues explores the recent advancements in supercapacitors utilizing bismuth oxide nanomaterials (BiO NMs). BiO NMs are promising materials for energy storage devices due to their redox behaviour, superior charge storage capacity, and eco-friendly nature. This chapter discusses the structure, properties, and applications of BiO-based materials, as well as the various chemical methods for synthesizing bismuth oxide. It also delves into the application of bismuth oxide in supercapacitors and highlights future research directions in this field.

Chapter 13 by Narendra P. Singh and colleagues discuss the green synthesis and applications of ceria nanomaterials. Green synthesis methods, using natural resources, offer a sustainable approach for producing ceria nanomaterials. These materials have diverse applications in fields like catalysis, water treatment, energy, photocatalysis, and biomedical areas. The chapter discusses the principles of green synthesis, characterization techniques, and the potential of ceria nanomaterials for sustainable and innovative solutions.

Chapter 14, led by Shivani R. Pandya and collaborators, provides a comprehensive overview of green synthesis techniques for gold nanoparticles (AuNPs). This chapter explores the historical development of green chemistry in nanoparticle synthesis, focusing on the mechanisms of stabilization and reduction facilitated by phytochemicals and biomolecules. It also discusses analytical methods for characterizing green-synthesized AuNPs and their applications. The chapter highlights the benefits of green synthesis, including improved biocompatibility, cost-effectiveness, and environmental friendliness. It explores the diverse applications of AuNPs in biomedicine, environmental remediation, and industrial catalysis. While green synthesis offers promising advantages, challenges such as scalability, repeatability, and regulatory barriers remain. The chapter discusses the potential of using machine learning and artificial intelligence to optimize green synthesis processes. By addressing these challenges and exploring innovative approaches, green synthesis can contribute to sustainable nanoparticle production and technological advancements in various fields.

In conclusion, this book provides a valuable resource for researchers, scientists, and students interested in the green synthesis and applications of nanomaterials. The chapters offer detailed information on the synthesis methods, characterization techniques, and potential applications of various nanomaterials, highlighting their significance in addressing environmental and societal challenges.

We extend our sincere gratitude to all contributors for their invaluable cooperation and dedication in providing high-quality chapters. The insightful comments from our esteemed reviewers played a crucial role in enhancing the overall quality of this book.

We would also like to acknowledge the unwavering support of Mr. Thomas Wohlbier and the entire team at Materials Research Forum LLC, Millersville PA, USA throughout the publication process. Their continuous cooperation and expertise were instrumental in bringing this book to fruition.

We hope that this comprehensive volume on the dynamic field of green nanomaterials will provide readers with a thorough understanding of the immense potential and diverse applications beyond the development of environmentally friendly nanomaterials.

Editors: Professor Martin F. Desimone

        Professor Rajshree B. Jotania

        Professor Ratiram G. Chaudhary

Green Synthesis and Emerging Applications of Frontier Nanomaterials   Materials Research Forum LLC
Materials Research Foundations 169 (2024) 1-34   https://doi.org/10.21741/9781644903278-1

Chapter 1

Eco-friendly synthesis of silica nanoparticles and their applications

Gabriel O. Ostapchuk[1,2,3], Natalia A. Scilletta[3,4], Hernán Levy[1,2],
Martín F. Desimone[3,5,6*], Paolo N. Catalano[1,2,3*]

[1] Instituto de Nanociencia y Nanotecnología (CNEA - CONICET), Nodo Constituyentes, Av. Gral. Paz 1499 (B1650KNA), San Martín, Buenos Aires, Argentina

[2] Departamento de Micro y Nanotecnología, Gerencia de Desarrollo Tecnológico y Proyectos Especiales, Gerencia de Área de Investigación, Desarrollo e Innovación, Centro Atómico Constituyentes, Comisión Nacional de Energía Atómica, Av. Gral. Paz 1499 (B1650KNA), San Martín, Buenos Aires, Argentina

[3] Universidad de Buenos Aires, Facultad de Farmacia y Bioquímica, Departamento de Ciencias Químicas, Cátedra de Química Analítica Instrumental, Junín 954 (1113), Buenos Aires, Argentina

[4] Laboratorio de Dosimetría Personal y de Área, Gerencia de Seguridad Radiológica y Nuclear, Gerencia de Área de Seguridad Nuclear y Ambiente, Centro Atómico Ezeiza, Comisión Nacional de Energía Atómica. Cam. Real Presbítero González y Aragón 15 (B1802) Ezeiza, Buenos Aires, Argentina

[5]CONICET - Universidad de Buenos Aires, Instituto de Química y Metabolismo del Fármaco (IQUIMEFA), Junín 954 (1113), Buenos Aires, Argentina

[6]Instituto de Ciências Biológicas (ICB), Universidade Federal do Rio Grande – FURG, Rio Grande, RS, Brazil

* M.F.D: desimone@ffyb.uba.ar, * P. N. C.: catalano@cnea.gov.ar

Abstract

Since their discovery, silica nanoparticles have become valuable materials with wide-ranging applications in biomedicine, catalysis, and environmental remediation. Their properties, including high surface area, variable pore size, surface reactivity, stability, and low toxicity, make them attractive for several fields. Methods such as laser ablation and sol-gel synthesis, which encompass both physical and chemical processes, can be employed for their synthesis. However, these conventional approaches have drawbacks, including the use of dangerous chemical and costly procedures, which can lead to biological and environmental concerns. Growing demand has been observed in recent years for eco-friendly synthesis of silica nanoparticles which is economically-viable, safe, reliable, scalable, and enables control over particle size distribution. Green synthesis methods have gained importance owing to their ability to minimize the adverse effects associated with conventional approaches. The key advantage of these sustainable strategies is the absence of harmful byproducts throughout the synthesis. Utilizing biological, renewable, and sustainable resources, green strategies offer simple and cost-effective routes for producing silica nanoparticles, removing or decreasing the need for complex and time-consuming procedures. This chapter discusses various green synthesis approaches for obtaining silica nanoparticles and explores their applications. The utilization of green methods promotes the sustainable production of silica nanoparticles and advances the field of green nanotechnology.

Keywords

Silica Nanoparticles, Green Synthesis, Sustainable Resources, Biomedical Applications, Catalysis, Environmental Remediation

Contents

1.   Introduction

Recently, nanotechnology has garnered increased interest, holding the potential to enhance life's quality by catalyzing scientific and technological progress. This phenomenon has captivated the global research community, resonating across diverse disciplines and kindling innovation [1]. Objects crafted on the nanoscale exhibit an array of distinctive properties, spanning optics, magnetism, and electricity, due to their large specific surface area, high surface energy, and quantum confinement effects [2]. Harnessing these exceptional physicochemical attributes, nanoparticles pave the way for boundless opportunities across domains such as medicine [3], electronics [4], food production [5], cosmetics [6], and the chemical sector [7].

In the realm of nanotechnology, silicon dioxide (SiO_2) or silica nanoparticles (referred to as SiNPs) have emerged as highly promising assets owing to their exceptional physicochemical attributes. Their distinctive traits, encompassing convenient functionalization, adjustable pore size, high surface area, and biocompatibility, position them as versatile contenders for a broad spectrum of applications [8,9]. Employing a range of methodologies and reactant combinations, these nanoparticles can be tailored into a wide variety of shapes and sizes [10]. This spectrum embraces amorphous structures [11], spherical and solid formations [12], as well as elongated rod-like [13] and intricate mesoporous configurations [14]. Notably, the pore sizes span from small to large dimensions, unveiling a myriad of possibilities [15].

The substantial surface area of nano-silica renders it an optimal candidate for a multitude of purposes, ranging from reinforcing fillers in the rubber industry to bolstering agricultural crop growth, enhancing concrete properties, catalyzing the insecticide sector, invigorating the paint industry, and revolutionizing the food and medical domains. Additionally, nano-silica is unveiling its potential across various biological and biotechnological pursuits, spanning cancer treatment, drug delivery, enzyme immobilization, DNA transfection, and environmental bioremediation [10-16].

The fabrication of these nanoparticles can be achieved through either 'top-down' [17] or 'bottom-up' methodologies [18]. The 'top-down' strategy involves employing physical or mechanical techniques to downsize bulk materials into nanoscale structures. Widely used methods encompass laser ablation [19, 20], ball-milling, electric arc discharge [21], and the radio frequency plasma method [22, 23]. However, implementing these techniques to achieve the desired synthesis often requires significant time due to the challenges linked with attaining and sustaining thermal stability. This consequently entails high energy consumption and necessitates the utilization of large space-consuming equipment, as exemplified by tube furnaces [6]. In contrast, the bottom-up approach, encompassing chemical techniques for synthesizing silica nanoparticles, often offers benefits such as precise control over size, shape, and surface properties. Sol-gel synthesis [24], hydrothermal synthesis [25], and microemulsion techniques [26] are representative methods under this category, adept at generating uniform nanoparticles with tailored attributes. However, this approach also presents a series of challenges related to supplies, environment, resources, time, and energy.

Chemical synthesis often necessitates specific precursors, reagents, and solvents, which may not always be readily available or could entail considerable expenses. Furthermore, the production of these chemicals could yield environmental implications. Many chemical methods involve hazardous substances, potentially resulting in toxic byproducts and waste, which could lead to environmental pollution if not handled properly. Elevated temperatures, pressures, and prolonged reaction times are often prerequisites for chemical synthesis processes, contributing to heightened energy consumption, increased costs, and elevated carbon emissions [27]. These processes can generate waste, including excess reagents and byproducts, posing disposal challenges, particularly in large-scale production.

To encounter these challenges, researchers are actively exploring greener and more sustainable synthesis pathways within the chemical approach [28]. By adopting eco-friendly or green synthesis methods, it is possible to address several of the disadvantages associated with traditional chemical and physical techniques for obtaining silica nanoparticles [29].

Green synthesis's methods follow an innovative paradigm rooted in sustainable principles. Green synthesis signifies a departure from conventional chemical and physical methodologies towards environmentally benign and resource-efficient approaches. This approach prioritizes the use of renewable raw materials, reduced energy consumption, and the avoidance of toxic solvents and hazardous byproducts. The essence of green synthesis lies in its commitment to harmonizing scientific advancement with ecological responsibility, aiming to minimize the environmental footprint associated with nanoparticle production [29, 30].

Green synthesis methods stand out for their ability to minimize or eliminate hazardous chemicals, mitigating the production of toxic waste. This directly addresses the environmental concerns linked to conventional chemical methods. Operating under milder conditions, these approaches necessitate lower temperatures and shorter reaction times, effectively reducing energy consumption, costs, and carbon emissions. The emphasis on sustainability within green synthesis results in fewer byproducts and waste, aligning with the principles of sustainable chemistry and facilitating a cleaner production process. Green synthesis strategies promote streamlined processes with fewer steps, leveraging renewable resources and simplified procedures, facilitating more straightforward scalability. By adhering to these practices, eco-friendly approaches not only resonate with the ethos of sustainable chemistry but also offer a means to mitigate the drawbacks inherent in traditional chemical and physical synthesis techniques for silica nanoparticles [29, 30]. This chapter comprehensively addresses a diverse spectrum of green synthesis methods employed in the production of silica nanoparticles (Fig. 1), thoroughly exploring their multifaceted applications. Additionally, it investigates the role of these green methodologies in augmenting the sustainable production of silica nanoparticles, thereby propelling advancements in the domain of green nanotechnology.

Figure 1. Schematic diagram of the different resources used for the eco-friendly synthesis of SiNPs. Plants, agricultural wastes, bacteria and yeast, algae and fungus, industrial wastes, and mineral clay are listed from upper left to right.

2.   Eco-friendly techniques for silica nanoparticles synthesis

2.1   Plant derivative sources

In recent literature, there is an abundance of reports elucidating eco-conscious methodologies for nanoparticle synthesis employing a rich variety of plant resources. These innovative processes harness extracts derived from diverse plant components, adeptly serving as both source materials and eco-friendly reducing agents, thereby obviating the customary reliance on synthetic chemicals in conventional nanomaterial synthesis. Plants are integral to the silicon biogeochemical cycle, absorbing silicic acid (H_4SiO_4) from their environment, which polymerizes into amorphous silica, often near transpiration channels [31]. Upon completing their life cycle, plants return silica to the soil, enriching it with humic acid. Phytoliths, composed mainly of amorphous SiO_2 with 5-15% H_2O, are the primary form for silicon accumulation in plants. Monocots such as *Cyperaceae*, *Gramineae*, and *Palmae*, dicots like *Myrtaceae*, *Proteaceae*, *Casuarinaceae*, *Mimosaceae*, and *Xanthorrhoeaceae*, conifers including *Taxodiaceae* and *Pinaceae*, as well as sphenophytes, are all recognized for their silicon accumulation [8]. Below, we will embark on an exploration of examples in the realm of SiNP fabrication based on plants, whether they function as abundant sources of silica (*Plant-based synthesis*) or play pivotal roles as mediators in the synthesis process by substituting essential reagents (*Plant-mediated synthesis*), offering sustainable alternatives to traditional methods.

2.1.1   Plant-based synthesis of SiNPs

Certain plant species, such as Equisetum (horsetail), particularly *Equisetum hyemale* and *Equisetum arvensis*, are quite intriguing owing to their high silica content. This silica, often termed 'biogenic silica', is accompanied by phenolic and flavonoid chemicals and various salt/mineral elements (Na, K, Mg, Al, Ca, Fe), making up about 10% of their composition. These plants can contain up to 25% silicon by dry weight, with a significant portion found in amorphous knobs and subcuticular layers. Silica is also distributed within cells, cell walls, and intercellular spaces, aiding in the absorption and transport of soil salicylic acid. Equisetum's global availability, rapid growth, and colonization of stagnant areas make it a cost-effective, silicon-rich resource for eco-friendly silicon nanoparticle synthesis [32, 33].

To obtain high-purity SiNPs from *Equisetum arvense*, Carneiro *et al.*, used a series of essential chemical and thermal processes [32]. The biomass underwent a sequence of treatments, including washing with deionized water to remove surface impurities and multiple washes with a 2% HCl acid solution to eliminate potential metallic residues. Subsequently, the samples were washed to attain the required pH level, either neutral or acidic, with pH levels known to impact material quality. After this, the material was dried and processed into powder using different-sized ceramic balls, resulting in 2.5 mm-sized powder samples. Finally, the powdered samples were subjected to different temperatures ranging from 773 K to 873 K in order to optimize their properties. Advanced techniques like Scanning Electron Microscopy (SEM) with energy dispersive X-ray analysis (EDAX) and confocal Raman microscopy have been used to analyze different forms of silica in these plants. They were able to obtain high-purity amorphous and porous SiNP agglomerates of around 300 nm with an average size for each particle of approximately 8 nm. Similarly, with subtle variations in their approach, Adach *et al.* successfully obtained SiNPs not only from *Equisetum arvense* and *Equisetum hyemale* but also from winter barley whiskers (*Hordeum vulgare L.*) and rice husk (*Oryza sativa L.*), showing the versatility of this methodology [33]. Utilizing SEM images, they confirmed that silica was present and distributed across the surface of the plant components. (Fig. 2).

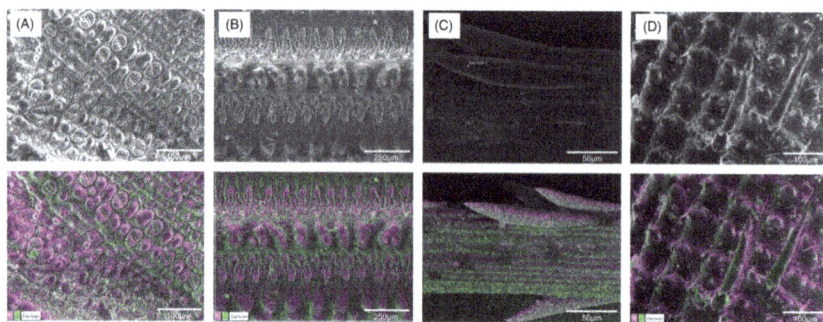

Figure 2. SEM images of the surfaces of the plant samples (A) Equisetum arvense, (B) Equisetum hyemale, (C) Hordeum vulgare, (D) Oryza sativa L. EDX analysis shows the distribution of the silicon (pink) and oxygen (green) atom on the plant surface. Reprinted from [33] with permission from Taylor & Francis Ltd.

Additionally, their study underscored the crucial role of pre-calcination treatment with HCl in obtaining high-purity nanoparticle samples, particularly for the removal of the aforementioned salts and minerals (Fig. 3).

In a different approach, Mehmood *et al.*, obtained silica from horsetail by refluxing the biomass in conjunction with sodium hydroxide (NaOH, 1.0 N), resulting in a mixture of sodium silicate (Na_2SiO_3) and water [34]. The subsequent addition of HCl 0.1 M encouraged the creation of NaCl and silicon dioxide (SiO_2) as products, which were purified by washing it multiple times with ethanol and water, followed by a 24-hour drying period at 50°C in an oven, obtaining the final SiNPs. SEM and Transmission electron microscopy (TEM) imaging, coupled with BET (Brunauer, Emmett, and Teller) analysis, unveiled nanoparticles of remarkably consistent size, averaging around 411.70 nm. BET and BJH (Barrett, Joyner, and Halenda) adsorption analyses indicated an average pore diameter of 10.261 nm and 15.01 nm, respectively, alongside evident aggregates (Fig. 4). An elemental analysis conducted via EDAX demonstrated the presence of carbon (C), oxygen (O), and silicon (Si) at composition percentages of 65.6%, 26.4%, and 8.0%, respectively. To assess the suspension's potential stability, Zeta Potential measurements were taken, yielding values of -32.9 ± 4.79 mV, indicative of excellent stability. The researchers suggested these SiNPs as efficient adsorbents for extracting carcinogenic hexavalent chromium (Cr(VI)) from aqueous solutions, addressing the pressing issue of heavy metal pollution in aquatic ecosystems and drinking water. Their porous structure, carbon, and oxygen content enhance adsorption capacity for various water pollutants. Heavy metals, particularly Cr(VI), threaten the environment due to non-biodegradability, high solubility, and widespread industrial use. While the use of green-synthesized nanoparticles for chromium removal is not new, the quest for nanoparticles that can be safely introduced into the environment without creating new ecological concerns remains a focal point of research. Researchers demonstrated SiNPs' efficient Cr(VI) capture under various pH conditions. Impressively, these green-synthesized SiNPs maintained high adsorption capability through six usage cycles, highlighting their recyclability offering cost savings in chemical processes and reducing environmental impact, fostering eco-friendly and sustainable long-lasting solutions.

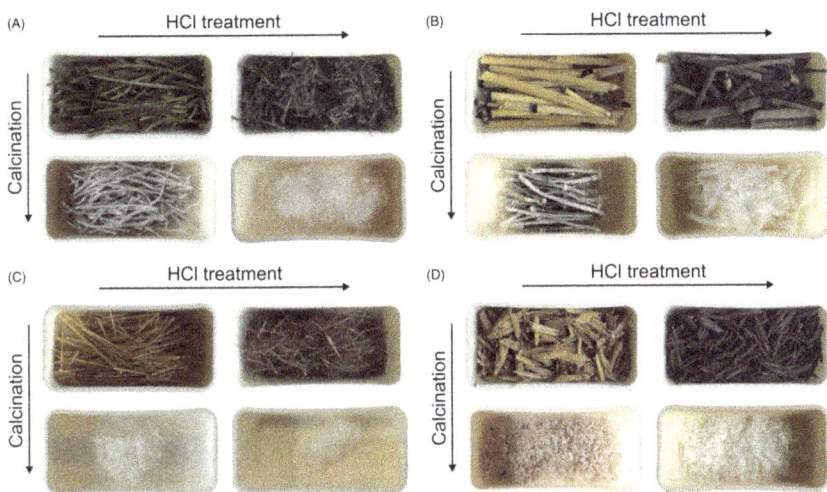

Figure 3. Images of examined plant samples (A) Equisetum arvense, (B) Equisetum hyemale, (C) Hordeum vulgare L., (D) Oryza sativa L. Each of the presented images shows the plant before and after chemical treatment (right arrow direction), and before and after calcination process (down arrow direction). Reprinted from [33] with permission from Taylor & Francis Ltd.

On the other hand, Adinarayana and colleagues proposed an extraction method involving microwaves to obtain highly dispersible SiNP from *Equisetum arvense* [35]. Their approach yielded nanoparticles with enhanced luminescence properties across various conditions. The process began with obtaining a plant extract in distilled water, followed by filtration through a 0.45 μm pore size syringe filter. Afterwards the extract was subjected to pyrolysis at high temperatures for 30 minutes (473 K) using a conventional household microwave oven. They determined, through TEM and Dynamic Light Scattering (DLS) analysis, an average nanoparticle size of approximately 2-3 nm, and their high crystallinity was confirmed using High-Resolution TEM (HRTEM). However, despite this protocol, the particles still exhibited a heterogeneous composition, as revealed by the results obtained from Energy Dispersive X-ray Fluorescence (EDXRF). These nanoparticles showed an absorption peak in the visible-UV range, specifically at a wavelength of 340 nm, and a fluorescence emission peak in the UV/blue region of the electromagnetic spectrum, with the maximum emission occurring at 420 nm. The presence of analytes and their subsequent adsorption onto the surface of these nanoparticles leads to a quenching effect on the initially observed fluorescence. This suggests that these environmentally friendly synthesized SiNP could serve as potential sensors. The researchers successfully tested this behavior in solutions containing Fe(III) ions, demonstrating the effectiveness of these particles in detecting and monitoring these ions in water, even up to the WHO-recommended level of 50 ppb.

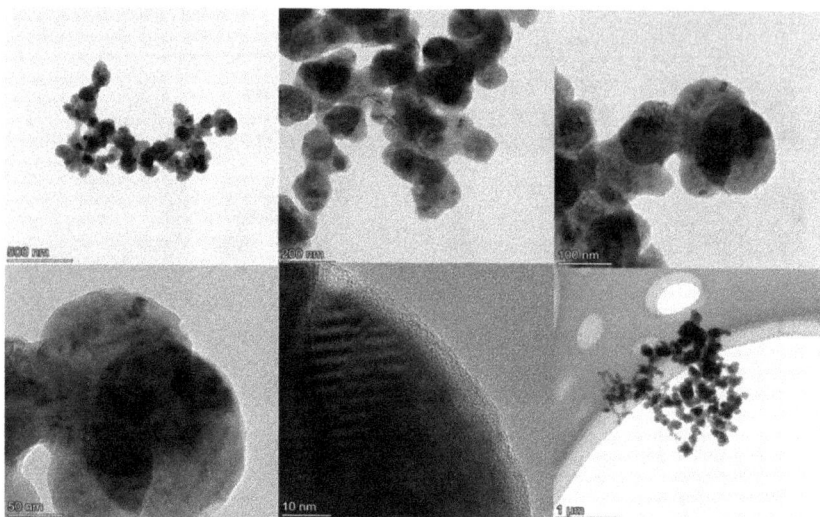

Figure 4. Transmission electron microscopy (TEM) images of green synthesized silicon nanoparticles (GS-SiNPs), magnified at the accelerating voltage of 200 KV, camera length of 520 mm, and electron wavelength of 0.0251 Å. Reprinted from [34], with permission from Elsevier.

Applications such as heavy metals or contaminants removal from water, as mentioned earlier, have been suggested for SiNP derived from extracts of various plants. Sachan *et al*. utilized plant leaves from weed (*Saccrum ravannae*), rice (*Oryza sativa*), and sugarcane *(Saccharum officinarum)* as silica sources [36]. They employed a procedure involving sample calcination at 700°C, followed by sodium silicate (Na_2SO_3) formation through reflux in 2.5 N NaOH and subsequent acidification with H_2SO_4, yielding SiNPs. These SiNPs, named SR-SNPs, OS-SNPs, and SO-SNPs according to their plant sources, exhibited distinct surface morphologies and sizes. SR-SNPs and OS-SNPs were spherical, while SO-SNPs had a hexagonal shape, with average sizes of 3.56 nm, 39.47 nm, and 29.13 nm, respectively. TEM analysis confirmed these features. SR-SNPs and OS-SNPs contained roughly 45 wt% Si, while SO-SNPs had around 35 wt% Si, as determined by EDAX. Field emission SEM and TEM (FESEM and FETEM, respectively) analyses highlighted topological variations among SiNPs from different sources. Powder X-Ray Diffraction analysis (PXRD) revealed crystalline characteristics with well-defined peaks, indicating the presence of silica phases. Thermogravimetric analysis (TGA) showed thermal stability, especially in SO-SNPs and OS-SNPs. BET analysis provided insights into pore features and surface areas. SR-SNPs had a surface area of 178.11 m^2/g, OS-SNPs exhibited 9.555 m^2/g, and SO-SNPs displayed 39.989 m^2/g. Regarding surface charge, SO-SNPs possessed the highest charge at -50.14 mV, followed by OS-SNPs at -33.76 mV, and SR-SNPs at -19.48 mV. These SiNPs were efficient adsorbents for Pb^{2+} and Cu^{2+} ions, removing over 95% of heavy metals from synthetic wastewater. Similarly, Sharma et al., synthesized spherical SiNPs (80 ± 5 nm) from bamboo leaves, further functionalizing them with APTES ((3-Aminopropyl) triethoxysilane), resulting in positively surface-charged A-SiNPs (85 ± 5 nm). These nanoparticles effectively adsorbed Cu^{2+} and NO_3^- ions, contaminants with long-term implications for human health and aquatic life [37]. Furthermore, certain studies underscored the additional value of nanoparticles produced via this method, leveraging the inherent properties of the source plant. For instance, Babu et al., prepared SiNPs from Cynodon dactylon L, an annual grass herb belonging to the Poaceae family [38]. This plant is renowned for its multifaceted biological effects, including anti-diabetic, anti-inflammatory, antimicrobial, antioxidant, and antiarthritic properties. The resulting SiNPs, ranging from 7 to 80 nm in size and characterized by minimal aggregation, exhibited remarkable antibacterial efficacy against Gram-negative bacteria (E. coli and P. aeruginosa). Importantly, this antimicrobial efficacy surpassed that of both plant extracts and chemically synthesized nanoparticles, with a concentration of 60 ug/ml showcasing the highest inhibitory effect. These findings underscored a synergistic interplay between the plant-derived components and silica structures.

2.1.2 Plant-mediated synthesis of SiNPs

In contrast to the methods discussed earlier, where plants serve as a source of silica, the literature presents alternative synthesis approaches employing plant extracts to replace one or more chemical reagents typically used in nanoparticle synthesis, such as catalysts and reducing agents. This substitution significantly reduces the potential environmental impact of these processes. For instance, Sankareswaran et al., utilized extracts from Phyllanthus emblica L., commonly known as Indian gooseberry [39]. This plant is renowned for its rich phenolic compound content in leaves, fruits, and barks, and it boasts diverse properties, including anti-inflammatory, anti-cancer, anti-microbial, and hypoglycemic activities. The extraction procedure followed Rajiv et al.'s methodology. The plant-derived extract was introduced into a solution containing Tetra Ethyl Ortho Silicate (TEOS) and ethanol, serving dual roles as both a reducing and capping agent. This enabled precise control over parameters like synthesis time and the final nanoparticle size. Following this, the nanoparticles underwent pre-characterization calcination. This method yielded spherical SiNPs, with an average size of 1.2 nm in size and a zeta potential of -40.5 mV, all while retaining biomolecules on their surface. In the realm of nanoparticle synthesis, the choice of reducing and capping agents is pivotal. It can profoundly influence SiNPs' properties, impacting critical attributes like size, shape, and surface functionalization, all of which determine their applicability. Such nuances assume heightened significance, especially in the context of deploying nanoparticles for cellular imaging, where the imperative is to minimize interference with cellular activity. To illustrate this point, Tiwari and collaborators embraced a green synthesis approach. They utilized APTES and vinyltrimethoxysilane (VTMS) as silica precursors, while Citrus limon (L.) performed a dual role of reducing and capping agent [40]. This innovative strategy yielded SiNPs boasting average sizes ranging from 20 to 57 nm. Notably, these nanoparticles exhibited impressive photoluminescence quantum yields of 7% and 25%, showcasing excitation-dependent blue-green luminescence. Their utility was further demonstrated through successful deployment in immunofluorescent cell imaging, proving particularly effective for imaging mouse epithelial cells and human white blood cells. These SiNPs stand out as promising luminous probes for cellular imaging, attributed to their well-defined size, high concentration threshold for eliciting antimicrobial effects, and robust surface stability. Similarly, extracts from *Euphorbia thymifolia* and *Punica granatum* (Periakaruppan *et al.*,) have been employed as both reducing and capping agents, resulting in SiNPs which demonstrated effects on the germination of *Sorghum bicolor* and antibacterial activity, respectively [41, 42]. Furthermore, Rahimzadeh and colleagues pushed green synthesis boundaries. They utilized *Rhus coriaria L.* extracts as a multifunctional agent and sodium metasilicate, a plant-derived silica form, instead of TEOS (Fig. 5) [43]. This innovative approach achieved fully green synthesis from Na_2SiO_3, ensuring environmental friendliness. The study successfully produced eco-friendly, nanometer-sized, and thermally stable SiNPs. This approach outperformed conventional synthesis in terms of efficiency, cost, and ecological soundness. In conclusion, green synthesis via plant extracts for SiNPs shows promise, offering eco-friendly, efficient routes to diverse nanomaterials and holding great potential for environmentally conscious nanotechnology applications and a sustainable future.

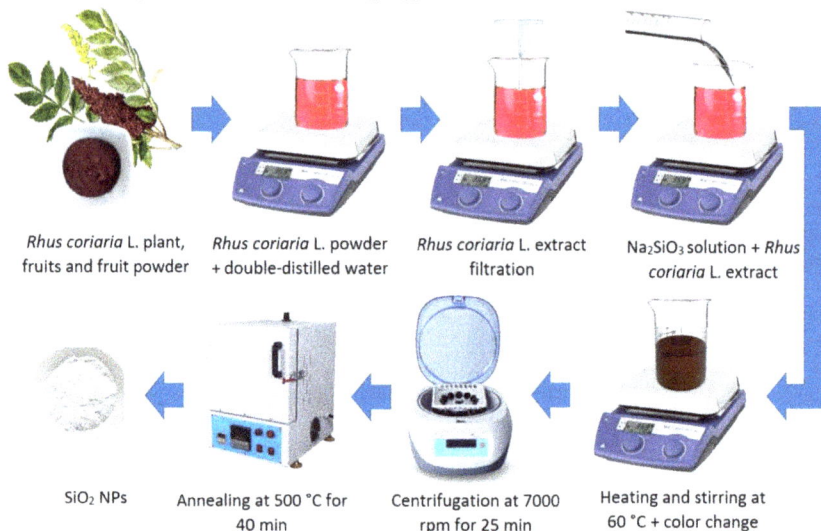

Figure 5. Schematic diagram of the green synthesis process of SiO₂ NPs using Rhus coriaria L. extract. Reprinted with permission from [43] licensed under the Creative Commons Attribution 4.0 International license.

2.2   Agricultural wastes

The rapid expansion of the global population is driving a swift increase in food production, consequently accelerating the accumulation of agricultural residues (AR) and agricultural wastes (AW). These byproducts are harnessed for the creation of eco-friendly fertilizers and as a renewable fuel source. Among the strategies employed to manage agricultural waste, a noteworthy avenue is the fabrication of SiNPs.

AR originates predominantly from four distinct sources. These sources encompass crops, food remnants, agro-industrial processes, and waste stemming from livestock (Fig. 6). These residues materialize after the productive parts of crops have been extracted or while managing livestock. Similar to other forms of waste, effective management protocols are indispensable for maintaining a clean environment. These protocols encompass the generation of novel products to amplify the value of the agricultural sector. The incineration of these agricultural residues results in the formation of agricultural waste ashes (AWAs) [44].

Agricultural Wastes

Crops
millet straws, corn stalks, sorghum straws, wheat straws, oat straw, barley straws,etc.

Agro-industrial processing
sugarcane bagasse, rice husks, palm shells, wheat bran, etc.

Food
banana peels, pumpkin shell, orange peels, apple peels, etc.

Livestock
cow dung poultry manure, camel manure, goat manure, swine manure, etc.

Figure 6. Different sources of agricultural wastes used for the synthesis of SiNPs.

Among the inventive products that can be derived from AWAs are SiNPs. Conventional techniques for producing these nanoparticles rely on non-renewable minerals, such as silica sand. However, a burgeoning emphasis on sustainability has spurred research into alternative sources. Consequently, exploration has centered on utilizing AWAs as an ecologically conscious resource for SiNP synthesis. Given the abundance of agricultural waste after each harvesting season, methodologies for nanoparticle synthesis that capitalize on these waste materials are not only economically viable but also ecologically sound. This methodology not only assists in the management of agricultural residues but also contributes to the overarching enhancement of value within the agricultural sector [44].

Rice husk (RH) constitutes about one-fifth of the entire weight of the rice grains. This residual material is obtained as a byproduct from the rice and paddy industry. Typically, these RH residues are disposed of by setting them on fire, making open burning the prevalent method for RH management. Given that rice is a dietary staple for over 3 billion individuals, especially in Asian regions, rice cultivation generates a significant volume of agricultural waste encompassing straw, husk, and ash. Approximately, every year, around 120 tons of RH are left as waste post-milling procedures. As the process of rice milling is crucial for producing high-quality edible rice kernels, the management of the resulting waste must align with principles of sustainable agriculture and circular economy. Rice-derived waste materials have been extensively examined as potential sources for generating SiNPs of desirable quality [45].

Since Jansomboon *et al.*, first prepared nanoparticles derived from rice hulls, many others followed them [46]. As an example, Deivaseeno *et al.*, employed a simple and economical sol-gel technique for the extraction of mesoporous SiNPs from RH through a bottom-up approach. Notably, they managed to achieve this without the requirement of surfactants to guide their structural development. The process involved treating RH with hydrochloric acid followed by heating, leading to the creation of high-purity RH ash with a silica content exceeding 98%. Through heating at 650°C for four hours, RH ash of exceptional purity devoid of both metal impurities and organic components was obtained. Their exploration utilized electron microscopy, revealing uniform clusters of spherical SiNPs displaying an average size below 20 nm. Furthermore, their investigation included BET analysis that unveiled pores with an average size of 8.5 nm, an area with specific surface measuring 300.2015 m^2/g, and a pore volume amounting to 0.659078 cm^3/g. Mesoporous SiNPs like those produced in this research, have the potential to act as carriers for delivering and releasing specific agricultural chemicals [45]. These nanoparticles were characterized by being very porous and lightweight, possessing a large outer surface area. These qualities make them highly suitable for various agricultural uses. These physical attributes are particularly beneficial when it comes to applying additional functionalities later on. This includes processes like enclosing agricultural chemicals such as regular fertilizers and pesticides, improving soil quality and the ability to hold water, serving as silicon-based fertilizers that aid plant growth and development, contributing to defenses against both living and environmental stresses, and ultimately boosting crop yield [47]. Farook *et al.*, developed an innovative method to extract silica from RH at room temperature, differentiating it from other high-temperature techniques. Researchers employed a sol-gel approach without templates or surfactants to create spherical SiNPs where calcination was not necessary. By adopting a sol-gel method that requires no template at room temperature, mesoporous SiNPs with spherical morphology were successfully produced. These nanoparticles were seen in TEM micrographs, demonstrating their 50.9 nm diameter. The nitrogen adsorption-desorption analysis revealed a high specific BET surface area of 245 m^2/g for RH silica. Importantly, these SiNPs exhibited a tight range of pore sizes, falling between 5.6 and 9.6 nm [48].

Another great example of AW is sugarcane subproducts. The global sugarcane industry yields vast quantities of sugarcane every year, with countries like Brazil as notable examples yielding over 600 million tons annually. During the process of extracting sugar and ethanol from sugarcane, significant amounts of AW are generated, including bagasse and straw. Around 200 kg of straw and tips and in between 250 to 270 kg of bagasse are produced from each ton of sugarcane.

Roughly 50% of these residues find use as an energy source in distillery plants, while the remainder is stockpiled. The incineration of this waste leads to the creation of 1-4% ash, resulting in an annual ash output of 3 to 12 million tons in Brazil alone. Typically, biomass ash from the sugarcane industry is either disposed of in landfills or utilized as plantation fertilizer. However, these practices pose health risks and environmental concerns, particularly related to soil and water [49].

Mohd et al., created SiNPs from sugarcane bagasse through a low-cost and environmentally friendly extraction and precipitation process. Sugarcane bagasse, a natural source rich in silica, served as the raw material. This silica content originates from orthosilicic acid present in the soil solution, which accumulates in sugarcane stems and leaves. In this work, the resulting amorphous SiNPs, had sizes of around 30 nm, 111 m^2/g specific surface area, and exhibited a spherical morphology. The adoption of this method offered advantages like reduced chemical consumption, economic feasibility, and the removal of metallic impurities [50]. Boonmee et al., also successfully produced SiNPs from sugarcane bagasse ash (SBA) using a sol-gel procedure. This method resulted in high-purity SiNPs with an average size of approximately 90 ± 10 nm. By sourcing sodium silicate from SBA, a process was undertaken. Initially, 100 g of SBA was refined by refluxing it with 1 M hydrochloric acid (HCl) at 100°C for 4 hours, removing impurities. The resulting material was washed to neutral pH, filtered, and dried. This acid-leached SBA was further processed by vigorously stirring it with 3 mol/l NaOH at 80°C for 4 hours, creating a sodium silicate solution. To generate the SiNPs, the solution was mixed with distilled water and slowly added drop by drop to an ethanol/ammonia mixture in a specific ratio. The resulting solution was aged into silica gel, after which it was centrifuged, washed, and subjected to different drying techniques, either heat drying in a vacuum oven or freeze drying, to study their impact on particle properties and size distribution. Finally, they studied how SiNPs affected the morphology, characteristics and mechanical properties of natural rubber composites [51]. These are mere examples of how using sugarcane wastes as a precursor for synthetizing SiNPs presents an approach to waste management and allows the valorization of valuable constituents from agricultural residues [52].

Many other AWs have been studied as sources for the production of SiNPs. Imoisili et al., prepared them using palm kernel shell ash (PKSA), a residual product from palm oil extraction. Oil palm trees are abundant in regions like America, Asia, and Africa, including Nigeria. The palm kernel shell (PKS) is the rigid endocarp of the palm fruit that remains after the kernel's removal for oil extraction. Employing a modified sol-gel extraction technique, researchers successfully produced these silica nanoparticles from PKSA. The extracted nanoparticles were characterized using X-ray diffraction (XRD), SEM with EDAX, Fourier transform infrared (FTIR), BET analysis, and TGA, among other techniques. The microstructural analysis demonstrated that the extracted SiNPs had unit sizes in the range of 50 to 100 nm, along with a notably specific surface area of 438 m^2/g [52]. Bananas rank second in global fruit production, contributing to 16% of the total. India accounts for 27% of the world's banana production, with cultivation spanning over 130 countries, primarily in tropical and subtropical regions. Some researchers, like Mohamad et al., employed banana peel ash to create, using the sol-gel method, mesoporous SiNPs with the capability to serve as an economical substitute adsorbent for addressing dye contamination in wastewater treatment [53]. Mohd Nazri et al., also utilized SiNPs as a helper to remove chemicals such as phenol. The researchers chose a sol-gel approach to transform bamboo leaf ashes into sodium silicate and ultimately into mesoporous SiNPs; Bamboo, widely cultivated for various applications, yields leaves that are yet another example of agricultural waste [54]. Lastly, another interesting example is the research made by Sivakumar *et al.* Employing cow dung ash, a byproduct of livestock farming, as a resource of value, the study achieved the extraction of pure silica. This was accomplished through the sol-gel process, where cow dung ash served as the starting material. The resulting particles had an approximate size of 200 nm and had a 100% purity level while exhibiting a spherical morphology [55]. In light of the cases discussed above in the literature, it becomes evident that AW stands out as a viable origin for the production of SiNPs.

2.3 Bacteria and yeast

Microbes play a key role in preserving silicon's geochemical availability under natural conditions. The complex process of weathering breaks down rocks and minerals physically, chemically, and biologically, converting minerals with complex structures into those with simpler ones. The most significant process is conceivably the hydrolysis of silicates, which is definitely related to biological weathering [56]. During biological weathering silica is extracted from insoluble silicates, enabling the release and uptake of new nutrients by plants. Therefore, the diversity of microbial species is crucial for silicon uptake in plants [57]. Given the abundance of bacteria that may perform the solubilization of silicates in the soil ecosystem, microbe-mediated extraction of silicate minerals becomes an attractive green synthesis technique for the production of SiNPs.

Several bio-based mechanisms are proposed for carrying out silicate dissolution [58]. They include:

Hydration of respiratory CO_2 and production of carbonic acid.

Carbonic acid is created when the respiratory CO_2 from soil bacteria is hydrated. Organic and inorganic acids promote silicate dissolution by providing protons (H^+) for protonation and silicate hydrolysis. In addition, they can act as chelating agents by forming complexes with the cationic components of silicates like aluminum, and iron, thereby leaving silicates soluble [59].

Production of Exopolysaccharides, Hydrogen Sulphide and Siderophores.

Microbe-produced exopolysaccharides have been related to both silicate disintegration and the erosion of rocks owing to their wetting characteristics [57]. Inorganic silicate is absorbed and bound to bacterial exopolysaccharide. Additionally, certain bacteria can reduce sulfate to hydrogen sulfide (H_2S), and the reduced sulfur can react with cations such as calcium and iron within silicate minerals, leading to the release of silica. Due to their ability to bind and transport iron, the siderophores produced by bacteria can contribute to the solubilization of silicates by scavenging iron from silicate minerals [60].

Enzyme based mechanism.

Previous studies have suggested that the secretion of enzymes by microorganisms, specifically the nitrate reductase enzyme which depends on nicotinamide adenine dinucleotide (NADH), is crucial for converting metallic ions into nanoparticles [61]. It has been suggested that this enzyme can also reduce inorganic silica substrates into nanoparticles, thus enabling their involvement in the environmentally friendly synthesis of SiNPs.

Pseudomonas, Staphylococcus, Mycobacterium, and Aminobacterium have been identified as silica-solubilizing bacteria [62]. Mohanraj and collaborators reported that actinomycetes isolated from contaminated soil were able to synthesize SiNPs from magnesium trisilicate hydrate, which acts as a substrate for silica [63]. The existence of functional groups was confirmed using FTIR and UV-vis spectrometry. The particle size analyser results showed that *Streptomyces* sp. culture was able to broadcast nanoparticles under 100 nm contrasted with alternative organisms. Researchers suggest that this phenomenon may be related to the reducing potential of secondary metabolites from this species. Afterwards, the viability and stability of the generated nanoparticles for treating effluents were further examined through encapsulation techniques. The immobilized SiNP treatment resulted in lower amounts of solids and toxic substances in the effluents. The research showed that encapsulated native SiNPs had a decolorization efficacy of 80%.

Natesan *et al.*, conducted another interesting study on *Trichoderma*, *Streptomyces*, and *Pseudomonas* spp. isolated from tea soils in India, investigating their ability to produce copper and SiNPs [64]. The researchers cultured the biocontrol agents in a basal medium enriched with copper sulfate and potassium silicofluoride as substrates. To confirm the elemental composition of the nanoparticles, XRD and EDAX studies were conducted. Analysis with FTIR indicated the presence of a functional siloxane group in the SiNPs. TEM results showed that the nanoparticle sizes ranged from 5 to 25 nm. The aim of this study was to employ antimicrobial nanoparticles to combat stem and root infections in tea plants. In this sense, tea fungal pathogens including *P. hypolateritia* and *P. theae* revealed highly effective growth inhibition when exposed to nanocopper and nanosilica. Furthermore, to ensure safety and efficacy, various nanoformulations of nanocopper and nanosilica were prepared using inert and environmentally friendly carrier materials. Subsequent tests on a zebrafish model demonstrated that the nanocopper and nanosilica formulations exhibited no toxicity. Therefore, this study provided a potential cutting-edge method for controlling root and stem diseases in tea agriculture, potentially enhancing indicators of tea quality.

In another approach, a green method for the synthesis of SiNPs utilizing a thermophilic bacteria (BKH1) as a template was described by Show *et al.*, [65]. Biosynthesis of SiNPs was conducted from inorganic (magnesium trisilicate) and organic (TEOS) silica precursor by the non-pathogenic thermophilic bacterium BKH1. The shape and distribution of the produced SiNPs were quite regular as verified by SEM and TEM. FTIR spectra and EDAX analysis showed that the BKH1 template method led to the effective synthesis of SiO_2. The zeta potential confirmed that nanoparticles were stable in dispersed medium, preventing aggregation. In addition, mechanistic studies were also conducted, revealing an increase in root mean square roughness through atomic force microscopy experiments, indicating silica deposition on the surface of bacterial cells. The TEM analysis displayed a rod-like template surface for BKH1 (Fig. 7). Besides, it could be observed that the SiO_2 nucleation process took place on the bacterial surface (BKH1). The researchers highlighted the potential for diverse biomedical applications of fluorescence emission spectra, spanning both UV and visible wavelengths. When utilized in this way, the UV radiation emitted can be used to eliminate undesirable germs, and the visible emission could be used for medical diagnosis as an optical probe.

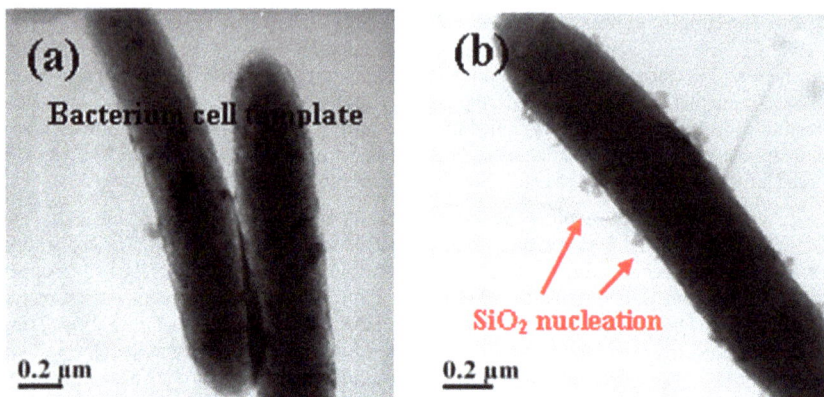

Figure 7. TEM micrograph of (a) bacterium cell template, (b) SiO₂ nucleation over the bacterium cell template. Reprinted from [65], with permission from Elsevier.

2.4 Algae and fungi

As explored in preceding sections, living organisms possess remarkable capabilities for nanoparticle production, and among them, algae and fungi stand out as exceptional resources. Their distinctive features set them apart from other biological agents, such as plants and bacteria, making them valuable contributors to the field of nanotechnology. They are easily cultivated, allowing for the rapid generation of substantial biomass in a short time frame, making them prime candidates for nanoparticle preparation. Moreover, their inherent capacity to produce various bioactive substances translates into stabilizing and reducing agents, effectively modifying nanoparticle properties.

2.4.1 Algae as nanofabricators

In the realm of green synthesis, there is a growing trend towards harnessing the potential of algae. Algae are rich sources of secondary metabolites, peptides, proteins, and pigments, effectively serving as nano-biofactories [66]. What makes them even more appealing are their rapid growth rates, cost-effective scalability, and ease of harvest, all of which render them ideal for the biological synthesis of nanoparticles.

Algae, the most primitive organisms inhabiting diverse ecosystems, also happen to be the dominant photosynthetic life forms on our planet [67]. Their remarkable ability to hyperaccumulate metals and convert them into nanoparticles positions them as prime candidates for green synthesis [67]. Across various classes of algae, including brown algae (*Phaeophyceae*), blue-green algae (*Cyanophyceae*), red algae (*Rhodophyceae*), and green algae (*Chlorophyceae*), a myriad of metallic and metal oxide nanoparticles were successfully prepared [68, 69].

Diatoms, a fascinating group of unicellular eukaryotic algae, stand out as a controlled and nanostructured source of silica. These remarkable organisms are found in almost all of Earth's aquatic habitats, with a dominant presence in oceanic phytoplankton [67]. Diatoms are renowned for their complex geometries and structural desings in their silica-based cell walls [70]. With over 10,000 identified species, diatoms exhibit diverse forms and structures, all under the genetic control of specific genes and proteins [71]. In vivo experiments have demonstrated the feasibility of controlling these structures, surpassing what can be achieved through conventional material engineering [72]. The unique deposition of silica in diatoms has been extensively studied through electron microscopy, revealing that they produce silica within intracellular compartments called as silica deposition vesicles (SDVs) [73]. These SDVs serve as 'cellular reaction vessels,' governing silica precipitation and pattern formation [74]. Notably, acid hydrolysis of isolated diatom cell walls has uncovered unusual amino acid derivatives, including N,N,N-trimethylhydroxylysine and dihydroxyproline, hinting at the composite nature of diatom silica [75]. The source of these derivatives was later identified as silaffins, a family of polypeptides covalently linked to long-chain polyamines [76]. These long-chain polyamines, whether attached to peptides or not, are distinctive constituents of diatom biosilica [77]. Diatom-derived porous silica microshells exhibit special potential in applications such as drug and gene delivery [78]. Their exceptional attributes, including unique architecture, mechanical resistance, high surface area (up to 200 m^2/g), thermal stability, adequate biocompatibility, ease of functionalization, optical and photonic properties, position them as intelligent platforms for drug and gene delivery [79, 80].

In the early stages of research, diatom frustules emerged as promising candidates for drug delivery without the need for surface modification. This approach primarily involved encapsulating therapeutic agents for oral delivery. Losic et al., innovatively utilized diatom microshells to deliver indomethacin, an extremely hydrophobic compound, demonstrating significant potential with substantial loading capacity and controlled sustained release over a two-week period [81]. To further refine drug delivery, researchers explored various surface functionalization strategies. For example, Terracino et al., achieved successful encapsulation and release of sorafenib, an anticancer drug with low water solubility, in MCF-7 and MDA-MB-231 cancer cells. They achieved this by using polyethylene glycol and a cell-penetrating peptide for the functionalization of diatomite nanoparticles, resulting in biofunctionalized nanoparticles with enhanced cellular uptake, minimal toxicity, and exceptional drug loading and release properties [82].

In another pioneering study, Sasirekha et al., introduced chitosan (Chi) to modify Amphora subtropica diatoms, enabling doxorubicin encapsulation within the diatom pores (Fig. 8). This approach yielded a biocompatible and biodegradable drug delivery system for cancer treatment with reduced toxicity compared to free doxorubicin [83].

Encapsulation versatility studies extend beyond cancer treatment. For instance, Vasani et al., engineered a stimulus-responsive system using diatom biosilica microcapsules derived from Aulacoseira sp. By anchoring oligo (ethylene glycol) copolymers to the surface, they achieved controlled drug release triggered by temperature changes, exhibiting antibacterial efficacy against Pseudomonas aeruginosa and Staphylococcus aureus [84].

Taking innovation a step further, Kroger and Voelcker embarked on a novel approach involving genetic engineering. They modified the genome of Thalassiosira pseudonana to exhibit an IgG-binding domain of protein G on the silica surface. This breakthrough allowed to attach cell-targeting antibodies, creating a platform for the selective delivery of the anticancer drug camptothecin. This platform showed remarkable potential in selectively locating and eliminating Neuroblastoma and B-lymphoma cells [85]. Diatom-derived biosilica has also found applications as a scaffold for immobilizing enzymes, antibodies, DNA aptamers, and drugs, facilitating the development of biosensors [86,87]. Several distinctive attributes set diatom-derived silica apart, including its intricate 3D structure, unparalleled by synthetically obtained nanoparticles, its optical and photoluminescent properties, ease of surface modification for diverse applications, and inherent biocompatibility. These qualities position it as a cost-effective material with a wide range of biomedical applications.

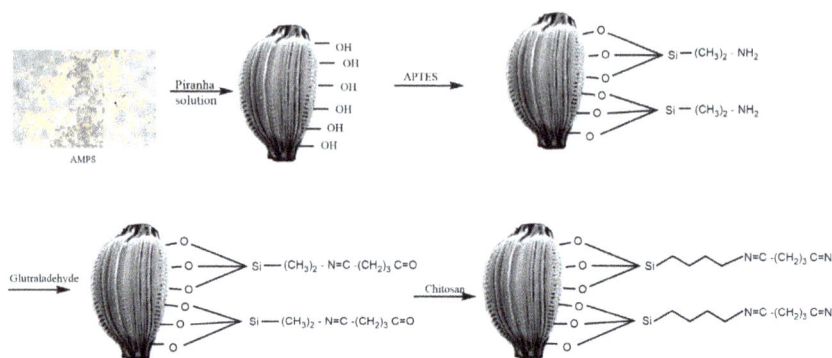

Figure 8. Steps involved in the functionalization of Amphora frustules (AF) are the demineralization to obtain Amino-AF (AF-NH2) using APTES, glutaraldehyde as crosslinker to link AF, and Chitosan to obtain (Chi@AF). Reprinted from [83], with permission from Elsevier.

2.4.2 Fungi: Nature's nanotechnicians

Fungi emerged as a vanguard in the realm of nanoparticle biosynthesis, offering a unique set of advantages. Their controlled cultivation capabilities render them highly adaptable to precise and consistent nanoparticle production, setting the stage for stable and reproducible results. Unlike bacteria, fungi exhibit remarkable resilience to genetic mutations, ensuring unwavering nanoparticle characteristics across extended timelines. Fungi, in particular, hold immense promise for large-scale nano-product development via bio-fermentation. Their track record of prolific extracellular enzyme production positions them at the forefront of commercial nanomaterial synthesis. This enzymatic prowess allows fungi to effectively reduce and stabilize metal ions, a fundamental step in nanoparticle fabrication. Moreover, the recent exploration of fungal organisms for deliberate nanomaterial synthesis highlights their potential as eco-friendly nanotechnological pioneers. While fungal organisms have long been recognized for their ability to leach carbonates and silicates from complex ores since the 1980s, their intentional application in synthesizing diverse oxide nanomaterials has only recently garnered attention. The literature unveils two primary facets of fungal-mediated nanoparticle synthesis: one involving the use of chemical precursors, and the other delving into the captivating realm of fungus-mediated bioleaching, which harnesses naturally available materials and agro-industrial by-products to yield oxide nanoparticles [88 - 93].

Fungi, particularly *Fusarium oxysporum*, a plant pathogenic fungus known to target crops like melon, tomato, cotton, banana, bean, chickpea, and more recently, *Arabidopsis thaliana*, have garnered attention in SiNPs synthesis [88]. Notably, Bansal *et al.*, observed that aqueous anionic complexes of SiF_6^{2-} and TiF_6^{2-} undergo hydrolysis in the presence of *Fusarium oxysporum*, yielding silica and titania nanoparticles at room temperature [89]. Intriguingly, this exposure prompts the release of complex proteins by the fungus, specifically 21 and 24 kDa molecular weight proteins, which were found to be responsible for the hydrolysis of the metal complexes into silica and titania. Remarkably, these fungi do not encounter these ionic complexes during their natural life cycle, making this capability unexpected. In a typical procedure, *Fusarium oxysporum* was cultured as previously described [90]. MGYP medium was used to inoculate the fungal mycelium, followed by incubation for ki72 hours at 27°C. Afterwards, the fungal mycelium was harvested and vigorously washed under sterile conditions. To obtain SiNPs, the obtained fungal mass was resuspended in aqueous solutions of K_2SiF_6 (pH 3.1) for 24 hours. The biotransformed products were collected by separating the fungal mycelium in suspension through filtration. Using this method, researchers obtained irregular, quasi-spherical nanoparticles with an average size of 9 ± 0.2 nm. Through XRD, FTIR, and SAED analyses, they determined the formation of crystalline SiO_2 structures with proteins in their structure, which were removed through calcination at 300°C for 3 hours. Notably, the hydrolysis of TEOS as a silica precursor also yielded similar results. What is particularly noteworthy is that these fungi can hydrolyze challenging metal halide precursors under acidic conditions. The regenerative capacity of biological systems, together with the finding of fungi's ability to hydrolyze metal complexes they never naturally encounter, holds enormous promise, especially for large-scale metal oxide semiconductor material synthesis. The sustainable and energy-saving nature of fungus-based biological processes for metal oxide preparation, compared to chemical methods cannot be overstated. Kannan *et al.*, conducted comprehensive optimization studies, exploring synthesis conditions involving temperature, pH, and incubation time. They assessed the performance of these nanoparticles using methyl orange, a common textile dye, to evaluate their efficacy as environmental contaminants removers. Additionally, they examined the stability of DNA molecules bound to the nanoparticle surfaces. Among various growth media tested, meal agar proved superior for *F. oxysporum* growth. They observed that the degradation of methyl orange by the biosynthesized SiNPs was pH-dependent, with pH 5 showing optimal results after 300 minutes of treatment. Furthermore, the SiNPs alone exhibited remarkable efficiency in protecting DNA against enzymatic cleavage. As discussed earlier, plants accumulate silicon from soil water in the form of water-soluble silicic acid (H_4SiO_4), which polymerizes and precipitates as amorphous silica. In this context, Bansal *et al.*, aimed to utilize an agroecological waste, RH, as a silica source for SiNPs production mediated by *F. oxysporum* [8]. This methodology integrates aspects previously mentioned in this chapter, with the fungus engaging in a bioleaching process of the plant material. In a simplified procedure, RH was suspended in sterile distilled water in the presence of harvested fungal biomass and agitated for 24 hours at 27°C. While samples were obtained at shorter times. It takes 24 hours to leach out 96% of the silica from RH. In this study, they showcased the effective biotransformation of amorphous silica present in RH into highly crystalline SiNPs mediated by this fungus. The resulting nanoparticles exhibited protein coatings, with sizes ranging from 2-6 nm, displaying apparent porosity and a cuboid structure. Interestingly, the use of only the extracellular cationic proteins from *F. oxysporum* yielded SiNPs, but not in the form of nanocrystallites. In a parallel study, they obtained SiNPs through the bioleaching of sand [91]. In essence, the process involved two key steps. First, the leaching of silica from the sand grains in the form of salicylic acid, and second, hydrolysis, both mediated by specific proteins within the biomass of *F. oxysporum*. Using this methodology, they successfully synthesized nanoparticles ranging from 2-5 nm, as observed through TEM. Building upon the pioneering work of these researchers who harnessed *F. oxysporum* as green nanotechnicians to produce SiNPs from various sources, numerous studies in the literature have highlighted the similar capabilities of different fungi. Piela *et al.*, for instance, employed a fungus from the same genus but a different species, *Fusarium culmorum*. Her work showcased this fungus's capacity to efficiently biotransform silica from raw corn cob husks into spherical SiNPs, measuring approximately 40-70 nm. This approach proved to be straightforward, cost-effective, and environmentally friendly [92]. The same research team extended their exploration to SiNP synthesis using *Aspergillus parasiticu*s strain and RH as the source material. Remarkably, they achieved a diverse array of nanoparticle shapes, including pyramidal, cubic, and spherical forms, measuring 400 nm, 85 nm, and 24 ± 8 nm, respectively. These variations in size and morphology were found to be intricately linked to the thermal treatments applied to the RH. Their SEM imaging revealed that the external structure of the substrate underwent significant changes under high-temperature conditions (specifically, a 121°C heat treatment), as illustrated in *Figure 9*, which could potentially enhance fungal enzyme access [93]. It is essential to underscore that the attainment of specific spatial nanoparticle configurations dictates their subsequent utilization. Given the broad utility of silica nanoparticles and the costliness and energy demands associated with traditional chemical and physical production, exploring diverse methods and alternatives for obtaining this valuable product is imperative.

In line with this, Vetchinkina *et al.*, achieved the synthesis of diverse SiNPs, alongside gold, silver, and selenium nanoparticles, employing xylotrophic and humus basidiomycetes, as well as soil bacteria [94]. They effectively controlled nanoparticle sizes and shapes by manipulating critical parameters, including mycelial extracts, isolated proteins, incubation duration, and culture medium, among others. They initially utilized two types of extracts: extracellular extracts obtained from fungal culture liquid filtration and intracellular extracts. After separating mycelia and fruiting bodies from wood substrates or the culture medium, they mechanically ground them at 18°C using a porcelain mortar with quartz sand to disrupt the cell envelope. Subsequently, an extraction process was conducted in a 20 mM Na/K-phosphate buffer (pH 6.0). This was followed by a series of steps, including centrifugation and filtration, ultimately yielding the sought-after final extract. SiNPs were then synthesized by incubating either of the two extracts in the presence of Na_2SiO_3 in a dark room. Regarding the resulting SiNPs, the majority exhibited spherical shapes. Employing the culture liquids of *P. ostreatus* and *G. frondosa* yielded small nanoparticles (5-15 nm) (Fig. 10a), while slightly larger ones (50-100 nm) were obtained with *L. edodes* and *G. lucidum* (Fig. 10b). Cultivating *A. bisporus* fungi in the presence of Na_2SiO_3 produced biogenic nanoparticles with diameters ranging from 30 to 100 nm. Interestingly, utilizing the culture liquid resulted in mesoporous SiNPs, with approximate sizes between 30 and 60 nm and 10 nm-sized pores (Fig. 10d).

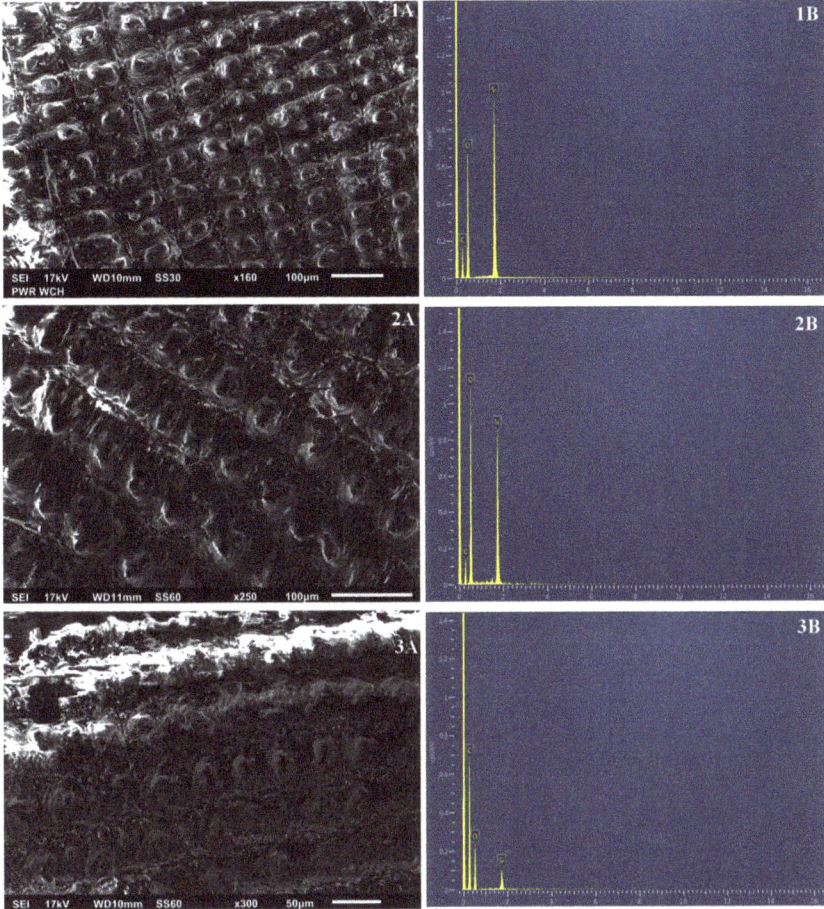

Figure 9. SEM images of RHs surfaces (after the thermal treatment or without it) and corresponding EDX spectra: (1A), (1B) untreated RHs; (2A), (2B) control sample (autoclavated RHs, dried in 200 C); (3A), (3B) samples after 11 d of biotransformation (autoclavated RHs, dried in 200 C). Left column presents SEM pictures of RHs outer surface. Right column presents EDX spectra of silica content in the areas shown in the pictures nearby. Reprinted from [93], with permission from Elsevier.

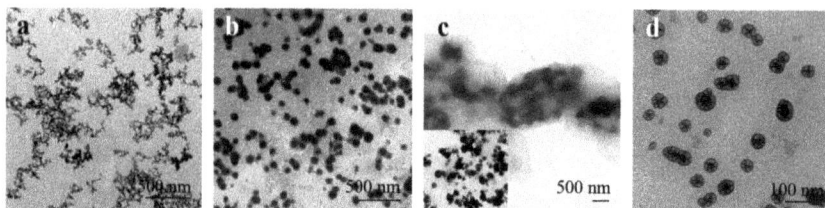

Figure 10. TEM of silica nanoparticles fabricated with G. frondosa (a) and L. edodes (b) culture liquids and with A. brasilense Sp245 (c); mesoporous silica nanoparticles obtained with A. bisporus (d). Bar markers are 100 and 500 nm. Reprinted with permission from [94]. Copyright 2019, American Chemical Society.

As an illustrative example, Albalawi *et al.*, biosynthesized SiNPs using the fungus *Aspergillus niger* [95]. Their study primarily focused on utilizing these nanoparticles as an alternative to pesticides in combating eggplant early blight disease produced by *Alternaria solani*. The mycosynthesized nanoparticles exhibited highly promising antifungal activity. Application of these SiNPs at varying concentrations (50 and 100 ppm) significantly enhanced early blight recovery in eggplants. Moreover, these mycosynthesized SiNPs demonstrated potential as growth-promoting agents, substituting synthetic fertilizers in both wholesome and infected eggplants. Exposure to these particles stimulated the plant's antioxidant defense system by inducing oxidative stress, enhancing antifungal activity while regulating the systemic immune response. The researchers propose that foliar application of Aspergillus niger mycosynthesized SiNPs, especially at 100 ppm, holds commercial promise as an antifungal and a robust enhancer of plant immunity against early blight disease. In summary, fungi stand out as invaluable nanotechnicians, demonstrating exceptional prowess in crafting green silica nanoparticles of varied shapes and sizes. Their remarkable capabilities beckon us toward a future where sustainable nanomaterial applications thrive, promising an eco-conscious, cutting-edge frontier in nanotechnology.

2.5   Industrial wastes

A vital long-term approach for achieving environmentally-friendly cycles in the industry is the reutilization of industrial waste. In this way, the synthesis of nanomaterials has proved to be an efficient method for handling and recycling waste [96]. Actually, it complies with the waste-to-wealth mission and minimal waste generation principles. In view of the constant waste formation and its high silica content, these wastes, especially their ashes, are being explored as possible raw materials for the synthesis of SiO_2.

Among solid-waste-based sources, fly ash is a low-cost and widely accessible option for making SiNPs owing to the large proportion of silica and alumina (more than 70%). Fly ash is a byproduct of combustion in thermal power plants that includes both crystalline phases like quartz and mullite and amorphous components including SiO_2, Al_2O_3, Fe_2O_3, TiO_2, and CaO. Due to the enormous amount of fly ash that is produced each year as waste throughout the world, fly ash is a viable and economical alternative supply material for the synthesis of SiNPs. In essence, fly ash comprises both crystalline and amorphous silica, though only amorphous silica may be transformed into SiNPs, as it is the only form that reacts with alkali. Many studies have produced SiNPs from leftover fly ashwith success. An efficient approach was created by Yadav *et al.*, to generate amorphous SiNPs from waste fly ash [97]. Their method entailed two steps. First, silica was extracted as sodium silicate by treatment with NaOH. Then, neutralization of the alkaline sodium silicate with diluted HCl allowed the formation of silica gel by the sol-gel method. According to FTIR analysis, siloxane and silanol were the major groups found in the resulting silica. Microscopic techniques revealed that the produced SiNPs were spherical, varying in size from 20 to 70 nm, and exhibiting purities ranging from 90 to 96%.

In an analogous investigation, Yan *et al.*, developed a technique to produce SiNPs from coal fly ash via the alkali dissolution method and CO_2-assistant precipitation technologies [98]. They utilized these nanoparticles to prepare nanosilica-supported CaO sorbents. Nanoparticles were able to enhance the sorption kinetics of CaO sorbents and the cyclic CO_2 uptakes. Therefore, these findings establish the novel material as an extraordinarily effective choice for upscaled CO_2 capture while offering an accessible method for high-value coal fly ash reuse.

2.6   Clay minerals

Among the innovative and greener approaches for synthesizing SiNPs, utilizing clay minerals (CM) as precursors stands out as a promising avenue. Many researchers have dived into using different types of clay and minerals owing to their amounts of silica. Sarikaya *et al.*, present a notable path wherein the alkaline treatment method yields a powder-form, highly pure amorphous nano-silica. Through this approach, pyrophyllite ore, extracted from Malatya, Turkey, emerges as a promising reservoir for nano-silica, imparting its latent potential by being transmuted into amorphous SiO_2 with a remarkable purity of 98%. The distinctive attribute of this transformation was its confinement to the nanoscale, with the resultant particle sizes being engineered to dimensions less than 50 nm [99]. Stopic and colleagues investigated the production of SiNPs using olivine minerals. They achieved this by subjecting the mineral to high-pressure carbonation within an autoclave, a controlled vessel for chemical reactions. As seen in *Figure 11*, they analyzed the oxide they produced using TEM and Scanning Transmission Electron Microscopy (STEM), and confirmed the rounded shape of the prepared silica particles. These nanoparticles had diameters ranging between about 400 and 500 nanometers, and are amorphous [100].

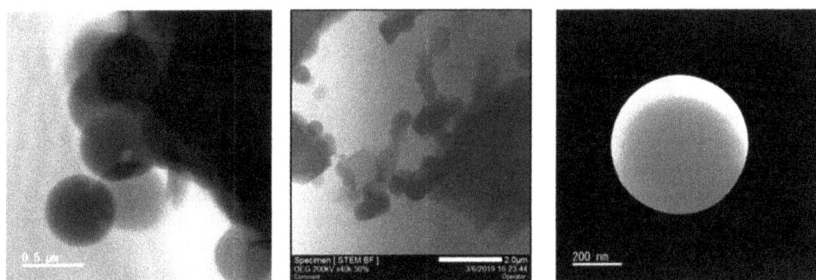

Figure 11. The TEM and STEM analysis of the round-shaped 400-500 nm silica particles obtained from olivine minerals. Reproduced by permission of Creative Commons Attribution (CC BY) license (http://creativecommons.org/licenses/by/4.0/) [100].

Numerous research studies for the production of silicon electrodes for lithium-ion batteries have illustrated that CM can serve as starting materials for producing structured SiNPs through a process called magnesiothermic reduction (MR). CM that contain a significant amount of aluminum, such as montmorillonite and kaolinite, constitute a significant portion of the spectrum of clay mineral varieties [101]. Aluminum and magnesium or silicon-oxygen compounds can potentially react during a MR, and lead to the formation of undesired byproducts like spinel or mullite. To mitigate this, a common practice involves transforming CM such as halloysite and palygorskite into amorphous SiO_2 through a specific chemical treatment, selective acid etching, to later engage in the MR process. This approach helps ensure the desired reaction outcomes and minimizes the formation of unwanted compounds [102, 103].

As the acid etching procedure introduces complexity and higher costs into the synthesis process, alternative research has taken the route of selecting CM without aluminum, as precursor materials. Adpakpang et al., fabricated novel electrode-worthy 2D nanoplates of elemental silicon using the MR. This pioneering synthesis leveraged exfoliated nanosheets from a magnesium silicate clay lacking aluminum, specifically Laponite, to yield porous structures with exceptional electrode capabilities [104]. However, adopting this approach significantly narrows down the range of CM available for utilization [101].

Mourhly et al. used an alkaline treatment and an acid-precipitation procedure instead of the MR to extract white, structureless SiNPs. This innovative approach utilized a porous mineral of volcanic origin called gray pumice. Pumice forms from combining SiO_2 and Al_2O_3 in varying proportions, creating an abundant and cost-effective resource found across the globe. This procedure involved dissolving silica from pumice using sodium hydroxide, then precipitating and purifying the silica through various steps like filtration and gel formation and using acid treatment to remove other minerals like aluminum, calcium, iron, and magnesium. The final product was a white silica powder with excellent yield and purity. Upon isolation, the amorphous silica was subjected to thorough characterization. The outcomes revealed that the nano-silica powder was successfully generated through an acid-base pathway. This synthesis yielded predominantly amorphous mesoporous structures possessing a substantial surface area of 422 m^2/g. When observed under TEM, the images portrayed uniformly distributed nano-silica particles with compact sizes ranging from 5 to 15 nanometers. The researchers suggest that the particle's inherent amorphous nature, its expansive surface area, and finely tuned dimensions, along with its economical manufacturing process, render it a prime contender for serving as a catalytic support in hydrogen production applications [105].

When it comes to producing nanostructured silicon through MR, molten salts could play a crucial role in managing the heat generated during the process. Utilization of molten salts can lower the reaction temperature and prevent high-temperature phases. This suggests that CM containing aluminum could be potentially used directly for synthesizing nanostructured silicon by combining molten salts and MR. Chen et al., selected three different types of Al-containing CM as precursors to produce porous SiNPs: layered montmorillonite, tubular halloysite, and chain-layered palygorskite. By using molten salt-assisted MR, they successfully synthesized porous SiNPs with enhanced porous structures and significant specific surface areas. This approach effectively managed heat, and prevented unwanted phases and the fusion of SiNPs. The molten salt also helped prevent nanoparticle agglomeration. And, the metal magnesium played a role in lowering the local reaction temperature through its vaporization. This method eliminates the need for complex pretreatment and broadens the amount of CM suitable as silicon sources [106].

The same team continued utilizing CM as starting materials in a modified MR reaction involving molten salt and successfully created three different silicon nanostructures. They found that the particular shape and structure of the resulting nanostructured silicon were influenced by the type of CM used. For instance, Si derived from montmorillonite retained a bidimensional porous structure resembling the original montmorillonite morphology. On the other hand, Si obtained from halloysite and palygorskite showcased distinct three-dimensional interconnected frameworks and loose textures composed of nanoparticles, which differed significantly from their precursor minerals.

Notably, these Si nanostructures displayed excellent electrochemical performance as anode materials for lithium-ion batteries due to their large specific surface area and hierarchical pore structure. Among them, Si(Mt) exhibited a more stable cycling performance, with a capacity retention of 78% over 200 cycles at 1.0 A g^{-1}, attributed to its limited surface area and enhanced transport properties [107].

Conclusions and future perspectives

In summary, the green synthesis of SiNPs is not only an innovative stride but a consequential benchmark in the narrative of sustainable development. It redefines the technological intervention into the uncharted territories of environmental exigencies by acknowledging the co-existence of scientific advancement and environmental conservation.

The future trajectory of green synthesis methods needs persistent reiteration, leading to the maturity of techniques, and paving the way for valuable insights into more refined and eco-compatible processes. The debate is no longer welling around the plausibility of eco-friendly synthesis but about enhancing the existing techniques to create less environmental impact.

The practical benefits of adopting green synthesis of silica nanoparticles are immense. It includes rational resource utilization, drastic reduction in waste generation, and avoidance of harmful by-products. At its core, it promotes sustainable production without jeopardizing the efficiency and effectiveness of SiNPs.

The myriad applications of SiNPs synthesized through eco-friendly methods cover vast domains like medicine, environmental science, agriculture, and electronics to name a few. The medical sphere hails its potential for its use in targeted drug delivery systems and precise biomedical imaging techniques. Their role in extracting hazardous metallic substances from wastewater showcases their importance in environmental conservation. In the agricultural sector, they enrich the soil fertility and consequently improve crop yield. Their utility extends to the electronics industry, where they are used in creating high-quality screens and conducting polymers. While incorporating green-synthesized nanoparticles into these domains is not novel, ongoing research centers on developing nanoparticles that can be introduced into the environment without triggering new ecological concerns.

Furthermore, the pursuit of recyclability takes center stage, offering potential cost savings in chemical processes while reducing environmental impact. This drive toward sustainable and eco-friendly solutions underscores the commitment to a greener future.

In the broader discourse surrounding nanoparticle synthesis, a dilemma persists - will we continue on the path of resource exploitation, or will we tilt our technologies toward sustainable innovation? The answer seems quite clear. The green synthesis of SiNPs, the multiple benefits they offer, and the wide application span invite us to a future steeped in sustainability and environmental responsibility.

References

[1] S. Malik, K. Muhammad, y Y. Waheed, «Nanotechnology: A Revolution in Modern Industry», *Molecules*, vol. 28, n.° 2, p. 661, ene. 2023. https://doi.org/10.3390/molecules28020661

[2] I. Hussain, N. B. Singh, A. Singh, H. Singh, y S. C. Singh, «Green synthesis of nanoparticles and its potential application», *Biotechnol. Lett.*, vol. 38, n.° 4, pp. 545-560, abr. 2016. https://doi.org/10.1007/s10529-015-2026-7

[3] H. M. E. Azzazy, M. M. H. Mansour, T. M. Samir, y R. Franco, «Gold nanoparticles in the clinical laboratory: principles of preparation and applications», *Clin. Chem. Lab. Med. CCLM*, vol. 50, n.° 2, ene. 2012. https://doi.org/10.1515/cclm.2011.732

[4] K. Maekawa *et al.*, «Drop-on-demand laser sintering with silver nanoparticles for electronics packaging», *IEEE Trans. Compon. Packag. Manuf. Technol.*, vol. 2, n.° 5, pp. 868-877, 2012

[5] P. Sanguansri y M. A. Augustin, «Nanoscale materials development – a food industry perspective», *Trends Food Sci. Technol.*, vol. 17, n.° 10, pp. 547-556, oct. 2006. https://doi.org/10.1016/j.tifs.2006.04.010

[6] F. K. Alanazi, A. A. Radwan, y I. A. Alsarra, «Biopharmaceutical applications of nanogold», *Saudi Pharm. J.*, vol. 18, n.° 4, pp. 179-193, oct. 2010. https://doi.org/10.1016/j.jsps.2010.07.002

[7] J. Virkutyte y R. S. Varma, «Green synthesis of metal nanoparticles: Biodegradable polymers and enzymes in stabilization and surface functionalization», *Chem Sci*, vol. 2, n.° 5, pp. 837-846, 2011. https://doi.org/10.1039/C0SC00338G

[8] V. Bansal, A. Ahmad, y M. Sastry, «Fungus-Mediated Biotransformation of Amorphous Silica in Rice Husk to Nanocrystalline Silica», *J. Am. Chem. Soc.*, vol. 128, n.° 43, pp. 14059-14066, nov. 2006. https://doi.org/10.1021/ja062113+

[9] S. Bettini, A. Santino, L. Valli, y G. Giancane, «A smart method for the fast and low-cost removal of biogenic amines from beverages by means of iron oxide nanoparticles», *RSC Adv.*, vol. 5, n.° 23, pp. 18167-18171, 2015. https://doi.org/10.1039/C5RA01699A

[10]   A. S. Taleghani *et al.*, «Mesoporous silica nanoparticles as a versatile nanocarrier for cancer treatment: A review», *J. Mol. Liq.*, vol. 328, p. 115417, abr. 2021. https://doi.org/10.1016/j.molliq.2021.115417

[11]   J. G. Croissant, K. S. Butler, J. I. Zink, y C. J. Brinker, «Synthetic amorphous silica nanoparticles: toxicity, biomedical and environmental implications», *Nat. Rev. Mater.*, vol. 5, n.° 12, pp. 886-909, sep. 2020. https://doi.org/10.1038/s41578-020-0230-0

[12]   D. L. Green, J. S. Lin, Y.-F. Lam, M. Z.-C. Hu, D. W. Schaefer, y M. T. Harris, «Size, volume fraction, and nucleation of Stober silica nanoparticles», *J. Colloid Interface Sci.*, vol. 266, n.° 2, pp. 346-358, oct. 2003. https://doi.org/10.1016/S0021-9797(03)00610-6

[13]   C. Xu, Y. Niu, A. Popat, S. Jambhrunkar, S. Karmakar, y C. Yu, «Rod-like mesoporous silica nanoparticles with rough surfaces for enhanced cellular delivery», *J Mater Chem B*, vol. 2, n.° 3, pp. 253-256, 2014. https://doi.org/10.1039/C3TB21431A

[14]   R. K. Kankala *et al.*, «Nanoarchitectured Structure and Surface Biofunctionality of Mesoporous Silica Nanoparticles», *Adv. Mater.*, vol. 32, n.° 23, p. 1907035, jun. 2020. https://doi.org/10.1002/adma.201907035

[15]   N. Pal, J.-H. Lee, y E.-B. Cho, «Recent Trends in Morphology-Controlled Synthesis and Application of Mesoporous Silica Nanoparticles», *Nanomaterials*, vol. 10, n.° 11, p. 2122, oct. 2020. https://doi.org/10.3390/nano10112122

[16]   S. Jafari, H. Derakhshankhah, L. Alaei, A. Fattahi, B. S. Varnamkhasti, y A. A. Saboury, «Mesoporous silica nanoparticles for therapeutic/diagnostic applications», *Biomed. Pharmacother.*, vol. 109, pp. 1100-1111, ene. 2019. https://doi.org/10.1016/j.biopha.2018.10.167

[17]   D. Xia, D. Li, Z. Ku, Y. Luo, y S. R. J. Brueck, «Top-Down Approaches to the Formation of Silica Nanoparticle Patterns», *Langmuir*, vol. 23, n.° 10, pp. 5377-5385, may 2007. https://doi.org/10.1021/la7005666

[18]   M. C. Gonçalves, «Sol-gel Silica Nanoparticles in Medicine: A Natural Choice. Design, Synthesis and Products», *Molecules*, vol. 23, n.° 8, p. 2021, ago. 2018. https://doi.org/10.3390/molecules23082021

[19]   S. Amoruso *et al.*, «Generation of silicon nanoparticles via femtosecond laser ablation in vacuum», *Appl. Phys. Lett.*, vol. 84, n.° 22, pp. 4502-4504, may 2004. https://doi.org/10.1063/1.1757014

[20]   M. Y. Kirillin *et al.*, «Laser-ablated silicon nanoparticles: optical properties and perspectives in optical coherence tomography», *Laser Phys.*, vol. 25, n.° 7, p. 075604, jul. 2015. https://doi.org/10.1088/1054-660X/25/7/075604

[21]   M. Kobayashi, S.-M. Liu, S. Sato, H. Yao, y K. Kimura, «Optical Evaluation of Silicon Nanoparticles Prepared by Arc Discharge Method in Liquid Nitrogen», *Jpn. J. Appl. Phys.*, vol. 45, n.° 8A, pp. 6146-6152, ago. 2006. https://doi.org/10.1143/JJAP.45.6146

[22]   B. M. Goortani, P. Proulx, S. Xue, y N. Y. Mendoza-Gonzalez, «Controlling nanostructure in thermal plasma processing: Moving from highly aggregated porous structure to spherical silica nanoparticles», *Powder Technol.*, vol. 175, n.° 1, pp. 22-32, jun. 2007. https://doi.org/10.1016/j.powtec.2007.01.014

[23]   Y. V. Kargina *et al.*, «Silicon Nanoparticles Prepared by Plasma-Assisted Ablative Synthesis: Physical Properties and Potential Biomedical Applications», *Phys. Status Solidi A*, vol. 216, n.° 14, p. 1800897, jul. 2019. https://doi.org/10.1002/pssa.201800897

[24]   R. S. Dubey, Y. B. R. D. Rajesh, y M. A. More, «Synthesis and Characterization of SiO2 Nanoparticles via Sol-gel Method for Industrial Applications», *Mater. Today Proc.*, vol. 2, n.° 4-5, pp. 3575-3579, 2015. https://doi.org/10.1016/j.matpr.2015.07.098

[25]   R. Sun, J. Zhou, y W. Wang, «A boric acid-assisted hydrothermal process for preparation of mesoporous silica nanoparticles with ultra-large mesopores and tunable particle sizes»,

Ceram. Int., vol. 49, n.° 12, pp. 20518-20527, jun. 2023.
https://doi.org/10.1016/j.ceramint.2023.03.181

[26]   C.-H. Lin, J.-H. Chang, Y.-Q. Yeh, S.-H. Wu, Y.-H. Liu, y C.-Y. Mou, «Formation of hollow silica nanospheres by reverse microemulsion», *Nanoscale*, vol. 7, n.° 21, pp. 9614-9626, 2015. https://doi.org/10.1039/C5NR01395J

[27]   S. D. Karande, S. A. Jadhav, H. B. Garud, V. A. Kalantre, S. H. Burungale, y P. S. Patil, «Green and sustainable synthesis of silica nanoparticles», *Nanotechnol. Environ. Eng.*, vol. 6, n.° 2, p. 29, ago. 2021. https://doi.org/10.1007/s41204-021-00124-1

[28]   A. Gour y N. K. Jain, «Advances in green synthesis of nanoparticles», *Artif. Cells Nanomedicine Biotechnol.*, vol. 47, n.° 1, pp. 844-851, dic. 2019.
https://doi.org/10.1080/21691401.2019.1577878

[29]   K. Parveen, V. Banse, y L. Ledwani, «Green synthesis of nanoparticles: Their advantages and disadvantages», presentado en 5TH NATIONAL CONFERENCE ON THERMOPHYSICAL PROPERTIES: (NCTP-09), Baroda (India), 2016, p. 020048. doi: 10.1063/1.4945168

[30]   S. Ying *et al.*, «Green synthesis of nanoparticles: Current developments and limitations», *Environ. Technol. Innov.*, vol. 26, p. 102336, may 2022.
https://doi.org/10.1016/j.eti.2022.102336

[31]   M. Shahrtash, «Silicon fertilization as a sustainable approach to disease management of agricultural crops», *J. Plant Prot. Res.*, vol. 58, n.° 4, 2018

[32]   M. E. C. Washington Le, «Preparation and Characterization of Nano Silica from Equisetum arvenses», *J. Bioprocess. Biotech.*, vol. 05, n.° 02, 2015.
https://doi.org/10.4172/2155-9821.1000205

[33]   K. Adach, D. Kroisova, y M. Fijalkowski, «Biogenic silicon dioxide nanoparticles processed from natural sources», *Part. Sci. Technol.*, vol. 39, n.° 4, pp. 481-489, may 2021.
https://doi.org/10.1080/02726351.2020.1758857

[34]   S. Mehmood *et al.*, «A green method for removing chromium (VI) from aqueous systems using novel silicon nanoparticles: Adsorption and interaction mechanisms», *Environ. Res.*, vol. 213, p. 113614, oct. 2022. https://doi.org/10.1016/j.envres.2022.113614

[35]   T. V. S. Adinarayana, A. Mishra, I. Singhal, y D. V. R. Koti Reddy, «Facile green synthesis of silicon nanoparticles from *Equisetum arvense* for fluorescence based detection of Fe(iii) ions», *Nanoscale Adv.*, vol. 2, n.° 9, pp. 4125-4132, 2020.
https://doi.org/10.1039/D0NA00307G

[36]   D. Sachan, A. Ramesh, y G. Das, «Green synthesis of silica nanoparticles from leaf biomass and its application to remove heavy metals from synthetic wastewater: A comparative analysis», *Environ. Nanotechnol. Monit. Manag.*, vol. 16, p. 100467, 2021

[37]   Sharma, P., Kherb, J., Prakash, J. et al. A novel and facile green synthesis of SiO2 nanoparticles for removal of toxic water pollutants. Appl Nanosci 13, 735–747 (2023).
https://doi.org/10.1007/s13204-021-01898-1

[38]   R. H. Babu, P. Yugandhar, y N. Savithramma, «Synthesis, characterization and antimicrobial studies of bio silica nanoparticles prepared from Cynodon dactylon L.: a green approach», *Bull. Mater. Sci.*, vol. 41, pp. 1-8, 2018

[39]   M. Sankareswaran, M. Vanitha, R. Periakaruppan, y A. Anbukumaran, «Phyllanthus emblica Mediated Silica Nanomaterials: Biosynthesis, Structural and Stability Analysis», *Silicon*, vol. 14, n.° 15, pp. 10123-10127, oct. 2022. https://doi.org/10.1007/s12633-022-01724-5

[40]   A. Tiwari, Y. L. Sherpa, A. P. Pathak, L. S. Singh, A. Gupta, y A. Tripathi, «One-pot green synthesis of highly luminescent silicon nanoparticles using Citrus limon (L.) and their applications in luminescent cell imaging and antimicrobial efficacy», *Mater. Today Commun.*, vol. 19, pp. 62-67, jun. 2019. https://doi.org/10.1016/j.mtcomm.2018.12.005

[41]   R. Periakaruppan, R. D. N, S. A. Abed, P. Vanathi, y J. S. Kumar, «Production of Biogenic Silica Nanoparticles by Green Chemistry Approach and Assessment of their Physicochemical Properties and Effects on the Germination of Sorghum bicolor», *Silicon*, vol. 15, n.° 10, pp. 4309-4316, jul. 2023. https://doi.org/10.1007/s12633-023-02348-z

[42]   R. Periakaruppan, M. P. S, P. C, R. P, G. R. S, y J. Danaraj, «Biosynthesis of Silica Nanoparticles Using the Leaf Extract of Punica granatum and Assessment of Its Antibacterial Activities Against Human Pathogens», *Appl. Biochem. Biotechnol.*, vol. 194, n.° 11, pp. 5594-5605, nov. 2022. https://doi.org/10.1007/s12010-022-03994-6

[43]   C. Y. Rahimzadeh, A. A. Barzinjy, A. S. Mohammed, y S. M. Hamad, «Green synthesis of SiO2 nanoparticles from Rhus coriaria L. extract: Comparison with chemically synthesized SiO2 nanoparticles», *PLOS ONE*, vol. 17, n.° 8, p. e0268184, ago. 2022. https://doi.org/10.1371/journal.pone.0268184

[44]   M. Aluga y K. Chewe, «AGRO-WASTE ASHES AS A FEEDER FOR THE SYNTHESIS OF SIO $_2$ NANOPARTICLES FOR ROAD CONSTRUCTION», presentado en WASTE MANAGEMENT AND ENVIRONMENTAL IMPACT 2022, Online, ago. 2022, pp. 53-63. doi: 10.2495/WMEI220051

[45]   D. Dorairaj, N. Govender, S. Zakaria, y R. Wickneswari, «Green synthesis and characterization of UKMRC-8 rice husk-derived mesoporous silica nanoparticle for agricultural application», *Sci. Rep.*, vol. 12, n.° 1, p. 20162, nov. 2022. https://doi.org/10.1038/s41598-022-24484-z

[46]   W. Jansomboon, K. Boonmaloet, S. Sukaros, y P. Prapainainar, «Rice Hull Micro and Nanosilica: Synthesis and Characterization», *Key Eng. Mater.*, vol. 718, pp. 77-80, nov. 2016. https://doi.org/10.4028/www.scientific.net/KEM.718.77

[47]   B. Sekhon, «Nanotechnology in agri-food production: an overview», *Nanotechnol. Sci. Appl.*, p. 31, may 2014. https://doi.org/10.2147/NSA.S39406

[48]   F. Adam, T.-S. Chew, y J. Andas, «A simple template-free sol–gel synthesis of spherical nanosilica from agricultural biomass», *J. Sol-Gel Sci. Technol.*, vol. 59, n.° 3, pp. 580-583, sep. 2011. https://doi.org/10.1007/s10971-011-2531-7

[49]   R. H. Alves, T. V. D. S. Reis, S. Rovani, y D. A. Fungaro, «Green Synthesis and Characterization of Biosilica Produced from Sugarcane Waste Ash», *J. Chem.*, vol. 2017, pp. 1-9, 2017. https://doi.org/10.1155/2017/6129035

[50]   N. K. Mohd, N. N. A. N. Wee, y A. A. Azmi, «Green synthesis of silica nanoparticles using sugarcane bagasse», presentado en 3RD ELECTRONIC AND GREEN MATERIALS INTERNATIONAL CONFERENCE 2017 (EGM 2017), Krabi, Thailand, 2017, p. 020123. doi: 10.1063/1.5002317

[51]   A. Boonmee y K. Jarukumjorn, «Preparation and characterization of silica nanoparticles from sugarcane bagasse ash for using as a filler in natural rubber composites», *Polym. Bull.*, vol. 77, n.º 7, pp. 3457-3472, jul. 2020. https://doi.org/10.1007/s00289-019-02925-6

[52]   P. E. Imoisili, K. O. Ukoba, y T.-C. Jen, «Green technology extraction and characterisation of silica nanoparticles from palm kernel shell ash via sol–gel», *J. Mater. Res. Technol.*, vol. 9, n.º 1, pp. 307-313, ene. 2020. https://doi.org/10.1016/j.jmrt.2019.10.059

[53]   D. F. Mohamad *et al.*, «Synthesis of Mesoporous Silica Nanoparticle from Banana Peel Ash for Removal of Phenol and Methyl Orange in Aqueous Solution», *Mater. Today Proc.*, vol. 19, pp. 1119-1125, 2019. https://doi.org/10.1016/j.matpr.2019.11.004

[54]   M. Khairul Hanif Mohd Nazri y N. Sapawe, «Effective performance of silica nanoparticles extracted from bamboo leaves ash for removal of phenol», *Mater. Today Proc.*, vol. 31, pp. A27-A32, 2020. https://doi.org/10.1016/j.matpr.2020.10.964

[55]   G. Sivakumar y K. Amutha, «Studies on Silica Obtained from Cow Dung Ash», *Adv. Mater. Res.*, vol. 584, pp. 470-473, oct. 2012. https://doi.org/10.4028/www.scientific.net/AMR.584.470

[56]   M. L. L. Formoso, «Some topics on geochemistry of weathering: a review», *An. Acad. Bras. Ciênc.*, vol. 78, n.º 4, pp. 809-820, dic. 2006. https://doi.org/10.1590/S0001-37652006000400014

[57]   M. Yadav, V. Dwibedi, S. Sharma, y N. George, «Biogenic silica nanoparticles from agro-waste: Properties, mechanism of extraction and applications in environmental sustainability», *J. Environ. Chem. Eng.*, vol. 10, n.º 6, p. 108550, dic. 2022. https://doi.org/10.1016/j.jece.2022.108550

[58]   N. Vasanthi, L. M. Saleena, y S. A. Raj, «Silica Solubilization Potential of Certain Bacterial Species in the Presence of Different Silicate Minerals», *Silicon*, vol. 10, n.º 2, pp. 267-275, mar. 2018. https://doi.org/10.1007/s12633-016-9438-4

[59]   H. Ehrlich, K. D. Demadis, O. S. Pokrovsky, y P. G. Koutsoukos, «Modern Views on Desilicification: Biosilica and Abiotic Silica Dissolution in Natural and Artificial Environments», *Chem. Rev.*, vol. 110, n.º 8, pp. 4656-4689, ago. 2010. https://doi.org/10.1021/cr900334y

[60]   M. M. Urrutia y T. J. Beveridge, «Formation of fine-grained metal and silicate precipitates on a bacterial surface (Bacillus subtilis)», *Chem. Geol.*, vol. 116, n.º 3-4, pp. 261-280, oct. 1994. https://doi.org/10.1016/0009-2541(94)90018-3

[61]   S. Abdeen, S. Geo, P. Praseetha, y R. Dhanya, «Biosynthesis of silver nanoparticles from Actinomycetes for therapeutic applications», 2014

[62]   S. Uroz, C. Calvaruso, M.-P. Turpault, y P. Frey-Klett, «Mineral weathering by bacteria: ecology, actors and mechanisms», *Trends Microbiol.*, vol. 17, n.° 8, pp. 378-387, 2009

[63]   R. Mohanraj *et al.*, «Decolourisation efficiency of immobilized silica nanoparticles synthesized by actinomycetes», *Mater. Today Proc.*, vol. 48, pp. 129-135, 2022. https://doi.org/10.1016/j.matpr.2020.04.139

[64]   K. Natesan *et al.*, «Biosynthesis of silica and copper nanoparticles from *Trichoderma* , *Streptomyces* and *Pseudomonas* spp. evaluated against collar canker and red root-rot disease of tea plants», *Arch. Phytopathol. Plant Prot.*, vol. 54, n.° 1-2, pp. 56-85, ene. 2021. https://doi.org/10.1080/03235408.2020.1817258

[65]   S. Show, A. Tamang, T. Chowdhury, D. Mandal, y B. Chattopadhyay, «Bacterial (BKH1) assisted silica nanoparticles from silica rich substrates: A facile and green approach for biotechnological applications», *Colloids Surf. B Biointerfaces*, vol. 126, pp. 245-250, feb. 2015. https://doi.org/10.1016/j.colsurfb.2014.12.039

[66]   V. Patel, D. Berthold, P. Puranik, y M. Gantar, «Screening of cyanobacteria and microalgae for their ability to synthesize silver nanoparticles with antibacterial activity», *Biotechnol. Rep.*, vol. 5, pp. 112-119, mar. 2015. https://doi.org/10.1016/j.btre.2014.12.001

[67]   D. Fawcett, J. J. Verduin, M. Shah, S. B. Sharma, y G. E. J. Poinern, «A Review of Current Research into the Biogenic Synthesis of Metal and Metal Oxide Nanoparticles via Marine Algae and Seagrasses», *J. Nanosci.*, vol. 2017, pp. 1-15, ene. 2017. https://doi.org/10.1155/2017/8013850

[68]   P. C. Sahoo, F. Kausar, J. H. Lee, y J. I. Han, «Facile fabrication of silver nanoparticle embedded $CaCO_3$ microspheres via microalgae-templated CO_2 biomineralization: application in antimicrobial paint development», *RSC Adv.*, vol. 4, n.° 61, p. 32562, jul. 2014. https://doi.org/10.1039/C4RA03623A

[69]   Z. Bao, J. Cao, G. Kang, y C. Q. Lan, «Effects of reaction conditions on light-dependent silver nanoparticle biosynthesis mediated by cell extract of green alga Neochloris oleoabundans», *Environ. Sci. Pollut. Res.*, vol. 26, pp. 2873-2881, 2019

[70]   R. C. Dugdale y F. P. Wilkerson, «Silicate regulation of new production in the equatorial Pacific upwelling», *Nature*, vol. 391, n.° 6664, pp. 270-273, ene. 1998. https://doi.org/10.1038/34630

[71]   M. Sumper y E. Brunner, «Learning from Diatoms: Nature's Tools for the Production of Nanostructured Silica», *Adv. Funct. Mater.*, vol. 16, n.° 1, pp. 17-26, ene. 2006. https://doi.org/10.1002/adfm.200500616

[72]   T. Nakajima y B. E. Volcani, «3,4-Dihydroxyproline: A New Amino Acid in Diatom Cell Walls», *Science*, vol. 164, n.° 3886, pp. 1400-1401, jun. 1969. https://doi.org/10.1126/science.164.3886.1400

[73]   R. W. Drum y H. S. Pankratz, «Post mitotic fine structure of Gomphonema parvulum», *J. Ultrastruct. Res.*, vol. 10, n.° 3-4, pp. 217-223, abr. 1964. https://doi.org/10.1016/S0022-5320(64)80006-X

[74]   N. Kröger, R. Deutzmann, y M. Sumper, «Polycationic Peptides from Diatom Biosilica That Direct Silica Nanosphere Formation», *Science*, vol. 286, n.º 5442, pp. 1129-1132, nov. 1999. https://doi.org/10.1126/science.286.5442.1129

[75]   T. Nakajima y B. E. Volcani, «ε-N-trimethyl-L-δ -hydroxylysine phosphate and its nonphosphorylated compound in diatom cell walls», *Biochem. Biophys. Res. Commun.*, vol. 39, n.º 1, pp. 28-33, abr. 1970. https://doi.org/10.1016/0006-291X(70)90752-7

[76]   N. Poulsen y N. Kröger, «Silica Morphogenesis by Alternative Processing of Silaffins in the Diatom Thalassiosira pseudonana», *J. Biol. Chem.*, vol. 279, n.º 41, pp. 42993-42999, oct. 2004. https://doi.org/10.1074/jbc.M407734200

[77]   N. Kröger, R. Deutzmann, C. Bergsdorf, y M. Sumper, «Species-specific polyamines from diatoms control silica morphology», *Proc. Natl. Acad. Sci.*, vol. 97, n.º 26, pp. 14133-14138, dic. 2000. https://doi.org/10.1073/pnas.260496497

[78]   M. S. Aw, S. Simovic, Y. Yu, J. Addai-Mensah, y D. Losic, «Porous silica microshells from diatoms as biocarrier for drug delivery applications», *Powder Technol.*, vol. 223, pp. 52-58, jun. 2012. https://doi.org/10.1016/j.powtec.2011.04.023

[79]   X. Lin, L. Tirichine, y C. Bowler, «Protocol: Chromatin immunoprecipitation (ChIP) methodology to investigate histone modifications in two model diatom species», *Plant Methods*, vol. 8, n.º 1, p. 48, dic. 2012. https://doi.org/10.1186/1746-4811-8-48

[80]   I. Rea, M. Terracciano, y L. De Stefano, «Synthetic vs Natural: Diatoms Bioderived Porous Materials for the Next Generation of Healthcare Nanodevices», *Adv. Healthc. Mater.*, vol. 6, n.º 3, p. 1601125, feb. 2017. https://doi.org/10.1002/adhm.201601125

[81]   I. Rea *et al.*, «Diatomite biosilica nanocarriers for siRNA transport inside cancer cells», *Biochim. Biophys. Acta BBA - Gen. Subj.*, vol. 1840, n.º 12, pp. 3393-3403, dic. 2014. https://doi.org/10.1016/j.bbagen.2014.09.009

[82]   L.-Z. Sun y Y.-J. Ying, «Moving dynamics of a nanorobot with three DNA legs on nanopore-based tracks», *Nanoscale*, p. 10.1039.D3NR03747A, 2023. https://doi.org/10.1039/D3NR03747A

[83]   R. Sasirekha *et al.*, «Surface engineered Amphora subtropica frustules using chitosan as a drug delivery platform for anticancer therapy», *Mater. Sci. Eng. C*, vol. 94, pp. 56-64, ene. 2019. https://doi.org/10.1016/j.msec.2018.09.009

[84]   R. B. Vasani, D. Losic, A. Cavallaro, y N. H. Voelcker, «Fabrication of stimulus-responsive diatom biosilica microcapsules for antibiotic drug delivery», *J. Mater. Chem. B*, vol. 3, n.º 21, pp. 4325-4329, 2015. https://doi.org/10.1039/C5TB00648A

[85]   B. Delalat *et al.*, «Targeted drug delivery using genetically engineered diatom biosilica», *Nat. Commun.*, vol. 6, n.º 1, p. 8791, nov. 2015. https://doi.org/10.1038/ncomms9791

[86]   K. Squire, X. Kong, P. LeDuff, G. L. Rorrer, y A. X. Wang, «Photonic crystal enhanced fluorescence immunoassay on diatom biosilica», *J. Biophotonics*, vol. 11, n.º 10, p. e201800009, oct. 2018. https://doi.org/10.1002/jbio.201800009

[87]   K. J. Squire *et al.*, «Photonic crystal-enhanced fluorescence imaging immunoassay for cardiovascular disease biomarker screening with machine learning analysis», *Sens. Actuators B Chem.*, vol. 290, pp. 118-124, jul. 2019. https://doi.org/10.1016/j.snb.2019.03.102

[88]   C. B. Michielse y M. Rep, «Pathogen profile update: *Fusarium oxysporum*», *Mol. Plant Pathol.*, vol. 10, n.º 3, pp. 311-324, may 2009. https://doi.org/10.1111/j.1364-3703.2009.00538.x

[89]   V. Bansal *et al.*, «Fungus-mediated biosynthesis of silica and titania particles», *J. Mater. Chem.*, vol. 15, n.º 26, p. 2583, 2005. https://doi.org/10.1039/b503008k

[90]   V. Bansal, D. Rautaray, A. Ahmad, y M. Sastry, «Biosynthesis of zirconia nanoparticles using the fungus Fusarium oxysporum», *J. Mater. Chem.*, vol. 14, n.º 22, p. 3303, 2004. https://doi.org/10.1039/b407904c

[91]   V. Bansal, A. Sanyal, D. Rautaray, A. Ahmad, y M. Sastry, «Bioleaching of Sand by the FungusFusarium oxysporum as a Means of Producing Extracellular Silica Nanoparticles», *Adv. Mater.*, vol. 17, n.º 7, pp. 889-892, abr. 2005. https://doi.org/10.1002/adma.200401176

[92]   A. Piela *et al.*, «Biogenic synthesis of silica nanoparticles from corn cobs husks. Dependence of the productivity on the method of raw material processing», *Bioorganic Chem.*, vol. 99, p. 103773, jun. 2020. https://doi.org/10.1016/j.bioorg.2020.103773

[93]   A. Zielonka, E. Żymańczyk-Duda, M. Brzezińska-Rodak, M. Duda, J. Grzesiak, y M. Klimek-Ochab, «Nanosilica synthesis mediated by Aspergillus parasiticus strain», *Fungal Biol.*, vol. 122, n.º 5, pp. 333-344, may 2018. https://doi.org/10.1016/j.funbio.2018.02.004

[94]   E. Vetchinkina, E. Loshchinina, M. Kupryashina, A. Burov, y V. Nikitina, «Shape and Size Diversity of Gold, Silver, Selenium, and Silica Nanoparticles Prepared by Green Synthesis Using Fungi and Bacteria», *Ind. Eng. Chem. Res.*, vol. 58, n.º 37, pp. 17207-17218, sep. 2019. https://doi.org/10.1021/acs.iecr.9b03345

[95]   M. A. Albalawi, A. M. Abdelaziz, M. S. Attia, E. Saied, H. H. Elganzory, y A. H. Hashem, «Mycosynthesis of Silica Nanoparticles Using Aspergillus niger: Control of Alternaria solani Causing Early Blight Disease, Induction of Innate Immunity and Reducing of Oxidative Stress in Eggplant», *Antioxidants*, vol. 11, n.º 12, p. 2323, nov. 2022. https://doi.org/10.3390/antiox11122323

[96]   P. Samaddar, Y. S. Ok, K.-H. Kim, E. E. Kwon, y D. C. W. Tsang, «Synthesis of nanomaterials from various wastes and their new age applications», *J. Clean. Prod.*, vol. 197, pp. 1190-1209, oct. 2018. https://doi.org/10.1016/j.jclepro.2018.06.262

[97]   V. Kumar Yadav y M. H. Fulekar, «Green synthesis and characterization of amorphous silica nanoparticles from fly ash», *Mater. Today Proc.*, vol. 18, pp. 4351-4359, 2019. https://doi.org/10.1016/j.matpr.2019.07.395

[98]   F. Yan *et al.*, «Green Synthesis of Nanosilica from Coal Fly Ash and Its Stabilizing Effect on CaO Sorbents for CO_2 Capture», *Environ. Sci. Technol.*, vol. 51, n.º 13, pp. 7606-7615, jul. 2017. https://doi.org/10.1021/acs.est.7b00320

[99]   M. Sarikaya, T. Depci, R. Aydogmus, A. Yucel, y N. Kizilkaya, «Production of Nano Amorphous SiO_2 from Malatya Pyrophyllite», *IOP Conf. Ser. Earth Environ. Sci.*, vol. 44, p. 052004, oct. 2016. https://doi.org/10.1088/1755-1315/44/5/052004

[100] S. Stopic *et al.*, «Synthesis of Nanosilica via Olivine Mineral Carbonation under High Pressure in an Autoclave», *Metals*, vol. 9, n.° 6, p. 708, jun. 2019. https://doi.org/10.3390/met9060708

[101] F. Bergaya y G. Lagaly, «General Introduction», en *Developments in Clay Science*, vol. 5, Elsevier, 2013, pp. 1-19. doi: 10.1016/B978-0-08-098258-8.00001-8

[102] X. Zhou, L. Wu, J. Yang, J. Tang, L. Xi, y B. Wang, «Synthesis of nano-sized silicon from natural halloysite clay and its high performance as anode for lithium-ion batteries», *J. Power Sources*, vol. 324, pp. 33-40, ago. 2016. https://doi.org/10.1016/j.jpowsour.2016.05.058

[103] J. Ryu, D. Hong, M. Shin, y S. Park, «Multiscale Hyperporous Silicon Flake Anodes for High Initial Coulombic Efficiency and Cycle Stability», *ACS Nano*, vol. 10, n.° 11, pp. 10589-10597, nov. 2016. https://doi.org/10.1021/acsnano.6b06828

[104] K. Adpakpang, S. B. Patil, S. M. Oh, J.-H. Kang, M. Lacroix, y S.-J. Hwang, «Effective Chemical Route to 2D Nanostructured Silicon Electrode Material: Phase Transition from Exfoliated Clay Nanosheet to Porous Si Nanoplate», *Electrochimica Acta*, vol. 204, pp. 60-68, jun. 2016. https://doi.org/10.1016/j.electacta.2016.04.043

[105] A. Mourhly, M. Khachani, A. E. Hamidi, M. Kacimi, M. Halim, y S. Arsalane, «The Synthesis and Characterization of Low-Cost Mesoporous Silica SiO_2 from Local Pumice Rock», *Nanomater. Nanotechnol.*, vol. 5, p. 35, ene. 2015. https://doi.org/10.5772/62033

[106] Q. Chen *et al.*, «From natural clay minerals to porous silicon nanoparticles», *Microporous Mesoporous Mater.*, vol. 260, pp. 76-83, abr. 2018. https://doi.org/10.1016/j.micromeso.2017.10.033

[107] Q. Chen *et al.*, «Clay minerals derived nanostructured silicon with various morphology: Controlled synthesis, structural evolution, and enhanced lithium storage properties», *J. Power Sources*, vol. 405, pp. 61-69, nov. 2018. https://doi.org/10.1016/j.jpowsour.2018.10.031

Green Synthesis and Emerging Applications of Frontier Nanomaterials   Materials Research Forum LLC
Materials Research Foundations 169 (2024) 35-58   https://doi.org/10.21741/9781644903278-2

Chapter 2

Green silver nanoparticles: Synthesis, characterization and applications

Fátima Ibarra[1], Sofia Municoy[1], Pablo E. Antezana[1,2], Fresia Silva Sofrás[1],
Pablo Santo Orihuela[1,3], Martín F. Desimone[1,4*]

[1]Universidad de Buenos Aires, Consejo Nacional de Investigaciones Científicas y Técnicas (CONICET), Instituto de la Química y Metabolismo del Fármaco (IQUIMEFA), Facultad de Farmacia y Bioquímica Junín 956, Piso 3 (1113), Buenos Aires, Argentina

[2]Universidad de Buenos Aires. Consejo Nacional de Investigaciones Científicas y Técnicas (CONICET). Instituto de Bioquímica y Medicina Molecular (IBIMOL). Facultad de Farmacia y Bioquímica. Buenos Aires, Argentina

[3]Centro de Investigaciones en Plagas e Insecticidas (CIPEIN), Instituto de Investigaciones Científicas y Técnicas para la Defensa CITEDEF/UNIDEF, Consejo Nacional de Investigaciones Científicas y Técnicas, Buenos Aires, Argentina (CONICET), Juan B. de La Salle 4397, Villa Martelli, Buenos Aires 1603, Argentina

[4]Instituto de Ciências Biológicas (ICB), Universidade Federal do Rio Grande – FURG, Rio Grande, RS, Brazil

* desimone@ffyb.uba.ar

Keywords

Silver Nanoparticles, Green Synthesis, Plant Extracts, Bacteria, Fungi

Abstracts

Since the first synthesis in the 19th century, silver nanoparticles (AgNPs) have been used for several applications, mainly for its bactericidal properties. Although physical and chemical methods are the most used approaches for preparing AgNPs, in recent years they have been replaced by eco-friendlier, cost-effective and simpler methods. Thus, green synthesis employing bacteria, plant extracts and fungi for the reduction of silver to AgNPs has emerged as an interesting alternative. The use of plants for synthesizing AgNPs involves the preparation of aqueous extracts from the roots, leaves or stems to obtain secondary metabolites with antioxidant activity that reduce silver salts (i.e.: $AgNO_3$) or act as capping agents of the AgNPs. In case of green synthesis based on bacteria, the microorganisms are first cultivated in a suitable medium and then silver salts are added. When fungi are used for the synthesis, the fungus is frequently cultured on agar, then transferred to a liquid media and the biomass generated placed in water to recover the active compounds. After filtering, the biomass is removed and the filtrate is mixed with $AgNO_3$. Depending on the synthesis procedures, AgNPs of different size and shape are generated. In the present chapter, different methods based on the green synthesis of silver nanoparticles and the multiple applications of the AgNPs thus prepared will be discussed.

Contents

1.   **Introduction**

Silver nanoparticles (AgNPs) are tiny silver particles that are typically 1 to 100 nanometers in diameter. They have unique properties that make them valuable for a wide range of applications. In this sense, AgNPs are highly efficient to combat a wide spectra of bacteria, including multidrug-resistant strains. This makes them valuable for use in medical devices, wound dressings, and other applications where it is important to prevent or treat infections [1–3]. In addition, AgNPs can be used as catalysts for a variety of chemical reactions [4]. This makes them useful in many industrial and environmental applications, such as water purification [5,6]. Furthermore, AgNPs have unique optical properties, such as the ability to absorb and scatter light. This makes them interesting for a great variety of applications, such as solar cells, optical sensors, and imaging devices. It is also worth to mention that AgNPs are used in different consumer products, such as clothing, cosmetics, and food packaging. They are also used in some cleaning products and air purifiers.

While silver nanoparticles offer many potential benefits, there are also some potential risks associated with their use. For example, silver nanoparticles can be toxic to some aquatic organisms [7–9]. Additionally, there is some concern that silver nanoparticles could penetrate into the human body and cause adverse health effects [10].

Green synthesis of nanoparticles is a process that uses natural resources, such as plant extracts, microorganisms, and agricultural wastes, to produce NPs. These methods are considered to be more environmentally friendly and sustainable than traditional methods, which often use toxic chemicals and harsh conditions [11,12]. There are a number of reasons why green synthesis methods are considered to be more environmentally friendly and sustainable than traditional methods. First, green synthesis methods do not use toxic chemicals. This makes them safer for the environment and for human health. Second, green synthesis methods use renewable resources, such as microorganisms and plant extracts. This makes them more sustainable than traditional methods, which often use non-renewable resources. Third, green synthesis methods are often less expensive than traditional methods. This is because green synthesis methods do not require the use of expensive chemicals, energy, or equipment [13].

In addition to these environmental and economic benefits, green synthesis methods also offer a number of other advantages. For example, green synthesized AgNPs are often more biocompatible than traditionally synthesized AgNPs. This is because green synthesized AgNPs are typically coated with natural biomolecules, such as proteins and polysaccharides. These biomolecules help to reduce the toxicity of the AgNPs and make them more compatible with living cells and organisms [14,15].

Overall, green synthesis is an outstanding approach to produce AgNPs. Green synthesis methods are more environmentally friendly, sustainable, and biocompatible than traditional methods. As the technology continues to develop, green synthesized AgNPs are expected to play an increasingly important role in many industries and aspects of our lives.

In this book chapter, we will discuss the green synthesis of AgNPs in detail. We will cover the various alternatives of green synthesis procedures, the advantages of green synthesis methods, and the potential applications of green synthesized AgNPs.

2. Methodologies for the green production of silver nanoparticles

2.1 Plant-mediated synthesis of silver nanoparticles

2.1.1 Phyto-nanosynthesis

According to Pubmed, the very first research into the utilization of Geranium (*Pelargonium graveolens*) for the plant-mediated production of AgNPs was reported in 2003, becoming the pioneer in the realm of Plant-mediated nanotechnology[16]. This innovative approach has gained prominence as a highly efficient, convenient, rapid, eco-friendly, and non-toxic technique for AgNPs synthesis.

The number of articles related to silver nanoparticles obtained with plant extracts has steadily increased since 2003 and the highest number of articles was reached in 2022 with 377. This year, 2023, 277 articles have already been published (Fig. 1).

Various plant components, including bark, roots, stems, fruits, seeds, callus, peels, leaves, and flowers, have been used for nanoparticle synthesis [17–20]. The key to this methodology lies in the crucial role played by phytochemicals produced by the plant. These naturally occurring water-soluble compounds, comprising terpenoids, polyphenols, flavones, carboxylic acids, carbohydrates, phenolic acids, proteins and alkaloids, are responsible for the bio-reduction of silver ions during the synthesis (Fig. 2)[21].

As it was mentioned before, plant extracts contain a great variety of phytocomponents with phenol, amino, carboxyl, hydroxyl and carbonyl groups, thus, can reduce metals such as silver. Research findings have underscored the significance of specific functional groups within these phytochemicals, notably alkaloids, anthracenes and flavones, characterized by their $-C-O-C-$, $-C=C-$, $-C=O-$, and $-C-O$ moieties [17]. Remarkably, in most instances, the reducing agents present in the plant extract simultaneously serve as both stabilizing and capping agents, eliminating the need for the external introduction of such agents [18–23].

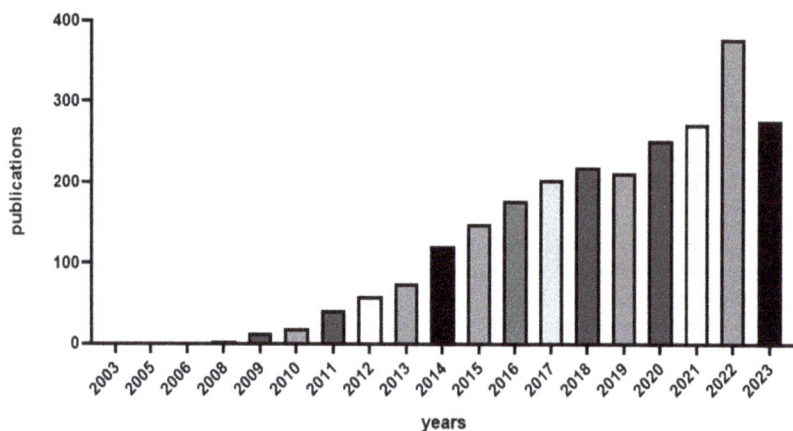

Fig. 1. Pubmed data describing the growth in number of publications since 2003 on synthesis of silver nanoparticles mediated by plant extracts.

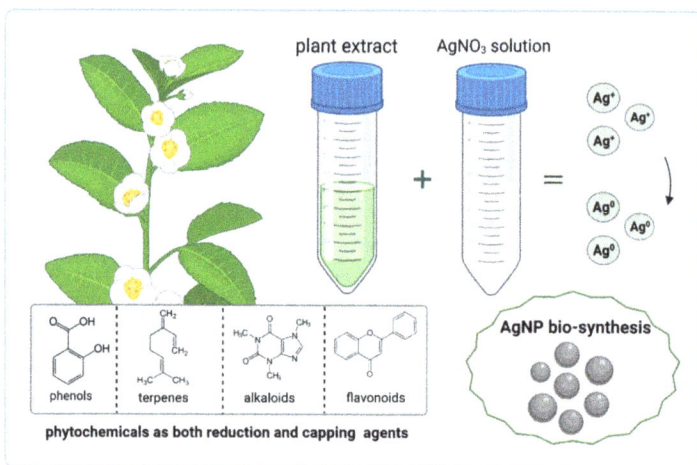

Fig 2. General scheme of AgNPs synthesis mediated by plant extracts. Silver is reduced by the action of phytochemicals.

Nanoparticles have been synthesized from more than 100 leaf extracts, more than 20 from roots, and more than 30 from fruits and seeds. Syntheses based on flower extracts were also cited. This indicates that a wide range of plant families have the capacity to promote the synthesis of diverse shapes and sizes for their employment in various applications. AgNPs synthesized with plant-extract posse sizes ranging from few nanometers to thousands of nanometers (*c.a.* 4,000 nm), whereas shapes range from spherical to rod and hexagonal. An exhaustive review of the plants used in nanoparticle synthesis, together with the variability in dimensions and shape, or their antibacterial, anticancer, antifungal, and antioxidant activity, have been reported in recent scientific works (Fig. 3) [17–19,22–24].

Specie	nm	shape	proposed reductor agent	activity	Ref
Chrysanthemum indicum	37–72	spherical	Flavonoids, terpenoids, and glycosides	antibacterial	a
Tagets erecta	10-90	spherical, hexagonal	Alkaloids, flavonoids, and cardiac glycoside	antifungal	b
Cassia angustifolia	10-80	spherical	Phenols, compounds, aromatics, alkane compounds	anticancer and antiantioxidant	c
Aloe vera	5-50	octahedron	flavanones, and terpenoids	antibacterial	d
Ocimum sanctum	5-50	spherical	proteins, terpenoids, ketones, carboxilic acid	antifungal	e
Acacia nilotica	10-50	spherical	proteins and phenolic compounds	antioxidant	f
Mukia maderaspatna	20-50	irregular	alkaloids, terpenoids flavonoids, glycosides	anticancer	g
Phoenix dactylifera	15-40	spherical	Alkenes, alcohols, carboxylic acid, esters and ethers	antifungal	h
Plumbago indica	50-70	spherical, oval	Plumbagin, flavonoids, tannins, alkaloids, phenolic compounds	anticancer	i
Trianthema decandra	36-74	spherical	saponins	antibacterial	j

*Fig 3. Example of nanoparticles obtained employing various parts of plants. Species name, size, shape and reported biological activity are detailed. (**a to f**[25–34]).*

2.1.2 Advantages of plant-extract as bio-precursor

This multifaceted approach, rooted in the rich chemistry of plant-derived compounds, not only exemplifies the elegance of green synthesis but also underscores its self-contained nature, minimizing the reliance on exogenous capping and stabilizing agents. Consequently, it stands as a pioneering technique in the burgeoning field of environmentally sustainable nanotechnology.

The green synthesis method is classified within the bottom-up methods, including synthesis from plant extract. In this methodology, atoms and molecules are assembled to form nanoparticles. It is desirable that these particles be nanometric in size, homogeneous and uniform and that they do not clump together forming larger complexes with a diversity of sizes. Thus, the role of a stabilizing agent becomes relevant. Some authors propose that the stabilizing function of plant extracts is due to the fact that their composition usually contains saponins [35]. Saponins are biosurfactants, which can replace synthetic surfactants.

The scalability of AgNPs production through plant-mediated synthesis is a notable advantage, making it well-suited for large-scale production, in comparison to alternative methods. AgNPs synthesized from plant extracts exhibit more stability and a huge variability of shapes and dimensions, these can be achieved by modifying various reaction conditions such as reaction time, pH, temperature and the amount of plant extract and silver source. This technique also offers an efficient and rapid route, with a higher yield of nanoparticles synthesis, avoiding the intricate and time-consuming processes associated with maintaining cell cultures, as often required in other biological methods [23].

On the other hand, the plant extract has antioxidant content that could be very useful to counter the own effect of the metal nanoparticles that produce free radical for their nature. Although sometimes this is a sought-after characteristic, as in the case of antimicrobials, it is also not entirely clear how toxic these effects can be on the environment and the human body. But, with this ecofriendly method, an amount of organic material is deposited over the surface of the nanoparticle, and it contributes to safety-by-design, improving their biocompatibility. Without this effect, a high oxygen free radicals amount can generate a lot of damage in living systems [36].

Notably, plant-synthesized nanoparticles exhibit some advantages regarding their application in human health due to the medicinal value coming from secondary metabolites present in plant extracts [18,20]. Green synthesized silver nanoparticles have been shown to possess an effective broad-spectrum antibacterial effect and have been shown to be effective to combat yeast and fungal pathogens. Nowadays, plant extract-synthesized AgNPs are rapidly becoming a key focus in nanobiotechnology due to their promising antitumor activity. Several research efforts are dedicated to confirming their potential as cancer-killing agents. Extensive practical studies are warranted to unlock the full spectrum of possibilities offered by plant-mediated synthesis of AgNPs in healthcare and beyond [18,20].

2.1.3 The in vitro plant-based reaction

During a metal nanoparticles green synthesis using plant extracts, there are five factors which should be chosen: the reaction temperature, the reaction duration, the metal salt (in this case is usually $AgNO_3$ in a particular concentration, commonly 0,1mM) [37], the plant extract, the reaction medium (just water or a buffer solution), and the necessary of a stabilizer agent. In this case, it must be chosen an ecofriendly one. But, as it was already mentioned, an advantage of plant extracts is that they have a role as reducing and stabilizer agent at the same time.

To prepare the plant extract, the part of the plant is introduced into a solvent to perform the extraction. Generally, the major extraction is possible at high temperature, for example 100°C, as Asif *et al.* mentioned in their work [38]. But it depends on the stability of the components to be extracted and if the solvent is volatile at low temperatures, like an alcohol. For example, in the work of Giri *et al.*, the extraction is carried out in methanol using liquid nitrogen [39]. However, the most used solvents are aqueous. The last step is very simple and involves separating the plant from the extract using a filtration method.

The most common way to synthesize nanoparticles using plants is to mix the salt solution with the extract of the desired plant in the appropriate reaction medium, for a determined time, at a defined temperature and under a constant stirring. The reaction is visible by the appearance of a yellow brown color in the solution due to the production of AgNPs by the conversion of Ag^+ to Ag^0. The longer the time, the more intense the color will be. At the end, the product is centrifuged to obtain the nanoparticles in the precipitated.

The extract components and the rest of the factors determine the features of the obtained nanoparticles. Major authors had reported that the best results are obtained under an alkaline pH: better stability and less dispersion in the size obtained. But an extremely alkaline pH allows the formation of agglomerates that could be attributed to the charge change over the surface [40]. About the temperature, most of the reported syntheses were performed at room temperature, but high temperatures seem to be useful to shorten the reaction time and to obtain smaller nanoparticles [40]. So, the pH along with temperature seems to determine the size of the nanoparticles. Furthermore, it is common that the amount of product obtained is proportional to the reaction time. Other treatments can be applied after centrifugation: the product can be dried to obtain a dry product or even it can be exposed to muffle, so the organic rest are eliminated. But in solution, it has been reported that silver nanoparticles obtained by plant-mediated synthesis show good stability for at least three months [35].

The shape is usually determined by the capping agents provided by the same plant extract, or by the addition of another stabilizing agent. Some works mentioned that NaCl salt is used to guide the formation of the nanoparticles due to absorption and stabilization of the surfaces of the product obtained, being this a green reagent[37]. Another example could be sodium dodecyl sulfate (SDS) [40].

Some works explained the effect of leaves extract concentration on the surface plasmon resonance (SPR) of the AgNPs obtained by plant-based synthesis, both in the appearance and intensity of the observed peak [38]. Considering that the SPR exposes characteristics of the size and morphology of the nanoparticles, it is likely that the concentration of the extract is decisive in establishing the properties of the AgNPs obtained. More intense signals indicate a greater surface area, which would indicate that smaller nanoparticles were formed, with a greater exposed surface area and, therefore, more stable. Most reported plant-based syntheses mention obtaining nanoparticles in the range of 1 to 30 nm and generally monodisperse with spherical shape. However, triangular, hexagonal, scales or rod-like silver nanoparticles can also be obtained [40]. These structures can be visualized by scanning or transmission electron microscopy (SEM and TEM, respectively).

2.1.4   Crucial keys in nanosynthesis mediated by plants

The choice of the plant to use as a reducing agent is crucial since the synthesis and its performance depend on it. Furthermore, it is vital to consider the presence of toxic compounds in some plants, to avoid their use in synthesis [23]. Moreover, the chosen plant extract must undergo through characterization and must exhibit remarkable stability to ensure the reproducibility of the process. It becomes imperative to establish facile, large-scale isolation and purification steps for plant metabolites responsible for nanosynthesis [23].

However, a significant problem raised during green synthesis is the aggregation of AgNPs, which leads to decreases surface area to volume ratio, altering AgNPs effectiveness and functionality. A strategic solution to mitigate this concern lies in finding the active molecule responsible for the reduction of silver ions and with the ability to work as capping agents. Consequently, genetic engineering techniques can be employed to enhance the production of the natural products of interest that generate dispersed and stable nanoparticles at a comparable cost to that of chemical synthesis [23].

The influence of temperature on the physicochemical environment merits careful consideration as well. It is plausible that temperature exerts a profound influence on the nucleation of AgNPs and plays a crucial role in controlling their aggregation, ultimately affecting the stability of the nanoparticles. However, it should be noted that the fate of biomolecules within the plant extract at elevated temperatures remains a relatively uncharted territory in current research [23].

Another facet of the plant-mediated synthesis of nanoparticles is its susceptibility to geographical and seasonal variations. These fluctuations can produce variations in the morphologies and properties of the synthesized nanoparticles. Consequently, a comprehensive investigation becomes indispensable to ascertain the robustness of the chosen plant extract, ensuring the production of uniform particles under diverse seasonal conditions, all within a single, standardized synthesis protocol. To achieve this, phytochemical analysis of the plant extract, involving the profiling of metabolites, is essential. Such an analysis will elucidate the variations in metabolite composition across different seasons and delineate their specific impacts on the nanosynthesis process [23].

In sum, the synthesis of AgNPs with plant-mediated approaches is a multifaceted endeavor. It hinges on the judicious selection of plant materials, the meticulous characterization of plant extracts, the mitigation of aggregation issues, the exploration of temperature effects, and a comprehensive understanding of the influence of geographical and seasonal factors. These nuances collectively constitute a rich domain for further academic exploration and practical application in the realm of nanotechnology.

There are more than 300,000 plants on earth. Such huge natural diversity is already present on earth offering potent biochemical tools. To unlock the secrets of plant-mediated AgNP synthesis, researchers must embark on a comprehensive survey of flora across diverse ecosystems.

2.2   Bacterial synthesis of AgNPs

Microorganisms have demonstrated to be a potent natural agent for simple, inexpensive, and non-toxic synthesis of AgNPs, avoiding costly reactants and the use of considerable energy needed for physiochemical methods [41]. Bacteria are frequently preferred for the green production of AgNPs due to their fast growth, easy cultivation and simple manipulation. Bacteria can also easily adapt to severe and toxic conditions produced by the presence of heavy metal ions [42]. Thus, biosynthesis of AgNPs by different bacterium is the result of their intrinsic response to survive under these stress environments.

Bacteria-based biosynthesis of AgNPs involves extracellular and intracellular redox reactions (Fig. 4)[43].

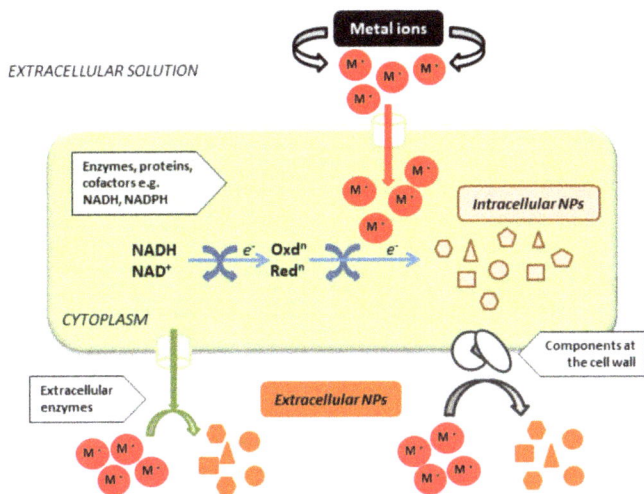

Fig 4. Schematic extracellular and intracellular mechanisms for bacterial AgNPs synthesis [42].
Copyright © 2022 Elsevier Inc.

In the intracellular method, bacteria are grown and multiplied in a medium containing the silver salt and organic and inorganic nutrients. Under these conditions, silver ions are delivered into the bacteria via active transport, ion channels, endocytosis, or diffusion across the lipid membrane [44]. To avoid toxic responses due to the entry of Ag^+, bacteria quickly reduce the ions to AgNPs, through intracellular reductase enzymes like the reduced form of nicotinamide adenine dinucleotide phosphate (NADPH) and the well-known nicotinamide adenine dinucleotide (NADH). An indication that AgNPs synthesis occurred is a color variation of the sample to a dark reddish-brown color. The resulting AgNPs can be accumulated either on the cytoplasmic membrane, cell wall or periplasm of bacteria which means that further procedures are necessary to separate the AgNPs from the interior of the cells. For example, after lysing bacteria, AgNPs can be recovered by a physical method like ultra-sonication and heat treatment, or chemical approaches such as using salt or surfactants [45]. Because these steps of extracting the AgNPs from inside bacteria are time-consuming and multi-step methods, extracellular synthesis methods are generally preferred [46].

One of the few reported works that study the intracellular biosynthesis of AgNPs by bacteria has been published by Kalimuthu *et al.* [47]. In this work, they used the bacterium *Bacillus licheniformis* isolated from sewage collected from municipal wastes. The preparation of the AgNPs involved the mix of a solution of 1 mM $AgNO_3$ with 2 g of harvested wet biomass and incubated at 37 °C under agitation. After 24 h, the color changed from whitish-yellow to brown denoting the obtaining of AgNPs so ultrasonic lysis of bacterial cells was carried out in order to recovery the trapped particles. The sonicated samples were then centrifugated to removed remains of cells and the supernatants containing the AgNPs were fully characterized by UV–visible spectrometry, X-ray diffraction (XRD), SEM and Energy-dispersive X-ray spectroscopy (EDX). As a result, the bacterial-synthesized AgNPs exhibited the characteristic strong peak at 440 nm assigned to the typical SPR of the silver NPs [48] and EDX confirmed that the NPs were made of silver. Furthermore, XRD spectrum revealed that biosynthesized AgNPs were in the form of nanocrystals and SEM images showed an average particle size of 50 nm. What is also important to highlight is the capability of the bacterium to resist exposure to the silver nitrate solution, as after incubating the biomass with AgNO₃ for 24 h, was still capable of forming colonies. An extensive protocol for the intracellular biosynthesis of AgNPs by *Bacillus licheniformis* was also described by Sriram *et al.* in 2012.

Rhodococcus spp has demonstrated to produced AgNPs intracellularly as well[49]. This actinobacteria is used for biotransformation and biodegradation of many organic substances in industry because of their large amount of enzymes[50]. In this sense, Otari *et al.* exposed *Rhodococcus* NCIM 2891 to a solution of AgNO₃ for the enzymatic reduction of Ag⁺ and obtention of monodispersed intracellular AgNPs. For this, they incubated the grown culture of bacterium with a solution of 1 mM AgNO₃ for 24 h and the formed AgNPs were then characterized by UV-visible spectroscopy, TEM, EDX, XRD and Fourier-transform infrared (FTIR). As a result, the UV–Visible spectrum of the suspension exhibited the characteristic SPR of AgNPs around 405 nm that also suggested a spherical shape of the particles. XRD pattern revealed the face-centered cubic crystal structure of the AgNPs and the full-width at half-maximum of the diffraction peaks permitted to calculate through the Debye–Sherrer formula a nanoparticle size of 25 nm, which is in agreement with the TEM images (Fig. 5). The FTIR analysis showed peaks of amides corresponding to peptides and proteins present on the surface of the AgNPs that favor their stabilization, preventing their aggregation. Finally, EDS corroborated the crystalline nature of AgNPs. Due to the antimicrobial activity of AgNPs is fully reported [51–53] they studied the bactericidal effect of the biosynthesized AgNPs against *Enterococcus faecalis*, *Escherichia coli*, *Staphylococcus aureus*, *Pseudomonas aeruginosa*, *Klebsiella pneumoniae* and *Proteus vulgaris*. They found that AgNPs obtained from *Rhodococcus* NCIM 2891 showed bacteriostatic and bactericidal activity against these Gram-positive and Gram-negative microorganisms when 50 µg/ml of AgNPs was used.

Bacterial intracellular synthesis of AgNPs has also been explored by Seshadri *et al.*, [54]. In this study, they obtained AgNPs by exposing the highly silver-resistant marine bacterium, *Idiomarina sp. PR58-8*, with AgNO₃. Firstly, the bacterium was isolated from soil samples at banks of the Mandovi River in Goa, India, and grown in Zobell Marine Broth 2216 in presence of 5 mM Ag⁺ for 48 h under agitation. After this period, the liquid media turned brownish-black indicating the formation of AgNPs. They found that *Idiomarina sp. PR58-8* can tolerate up to 7 mM of AgNO₃ and stored more than 90 % of silver intracellularly. Characterization of AgNPs by UV-visible spectrophotometry, XRD, and TEM was performed. UV- visible spectrum revealed a broad peak at 450 nm due to the SPR of AgNPs and the XRD analysis confirmed the presence of Ag. Furthermore, the crystallite size of 25 nm was calculated by the Debye–Sherrer and agreed with the size observed in TEM images. Because bacteria respond to metal stress by intracellularly promoting the expression of thiol peptides that capture the Ag ions, they also measured the intracellular total thiol and non-protein thiol concentration in *Idiomarina sp. PR58-8* after exposure to Ag solution. They found that the Ag-exposed bacterium reacted to the metal stress by raising it intracellular content of total thiol as a result of the induction of non-protein thiols. Thus, they demonstrated that non-protein thiols are crucial in Ag resistance of *Idiomarina sp. PR58-8.*

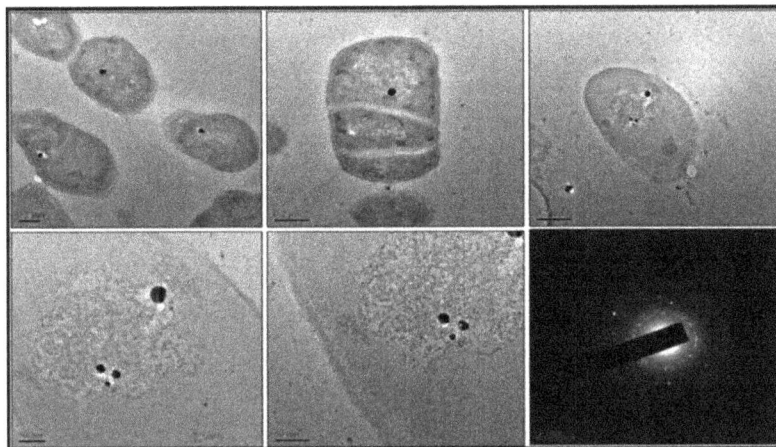

Fig 5. TEM images of AgNPs formed in the cytoplasm Rhodococcus sp. and their Selected Area Electron Diffraction (SAED) pattern showing the crystalline nature of nanoparticles. Reprinted from [49]. Copyright © 2014 Elsevier B.V.

Likewise, *Vibrio alginolyticus* is a marine bacterium that was tested to produced AgNPs intracellularly [55]. Rajeshkumar *et al.* isolated the marine bacteria from marine water and after growing the culture for 24 h, 1 M AgNO₃ was added to the broth. After 4 h, the clear yellowish of the bacterial culture changed to brown color showing a high peak at 440 nm of the SPR, confirming the formation of AgNPs. SEM pictures exposed predominantly spherical NPs of 50-100 nm and EDS results showed a strongest signal in the Ag region at 3 keV proving the reduction of Ag^+ to Ag^0.

On the other hand, for the extracellular synthesis, some microorganisms have the capability to release reductase enzymes into the solution. These reductase enzymes play a crucial role in reducing silver ions, converting them into AgNPs. Simultaneously, certain proteins associated with the microorganism acts as stabilizers for these AgNPs. Consequently, the AgNPs become attached to the surface of the microorganisms, resulting in the formation of extracellular AgNPs [56,57]. However, there are other pathway which involves the separation of microorganisms from the solution after they have released a significant amount of reductase enzymes. In this scenario, the reductase enzymes continue to reduce silver ions into AgNPs. Notably, the particle size of the AgNPs generated through this method tends to be relatively smaller when compared to the previous approach [58,59].

In this vein, Kumar and Mamidyala employed the supernatant of *Pseudomonas aeruginosa* cultures to facilitate a cost-effective and straightforward green synthesis of AgNPs [60]. The authors achieved the reduction of silver ions when they treated a silver nitrate solution with the culture supernatant of *P. aeruginosa* at room temperature. In addition, they characterized the nanoparticles with various techniques such as UV–visible, TEM, EDX, FTIR spectroscopy, and XRD spectroscopy. The results revealed that AgNPs were monodisperse, spherical and with a size of 13 nm and they confirmed the characteristic SPR band at 430 nm by UV-visible analysis. Moreover, the FTIR analysis indicated that the protein component in the form of nitrate reductase enzyme, along with the rhamnolipids produced by the isolate in the culture supernatant, played a role both in the reduction process and as a capping material for the nanoparticles. The XRD spectrum confirmed the crystalline nature. Finally, the AgNPs exhibited antimicrobial activity over a wide variety of microorganisms, including various *Candida* species and Gram-positive or Gram-negative bacteria.

Ameen *et al.* have also used extracellular synthesis to produce AgNPs [61]. In this sense, they used *Cuprividus sp.*, a bacterium isolated from urban soil and presumed to have resistance to heavy metals. The initial step in the process involved incubating the isolated bacterium, which had initially shown the ability to synthesize AgNPs, in a rotary shaker at 37°C for 24 h. Following this incubation period, the bacterial cells were separated from the culture by centrifugation. Then, the supernatant was utilized for the extracellular procedure for AgNPs synthesis. Briefly, it was mixed with a AgNO₃ solution until a noticeable change in color indicated the successful synthesis of AgNPs. The effective formation of spherical crystalline nanoparticles with sizes ranging from 10 to 50 nanometers was confirmed through diffractometric, microscopic and spectroscopic analyses. Additionally, the FTIR spectroscopy revealed the presence of amides, supporting the previously proposed enzymatic mechanism for the extracellular reduction of silver ions into atomic silver nanoparticles. Antimicrobial activity of these AgNPs was confirmed by the disk diffusion assay against *Aeromonas enteropelogenes*, *Stenotrophomonas pavanii*, *Proteus mirabilis*, and *Enterobacter xiangfangensis*, which are human clinical pathogenic bacteria.

In addition, Iqtedar *et al.*, successfully synthesized 105 nm AgNPs employing *Bacillus mojavensis* [62]. They characterized the nanoparticles through atomic force microscopy (AFM), UV-visible spectroscopy, FTIR and XRD. It is interesting to remark that authors were able to reduce the AgNPs size up to 2.3 nm under optimized conditions, which included a temperature of 55 °C, pH of 8, the addition of the surfactant and the presence of K₂SO₄. On the other hand, the synthetized nanoparticles displayed antibacterial activity against *E. coli*, *K. pneumonia*, *Acinetobacter sp.*, and *P. aeruginosa*, but had no effect on *S. aureus*. In conclusion, this work shows that under optimized conditions, *Bacillus mojavensis* BTCB15 was able to produce extremely small AgNPs that exhibited potent antibacterial activity against multi-drug resistant microorganism.

Extracellular synthesis of AgNPs was also achieved by Huq and Akter using the supernatant of the bacterial culture from the *Paenarthrobacter nicotinovorans* strain [63]. The characterization of the AgNPs showed, using UV-Vis spectral analysis, an absorption maximum at 466 nm, confirming the successful synthesis of AgNPs. In addition, field emission-transmission electron microscopy (FE-TEM) analysis unveiled spherical-shaped nanoparticles with sizes ranging from 13 to 27 nm. Both EDX and XRD analyses confirmed the crystalline nature of the biosynthesized AgNPs. FTIR analysis provided insights into the involvement of various biomolecules acting as reducing and capping agents during the AgNPs synthesis process. Significantly, the bacterial-mediated synthesized AgNPs demonstrated the ability to inhibit the growth of pathogenic strains

P. aeruginosa and *B. cereus*, as evidenced by the development of clear zones of inhibition. Furthermore, FE-TEM analysis revealed that the synthesized AgNPs could disrupt the outer membrane and induce alterations in the cell morphology of the treated pathogens, ultimately leading to cell death (Fig.6). In summary, this study highlights the environment friendly, straightforward, and fast production of AgNPs using *P. nicotinovorans* and, since the AgNPs exhibited antimicrobial activity against both Gram-negative and Gram-positive pathogens, they could be a promising candidate for various applications in biotechnology and medicine.

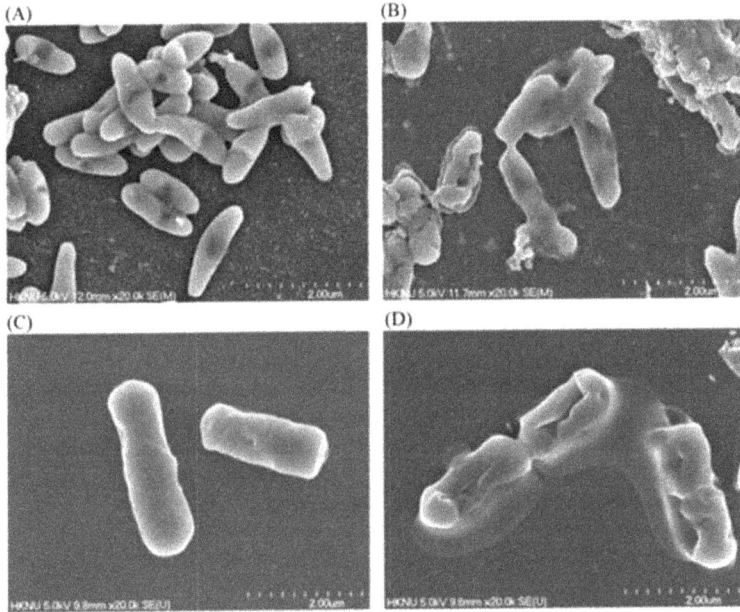

Fig 6. SEM images of normal P. aeruginosa cells (A), 1 × MBC AgNPs-treated P. aeruginosa cells (B), normal B. cereus cells (C), 1 × MBC AgNPs-treated B. cereus cells (D). Reprinted from [63]. Copyright © 2021MDPI https://creativecommons.org/licenses/by/4.0/

Finally, Matei *et al.*, were able to synthetize AgNPs using the filtrate from the culture of lactic acid bacteria, specifically *Lactobacillus sp.* strain LCM5 [64]. TEM analysis confirmed that the diameter of the produced AgNPs was in the range of 3 to 35 nm, with an average particle size of 13.84 ± 4.56 nm. These AgNPs exhibited a highly dispersion, with spherical shape, and parallel stripes, confirming their crystal structure. The antimicrobial activity of the AgNPs varied depending on the species and group of microorganisms tested (bacteria or fungi). For instance, the inhibition zones for *Aspergillus flavus* and *Aspergillus ochraceus* were similar. However, stronger inhibition was observed against *Penicillium expansum*. The effectiveness of the green synthesized AgNPs was more pronounced against Gram-negative bacteria, particularly *Chromobacterium violaceum*, where a broader inhibition area was observed compared to the fungi. In conclusion, the results obtained by these authors suggest that biosynthesized AgNPs using the culture filtrate of the lactic acid *Lactobacillus sp.* bacteria hold promise for various biotechnological applications as potent antimicrobial agents.

As it was mentioned, microbial production of AgNPs can take place through either intracellular or extracellular processes. Intracellular synthesis typically involves the use of acoustic waves and cell lysis agents to extract AgNPs. In contrast, extracellular synthesis of the nanoparticles does not require the use of such agents or acoustic waves. Consequently, extracellular synthesis is generally preferred and employed [41,50].

2.3   AgNPs green synthesis using fungi

Nowadays the synthesis of AgNPs using fungi is an interesting alternative. These microorganisms can serve as biofactories representing a simple and efficient way to create AgNPs. In addition, synthesis by fungi presents some advantages over other methods as production of different size and chemical composition nanoparticles. In this sense, fungi have other abilities such as metal bioconcentration, viability in metal rich environmental, wide range of extracellular enzyme secretion, and low-cost pathway.

When compared with bacteria, fungi are more tolerant to the agitation and flow pressure of the bioprocesses, leading to choose them for industrial-scale synthesis of metal-based nanomaterials [66,67]. Due to these factors, biosynthesis of metal and metal oxide nanoparticles using biological agents such as fungi, has gained popularity in area of nanotechnology, specifically myco-nanotechnology is an exciting frontier scientific field that studies how fungi unique metabolic pathways can be employed to synthesize nanostructures with precise dimensions and functionalities [58].

Generally, the process to synthesize AgNPs by fungi consists in the initial culture of the fungus on agar, and subsequently transfer it to a suitable liquid medium. The next step is to transfer the biomass produced to a water solution for the release of the molecules that play the role in the production of nanoparticles. After filtration, the biomass is separated, and silver nitrate is added to the filtrate [68–70] (Fig. 7).

Although the biogenic specific mechanisms have not been clarified, the consensus to this day indicates that extracellular synthesis of nanoparticles carry out between enzymes present in the fungal filtrate and silver ions, leading to production of elemental silver (Ag^0) at a nanometric scale. Products of the reaction change the UV-visible spectrum of the filtrate allowing to observe bands with absorbance wavelengths in the range from 400 to 450 nm through SPR bands reflecting alteration of the optical properties of the material [71]. The properties of obtained nanoparticles, like the dimensions or morphology, depends on synthesis conditions such as temperature, pH, fungus species among other variables [72].

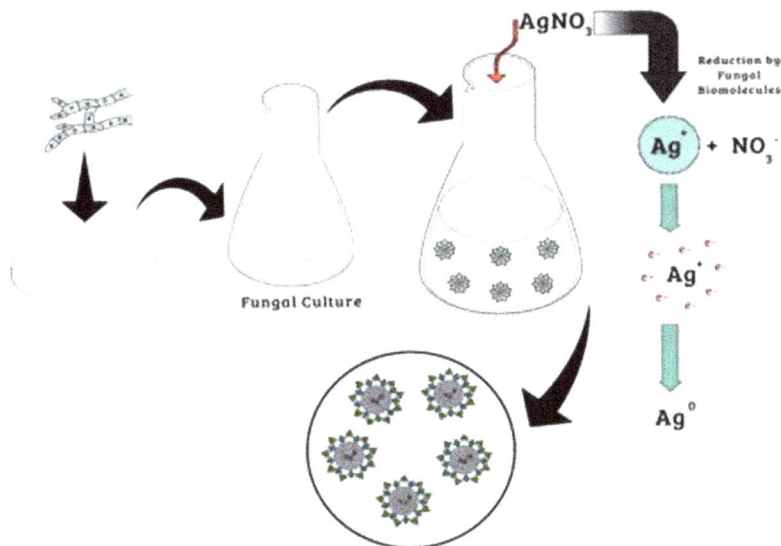

Fig 7: Mechanisms of biogenic synthesis of silver nanoparticles. Reprinted from [68]. Copyright © 2019 Guilger-Casagrande and Lima. Frontiers. Creative Commons Attribution License https://creativecommons.org/licenses/by/4.0/

Fungi can generate nanoparticles by biogenic synthesis and this process may carry out intra or extra cellular path. In case of intracellular synthesis, mycelia and fungal cell-free filtrate, respectively, can be used in fungi-mediated synthesis to produce intracellular AgNPs. In this case, the metal precursor is added to the mycelial culture and is internalized in the biomass. For this reason, the nanoparticles obtained by this kind of synthesis require several steps like extraction employing chemical treatment, centrifugation, and filtration to decompose the biomass and release the nanoparticles [67,68,73].

On the other hand, extracellular synthesis requires the addition of metal precursor to the aqueous filtrate containing only the fungal biomolecules, resulting in the formation of free nanoparticles in dispersion. Because of its simplicity and non-requirement liberation of nanoparticles from cells, this last method is most widely used [71,74]. In addition, fungi are easy to handle and they generate extracellular proteins with minimal cytotoxicity of the residues [68]. Several authors have demonstrated that biomolecules and proteins synthesized and released to extracellular media by fungi act by capped and stabilization of nanoparticles [75].

For instance, Abdel-Hadi *et al.*, focused on producing silver nanoparticles by employing extracellular proteins released from the culture of *Fusarium oxysporum* and subsequently evaluate their antibacterial effect over pathogenic microorganism [66]. They demonstrated that proteins can play an import role as a reducing and stabilizing agent through interaction with AgNPs and cysteine sites, free amino acids or by electrostatic interaction of negative charges of carboxyl groups within the extracellular enzyme (Fig. 8).

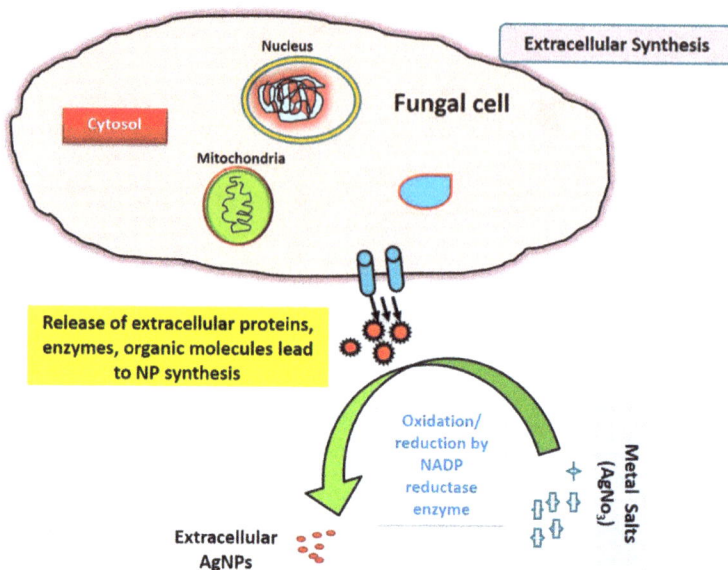

Fig 8. Schematic representation of extracellular mode of reduction of metallic salt/ions by fungal extracellular secretion. Reprinted from [66] . Copyright © 2023 MDPI
https://creativecommons.org/licenses/by/4.0/

As the rest of nanoparticles, the AgNPs obtained by fungi biogenic synthesis have several interesting uses in different fields, including biomedical, agriculture, and insect control. As we mentioned before, the production based on fungi offers significant advantages in terms of yield, based on high production and fungi environmental resistance. In addition, many authors showed results that could be taken as basis for new lines of controlling microorganisms as virus or bacteria and other organisms such as insects and their larvae [68]. For instance, Tyagi *et al.*, developed a rapid, simple and non-toxic *in vitro* extracellular silver nanoparticles synthesis using Entomopathogenic fungus (*Beauveria bassiana*) [76]. The authors obtained silver nanoparticles in fungal supernatants that they were confirmed by the absorbance peak at 450 nm in UV–Vis spectrophotometer and EDS, XRD and TEM/SEM microscopy. The microscopy analysis revealed the presence of nanoparticles with different shape (circular, triangular, hexagonal) with size ranging from 10 to 50 nm. The toxic effect of these nanoparticles was evaluated against *E. coli*, *P. aeruginosa* and *S. aureus* through growth kinetic studies and ciprofloxacin combination. The fungal assisted synthesized AgNPs presented a time dependent increase in antibacterial effect over

P. aeruginosa, S. aureus and *E. coli*. In addition, a clear synergistic action with the antibiotic ciprofloxacin was confirmed against the evaluated bacterial strains.

Wang *et al.*, [77] have biosynthesized AgNPs extracellularly using the culture supernatants of *Aspergillus sydowii*. The nanoparticles were analyzed employing UV absorption peak at 420 nm, TEM and XRD. The authors controlled and optimized three main synthesis factors (temperature, pH and substrate concentration). The biological assessment of AgNPs, showed effective antifungal activity against many clinical pathogenic fungi and *in vitro* antiproliferative activity to HeLa cells and MCF-7 cells.

Talaromyces funiculosus, wild endolichenic fungus, was employed to produce silver nanoparticles (*Talaromyces funiculosus*–AgNPs). The investigators carried out an evaluation of NPs by UV visible spectroscopy, dynamic light scattering (DLS), ATR-FTIR, XRD, SEM, and TEM. They demonstrated microbiological activity against *S. aureus, Streptococcus faecalis, Listeria innocua,* and *Micrococcus luteus* and nosocomial *P. aeruginosa, E. coli, Vibrio cholerae,* and *Bacillus subtilis* bacterial strains assessed by using percentage of inhibition (IC_{50}). Moreover, synthesized *TF*-AgNPs displayed cytotoxicity in a dose-dependent manner against MDA-MB-231 breast carcinoma cells. On the other hand, the combination of predation by the cyclopoid copepod *Mesocyclops aspericornis* and synthetized *TF*-AgNPs for the larval management of dengue vectors (*Aedes aegypti*) were carried out. The *TF*-AgNPs affected the condition of the larvae make them more susceptible to attack by the predators [78].

It is worth to highlight that the choice of fungi for biosynthesis of nanoparticles may present some disadvantages. They include the correct selection of fungi specie, evaluation of its growth parameters, sterile conditions, fungal growth and completed synthesis time. Lastly, difficulties may arise in relation to the scaling-up process. Nevertheless, green production of silver nanoparticles by fungi can lead to a significant diversity of interesting uses [68].

Conclusions

Silver nanoparticles are a rapidly growing area of research and development. New applications for AgNPs are being discovered all the time. As the technology continues to develop, silver nanoparticles are expected to play an even greater role in many industries and aspects of our lives. Although more research is needed to fully understand the potential risks and benefits of AgNPs, they are likely to play an increasingly important role in our lives in the future.

The advancement of green synthesis techniques has flourished due to the efforts of researchers from various backgrounds. The adaptability of green chemistry enables the production of a diverse array of silver nanomaterials with a multitude of potential applications. It is crucial to thoroughly characterize the resulting nanomaterials in every instance, as their properties will dictate their effectiveness in fulfilling their intended purpose and potentially reveal any harmful effects, such as nanotoxicological behavior. Moreover, the inherent intricate nature of biological sources (fungi, bacteria, plants, organisms) and the conditions manipulated during synthetic processes give rise to a diverse array of nanomaterials with varying surface properties, sizes and morphologies, which significantly influence their functionality and potential applications. A comprehensive understanding of the mechanism played by biological molecules, the pathways governing nanoparticle production and growth, and the impact of biogenic synthesis procedures is essential for tailoring nanomaterials to meet specific application requirements. Various eco-friendly sources and synthesis methods have been explored to develop less harmful, environmentally benign, and sustainable processes. The judicious utilization of natural resources in green synthesis approaches lessens our dependence on non-renewable energy sources and marks an advancement over conventional procedures. This enables the realization of cost-effective, simple, non-toxic, and sustainable procedures. Consequently, these approaches offer environmental and human health benefits, which should serve as the primary drivers of future efforts. These benefits encompass safer food and consumer products, cleaner water and air, enhanced worker security due to reduced chemical exposure, minimized harm to animals and plants, the employment of biodegradable chemicals, reduced reliance on landfills, and mitigated disruption of ecosystems, to name a few.

References

[1]A. Kaushal, I. Khurana, P. Yadav, P. Allawadhi, A.K. Banothu, D. Neeradi, S. Thalugula, P.J. Barani, R.R. Naik, U. Navik, K.K. Bharani, A. Khurana, Advances in therapeutic applications of silver nanoparticles, Chem. Biol. Interact., 382 (2023) 110590. https://doi.org/10.1016/j.cbi.2023.110590

[2]K.G. Kaiser, V. Delattre, V.J. Frost, G.W. Buck, J. V Phu, T.G. Fernandez, I.E. Pavel, Nanosilver: An Old Antibacterial Agent with Great Promise in the Fight against Antibiotic Resistance, Antibiotics, 12 (2023). https://doi.org/10.3390/antibiotics12081264

[3]A. Balestri, J. Cardellini, D. Berti, Gold and silver nanoparticles as tools to combat multidrug-resistant pathogens, Curr. Opin. Colloid Interface Sci., 66 (2023) 101710. https://doi.org/10.1016/j.cocis.2023.101710

[4]J.R. Ansari, N. Singh, S. Anwar, S. Mohapatra, A. Datta, Silver nanoparticles decorated two dimensional MoS2 nanosheets for enhanced photocatalytic activity, Colloids Surfaces A Physicochem. Eng. Asp., 635 (2022) 128102. https://doi.org/10.1016/j.colsurfa.2021.128102

[5]H. Tabassum, I.Z. Ahmad, Applications of metallic nanomaterials for the treatment of water, Lett. Appl. Microbiol., 75 (2022) 731–743. https://doi.org/10.1111/lam.13588

[6]A.K. Bhardwaj, S. Sundaram, K.K. Yadav, A.L. Srivastav, An overview of silver nano-particles as promising materials for water disinfection, Environ. Technol. Innov., 23 (2021) 101721. https://doi.org/10.1016/j.eti.2021.101721

[7]V.S. Andrade, A. Ale, P.E. Antezana, M.F. Desimone, J. Cazenave, M.F. Gutierrez, Ecotoxicity of nanosilver on cladocerans and the role of algae provision, Environ. Sci. Pollut. Res., 30 (2023) 27137–27149. https://doi.org/10.1007/s11356-022-24154-7

[8]A. Ale, G. Liberatori, M.L.M.L. Vannuccini, E. Bergami, S. Ancora, G. Mariotti, N. Bianchi, J.M.J.M.J.M. Galdopórpora, M.F.M.F. Desimone, J. Cazenave, I. Corsi, Exposure to a nanosilver-enabled consumer product results in similar accumulation and toxicity of silver nanoparticles in the marine mussel Mytilus galloprovincialis, Aquat. Toxicol., 211 (2019) 46–56. https://doi.org/10.1016/j.aquatox.2019.03.018

[9]A. Ale, C. Bacchetta, A.S. Rossi, J. Galdopórpora, M.F. Desimone, F.R. de la Torre, S. Gervasio, J. Cazenave, Nanosilver toxicity in gills of a neotropical fish: Metal accumulation, oxidative stress, histopathology and other physiological effects, Ecotoxicol. Environ. Saf., 148 (2018). https://doi.org/10.1016/j.ecoenv.2017.11.072

[10]   M. Garcés, N.D. Magnani, A. Pecorelli, V. Calabró, T. Marchini, L. Cáceres, E. Pambianchi, J. Galdoporpora, T. Vico, J. Salgueiro, M. Zubillaga, M.A. Moretton, M.F. Desimone, S. Alvarez, G. Valacchi, P. Evelson, Alterations in oxygen metabolism are associated to lung toxicity triggered by silver nanoparticles exposure, Free Radic. Biol. Med., 166 (2021) 324–336. https://doi.org/10.1016/j.freeradbiomed.2021.02.008

[11]   P.E. Antezana, S. Municoy, M.F. Desimone, 1 - Building nanomaterials with microbial factories, in: R. Pratap Singh, A.R. Rai, A. Abdala, R.G.B.T.-B.S.N. Chaudhary (Eds.), Micro Nano Technol., Elsevier, 2022: pp. 1–39. https://doi.org/10.1016/B978-0-323-88535-5.00012-3

[12]   R.G. Chaudhary, M.F. Desimone, Synthesis, Characterization, and Applications of Green Synthesized Nanomaterials (Part 1), Curr. Pharm. Biotechnol., 22 (2021) 722–723. https://doi.org/10.2174/138920102206210521165455

[13]   P.R. Bhilkar, A.S. Bodhne, S.T. Yerpude, R.S. Madankar, S.R. Somkuwar, A.R. Daddemal-Chaudhary, A.P. Lambat, M. Desimone, R. Sharma, R.G. Chaudhary, Phyto-derived metal nanoparticles: Prominent tool for biomedical applications, OpenNano, 14 (2023) 100192. https://doi.org/10.1016/j.onano.2023.100192

[14]   J.M. Galdopórpora, A. Ibar, M.V. Tuttolomondo, M.F. Desimone, Dual-effect core–shell polyphenol coated silver nanoparticles for tissue engineering, Nano-Structures & Nano-Objects, 26 (2021) 100716. https://doi.org/10.1016/j.nanoso.2021.100716

[15]   H. Singh, M.F. Desimone, S. Pandya, S. Jasani, N. George, M. Adnan, A. Aldarhami, A.S. Bazaid, S.A. Alderhami, Revisiting the Green Synthesis of Nanoparticles: Uncovering Influences of Plant Extracts as Reducing Agents for Enhanced Synthesis Efficiency and Its Biomedical Applications, Int. J. Nanomedicine, 18 (2023) 4727–4750. https://doi.org/10.2147/IJN.S419369

[16]   S.S. Shankar, A. Ahmad, M. Sastry, Geranium leaf assisted biosynthesis of silver nanoparticles, Biotechnol. Prog., 19 (2003) 1627–1631. https://doi.org/10.1021/bp034070w

[17]   S. Kumar, I.B. Basumatary, H.P.K. Sudhani, V.K. Bajpai, L. Chen, S. Shukla, A. Mukherjee, Plant extract mediated silver nanoparticles and their applications as antimicrobials and in sustainable food packaging: A state-of-the-art review, Trends Food Sci. Technol., 112 (2021) 651–666. https://doi.org/10.1016/j.tifs.2021.04.031

[18]   H.B. Habeeb Rahuman, R. Dhandapani, S. Narayanan, V. Palanivel, R. Paramasivam, R. Subbarayalu, S. Thangavelu, S. Muthupandian, Medicinal plants mediated the green synthesis of silver nanoparticles and their biomedical applications, IET Nanobiotechnology, 16 (2022) 115–144. https://doi.org/10.1049/nbt2.12078

[19]   C. Vanlalveni, S. Lallianrawna, A. Biswas, M. Selvaraj, B. Changmai, S.L. Rokhum, Green synthesis of silver nanoparticles using plant extracts and their antimicrobial activities: a review of recent literature, RSC Adv., 11 (2021) 2804–2837. https://doi.org/10.1039/D0RA09941D

[20]   S. Simon, N.R.S. Sibuyi, A.O. Fadaka, S. Meyer, J. Josephs, M.O. Onani, M. Meyer, A.M. Madiehe, Biomedical Applications of Plant Extract-Synthesized Silver Nanoparticles, Biomedicines, 10 (2022). https://doi.org/10.3390/biomedicines10112792

[21]   . Ritu, K. Verma, A. Das, P. Chandra, Phytochemical-Based Synthesis of Silver Nanoparticle: Mechanism and Potential Applications, Bionanoscience, 13 (2023) 1–22. https://doi.org/10.1007/s12668-023-01125-x

[22]   R. Rajan, K. Chandran, S.L. Harper, S.-I. Yun, P.T. Kalaichelvan, Plant extract synthesized silver nanoparticles: An ongoing source of novel biocompatible materials, Ind. Crops Prod., 70 (2015) 356–373. https://doi.org/10.1016/j.indcrop.2015.03.015

[23]   H.P. Borase, B.K. Salunke, R.B. Salunkhe, C.D. Patil, J.E. Hallsworth, B.S. Kim, S. V Patil, Plant extract: a promising biomatrix for ecofriendly, controlled synthesis of silver nanoparticles, Appl. Biochem. Biotechnol., 173 (2014) 1–29. https://doi.org/10.1007/s12010-014-0831-4

[24]   F. Arshad, G.A. Naikoo, I.U. Hassan, S.R. Chava, M. El-Tanani, A.A. Aljabali, M.M. Tambuwala, Bioinspired and Green Synthesis of Silver Nanoparticles for Medical Applications: A Green Perspective, Appl. Biochem. Biotechnol., (2023) 1–34. https://doi.org/10.1007/s12010-023-04719-z

[25]   S. Arokiyaraj, M.V. Arasu, S. Vincent, N.U. Prakash, S.H. Choi, Y.-K. Oh, K.C. Choi, K.H. Kim, Rapid green synthesis of silver nanoparticles from Chrysanthemum indicum L and its antibacterial and cytotoxic effects: an in vitro study, Int. J. Nanomedicine, 9 (2014) 379–388. https://doi.org/10.2147/IJN.S53546

[26]   H. Padalia, P. Moteriya, S. Chanda, Green synthesis of silver nanoparticles from marigold flower and its synergistic antimicrobial potential, Arab. J. Chem., 8 (2015) 732–741. https://doi.org/10.1016/j.arabjc.2014.11.015

[27]   B. Devaraj, V. Bhuvaneshwari, Evaluation of the Cytotoxic and Antioxidant Activity of Phyto-synthesized Silver Nanoparticles Using Cassia angustifolia Flowers, Bionanoscience, 9 (2019) 155–163. https://doi.org/10.1007/s12668-018-0577-5

[28]   K. Logaranjan, A.J. Raiza, S.C.B. Gopinath, Y. Chen, K. Pandian, Shape- and Size-Controlled Synthesis of Silver Nanoparticles Using Aloe vera Plant Extract and Their Antimicrobial Activity, Nanoscale Res. Lett., 11 (2016) 520. https://doi.org/10.1186/s11671-016-1725-x

[29]   Y. Rout, S. Behera, A.K. Ojha, P.L. Nayak, Green synthesis of silver nanoparticles using Ocimum sanctum (Tulashi) and study of their antibacterial and antifungal activities, J. Microbiol. Antimicrob., 4 (2012) 103–109. https://doi.org/10.5897/JMA11.060

[30]   A. Rwalinda, S.Ravikumar, Green Synthesis of Silver Nanoparticles Using Acacia Nilotica Leaf Extract and Its Antibacterial and Anti Oxidant Activity, Int. J. Pharm. Chem. Sci., 4 (2015) 433–444.

[31]   G.K. Devi, K. Sathishkumar, Synthesis of gold and silver nanoparticles using Mukia maderaspatna plant extract and its anticancer activity, IET Nanobiotechnology, 11 (2017) 143–151. https://doi.org/10.1049/iet-nbt.2015.0054

[32]   M. Oves, M. Aslam, M.A. Rauf, S. Qayyum, H.A. Qari, M.S. Khan, M.Z. Alam, S. Tabrez, A. Pugazhendhi, I.M.I. Ismail, Antimicrobial and anticancer activities of silver nanoparticles synthesized from the root hair extract of Phoenix dactylifera, Mater. Sci. Eng. C. Mater. Biol. Appl., 89 (2018) 429–443. https://doi.org/10.1016/j.msec.2018.03.035

[33]   T. Sujin Jeba Kumar, C.K. Balavigneswaran, R. Moses Packiaraj, A. Veeraraj, S. Prakash, Y. Natheer Hassan, K.P. Srinivasakumar, Green Synthesis of Silver Nanoparticles by Plumbago indica and Its Antitumor Activity Against Dalton's Lymphoma Ascites Model, Bionanoscience, 3 (2013) 394–402. https://doi.org/10.1007/s12668-013-0102-9

[34]   R. Geethalakshmi, D.V.L. Sarada, Gold and silver nanoparticles from Trianthema decandra: Synthesis, characterization, and antimicrobial properties, Int. J. Nanomedicine, 7 (2012) 5375–5384. https://doi.org/10.2147/IJN.S36516

[35]   U.K. Sur, B. Ankamwar, S. Karmakar, A. Halder, P. Das, Green synthesis of Silver nanoparticles using the plant extract of Shikakai and Reetha, Mater. Today Proc., 5 (2018) 2321–2329. https://doi.org/10.1016/j.matpr.2017.09.236

[36]   M.A. Quinteros, V. Cano Aristizábal, P.R. Dalmasso, M.G. Paraje, P.L. Páez, Oxidative stress generation of silver nanoparticles in three bacterial genera and its relationship with the antimicrobial activity, Toxicol. Vitr. an Int. J. Publ. Assoc. with BIBRA, 36 (2016) 216–223. https://doi.org/10.1016/j.tiv.2016.08.007

[37]   J.M. Galdopórpora, S. Municoy, F. Ibarra, V. Puente, P.E. Antezana, M.I.A. Echazú, M.F. Desimone, A Green Synthesis Method to Tune the Morphology of CuO and ZnO NanostructuresNo Title, Curr. Nanosci., 17 (2022).

[38]   M. Asif, R. Yasmin, R. Asif, A. Ambreen, M. Mustafa, S. Umbreen, Green Synthesis of Silver Nanoparticles (AgNPs), Structural Characterization, and their Antibacterial Potential, Dose. Response., 20 (2022) 15593258221088708. https://doi.org/10.1177/15593258221088709

[39]   A.K. Giri, B. Jena, B. Biswal, A.K. Pradhan, M. Arakha, S. Acharya, L. Acharya, Green synthesis and characterization of silver nanoparticles using Eugenia roxburghii DC extract and activity against biofilm-producing bacteria, Sci. Rep., 12 (2022) 8383. https://doi.org/10.1038/s41598-022-12484-y

[40]   S.K. Srikar, D.D. Giri, D.B. Pal, P.K. Mishra, S.N. Upadhyay, Green Synthesis of Silver Nanoparticles: A Review, Green Sustain. Chem., 6 (2016) 34–56. https://doi.org/10.4236/gsc.2016.61004

[41]   M.A. Huq, M. Ashrafudoulla, M.M. Rahman, S.R. Balusamy, S. Akter, Green Synthesis and Potential Antibacterial Applications of Bioactive Silver Nanoparticles: A Review, Polymers (Basel)., 14 (2022) 1–22. https://doi.org/10.3390/polym14040742

[42]   P.E. Antezana, S. Municoy, M.F. Desimone, Building nanomaterials with microbial factories, in: Raghvendra Pratap Singh, A. Abdala, A.R. Rai, Ratiram G. Chaudhary (Eds.), Biog. Sustain. Nanotechnol., Elsevier, 2022: pp. 1–39. https://doi.org/10.1016/B978-0-323-88535-5.00012-3

[43]   N. Alfryyan, M.G.M. Kordy, M. Abdel-Gabbar, H.A. Soliman, M. Shaban, Characterization of the biosynthesized intracellular and extracellular plasmonic silver nanoparticles using Bacillus cereus and their catalytic reduction of methylene blue, Sci. Rep., 12 (2022) 1–14. https://doi.org/10.1038/s41598-022-16029-1

[44]   A. Javaid, S.F. Oloketuyi, M.M. Khan, F. Khan, Diversity of Bacterial Synthesis of Silver Nanoparticles, Bionanoscience, 8 (2018) 43–59. https://doi.org/10.1007/s12668-017-0496-x

[45]   E. Janeeshma, P. Sameena, J.T. Puthur, Application of biogenic nanoparticles in the remediation of contaminated water, in: P. Singh, V. Kumar, M. Bakshi, C.M. Hussain, M. Sillanpää (Eds.), Environ. Appl. Microb. Nanotechnol. Emerg. Trends Environ. Remediat., Elsevier Science, 2022: pp. 33–41. https://doi.org/10.1016/B978-0-323-91744-5.00023-0

[46]   S.I. Tsekhmistrenko, V.S. Bityutskyy, O.S. Tsekhmistrenko, L.P. Horalskyi, N.O. Tymoshok, M.Y. Spivak, Bacterial synthesis of nanoparticles: A green approach, Biosyst. Divers., 28 (2020) 9–17. https://doi.org/10.15421/012002

[47]   K. Kalimuthu, R. Suresh Babu, D. Venkataraman, M. Bilal, S. Gurunathan, Biosynthesis of silver nanocrystals by Bacillus licheniformis, Colloids Surfaces B Biointerfaces, 65 (2008) 150–153. https://doi.org/10.1016/j.colsurfb.2008.02.018

[48]   S. Municoy, P.E. Antezana, C.J. Pérez, M.G. Bellino, M.F. Desimone, Tuning the antimicrobial activity of collagen biomaterials through a liposomal approach, J. Appl. Polym. Sci., n/a (2020) 50330. https://doi.org/10.1002/app.50330

[49]   S. V. Otari, R.M. Patil, S.J. Ghosh, N.D. Thorat, S.H. Pawar, Intracellular synthesis of silver nanoparticle by actinobacteria and its antimicrobial activity, Spectrochim. Acta - Part A Mol. Biomol. Spectrosc., 136 (2015) 1175–1180. https://doi.org/10.1016/j.saa.2014.10.003

[50]   M. Cappelletti, A. Presentato, E. Piacenza, A. Firrincieli, R.J. Turner, D. Zannoni, Biotechnology of Rhodococcus for the production of valuable compounds, Appl. Microbiol. Biotechnol., 104 (2020) 8567–8594. https://doi.org/10.1007/s00253-020-10861-z

[51]   S. Municoy, P.E. Antezana, M.G. Bellino, M.F. Desimone, Development of 3D-Printed Collagen Scaffolds with In-Situ Synthesis of Silver Nanoparticles, Antibiotics, 12 (2023) 1–19. https://doi.org/10.3390/antibiotics12010016

[52]   P.E. Antezana, S. Municoy, C.J. Perez, M.F. Desimone, Collagen Hydrogels Loaded with Silver Nanoparticles and Cannabis Sativa Oil, Antibiotics, 10 (2021) 1–18. https://doi.org/10.3390/antibiotics10111420

[53]   T. Bruna, F. Maldonado-Bravo, P. Jara, N. Caro, Silver Nanoparticles and Their Antibacterial Applications, Int. J. Mol. Sci., 22 (2021) 7202. https://doi.org/10.3390/ijms22137202

[54]   S. Seshadri, A. Prakash, M. Kowshik, Biosynthesis of silver nanoparticles by marine bacterium, Idiomarina sp PR58-8, Bull. Mater. Sci., 35 (2012) 1201–1205. https://doi.org/10.1007/s12034-012-0417-0

[55]   Rajeshkumar S, Malarkodi C, Paulkumar K, Vanaja M, Gnanajobitha G, Annadurai G, Intracellular and Extracellular Biosynthesis of Silver Nanoparticles By Using Marine Bacteria Vibrio Alginolyticus, An Int. J., 3 (2013) 21–25.

[56]   A. Banik, M. Vadivel, M. Mondal, N. Sakthivel, Molecular Mechanisms that Mediate Microbial Synthesis of Metal Nanoparticles, (2022) 135–166. https://doi.org/10.1007/978-3-030-97185-4_6

[57]   S. Gurunathan, J.W. Han, V. Eppakayala, M. Jeyaraj, J.H. Kim, Cytotoxicity of biologically synthesized silver nanoparticles in MDA-MB-231 human breast cancer cells, Biomed Res. Int., 2013 (2013). https://doi.org/10.1155/2013/535796

[58]   T.A. Jorge de Souza, L.R. Rosa Souza, L.P. Franchi, Silver nanoparticles: An integrated view of green synthesis methods, transformation in the environment, and toxicity, Ecotoxicol. Environ. Saf., 171 (2019) 691–700. https://doi.org/10.1016/j.ecoenv.2018.12.095

[59]   D. Garg, A. Sarkar, P. Chand, P. Bansal, D. Gola, S. Sharma, S. Khantwal, Surabhi, R. Mehrotra, N. Chauhan, R.K. Bharti, Synthesis of silver nanoparticles utilizing various biological systems: mechanisms and applications—a review, Prog. Biomater., 9 (2020) 81–95. https://doi.org/10.1007/s40204-020-00135-2

[60]   C.G. Kumar, S.K. Mamidyala, Extracellular synthesis of silver nanoparticles using culture supernatant of Pseudomonas aeruginosa, Colloids Surfaces B Biointerfaces, 84 (2011) 462–466. https://doi.org/10.1016/j.colsurfb.2011.01.042

[61]   F. Ameen, S. AlYahya, M. Govarthanan, N. ALjahdali, N. Al-Enazi, K. Alsamhary, W.A. Alshehri, S.S. Alwakeel, S.A. Alharbi, Soil bacteria Cupriavidus sp mediates the extracellular synthesis of antibacterial silver nanoparticles, J. Mol. Struct., 1202 (2020) 127233. https://doi.org/10.1016/j.molstruc.2019.127233

[62]   M. Iqtedar, M. Aslam, M. Akhyar, A. Shehzaad, R. Abdullah, A. Kaleem, Extracellular biosynthesis, characterization, optimization of silver nanoparticles (AgNPs) using Bacillus mojavensis BTCB15 and its antimicrobial activity against multidrug resistant pathogens, Prep. Biochem. Biotechnol., 49 (2019) 136–142. https://doi.org/10.1080/10826068.2018.1550654

[63]   M.A. Huq, S. Akter, Bacterial mediated rapid and facile synthesis of silver nanoparticles and their antimicrobial efficacy against pathogenic microorganisms, Materials (Basel)., 14 (2021) 2615. https://doi.org/10.3390/ma14102615

[64]   A. Matei, S. Matei, G.M. Matei, G. Cogǎlniceanu, C.P. Cornea, Biosynthesis of silver nanoparticles mediated by culture filtrate of lactic acid bacteria, characterization and antifungal activity, Eurobiotech J., 4 (2020) 97–103. https://doi.org/10.2478/ebtj-2020-0011

[65]   U. Farooq, X. Liu, W. Zhou, M. Hassan, L. Niu, L. Meng, Cell lysis induced by nanowire collision based on acoustic streaming using surface acoustic waves, Sensors Actuators B Chem., 345 (2021) 130335. https://doi.org/10.1016/j.snb.2021.130335

[66]   A. Abdel-Hadi, D. Iqbal, R. Alharbi, S. Jahan, O. Darwish, B. Alshehri, S. Banawas, M. Palanisamy, A. Ismail, S. Aldosari, M. Alsaweed, Y. Madkhali, M. Kamal, F. Fatima, Myco-Synthesis of Silver Nanoparticles and Their Bioactive Role against Pathogenic Microbes, Biology (Basel)., 12 (2023). https://doi.org/10.3390/biology12050661

[67]   A. Dhaka, S. Chand Mali, S. Sharma, R. Trivedi, A review on biological synthesis of silver nanoparticles and their potential applications, Results Chem., 6 (2023) 101108. https://doi.org/10.1016/j.rechem.2023.101108

[68]   M. Guilger-Casagrande, R. de Lima, Synthesis of Silver Nanoparticles Mediated by Fungi: A Review , Front. Bioeng. Biotechnol. 7 (2019). https://doi.org/10.3389/fbioe.2019.00287

[69]   B.F. Costa Silva LP, Oliveira JP, Keijok WJ, da Silva AR, Aguiar AR, Guimarães MCC, Ferraz CM, Araújo JV, Tobias FL, Extracellular biosynthesis of silver nanoparticles using the

cell-free filtrate of nematophagous fungus Duddingtonia flagrans, Int J Nanomedicine, 12 (2017) 6373–6381. https://doi.org/10.2147/IJN.S137703

[70]   E.M. Mekkawy AI, El-Mokhtar MA, Nafady NA, Yousef N, Hamad MA, El-Shanawany SM, Ibrahim EH, In vitro and in vivo evaluation of biologically synthesized silver nanoparticles for topical applications: effect of surface coating and loading into hydrogels, Int J Nanomedicine, 12 (2017) 759–777. https://doi.org/10.2147/IJN.S124294

[71]   R.M. Elamawi, R.E. Al-Harbi, A.A. Hendi, Biosynthesis and characterization of silver nanoparticles using Trichoderma longibrachiatum and their effect on phytopathogenic fungi, Egypt. J. Biol. Pest Control, 28 (2018) 28. https://doi.org/10.1186/s41938-018-0028-1

[72]   S.H. Lee, B.-H. Jun, Silver Nanoparticles: Synthesis and Application for Nanomedicine, Int. J. Mol. Sci., 20 (2019). https://doi.org/10.3390/ijms20040865

[73]   S. Rajput, R. Werezuk, R.M. Lange, M.T. McDermott, Fungal Isolate Optimized for Biogenesis of Silver Nanoparticles with Enhanced Colloidal Stability, Langmuir, 32 (2016) 8688–8697. https://doi.org/10.1021/acs.langmuir.6b01813

[74]   K. Gudikandula, P. Vadapally, M.A. Singara Charya, Biogenic synthesis of silver nanoparticles from white rot fungi: Their characterization and antibacterial studies, OpenNano, 2 (2017) 64–78. https://doi.org/10.1016/j.onano.2017.07.002

[75]   R.G. Saratale, I. Karuppusamy, G.D. Saratale, A. Pugazhendhi, G. Kumar, Y. Park, G.S. Ghodake, R.N. Bharagava, J.R. Banu, H.S. Shin, A comprehensive review on green nanomaterials using biological systems: Recent perception and their future applications, Colloids Surfaces B Biointerfaces, 170 (2018) 20–35. https://doi.org/10.1016/j.colsurfb.2018.05.045

[76]   S. Tyagi, P.K. Tyagi, D. Gola, N. Chauhan, R.K. Bharti, Extracellular synthesis of silver nanoparticles using entomopathogenic fungus: characterization and antibacterial potential, SN Appl. Sci., 1 (2019) 1545. https://doi.org/10.1007/s42452-019-1593-y

[77]   D. Wang, B. Xue, L. Wang, Y. Zhang, L. Liu, Y. Zhou, Fungus-mediated green synthesis of nano-silver using Aspergillus sydowii and its antifungal/antiproliferative activities, Sci. Rep., 11 (2021) 10356. https://doi.org/10.1038/s41598-021-89854-5

[78]   Y.K. Mohanta, D. Nayak, A.K. Mishra, I. Chakrabartty, M.K. Ray, T.K. Mohanta, K. Tayung, R. Rajaganesh, M. Vasanthakumaran, S. Muthupandian, K. Murugan, G. Sharma, H.-U. Dahms, J.-S. Hwang, Green Synthesis of Endolichenic Fungi Functionalized Silver Nanoparticles: The Role in Antimicrobial, Anti-Cancer, and Mosquitocidal Activities, Int. J. Mol. Sci., 23 (2022). https://doi.org/10.3390/ijms231810626

Green Synthesis and Emerging Applications of Frontier Nanomaterials   Materials Research Forum LLC
Materials Research Foundations 169 (2024) 59-80   https://doi.org/10.21741/9781644903278-3

Chapter 3

Green Synthesis and emerging applications of iron based nanomaterials

Hina N. Chaudhari[1], Dipti D. Parmar[1], Charmi D. Patel[1], Rajshree B. Jotania[1*]

[1]Department of Physics, Electronics and Space sciences, University School of Sciences, Gujarat University, Ahmedabad, 380009, India

rajshree_jotania@yahoo.co.in

Abstract

A vital motivation to investigate new synthesis methods for green-based iron nanoparticles (NPs) is their growing significance over the past ten years and the variety of uses for which they are used (biomedical, microwave absorbers, EMI shielding, etc.). Typically, they are magnetite (Fe_3O_4), maghemite (γ-Fe_2O_3), or a combination of the two, consisting of magnetic iron oxide nanoparticles. The term "green synthesis of nanomaterials" refers to the process of creating different metal nanoparticles using bioactive elements, including plant matter, microorganisms, and a range of bio-wastes like vegetable waste, fruit peel waste, eggshells, agricultural waste, etc. Nanoparticles synthesized by organic bioactive substances can reduce environmental damage. It is possible to produce less harmful by-products and dispose of organic waste and reagents by synthesizing iron nanoparticles in an aqueous medium under standard conditions (temperature and pressure).

Keywords

Iron Nanomaterials, Green Synthesis, Bio-Waste, Biomedical Application, EMI Shielding

Contents

1.   Introduction

Various materials are studied in the multidisciplinary field of material science. Recently, interest in innovative inorganic multifunctional nanomaterials has emerged as a result of the breadth of nanotechnology expanding to various application sectors [1]. The prospective uses of nanoparticles in a variety of technical domains, including electronics, optics, magnetic data storage, magnetic resonance imaging, catalysis, etc., attract considerable study interest [2-4]. Due to their biocompatibility, non-toxicity, and inducible magnetic moments, magnetic nanoparticles (MNPs) are a significant class of functional materials for applications in biomedical research [5-7].

Depending on their various oxidation states, iron oxides can be classified as [8]:

> ➢ Iron (II) ferrous oxide (wüstite-FeO)

> ➢ Iron (ferrous ferric) oxide (magnetite-Fe_3O_4)

> ➢ Iron (ferric) oxides Fe_2O_3

> • Hematite (α-Fe_2O_3)

> • β-Fe_2O_3

Since the 1990's, magnetic oxides, which are essentially always utilized as magnetic nanoparticles (NPs), have been the most widely used oxide materials in general/everyday biomedicine. Typically, these are spinel-structured magnetic iron oxides such as magnetite Fe_3O_4 or maghemite -Fe_2O_3 [9]. The enhancement of NPs production techniques, making them more effective, straightforward, and clean, hence lowering environmental contamination, is a current trend in nanotechnology. Humans require iron for the production of hemoglobin, and they also utilize it to make thermite, a catalyst, as well as coatings, paint, and colored concrete [8].

For a variety of biomedical applications, researchers have created MNPs with a particular chemical composition, shape, size, and surface coating (core-shell nanostructure). Most often, 100 nm or smaller iron oxide-based nanoparticles with an organic or inorganic shell are utilized. For example, Anand *et al.,* [10] prepared ZnO NPs by the technique of simple mixing and stirring of the metal salt and extract with Prunus dulcis (Almond Gum) green route for antimicrobial and supercapacitor purposes. Khalil *et al.,* [11] reported stable iron oxide NPs with *Sageretia thea* (Osbeck) for wastewater treatment applications. ZnO nanoparticles were studied by Paul *et al.,* [12] utilizing seeds extract from the tender pots of *Parkia roxburghii* (traditional vegetable abundantly in different areas of north-east India) by the one-pot green route method for photocatalytic degradation of organic dyes. Nasrollahzadeh *et al.*, reported CuO nanoparticles synthesized via the biosynthetic route using *Anthemis nobilis* flowers and *Thymus vulgaris* L leaves [13]. This chapter includes the development of iron NPs prepared via the green synthesis route and their applications in various fields.

2.   Green inspired synthesis of iron-based nanomaterials

Magnetic nanoparticles synthesis employing environmentally friendly techniques aids in a better understanding of the difficulties posed by chemical and physical processes. In recent era, green - inspired nanotechnology has attracted a lot of attention because it can eliminate hazardous organic solvents and help with environmental restoration. The biocompatibility of many MNPs, including magnetite, maghemite, and zero-valent iron, has been carefully investigated [14, 15]. Furthermore, plant extract, plant tissue, exudates, microorganisms, fungi, vitamins, enzymes, etc., have been mixed as effectual substitutes for MNPs synthesis [16]. It is beneficial to synthesize NPs from plants over biological entities as it saves time and effort on maintaining microbial cultures because plants have ascorbic acid, reductases, flavonoids, and dehydrogenase, which work like reducing agents [17, 18]. Carbon leaf extract was used by Awwad *et al.*, [19] to prepare magnetite NPs (Fe_3O_4) at a lower temperature of 80-85 °C. Obtained particles show advanced properties with an average diameter between 4-8 nm. In order to create magnetite NPs, Eatemadi *et al.*, [20] use poly (ethylene glycol), PEO, as a surfactant and solvent at the same time. Due to its greater boiling point (300 ℃ – 400 ℃) and lower toxicity, PEO has been widely used as an environmentally friendly solvent for several chemical synthesis processes.

According to Singh *et al.*, (2018) [21], several microorganisms, including bacteria, yeast, and fungi, have the ability to absorb metal ions on cell surfaces and then create active intracellular or extracellular enzymes to create iron NPs. These procedures can happen inside or outside of cells. Since the cell wall is not destroyed, extracellular synthesis is favored for iron NPs, which are made by microorganisms. Many different microbes have been used so far to produce Pb, Au, Cu, and Ag NPs. The microorganisms create proteins and peptides that act as natural stabilizers to prevent nanoparticle agglomeration [22, 23]. Microorganisms can also help reduce heavy metal toxicity through metal-organic complexation, metal oxidation, intracellular and extracellular metal sequestration, and metal chelators [24]. The best way to biosynthesize NPs with high monodispersed and controlled sizes must be ascertained [25].

Sachin Kumar *et al.*, [26] used tomato pulp to produce barium (BaM) hexaferrites (T0-T5) with different concentrations (0 g, 0.5 g, 1 g, 2 g, 3 g, 5 g per 100 mL of deionized solution) via sustainable and eco-friendly sol–gel auto-combustion approach. The obtained nanoparticles have average particle size of 75 nm and 22 nm. In the "green" sol-gel synthesis of cobalt ferrite NPs, an olive leaf extract containing phenolic compounds (oleuropein, phenolic acids, phenolic alcohols, and flavonoids) was utilized as a chelating agent, resulting in grain sizes ranging from 15 to 30 nm. [27]. The quantity of plant extract affects the acquired saturation magnetization and coercivity values. Magnesium ferrite NPs were created employing a *Solanum Lycopersicum* extract having average grain sizes between 18 - 65 nm [28]. Vitamin C, found in tomato extracts, plays a crucial part in the chelating process that results in spinel ferrite production.

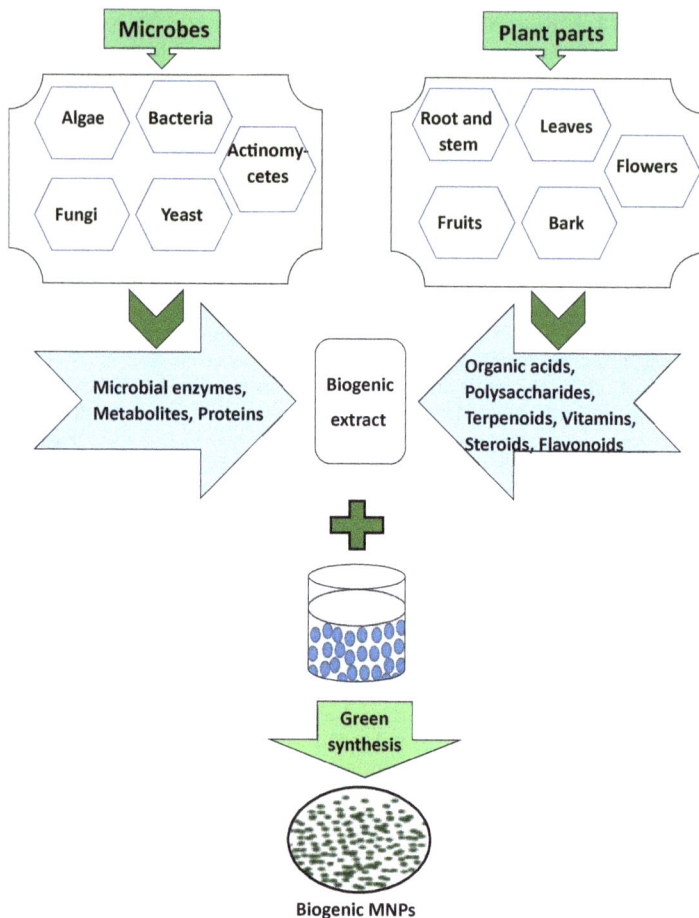

Fig. 1 Schematic diagram to synthesis NPs using green route [29-38]

Fig. 1 shows the flowchart to prepare green- inspired iron NPs using a sol-gel auto-combustion method. A stoichiometric amount of metal nitrate was dissolved separately in distilled water and then mixed well by continuous stirring on a magnetic stirrer to prepare a nitrate solution. Thereafter, green extract (plant tissue, fruits, leaves, flowers, fungi, etc.) was added dropwise into the prepared nitrates solution. This solution was stirred continuously on a hot plate. Then ammonia solution was slowly added to neutralize the solution. The dark greenish solution was kept for heating on a hot plate at 100 °C until combustion started and it was transformed into a fluffy powder. At last, the combusted powders were ground using a mortar and pestle to obtain a fine powder and heated in a muffle furnace at an appropriate temperature to obtain the final product of nanomaterials.

2.1   Advantages of green synthesis method [39-41]

> Non-toxic process

> Pollution free

> Eco-friendly

> Low cost

> Sustainable

3.   Applications of iron based nanomaterials

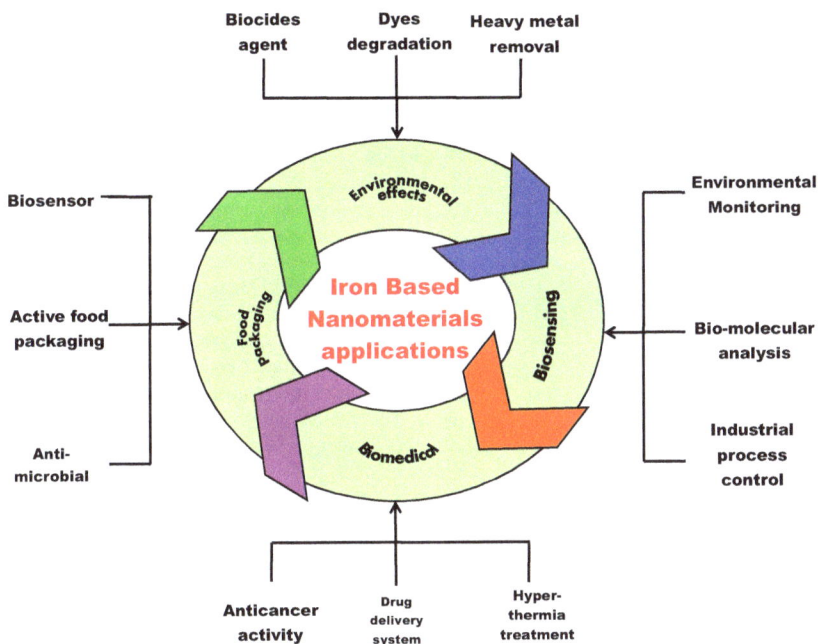

Fig. 2 Illustration of applications of iron based nanomaterials [42-47]

3.1 Environmental effects

Environmental contamination has become a global issue that is being watched closely by many people. Pollutant source and sink systems are produced by the constant exchange of pollutants between the three media of water, air, and soil. Due to their small particle size, high surface-to-volume ratio, and strong redox power, iron NPs have demonstrated remarkable performance for environmental remediation [48]. Researchers have studied the pH, ionic strength, and presence of other ions that may affect the efficiency of iron NPs in environmental cleanup [49].

The decomposition of iron NPs influences not only the surface but also its chemical characteristics. It also has an impact on the speciation of heavy metals and the pH of the environment when they are hazardous to living things. During the reaction step, iron NPs often directly alters the pH and pollutant specificity [50-52].

3.2 Food packaging

The penetrating qualities, barrier properties, heat resistance, and antibacterial activity are all changed or improved when nanomaterials are used in food packaging [53]. The enormous surface area and high surface energy of NPs define them. As a result, the NPs and polymer bonds interact strongly at the interface. The characteristics of the bio/polymers that utilized for packaging are greatly enhanced by this [54]. Food contact with oxygen during preservation is one of the key issues. Numerous food degradation processes, including lipid oxidation, microbial growth, enzymatic browning, and nutritional deterioration, are brought on by biopolymers. When food is in contact with oxygen, it leads to degradation of food which results in a reduction in shelf life, particularly for oxygen-sensitive foods [55]. As a solution, food is frequently packaged using materials that can reduce oxygen penetration [56]. Iron NPs can be used to prevent food from oxygen. There are two approaches that could be used to use nano-iron in packaging. The first technique involves adding nano- zerovalent iron (nZVI) containing sachets to the packaging; the second is to use a monolayer or multilayer structure to include the NPs in the packaging film [57, 58]. One such example is the use of FeO to modify bentonite and kaolinite. The European Food Safety Authority (EFSA) has accepted these compounds as non-nanoparticles [59]. Khalay et al., [60] reported that the PP (polypropylene) nanocomposites, which contained montmorillonite (OMMT) with iron NPs modification, were suitable for use in food packaging. They explored that the NPs can act as active oxygen and are capable of being removed by chemically reacting with it. Also, Busolo and Lagaron [58] observed that iron kaolinite present in active composites functions as a passive barrier because of its winding channel, which prevents gas transport and gas traps; moreover, it interacts with molecular oxygen. Microorganisms spoil food throughout the storage process, which is the main issue. In order to reduce the influence of microorganisms on food during storage, utilize IONPs and nanocomposites with antibacterial properties. They increase food safety and lengthen shelf life of food [61].

3.3 Biosensing

Forensic investigation, environmental monitoring, and biomedical diagnosis all have a strong interest in the detection of biological agents, diseases, and harmful substances [62]. A sensor typically has two parts: a transduction element for signaling the binding event and a recognition element for binding to the target. These systems are interesting prospects for sensing applications [63] due to the distinctive physicochemical characteristics of NPs [64] and the natural improvement in signal-to-noise ratio given by miniaturization [65]. Gold NPs are an example that displays distinctive electronic and optical characteristics depending on the size and shape of the NPs. Gold NPs about 500 – 550 nm exhibited a strong absorption peak occurring from surface Plasmon resonance (SPR), as shown in Fig. 3 [66-69]. Although the fundamental physical principles of SPR are highly complex, they result from the collective oscillation of the conductive electrons caused by the resonant excitation by the incident photons. SPR signals changes in solvent and binding by being sensitive to the environment. This phenomenon gives rise to the popular and frequently used colorimetric sensing. Under some circumstances, metallic NPs also exhibit excellent quenching ability [70-72] and photoluminescence [73-75].

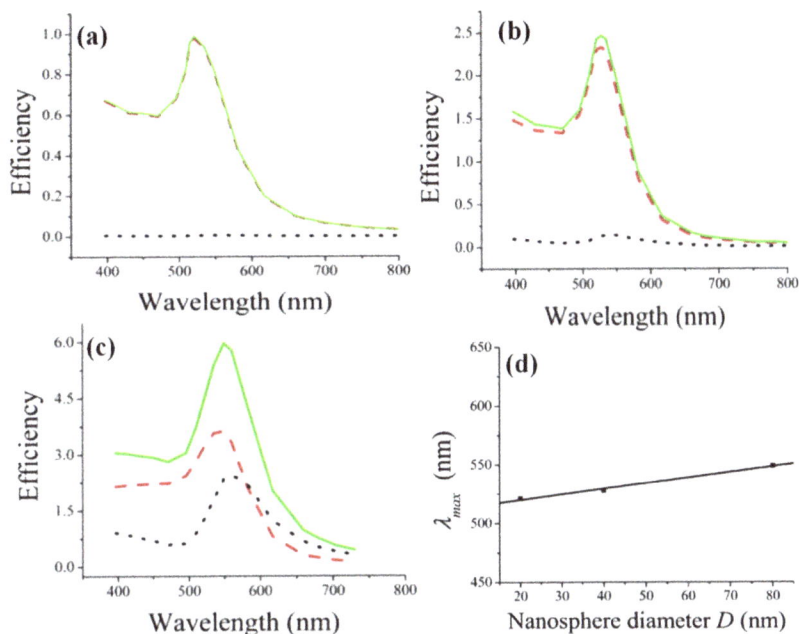

Fig 3. Calculated spectra of the efficiency of absorption (red dashed), scattering (black dotted), and extinction (green solid) for gold nanospheres (a) D =20 nm, (b) D =40 nm, (c) D =80 nm, (d) variation of surface plasmon extinction maximum λ_{max} with nanosphere diameter D (nm) [66]

3.4   Biomedical applications

3.4.1   Role of iron nanoparticle on anticancer activity

By using an eco-friendly biological method, K. Rajendran et al., [76] investigated hematite NPs from ferric chloride in the culture supernatant of B. cereus SVK1 at physiological temperature (98.6 °F) and pH. The peaks of Fe, O, and P are visible in the EDX spectrum (Fig. 4).

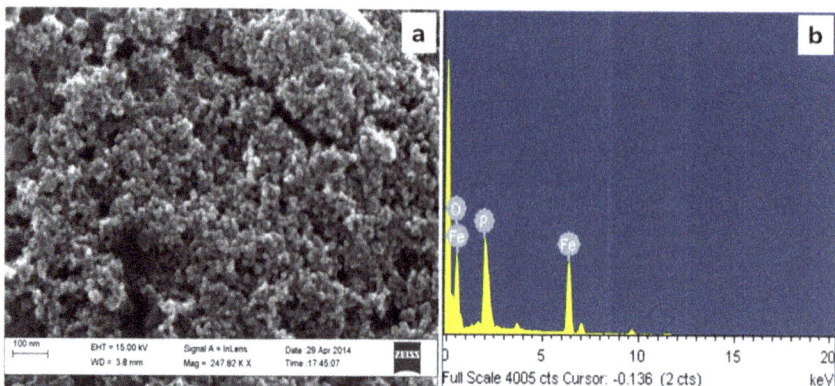

Fig. 4 Characterization of hematite nanoparticles (a) SEM and (b) EDX spectrum [76]

According to their study, the eco-friendly biological method for synthesizing hematite NPs is more effective and economical compared to the conventional chemical method (Hepatocellular carcinoma cell line). They also suggest that prepared hematite NPs are less harmful and can be used in cancer treatment. The cytotoxicity effect (Fig. 5) of iron oxide NPs on human HepG2 liver cancer cells was investigated by K. Rajendran *et al.*, They also experimented with hematite nanoparticle concentrations of 50 ng/ml, 100 ng/ml, 250 ng/ml, 500 ng/ml, and 1000 ng/ml (nanograms). At very low concentrations (704 ng/ml), hematite nanoparticles with an average diameter of ~30 nm demonstrated considerable anticancer action against HepG2 cell lines. The size of nanoparticles are important factors in causing toxicity [76].

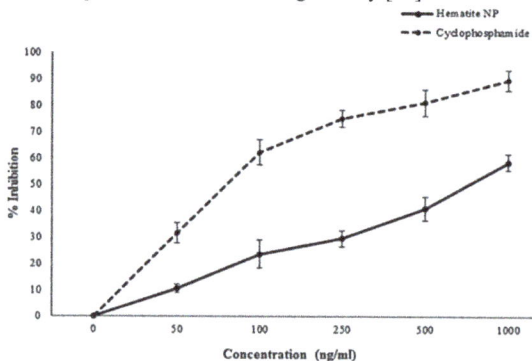

Fig. 5 Cytotoxic effect of hematite nanoparticles and cyclophosphamide (CYP) against HepG2 cancer cells [76]

Another study, reported by S. Kossatz *et al.*, [77], shows the effect of iron oxide NPs on tumor cells. They claimed that magnetic hyperthermia efficiently killed tumor cells through the heating process. The biggest difficulties in this area are the invention of healing temperatures that are targeted to the entire tumor area. Therefore, they proposed to enhance magnetic hyperthermia in breast cancer by using novel NPs. They also suggest that cell internalization and a chemotherapeutic agent with high heating potential are required to increase cell death. Their finding also noted that the composite of MF66 performed with N6L (Nucant multivalent pseudopeptide), doxorubicin (DOX), and magnetic hyperthermia might significantly boost the therapeutic effect of breast cancer magnetic hyperthermia. They proposed the research on merging two effective methods for destroying tumor cells as a clinical practice, such as intratumor nanoparticles used in chemotherapy injection and magnetic hyperthermia.

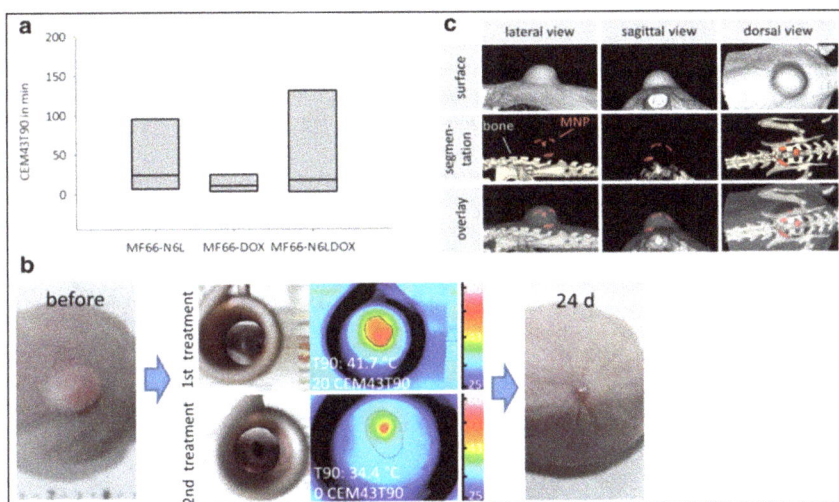

Fig. 6 Individual temperature dosages over tumor areas (a) By using tumor surface temperature during hyperthermia treatment, median temperature dosages were calculated as cumulative equivalent minutes (CEM43T90) and displayed as box plots. (b) Example of a treatment sequence within the alternating magnetic field (AMF), the corresponding temperature distribution over the tumor surface and the effect on tumor volume (c) Intratumoral distribution of magnetic nanoparticles (MNP) (MF66-N6LDOX) was determined using micro computed tomography 24 h prior to the first hyperthermia treatment. N6L Nucant multivalent pseudopeptide, DOX doxorubicin [77]

Iron oxide NPs synthesized using *Sargassum myriocystum* leaf extract were investigated by B. Sangeetha and A.K. Kumaraguru [78]. They found that the biosynthesized iron oxide with a size of 2.8 nm was utilized in the cure of cancer, and the experiments were conducted for both *in vitro* and *in vivo* testing. The flutamide having iron oxide NPs resulted in the growth in life of the tumor-bearing mice. According to their findings, flumide nanoparticles having a concentration of 10 mg/kg body weight by reducing the volume of nutritive fluid resulted controlling in tumor growth and expanded the life of Dalton's lymphoma ascite-bearing mice. As a result, flutamide nanoparticles at a dose of 10 mg/kg body weight have tumor-killing effects against Daltons's lymphoma ascite-bearing mice.

3.4.2 Drug delivery system

Due to their impressive properties, magnetic NPs are well suited for drug delivery devices that are highly targeted and organ specific. The utilization of magnetic NPs as a transporter in drug delivery was described by Widder *et al.,* in the 1970s [79]. The sustainability of NPs depends on their size and surface area [80]. As it focuses on a specified area, the drug is bombarded and absorbed by the targeted cells. Due to the targeted cells focus on a specific area, the drug is discharged and absorbed by them. The molecular structure and nature of supporting materials, the distribution mechanism, as well as the release temperature and pH, all depend on rate of the targeted drug. Magnetic-loaded systems and complicated systems like core shells and their composites have been tested for drug delivery. The test drug is chemically attached to the iron surface by various functional groups, including carboxyl, amino, and thiol, etc. [81-84]. A Scanning electron microscopy study showed small spheres with a diameter of 0.25 to 1.35 μm and an average size of 1.0 μm. Each albumin matrix is observed to have clumps of Fe_3O_4 dispersed mostly around the periphery of the particle by transmission electron microscopy. (Fig. 7, main panel and inset (i)).

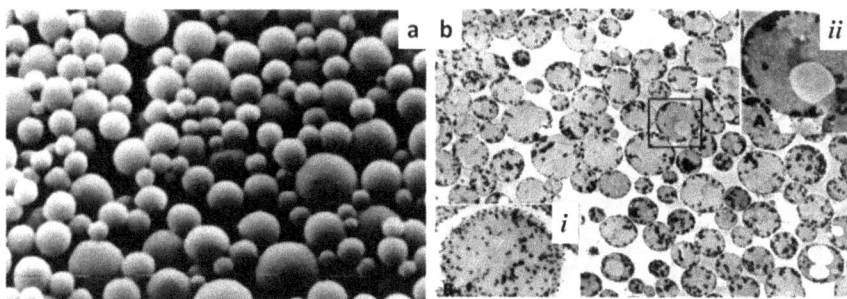

Fig. 7 (a) Scanning electron micrograph of magnetically responsive small albumin microspheres following evaporation from an ether suspension. magnification = × 8500. Calibration line = 1 μm. (Reproduced from Widder et al., 1978a. by permission of Academic Press.) (b) Transmission electron micrograph of microspheres prepared as in Main panel, standard preparation of microspheres (magnification = × 8500). Inset (i), enlargement of a standard microsphere containing peripherally oriented electron-dense clumps of a Fe, O, a medium dense albumin matrix, and an eccentric electron-lucent vacuole (magnification = × 38,250). Inset (ii), enlargement of a microsphere prepared with cationic surfactant, exhibiting a homogeneous distribution of Fe90, (magnification = × 19,040). (Adapted from Widder et. al., 1978a.) [80].

Magnetic NPs for the drug doxorubicin were utilized by Yuan *et al.,* in 2008. The temperature-sensitive polymer derivative encapsulated the magnetic nanoparticles, which acted as a core and were attached to doxorubicin by an acid-labile linker. Both pH~5.3 for cancerous cells and pH~7.4 for blood cells were used in this practice. Additionally, the effectiveness of the NPs was tested at temperatures of 20°C, 37°C, and 40°C (hyperthermal temperature). According to their findings, a pH value of 5.3 is more preferable to a pH value equal to 7.4 for the release of doxorubicin [85]. Transmission electron microscopy (TEM) measurements ((Fig. 8 (a)) showed that the average size of the uncoated MNPs was 9.6 $\pm$ 1.3 nm, but after coating with PLGA biopolymer poly (DL-lactic-co-glycolic acid), this size increased to 53.2 $\pm$ 6.9 nm. A magnetic property was shown by the M-H loop of Fe_2O_3 MNPs ((Fig. 8 (a)).

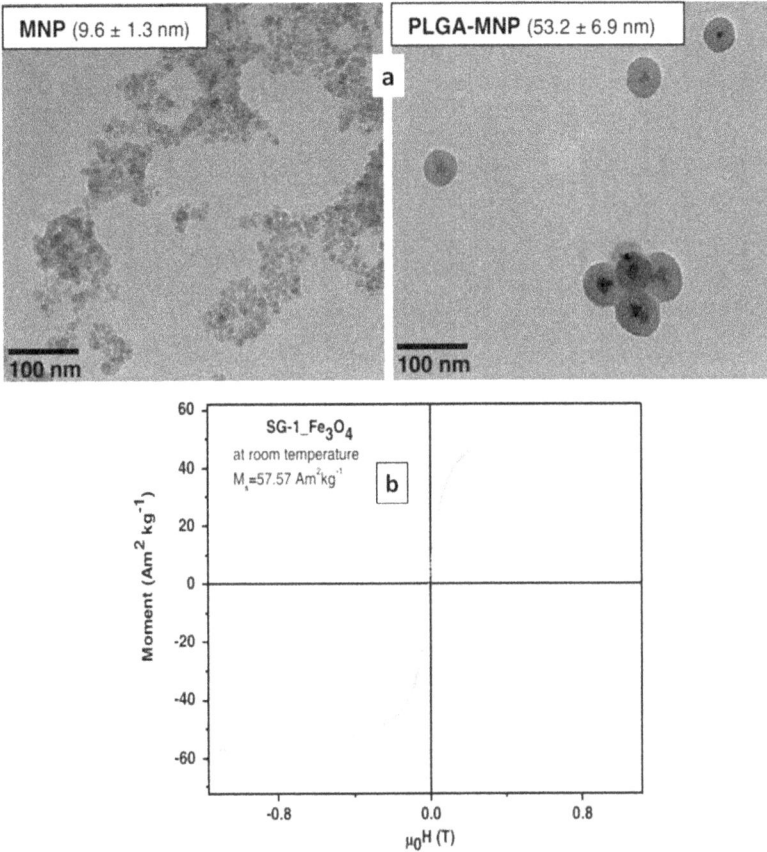

Fig. 8 *(a) TEM images of* MNPs or PLGA coated MNPs (PLGA-MNP) (b) Magnetization curve of initial Fe_3O_4 MNP at room temperature [85].

Additionally, chitosan, which is coated with magnetic NPs, has been appropriately tested and is utilized in the release of drugs (such as Doxorubicin and Bortezomib) depending on pH values [86,87]. Cisplatin can be utilized as a carrier of anti-cancer drug delivery, but due to its extreme toxicity, it is essential to target the drugs extremely precisely to the target cells [88,89]. Ibuprofen (an anti-inflammatory drug) has also been investigated as a drug carrier for the targeted delivery of drugs using magnetic NPs [90].

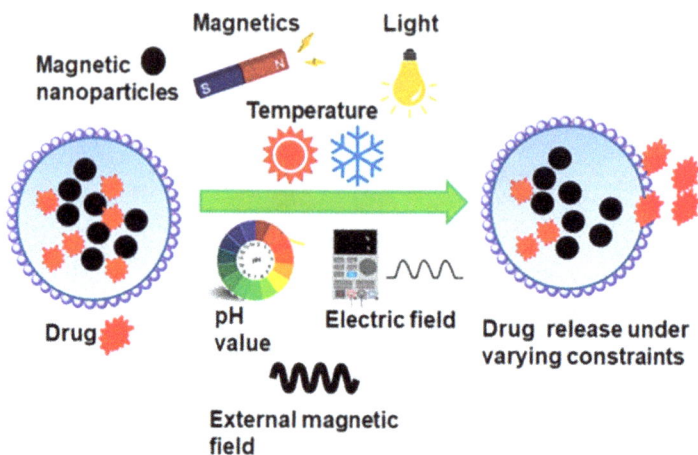

Fig. 9 Schematic representation of drug release using a MNP nanocomposite with applied external magnetic field under varying constraints [91].

3.4.3 Hyperthermia treatment

Gordon *et al.,* [92] employing locally generated heat energy produced by treating colloidally suspended submicron particles (magnetite nanoparticles) that were injected into tumor cells by applying an external magnetic field in 1979. For hyperthermia treatment, the most frequently studied spinel ferrite (NP) is magnetite, which is frequently surface-coated with an organic or inorganic coating agent [93]. Beyond the treatment of cancer, hyperthermia has also been utilized to treat cardiac arrhythmia, endometrial bleeding, varicose veins, excess body fat, etc. An approach to treating localized cancer called magnetic hyperthermia uses heated magnetic NPs that are injected into the tumor cells. The cancer cell death occurs at a temperature of about 43°C [94]. As tumor cells have a higher metabolic rate, they are more affected by hyperthermia effects compared to healthy cells [95]. In depth studies by Chang *et al.,* show that hyperthermia treatment can result in cell death due to various mechanisms [96]. According to research, using MNPs with hyperthermia during cancer treatment increases the effectiveness when mixed with other forms of therapy including, chemotherapy or radiotherapy. In-depth research has been conducted on heat-mediated drug delivery for the treatment of cancer [97].

For hyperthermia treatment, heat release of MNPs was utilized, depending on the morphology, synthesis procedure, particle dimensions, applied surface modification, and chemical compositions [98]. As a result, the heating efficiency varied for the cubic, spherical, and elongated magnetite (Fe_3O_4) NPs prepared by the solvothermal method [99]. The shape of the prepared NPs is also affected by the distribution of temperature [100]. Magnetite nanorods were prepared with an iron oleate precursor using the thermal decomposition technique. Amphilic surfactant- Tetramethy l ammonium 11-aminoundecanoate (TMAAD), was used to make them more appropriate for biological applications and for hyperthermia. This caused the hydrophobic oleate-coated iron oxide nanorods to become hydrophilic [101]. Hydrophilic rods were tested *in vitro* against MCF-7 breast cancer cell lines, and the results showed that they had anticancer action with reduced hemolysis because of minor erythrocyte damage.

3.4.4 Magnetofection

Magnetofection is a common use for magnetic NPs in biomedical applications. DNA-vectored magnetic NPs are injected into the cells by an external magnetic field using a very efficient transfection technique. Additionally, it is a quick and easy technique that allows transfection at a low dose up to saturation level. As they don't require receptors or other cell membrane proteins, nonpermissive cells may be transfected by the magnetic particles for cell uptake. For the vectorization of negatively charged DNA, the surfaces of the magnetic nanoparticles utilized for this goal are typically coated with polycationic polymers. The vector competence is dramatically increased by viral and nonviral vector magnetofection of nucleic acids [102]. A positively charged chemical creates nucleic acid/chemical complexes with negatively charged nucleic acids in a process known as lipofection that are subsequently attracted to the negatively charged cell membrane; using electroporation, a brief electrical pulse damages cell membranes and creates holes in them through so that nucleic acids can flow; or microinjection, which is a time-consuming method that efficiently injects nucleic acids straight into the cytoplasm or nucleus [103].

Conclusions and future scope

Iron-based nanoparticles have expanding range of applications in different fields, the major problem is still to synthesize nanoparticle or its composites with a well morphology and hysteresis properties using environmental friendly, simplest, economical, and repeatable synthesis techniques. Coating of prepared magnetic NPs must fulfill the needs of the intended application while also optimizing magnetic performance and biocompatibility. Iron nanoparticles can help to minimize environment pollution by altering the pH of dangerous contaminants. Hematite nanoparticles were tested for anticancer activity in HepG2 cells. Magnetic NPs are particularly suited for highly focused and organ-specific drug delivery devices due to their outstanding features. MNPs with organic or inorganic coating agents are utilized to treatment of hyperthermia.

Over the last few decades, research into the use of MNPs in biomedical applications has advanced rapidly, leading in the advancement of NP medicinal applications. However, more research into how NPs and MNPs interact with biological systems in the human body is required. Knowing this allows one to comprehend NP toxicity and mapping. The importance of MNPs in nature will be highlighted in the near future. The importance of magnetic navigation for various animals has lately been recognized. Understanding its biological system mechanism is an absorbing scientific task.

References

[1] L.A. Paramo, A.A. Feregrino-Perez, R. Guevara, S. Mendoza, K. Esquivel, Nanoparticles in agroindustry: applications, toxicity, challenges, and trends, Nanomaterials. 10 (2020) 1654.https://doi.org/10.3390/nano10091654

[2] D.E. Laughlin, D.N. Lambeth, Microstructural and crystallographic aspects of thin film recording media, IEEE Transactions on Magnetics. 36 (2000) 48-53.https://doi.org/10.1109/20.824424

[3] B. Issa, I.M. Obaidat, B.A. Albiss, Y. Haik, Magnetic nanoparticles : surface effects and properties related to biomedicine applications, Int. J. Mol. Sci. 14 (2013) 21266-21305.https://doi.org/10.3390/ijms141121266

[4] S. Schrittwieser, D. Reichinger, J. Schotter, Applications, Surface modification and functionalization of nickel nanorods, Materials. 11 (2018) 45. https://doi.org/10.3390/ma11010045

[5] V.F. Cardoso, A. Francesko, C. Ribeiro, M. Bañobre-López, P. Martins, S. LancerosMendez, Advances in Magnetic Nanoparticles for Biomedical Applications, Adv. Healthc. Mater. 7 (2018) 1700845. https://doi.org/10.1002/adhm.201700845

[6] M.I. Anik, M.K. Hossain, I. Hossain, A.M.U.B. Mahfuz, M.T. Rahman, I. Ahmed, Recent progress of magnetic nanoparticles in biomedical applications: A review, Nano Sel. 2 (2021) 1146–1186.https://doi.org/10.1002/nano.202000162

[7] N. Tran, T.J. Webster, Magnetic nanoparticles: biomedical applications and challenges, J. Mater. Chem. 20 (2010) 8760.https://doi.org/10.1039/c0jm00994f

[8] S.R. Pandya, H. Singh, Surface-Tailored Iron Oxide Magnetic Nanomaterials for Biomedical Applications, Materials Research Foundations 143 (2023) 1-40. https://doi.org/10.21741/9781644902332-1

[9] R.C. Pullar, Applications of Magnetic Oxide Nanoparticles in Hyperthermia, Materials Research Foundations 143 (2023) 76-101. https://doi.org/10.21741/9781644902332-3

[10] Theophil, Renuka D, R. Radhakrishnan, L. Anandaraj, J.S Savarenathan, G. Ramalingam, C.M. Magdalane, A.K. Bashir, M. Maaza, K. Kaviyarasu, Green synthesis of ZnO nanoparticle using prunus dulcis (almond gum) for antimicrobial and supercapacitor applications. Surf Interfaces. 17 (2019) 100376.17:100376.https://doi:10.1016/j.surfin.2019.100376

[11] A.T. Khalil, M. Ovais, I. Ullah, M. Ali, Z.K. Shinwari, M. Maaza, Biosynthesis of iron oxide (Fe_2O_3) nanoparticles via aqueous extracts of sageretia thea (osbeck.) and their pharmacognostic properties. Green Chem Lett Rev. 10 (2017) 186–201.https://doi:10.1080/17518253.2017. 1339831

[12] B. Paul, S. Vadivel, S. Dhar, S. Debbarma, M. Kumaravel, One-pot green synthesis of zinc oxide nano rice and its application as sonocatalyst for degradation of organic dye and synthesis of 2-benzimidazole derivatives. J Phys Chem Solids. 104 (2017) 152–159. https://doi:10.1016/j. jpcs.2017.01.007

[13] M. Nasrollahzadeh, S.M. Sajadi, A. Vartooni, Green synthesis of cuo nanoparticles by aqueous extract of anthemis nobilis flowers and their catalytic activity for the A(3) coupling reaction. J Colloid Interface Sci. 459 (2015) 183–188.https://doi:10.1016/j.jcis.2015.08.020

[14] N. Joshi, J. Filip, V.S. Coker, J. Sadhukhan, I. Safarik, H. Bagshaw and J.R. Lloyd, Microbial reduction of natural Fe (III) minerals ; toward the sustainable production of functional magnetic nanoparticles, Front. Environ. Sci., 6 (2018) 1–11. https://doi.org/10.3389/fenvs.2018.00127

[15] S. Shukla, R. Khan, A. Daverey, Environmental Technology & Innovation Synthesis and characterization of magnetic nanoparticles , and their applications in wastewater treatment : A review, Environmental Technology and Innovation, 24 (2021) 101924. https://doi.org/10.1016/j.eti.2021.101924

[16] P. Singh, Y. Kim, D. Zhang, D. Yang, Biological synthesis of nanoparticles from plants and microorganisms, Trends in Biotechnology, 34 (2016) 588–599. https://doi.org/10.1016/j.tibtech.2016.02.006

[17] R.G. Chaudhary, A.K. Potbhare, P.B. Chouke, A.R. Rai, R. Mishra, M.F. Desimone, A.A. Abdala, Graphene-based materials and their nanocomposites with metal oxides : biosynthesis, electrochemical, photocatalytic and antimicrobial applications, Material Research Forum. 83 (2020) 79-116. https://doi.org/10.21741/9781644900970-4

[18] P.B. Chouke, K.M. Dadure, A.K. Potbhare, G.S. Bhusari, A. Mondal, K. Chaudhary, V. Singh, M.F. Desimone, R.G. Chaudhary, D.T. Masram, Biosynthesized δ-Bi₂O₃ Nanoparticles from Crinum viviparum flower extract for photocatalytic dye degradation and molecular docking, ACS Omega, 2022, 7 (24),20983–20993.https://doi.org/10.1021/acsomega.2c01745d

[19] A.M. Awwad, N.M. Salem, A Green and Facile Approach for Synthesis of Magnetite Nanoparticles, nanoscience and nanotechnology, 2 (2012) 208-213. https://doi.org/10.5923/j.nn.201206.09

[20] A. Etemadi, H. Daraee, N. Zarghami, H.M. Yar, A. Akbarzadeh, Nanofiber: Synthesis and biomedical applications, Artificial Cells, Nanomedicine and Biotechnology, 44 (2011) 111-121.https://doi.org/10.3109/21691401.2014.922568

[21] J. Singh, T. Dutta, K.H. Kim, M. Rawat, P. Samddar, P. Kumar, "Green' synthesis of metals and their oxide nanoparticles: applications for environmental remediation, J. Nanobiotechnol. 16 (2018) 48. https://doi.org/10.1186/s12951-018-0408-4

[22] D.R. Lovley, J.F. Stolz, G.L. Nord, E.J.P. Phillips, Anaerobic production of magnetite by a dissimilatory iron-reducing microorganism, Nature, 330 (1987) 252–254.https://doi.org/10.1038/330252a0

[23] P. Singh, Y.J. Kim, D. Zhang, D.C. Yang, Biological synthesis of nanoparticles from plants and microorganisms, Trends Biotechnol. 34 (2016) 588–599. https://doi. org/10.1016/j.tibtech.2016.02.006

[24] S. Iravani, R.S. Varma, Bacteria in heavy metal remediation and nanoparticle biosynthesis, ACS Sustainable Chem. Eng. 8 (2020) 5395–5409.https://doi.org/ 10.1021/acssuschemeng.0c00292

[25] G.S. Dhillon, S.k. Brar, S. Kaur, M. Verma, Green approach for nanoparticle biosynthesis by fungi: current trends and applications, Crit. Rev. Biotechnol. 32 (2012) 49–73.http://doi.org/10.3109/07388551.2010.550568

[26] S. Godara, J. Prakash, R. Jasrotia, J. Ahmed, A. M. Tamboli, A. Hossain, Suman, A. Verma, P. Kumar, M. Singh, S. Verma, Rahul K. Dhaka, A. Kandwal, Green synthesis of magnetic nanoparticles of $BaFe_{12}O_{19}$ hexaferrites using tomato pulp: structural, morphological, optical, magnetic and dielectric traits, J Mater Sci: Mater Electron. 34 (2023) 1516. https://doi.org/10.1007/s10854-023-10859-z

[27] S. Banifatemi, E. Davar, B. Aghabaran, J. A. Segura, F. J. Alonso, S. M. Ghoreishi, Green synthesis of $CoFe_2O_4$ nanoparticles using olive leaf extract and characterization of their magnetic properties, Ceram. Int. 47 (2021) 19198-19204. https://doi.org/10.1016/j.ceramint.2021.03.267

[28] H. N. Chaudhari, P. N. Dhruv, C. Singh, S. S. Meena, S. Kavita, R. B. Jotania, Effect of heating temperature on structural, magnetic and dielectric properties of magnesium ferrites prepared in the presence of Solanum lycopersicum fruit extract, J. Mater. Sci.: Mater. Electron. 31 (2020) 18445-18463. https://doi.org/10.1007/s10854- 020-04389-1

[29] Dhillon, G. S., Brar, S. K., Kaur, S., & Verma, M. (2011). Green approach for nanoparticle biosynthesis by fungi: current trends and applications. Critical Reviews in Biotechnology, 32(1), 49–73. https://doi.org/10.3109/07388551.2010.550568

[30] https://www.news-medical.net/life-sciences/What-are-Algae.aspx

[31] https://jpt.spe.org/bacteria-real-water-issue-hydraulic-fracturing

[32] https://www.sciencenews.org/article/aspergillus-fungi-mycotoxins-aflatoxin-food-contamination

[33] https://stock.adobe.com/search?k=yeast+cell

[34] https://www.aakash.ac.in/blog/web-stories/biology-modifications-of-root-stem-and-leaf-with-examples/

[35] https://www.healthyeating.org/nutrition-topics/general/food-groups/fruits

[36] https://www.healthline.com/nutrition/pine-bark-extract

[37] https://www.1800flowers.com/blog/flower-facts/flower-color-meanings/

[38] https://www.chuka.ac.ke/portfolio/bioprospecting-of-actinomycetes-for-production-of-bioactive-compounds-with-antibacterial-properties-from-various-soils/

[39] https://www.alamy.com/stock-photo-laboratory-conical-flask-with-green-liquid-isolated-on-white-117645937.html

[40] F.K. Alsammarraie, W. Wei, Z. Peng, A. Mustapha, L. Mengshi, Green synthesis of silver nanoparticles using turmeric extracts and investigation of their antibacterial activities, Colloids Surf. B 171 (2018) 398–405. https://doi:10.1016/j.colsurfb.2018.07.059

[41] N. Kataria, V.K. Garg, Green synthesis of Fe_3O_4 nanoparticles loaded sawdust carbon for cadmium (II) removal from water: Regeneration and mechanism, Chemosphere, 208 (2018) 818–828.https://doi:10.1016/j.chemosphere.2018.06.022

[42] M. Nasrollahzadeh, M. Sajadi, Pd nanoparticles synthesized in situ with the use of Euphorbia granulate leaf extract: Catalytic properties of the resulting particles, J. Colloid. Interface Sci. 462 (2016) 243–251. https://doi:10.1016/j.jcis.2015.09.065

[43] https://www.hydrogenlink.com/dyedecoloration

[44] https://www.biolinscientific.com/blog/how-to-characterize-biomolecular-interactions-with-qcm-d

[45] https://www.dreamstime.com/pressure-transmitter-monitor-sent-measuring-value-to-programmable-logic-controller-plc-to-control-oil-gas-process-image119734043

[46] https://www.verywellhealth.com/hyperthermia-and-cancer-5076038

[47] https://www.medindia.net/patients/patientinfo/drug-delivery-system.htm#about

[48] https://nano-magazine.com/news/2019/2/20/study-says-consumers-prefer-nanotechnology-in-active-food-packaging

[49] S.C. Tang, I.M. Lo, Magnetic nanoparticles: essential factors for sustainable environmental applications. Water Res. 47 (2013) 2613–2632. https://doi.org/ 10.1016/j.watres.2013.02.039

[50] M. Gil-Diaz, J. Alonso, E. Rodriguez-Valdes, J.R. Gallego, M.C. Lobo, Comparing different commercial zero valent iron nanoparticles to immobilize As and Hg in brownfeld soil, Sci. Total Environ. 584 (2017) 1324–1332. https://doi.org/ 10.1016/j.scitotenv.2017.02.011

[51] X. Zhao, W. Liu, Z. Cai, B. Han, T. Qian, D. Zhao, An overview of preparation and applications of stabilized zero-valent iron nanoparticles for soil and groundwater remediation, Water Res. 100 (2016) 245–266. https://doi.org/10.1016/j. watres.2016.05.019

[52] L.G. Cullen, E.L. Tilston, G.R. Mitchell, C.D. Collins, L.J. Shaw, Assessing the impact of nano- and micro-scale zerovalent iron particles on soil microbial activities: particle reactivity interferes with assay conditions and interpretation of genuine microbial effects, Chemosphere, 82 (2011) 1675–1682. https://doi.org/10.1016/j. chemosphere.2010.11.009

[53] A. Ronavari, M. Balazs, P. Tolmacsov, C.Molnar, I. Kiss, A. Kukovecz, Z.Konya, Impact of the morphology and reactivity of nanoscale zero-valent iron (nZVI) on dechlorinating bacteria, Water Res. 95 (2016) 165–173. https://doi.org/10.1016/j. watres.2016.03.019

[54] S.M.A.S. Keshk, S. Bondock, M. Abu Haija, Synthesis of a Magnetic Nanoparticles/Dialdehyde Starch-Based Composite Film for Food Packaging, Starch-Starke, 71 (2018), 1800035.https://doi: 10.1002/star.201800035

[55] S. Jafarzadeh, A. Salehabadi, A.M. Nafchi, N. Oladzadabbasabadi, S.M. Jafari, Cheese packaging by edible coatings and biodegradable nanocomposites; improvement in shelf life, physicochemical and sensory properties, Trends Food Sci. Technol. 116 (2021) 218–231.https://doi:10.1016/j.tifs.2021.07.021

[56] M. Honglei, G. Haiyan, Ch. Hangjun, T. Fei, F. Xiangjun, G. Linmei, A nanosised oxygen scavenger: Preparation and antioxidant application to roasted sunflower seeds and walnuts, Food Chem. 136 (2013) 245–250. https://doi: 10.1016/j.foodchem.2012.07.121

[57] Z. Foltynowicz, A. Bardenshtein, S. Sängerlaub, H. Antvorskov, W. Kozak, Nanoscale, zero valent iron particles for application as oxygen scavenger in food packaging, Food Packag. Shelf Life. 11 (2017) 74–83. https://doi.org/10.1016/j.fpsl.2017.01.003

[58] C. Vilela, M. Kurek, Z. Hayouka, B. Röcker, S. Yildirim, M.D.C. Antunes, J. Nilsen-Nygaard, M.K. Pettersen, C.S.R. Freire, A concise guide to active agents for active food packaging, Trends Food Sci. Technol. 80 (2018) 212–222.https://doi.org/10.21256/zhaw-10457

[59] M.A. Busolo, J.M. Lagaron, Oxygen scavenging polyolefin nanocomposite films containing an iron modified kaolinite of interest in active food packaging applications, Innov. Food Sci. Emerg. Technol. 16 (2012) 211–217. https://doi: 10.1016/j.ifset.2012.06.008

[60] EFSA Panel on Food Contact Materials, Enzymes, Flavourings and Processing Aids (CEF). Scientific Opinion on the safety assessment of the active substances iron, iron oxides, sodium chloride and calcium hydroxide for use in food contact materials. EFSA J. 11(10) (2013) 3387. https://doi.org/10.2903/j.efsa.2013.3387

[61] M.J. Khalaj, H. Ahmadi, R. Lesankhosh, G. Khalaj, Study of physical and mechanical properties of polypropylene nanocomposites for food packaging application: Nano-clay modified with iron nanoparticles, Trends Food Sci. Technol.,51 (2016) 41–48. http://dx.doi.org/10.1016/j.tifs.2016.03.007

[62] T.R.N. Mary, R. Jayavel, Fabrication of chitosan/Cashew Nut Shell Liquid/plant extracts-based bio-formulated nanosheets with embedded iron oxide nanoparticles as multi-functional barrier resist eco-packaging material, Appl. Nanosci. 12 (2022) 1719–1730.https://doi:10.1007/s13204-022-02377-x

[63] D. Diamond, Principles of Chemical and Biological Sensors (Ed.: D.Diamond), John Wiley & Sons, New York, NY 1998, 1–18. https://www.wiley.com/en-us/9780471546191

[64] N.L. Rosi, C.A. Mirkin, Nanostructures in Biodiagnostics, Chem. Rev. 105 (2005) 1547-1562. https://doi.org/10.1021/cr030067f

[65] P.E. Sheehan, L.J. Whitman, Detection limits for nanoscale biosensors, Nano Lett. 5 (4) (2005) 803-7.https:// doi: 10.1021/nl050298x

[66] M.C. Daniel, D. Astruc, Gold Nanoparticles: Assembly, Supramolecular Chemistry, Quantum-Size-Related Properties, and Applications toward Biology, Catalysis, and Nanotechnology, Chem. Rev. 104 (2004) 293-346. https://doi.org/10.1021/cr030698+

[67] P.K. Jain, K.S. Lee, I.H. El-Sayed, M.A. El-Sayed, Calculated Absorption and Scattering Properties of Gold Nanoparticles of Different Size, Shape, and Composition: Applications in Biological Imaging and Biomedicine, J. Phys. Chem. B, 110 (2006) 7238-7248. https://doi.org/10.1021/jp057170o

[68] G. Mie, Contributions to the optics of turbid media, particularly of colloidal metal solutions, Ann. Phys. 25 (1908) 377-445. https://doi.org/10.1002/andp.19083300302

[69] K.L. Kelly, E. Coronado, L.L. Zhao, G.C. Schatz, The Optical Properties of Metal Nanoparticles: The Influence of Size, Shape, and Dielectric Environment, J. Phys. Chem. B, 107 (2003) 668-677. https://doi.org/10.1021/jp026731y

[70] S. Eustis, M.A. El-Sayed, Gold nanoparticles are more precious than pretty gold: Noble metal surface plasmon resonance and its enhancement of the radiative and nonradiative properties of nanocrystals of different shapes, Chem. Soc. Rev. 35 (2006) 209-217. https://doi.org/10.1039/B514191E

[71] K.H. Su, Q.H. Wei, X. Zhang, J.J. Mock, D.R. Smith, S. Schultz, Interparticle Coupling Effects on Plasmon Resonances of Nanogold Particles, Nano Lett. 3 (2003) 1087-1090. https://doi.org/10.1021/nl034197f

[72] K.E. Sapsford, L. Berti, I.L. Medintz, Materials for Fluorescence Resonance Energy Transfer Analysis: Beyond Traditional Donor–Acceptor Combinations, Chem. Int. Ed. 45 (2006) 4562. https://doi.org/10.1002/anie.200503873

[73] K.G. Thomas, P.V. Kamat, Chromophore-Functionalized Gold Nanoparticles, Acc. Chem. Res. 36 (2003) 888-898. https://doi.org/10.1021/ar030030h

[74] J. Zheng, C. Zhang, R.M. Dickson, Highly Fluorescent, Water-Soluble, Size-Tunable Gold Quantum Dots, Phys. Rev. Lett. 93 (2004) 077402. https://doi.org/10.1103/PhysRevLett.93.077402

[75] M.A. van Dijk, M. Lippitz, M. Orrit, Far-Field Optical Microscopy of Single Metal Nanoparticles, Acc. Chem. Res. 38 (2005) 594-601. https://doi.org/10.1021/ar0401303

[76] J.R. Lakowicz, Radiative decay engineering 5: metal-enhanced fluorescence and plasmon emission, Anal. Biochem. 337 (2005) 171-194. https://doi.org/10.1016/j.ab.2004.11.026

[77] K. Rajendran, V. Karunagaran, B. Mahanty, S. Sen, Biosynthesis of hematite nanoparticles and its cytotoxic effect on HepG2 cancer cells, Int. J. Biol. Macromol. 74 (2015) 376–381. https://doi.org/10.1016/j.ijbiomac.2014.12.028

[78] S. Kossatz, J. Grandke, P. Couleaud, A. Latorre, A. Aires, K. Crosbie-Staunton, R.Ludwig, H. Dahring, V. Ettelt, A. LazaroCarrillo, M. Calero, M. Sader, J. Courty, Y. Volkov, A. Prina-Mello, A. Villanueva, A. Somoza, A.L. Cortajarena, R. Miranda, I. Hilger, Efficient treatment of breast cancer xenografts with multifunctionalized iron oxide nanoparticles combining magnetic hyperthermia and anti-cancer drug delivery, Breast Cancer Res. 17 (2015) 1–17. https://doi.org/10.1186/s13058-015-0576-1

[79] N. Sangeetha, A.K. Kumaraguru, Antitumor effects and characterization of Biosynthesized Iron Oxide nanoparticles using Seaweeds of Gulf of mannar, Int. J. Pharm. Pharm. Sci. 7 (2015) 469-476.https://journals.innovareacademics.in/index.php/ijpps/article/view/3755

[80] Z. Chen, B. Li, J. Zhang, L. Qin, D. Zhou, Y. Han, Z. Du, Z. Guo, Y. Song, R. Yang, Quorum sensing affects virulence associated proteins F1, LcrV, KatY and pH6 etc. of Yersinia pestis as revealed by protein microarray-based antibody profiling, Microbes Infect. 8 (2006) 2501-2508. https://doi.org/10.1016/j.micinf.2006.06.007

[81] K.J. Widder, A.E. Senyei, D.F. Ranney, Magnetically responsive microspheres and other carriers for the biophysical targeting of antitumour agents, J. Adv. Pharmacol. Chemother. 16 (1979) 213-271. https://doi.org/10.1016/S1054-3589(08)60246-X

[82] G. Unsoy, U. Gunduz, O. Oprea, D. Ficai, M. Sonmez, M. Radulescu, M. Alexie, A. Ficai, Magnetite: from synthesis to applications, Curr Top Med Chem. 15 (2015) 1622-1640. https://doi.org/10.2174/1568026615666150414153928

[83] P. Theamdee, R. Traiphol, B. Rutnakornpituk, U. Wichai, M. Rutnakornpituk, Surface modification of magnetite nanoparticle with azobenzene-containing water dispersible polymer, J. Nano. Part. Res. 13 (2011) 4463-4477. https://doi.org/10.1007/s11051-011-0399-7

[84] D. Dorniani, A.U. Kura, S.H. Hussein-Al-Ali, M.Z. Bin Hussein, S. Fakurazi, A.H. Shaari, Z. Ahmad, Release Behavior and Toxicity Profiles towards Leukemia (WEHI-3B) Cell Lines of 6-Mercaptopurine-PEG-Coated Magnetite Nanoparticles Delivery System, Sci. World J. 2014 (2014) 1-11. https://doi.org/10.1155/2014/972501

[85] A.F. Wang, W.X. Qi, N. Wang, J.Y. Zhao, F. Muhammad, K. Cai, H. Ren, F.X. Sun, L. Chen, Y.J. Guo, M.Y. Guo, G.S. Zhu, A smart nanoporoustheranostic platform for simultaneous enhanced MRI and drug delivery, Micropor. Mesopor. Mat. 180 (2013) 1-7. https://doi.org/10.1016/j.micromeso.2013.06.015

[86] N.K. Verma, K. Crosbie-Staunton, A. Satti, S. Gallagher, K.B. Ryan, T. Doody, C. McAtamney, R. MacLoughlin, P. Galvin, C.S. Burke, Y. Volkov, Y.K. Gun'ko, Magnetic core-shell nanoparticles for drug delivery by nebulization, J. Nanobiotechnol. 11 (2013) 1-12. https://doi.org/10.1186/1477-3155-11-1

[87] Q. Yuan, R. Venkatasubramanian, S. Hein, R.D.K. Misra, A stimulus-responsive magnetic nanoparticle drug carrier: Magnetite encapsulated by chitosan-graftedcopolymer, Acta. Biomater. 4 (2008) 1024-1037. https://doi.org/10.1016/j.actbio.2008.02.002

[88] G. Unsoy, S. Yalcin, R. Khodadust, P. Mutlu, O. Onguru, U. Gunduz, Chitosan magnetic nanoparticles for pH responsive Bortezomib release in cancer therapy, Biomed. Pharma. 68 (2014) 641-648. https://doi.org/10.1016/j.biopha.2014.04.003

[89] G. Unsoy, R. Khodadust, S. Yalcin, P. Mutlu, U. Gunduz, Synthesis of Doxorubicin loaded magnetic chitosan nanoparticles for pH responsive targeted drug delivery, Eur. J Pharm. Sci. 62 (2014) 234-250. https://doi.org/10.1016/j.ejps.2014.05.021

[90] M. Konishi, Y. Tabata, M. Kariya, H. Hosseinkhani, A. Suzuki, K. Fukuhara, M. Mandai, A. Takakura, S. Fujii, In vivo anti-tumor effect of dual release of cisplatin and adriamycin from

biodegradable gelatin hydrogel, J. Cont. Rel. 103 (2005) 7-19.
https://doi.org/10.1016/j.jconrel.2004.11.014

[91] J.H. Kim, Y.S. Kim, K. Park, S. Lee, H.Y. Nam, K.H. Min, H.G. Jo, J.H. Park, K. Choi, S.Y. Jeong, R.W. Park, I.S. Kim, K. Kim, I.C. Kwon, Antitumor efficacy of cisplatin-loaded glycol chitosan nanoparticles in tumour-bearing mice, J. Cont. Rel. 127 (2008) 41-49. https://doi.org/10.1016/j.jconrel.2007.12.014

[92] S. Khizar, N.M. Ahmad, N. Line, N. Jaffrezic-Renault, A. Errachid-el-salhi, A. Elaissari, Magnetic nanoparticles: from synthesis to therapeutic applications, ACS Appl. Nano. Mater. 4 (2021) 4284-4306. https://doi.org/10.1021/acsanm.1c00852

[93] R.T. Gordon, J.R. Hines, D. Gordon, Intracellular hyperthermia: a biophysical approach to cancer treatment via intracellular temperature and biophysical alterations, Med. Hypothesis 5 (1979) 83-102. https://doi.org/10.1016/0306-9877(79)90063-X

[94] P.M. Martins, A.C. Lima, S. Ribeiro, S. Lanceros-Mendez, P. Martins, Magnetic nanoparticles for biomedical applications: from the soul of the earth to deep history of ourselves, ACS Appl. Bio. Mater. 4 (2021) 5839-5870. https://doi.org/10.1021/acsabm.1c00440

[95] K. Tofani, S. Tiari, Magnetic nanoparticle hyperthermia for cancer treatment: a review on nanoparticle types and thermal analyses, ASME of Medicinal Diagnostics 4(2021) 030801. https://doi.org/10.1115/1.4051293

[96] K.K. Kefeni, T.A.M. Msagati, T.T.I. Nkambule, B.B. Mamba, Spinel ferrite nanoparticles and nanocomposites for biomedical applications and their toxicity, Mater. Sci. Eng. C 107 (2020) 110314. https://doi.org/10.1016/j.msec.2019.110314

[97] D. Chang, M. Lim, J.A.C.M. Goos, R. Qiao, X.Y. Ng, F.M. Mansfeld, M. Jackson, T.P. Davis, M. Kavallaris, Front. Pharmacol. 9 (2018) 831. https://doi.org/10.3389/fphar.2018.00831

[98] P. Das, M. Colombo, D. Prosperi, Recent advances in magnetic fluid hyperthermia for cancer therapy, Colloids Surf. B 174 (2019) 42-55. https://doi.org/10.1016/j.colsurfb.2018.10.051

[99] Y. Wang, Y. Miao, M. Su, X. Chen, H. Zhang, Y. Zhang, W. Jiao, Y. He, J. Yi, X. Liu, H. Fan, Engineering ferrite nanoparticles with enhanced magnetic response for advanced biomedical applications, Mater. Today Adv. 8 (2020) 100119. https://doi.org/10.1016/j.mtadv.2020.100119

[100]   O. Polozhentsev, A.V. Soldatov, Efficiency of heating magnetite nanoparticles with different surface morphologies for the purpose of hyperthermia, J. Surf. Investig. 15 (2021) 799-805. https://doi.org/10.1134/S1027451021040364

[101]   Y. Tang, R.C.C. Flesch, T. Jin, Y. Gao, M. Ho, Effect of nanoparticle shape on therapeutic temperature distribution during magnetic hyperthermia, J. Phys. D: Appl. Phys. 54 (2021) 165401. https://doi.org/10.1088/1361-6463/abdb0e

[102]　A.J. Rajan, N.K. Sahu, Hydrophobic-tohydrophylic transition of Fe_3O_4 nanorods for magnetically induced hyperthermia, ACS Appl. Nano. Mater. 4 (2021) 4642-4653. https://doi.org/10.1021/acsanm.1c00274

[103]　S. Rahim, F. Jan Iftikhar, M.I. Malik, Biomedical applications of magnetic nanoparticles, in: M.R. Shah, M. Imran, S.B.T.-M.N. for D.D. and D.A. Ullah (Eds.), Micro Nano Technol., Elsevier, (2020) 301-328. https://doi.org/10.1016/B978-0-12-816960-5.00016-1

Chapter 4

Bioinspired zinc-based nanomaterials: Synthesis, biomedical, environmental and agricultural applications

P.R. Bhilkar[1], K. Shrivastav[1], A.S. Kahate[1], S. Tripathy[1], R.S. Madankar[1], A.R. Chaudhary[2,3], A.A. Balki[2], A.P. Lambat[4*], M.S. Nagmote[1], S. Somkuwar[5] and R.G. Chaudhary[1*]

[1]Post Graduate Department of Chemistry, S. K. Porwal College, Kamptee-441001, RTM Nagpur University, Nagpur, India

[2]Lady Amritabai Daga College for Women of Arts, Commerce and Science, Nagpur-440010, India

[3]Post Graduate Department of Botany, R.T.M. Nagpur University, Nagpur-440033, India

[4]Sevadal Mahila Mahavidyalaya, Umred Road, Nagpur-440024, India

[5]Dr. Ambedkar College, Deekshabhoomi, Nagpur-440024, India

chaudhary_rati@yahoo.com and lambatashish@gmail.com

Abstract

Metal-based nanomaterials play a vital role in the progress of solid-state nanomaterial science as they have numerous benefits in various areas of human life from health sciences to nanotechnology. Specially, Zinc-based nanomaterials (Zn-based NMs) drawn the attention worldwide due to their biocompatible, sustainable, and economical nature. Zn-based NMs have been widely used in various sectors viz., biomedical, clinical, semiconductors, photoluminescence, photocatalysis, agriculture, water treatment and many more because of their high surface area, excellent porosity, stability, scaffold morphology, good optical and electrochemical property. Because of this they can emerge as potential material in a variety of potential fields such as optics, biomedicine and environment. Since, many conventional routes have been used to synthesize Zn-based NMs, however, biogenic approaches are widely employed recently. By keeping this scenario in mind, the present chapter focuses on biogenic synthesis of Zn-based NMs with their excellent properties and its utility in biomedical, environmental and agriculture sector.

Keywords

Zn-Based NMs, Bioinspired Synthesis, Optical Properties, Biomedical Applications, Photocatalytical Activity

Contents

1.   Introduction

Nanomaterials (NMs) are the key pillar of the nanoscience and nanotechnology field which emerge as a fast-developing research area with numerous applications in different potential fields. Nanotechnology has enabled the controlled synthesis of NMs with a structure having at least one dimension less than 100 nm, which is one of the reasons why it is receiving attention. Also, NMs represents a novel and permitting that potentials to provide wide range of enhanced technologies in variety of field including biomedical, environmental and agriculture industries. Recently metal-based NMs fascinated in the NMs world because of its excellent physiochemical and biological behaviors. Extensive production of metal-based NMs made it easier for their product to use in household, healthcare and technical applications [1-3]. Among the various metal-based NMs, zinc oxide (ZnO) and zinc-based NMs (Zn-based NMs) are often preferred in biological and ecological area due to biodegradability, biocompatibility, and nontoxic nature [4]. Further, many implausible functions such as photocatalytic activity, physiochemical stability, antibacterial activity, UV-protection property, optimum band gap and high excitation binding energy exhibited by Zn-based NMs which made it star in the nanomaterial arena (Fig. 1) [5]. Zinc-based NMs, notably zinc oxide

(ZnO), demonstrates unparalleled optical, chemical sensing, semiconducting, electric and thermal conductivity, and piezoelectric attributes. Additionally, these NMs are commonly referred to as II-VI semiconductors [6]. Due to the wide bandgap, ZnO's properties are greatly affected, such as its electrical conductivity and its optical absorption. In combination with other metals, ZnO can persist at higher temperatures and have a higher conductivity. The durability, selectivity and heat resistance property can be enhanced of ZnO NMs by modification with inorganic or organic moiety [7, 8].

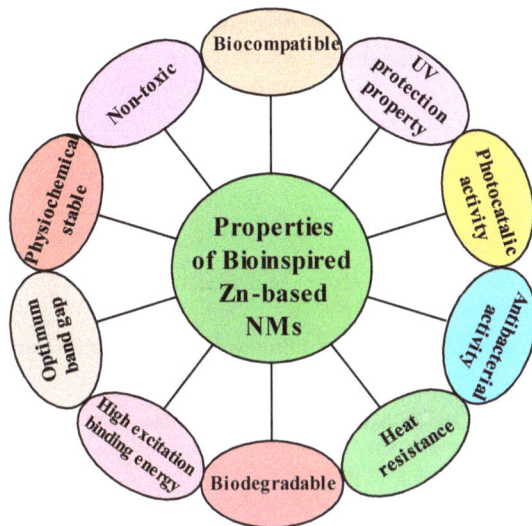

Figure 1: Promising properties shown by bioinspired Zn-based NMs

Zinc is an important co-factor present in the human anatomy and plays vital roles in functioning of large number of macromolecules and enzymes. The function of metalloproteins is also dependent on zinc, though it is considered relatively harmless, there is indication that it can be toxic to cells when it is dissolved as a free ion. Therefore, the biocompatible and innoxious nature made Zn-based NMs good for biomedical applications [9]. The ZnO based NMs exhibited high antibacterial and antifungal properties along with anticancer, biosensing and bioimaging activities. Apart from its exceptional optical absorption in UVA and UVB zones (315 to 400 nm), ZnO is also employed as a UV protector in the cosmetic industry [10]. It has also been shown that ZnO nanoparticles (NPs) cause cytotoxicity by generating reactive oxygen species (ROS) and destroying mitochondrial membrane potential, resulting in caspase cascade activation followed by cell apoptosis. A variety of plant bioactive drugs and chemotherapeutic anticancer drugs have been delivered to tumor cells using Zn-based NMs as effective carriers. Subsequently, Zn-based NMs are relatively inexpensive and harmless than other metal-based NMs and offer broad range of many medicinal uses such as anti-inflammatory, anti-aging, anti-diabetic and wound healing (Fig. 2) [11-14].

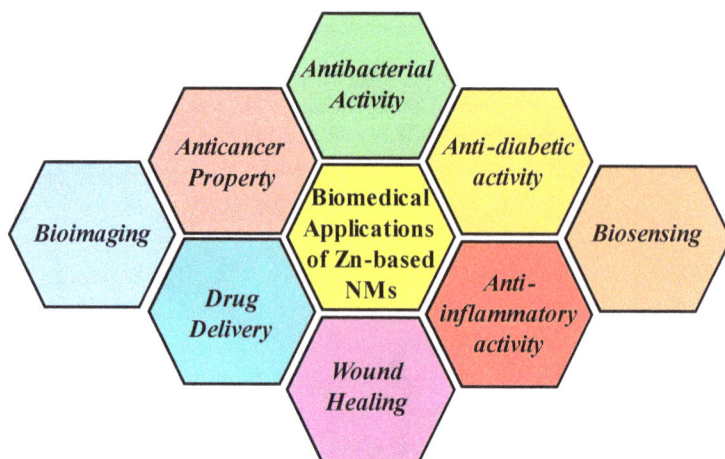

Figure 2: Biomedical applications of Bioinspired Zn-based NMs

The superiority of ZnO and Zn-based NMs becomes evident when compared to other metal-based NMs. This is primarily attributed to their abundance of point defects, particularly oxygen vacancies, as well as their higher production of hydroxyl ions, leading to a significantly faster reaction rate. Further, due to their optimum bandgap, similar to TiO_2, they can be the better choice for photocatalyst which can be utilized for the breakdown of many hazardous dyes and non-degradable pesticide and drugs in aqueous and soil system [15]. Herein this chapter, the synthesis of bioinspired Zn-based NMs from different biological sources are discussed in detailed manner. Also, their biomedical, environmental and agricultural applications are discussed along with futuristic approaches.

2.   Bioinspired synthesis of Zn-based NMs

The synthesis method plays a crucial role in the formation and development of NMs as it affects the size, surface area and surface modification. There are many conventional methods for the synthesis of Zn-based NMs but having certain disadvantages such as critical handling, high energy consumption, and hazardous byproduct. The bioinspired method is another alternative for the production of Zn-based NMs as it fulfills all the goals of green approach *viz.*, safe, cheap, less-hazardous ecofriendly and covered all the limitations of conventional methods [16]. Use of biological sources like different parts of plants, microbes and algae produce in controlled manner diffrent shape and sized Zn-based NMs (Fig. 3). In this portion, the different biogenic sources *viz.*, plant, microbes and biomolecules.

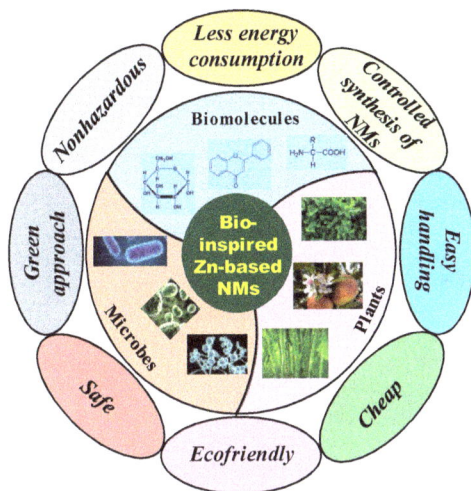

Figure 3. Biogenic sources and their advantages of bioinspired method of synthesis of Zn-based NMs

2.1   Plants

Among the different biogenic sources, plant-mediated synthesis of Zn-based NMs is chosen mostly due to its availability, cost-effectiveness and safe for human therapeutic use. For biosynthesis purpose, the extract is prepared from various element of the plant *viz.*, leaf, flower, fruits, stem, bark and roots in suitable solvent (broadly use water or ethanol). Numerous secondary metabolites such as alkaloids, flavonoids, carotenoids, polyphenols, sugars, acids, tannins and many more, the plant extracts which responsible for the production of controlled shape and size of Zn-based NMs. Further, the extract comprises of surfactants and stabilizing agents which play crucial role for the synthesis desired NMs [17]. Chouke et al. [18] synthesized ZnO nanospheres using the leaves extract of *Aspidopterys Cordata*. The XRD study reveals hexagonal-wurtzite structure of ZnO with average crystal size 16-17 nm and 3.11 eV bandgap. Further, Chouke et al. [19] attempted to prepared the hexagonal ZnO NPs using the aqueous extract of *Bauhinia racemosa* via co-precipitation method. For this purpose, 20 g of plant powder were used to make 20 mL extract which mix with zinc sulphate at room temperature. Formation of ZnO NPs indicated by changing faint to dark green color. Morphological examination shows the spherical shaped of synsized NPs with average crystalline size 11 to 25 nm and band gap 3.55 eV. Similarly, Kaliyamoorthy et al. [20] reported a synthesis of *Pisonia grandis* assisted preparation of ZnO NPs using co-precipitation technique at 60 °C for 2 hrs. Kaliyamoorthy et al. prepared plant extract using leaves powder in aqueous medium and stirred it with zinc acetate to produce ZnO NPs. The synthesized NPs exhibited semi-spherical shaped with an average crystalline size 30.32 nm.

Further, it was found that ZnO-doped in graphene-based NMs exhibited scaffold structures. The numerous reports revealed marvelous structures and shape of Zinc-based nanocomposites (NCs), and one of report by Tariq et al. [21] demonstrated an easy preparation of ZnO loaded graphene oxide (ZnO/GO) NCs using the *Sonchus Oleraceus* aqueous extract. The 30 g of plant fine powder was used to prepare the extract followed by mixing it with zinc acetate for 4 hrs at 55 °C to produces the trigonal shape ZnO NPs. Further, as-synthesized NPs were loaded on GO sheet with the aid of thermal condition and continuous stirring. Likewise, Sánchez-Albores et al. [22] demonstrated successfully biosynthesized ZnO/GO NCs using powder of orange peel for the photocatalytic degradation of methylene blue. By promoting the impregnation of organic compounds, the phytochemicals present in the extract facilitated their conversion into graphene oxide. The study found that reducing zinc source and increasing annealing temperature improved the photocatalytic performance. The performance was positively impacted by the optimization of catalyst dosage and pH in the dye solution. A mesoporous phytoreduced graphene oxide-ZnO (rGO-ZnO) nanohybrid was fabricated by Umekar et al. [23] employing *Clerodendrum infortunatum* leaf extract as a green reducer at ambient temperature. The rGO-ZnO nanohybrid absorption spectra displayed a minor blueshift under UV-DRS, indicating that zinc oxide was successfully anchored to the rGO surface. Similarly, Chaudhary et al. [24] reported a fabrication of rGO-ZnO NCs using the aqueous leaf extract of *Sesbania bispinosa* via hydrothermal method. The hexagonal wurtzite structure was observed in the rGO-ZnO NCs with average crystal sizes of 15-20 nm, and exhibited excellent photocatalytical performance under UV-light in less time.

Many metallic-ZnO based NCs were efficiently synthesized with the aid of plant extracts. For instance, Pandiyan et al. [25] demonstrates the hydrothermal method for preparation of silver and gold-doped ZnO NPs using $ZnCl_2$, $HAuCl_4$, and AgCl as precursors. The process involves adding 0.1 M of these solutions to 20 mL of *Justicia adhatoda* leaf extract, stirring for 1 hour, and heating in a stainless-steel autoclave for 48 hrs. The average crystalline size of synthesized NCs found to be 25 nm with band gap 2.98 eV. Similarly, Jan et al. [26] reported the single step biosynthesis of Fe doped ZnO NPs with aid of 100 mL aqueous extract of *Myrtus communis* L. via co-precipitation method. Zinc nitrate and iron nitrate were mixed along with plant extract for 3 hrs at 60 °C to produced irregular spherically shaped NCs. Also, Subbiah et al. [27] reported the synthesis of pure ZnO, $Zn_{0.98}Mn_{0.02}O$, and $Zn_{0.96}Mn_{0.02}Mg_{0.02}O$ nanostructures using *Carica papaya* leaves extract as a reducing and capping agent. The nanostructures show none of secondary phase detection and none of alteration in the hexagonal-wurtzite structure of ZnO NPs. Crystalline size reduces due to the doping of MgCo from 42 to 36 nm. By virtue of its wide energy gap, the synthesized ZnO NPs can effectively operate at higher temperatures and frequencies, surpassing the capabilities of conventional semiconductors.

Similarly, polymeric-doped ZnO NPs enhance nanocomposite's properties. Noticeably, Dejen et al. [28] demonstrated the synthesis of ZnO/PVA NCs with different ratio with the aid of using *Moringa oleifeira* leaves extract template. The NPs exhibited rod shaped morphology and particle sizes of 7.57 and 15.27 nm. Micro-structure examination confirmed successful loading of ZnO NPs on polyvinyl alcohol (PVA), with bandgap decreased. Similarly, Subashini et al. [29] reported synthesis of ZnO NPs using *Sterculia foetida* leaf extract, and introduced into poly(glutaric acid-co-ethylene glycol-co-acrylic acid) hydrogel to get polymer composites (GEA-ZnO hydrogel NCs).

2.2   Microbes

The implementation of unique strains of bacteria and the advancement of microbial methods for NPs production could reduce the harmful effects on the environment and remove issues and challenges associated with current methods. To tackle this issue, Barani et al. [30] successfully synthesized ZnO NPs *via* two new aquatic bacterium, *Marinobacter* sp. 2C8 and *Vibrio* sp. VLA. The FT-IR spectrum showed successfully biofabrication of NPs with obtaining peak at 482 cm^{-1}, while XRD analysis showed hexagonal wurtzite structured and spherical ZnO NPs. TEM and AFM techniques confirmed the formation of highly stable ZnO NPs with acceptable particle sizes. Similarly, Dhandapani et al. [31] elucidated the process of synthesizing ZnO NPs on cotton fabric through the utilization of urea broth and the inclusion of the ureolytic microbe, *Serratia ureilytica* (HM475278). This involved the conversion of urea into biologically engaged ammonia, which served as a precursor for the formation of ZnO NPs. For this purpose, biogenic ammonia is estimated from bacterial culture and centrifuging samples, filtering through a membrane, and testing for ammonia presence using the indophenols method, then incubating for 30 minutes. The synthesis process involved dissolving zinc acetate in de-ionized water, stirring it, and heating the medium at 50 °C for various durations. Azizi et al. [32] synthesized the ZnO NPs using the extract of marine microalgae *Sargassum muticum*. The study involved mixing dried algae powder with distilled water, heating it to 100 °C, and filtering it. Zinc acetate dehydrate was then mixed with the extract, resulting in a pale white solid which on heated at 450 °C produced hexagonal ZnO NPs. The sulphate and hydroxyl functionalities present in polysaccharide were shown to be involved in the production of ZnO NPs by analyzing FTIR spectra.

For the first time Selvarajan and Mohanasrinivasan [33] reported synthesis of ZnO NPs using probiotic bacterial species *Lactobacillus plantarum* VITES07. Following inoculation into sterile de Man, Rogosa and Sharpe broth, the bacterial culture was incubated for a duration of 24 hrs. The pH was adjusted to 6, and then zinc sulfate solution was added. The culture solution was heated, and a white precipitate of ZnO NPs formed. The production of ZnO NPs was successfully carried out by Ganesan et al. [34] employing the extract of *Periconium* species, an endophytic fungus through sol gel method. The extract was prepared by *Periconium* species was cultured in potato dextrose broth for 21 days, filtered, dried, pulverized into powder, and boiled in de-ionized water. Solution of zinc nitrate was mixed with fungal extract and then evaporated to create a sol. A gel with homogeneous Zn^{2+} ions formed when the solution was heated up in a hot air oven for 24 hours. To generate ultrafine ZnO NPs, the dried gel was calcined at 700 °C for 4 hrs after being dried at 125 °C for 12 hours. The synthesized nanoparticles are polydispersed with quasi spherically shaped with average crystalline size of 40 nm. Likewise, Jayaseelan et al. [35] delivers an reasonable and straightforward process for production of ZnO NPs employing the ecological reducing and capping agent with the aid of bacterium, *Aeromonas hydrophila*. The bacterial cells were grown in sterile distilled water for 36 hours, then diluted four times with nutrient broth media. The culture solution was subsequently supplemented with 0.1 g of ZnO and left to incubate for a duration of 24 hours. Within a time frame of 12-48 hours, noticeable white clusters, which merged together, were observed at the flask's base. However, the scalability of this method is minimum compared with other method. Additionally, using a microbiological technique makes it difficult to fabricate nanocomposites of ZnO correctly and uniformly with an inorganic or organic moiety, another metallic or non-metallic group or an atom.

2.3 Biomolecules

The most potent alternatives for hazardous chemicals reducing agents are biomolecules, and biocompatible reductant can be employed to generate Zn-based NMs. Different classes of biomolecules *viz.,* Amino acids, polyphenols, reducing sugar, acids flavonoids, etc., can be used in the nanomaterial's synthesis. Tanna et al. [36] synthesized the histidine capped ZnO NPs in ethanolic medium with high efficiency using solvothermal method. The synthesis of nanosized ZnO NPs involved dissolving histidine in ethanol, adding zinc acetate dihydrate and NaOH, heating in an autoclave. The presence of histidine-capped ZnO NPs gives rise to a droplet-shaped structure owing to the chaotic arrangement of individual beads, ultimately resulting in a substantial aggregation of particles. The presence of L-histidine significantly modifies ZnO superficial, forming smaller nanocrystals. Similarly, Ramani et al. [37] employed L-alanine, L-threonine, and L-glutamine for the synthesis of ZnO nanostructures (ZnO-NSs) with various morphologies. The amino acids play a crucial role as capping agents. Likewise, Tasdemir et al. [38] presents a green and economical approach for reducing natural L-ascorbic acids to produce stable ZnO thin films. The results of the investigation demonstrated the structural, morphological, electrical, and antibacterial characteristics of ZnO samples are strongly affected by the concentrations of ascorbic acid. The average particle size and thickness of the ZnO materials decreased after the addition of ascorbic acid from 58.29 to 48.68 nm, which improved their physical properties. The process of absorption and morphological change during the deposition phase were both impacted by the presence of ascorbic acid. Marinho et al. [39] utilized zinc acetate and urea as precursor for preparation of nano-sized ZnO NPs by utilizing a microwave hydrothermal approach in addition to traditional water bath heating. The formation of ZnO NPs using this approach is fast and homogenous.

Further, Hollow spheres structures ZnO and CuO-ZnO NCs were fabricated by Kim et al. [40] in a single pot via a hydrothermal reaction involving glucose and subsequent heating. The process involved dissolving D(+)-glucose monohydrate in distilled water, adding zinc nitrate hexahydrate, and stirring for few minutes to prepare ZnO and for CuO-ZnO NCs, copper nitrate was added extra in the solution. After being transferred to a Teflon lined autoclave, the solutions were subjected to a temperature of 180 °C for a duration of 24 hrs. Following this, the dried reaction mixtures were exposed to a heat of 500 °C for three hours, leading to the formation of hollow spheres consisting of both pure ZnO and CuO-ZnO composite material. Peychev and Vasileva [41] reported the formation of Au/ZnO NCs employing unique solution-solid synthesis method. A warm deionized water was used to dissolve a soluble starch solution, followed by the addition of zinc nitrate and chloroauric acid. The solution was heated, cooled, and incubated for 40 hours. After replacing the solvent with ethanol, the solid was collected and subsequently dried. The solid was then calcined in an air atmosphere at 600 °C for 4 hours. The hydrothermal development of amorphous ZnO or nanocrystalline ZnS shells on gold NPs was effectively controlled by Klug et al. [42] through the utilization of free-base amino acids. Among the amino acids investigated, L-histidine exhibited exceptional performance, yielding unaggregated and monodisperse sols of superior quality compared to other amino acids suitable for this technique.

3.　Potential applications of Zn-based NMs

Recently, numerous research has been conducted in a diversity of potential domains, including biomedicine, environment issue, and agriculture using the strength of an extraordinary features of bioinspired Zn-based NMs.

3.1 Biomedical applications

The biology, biotechnology, and medicine stand to benefit greatly from the use of nanoscience and nanotechnologies, both those now underutilized and those still to be discovered. Because of their similar size scale to biological components like proteins and antibodies, nanoparticles can be advantageously applied in a variety of biological and medical applications [43]. Recently, an important finding on Zn-based NMs has been made by the biomedical community. The Zn-based NMs has unique physical properties including a high surface area to volume ratio, optical emission, remarkable electrical and magnetic activities and variety of biological applications viz. drug delivery and biosensors.

3.1.1 Antimicrobial activity

There are several metal-based NMs like silver, gold, platinum, iron, copper, titanium, tungsten, manganese, cobalt, nickel, zinc etc. having potential applications in various sectors. Among these, Zn-based NMs like ZnO, ZnS, Ag-ZnO, $NiO–ZnO/TiO_2$, $ZnFe_2O_4$, $ZnAl_2O_4$, $ZnAl_2O_4–TiO_2$, Mg-doped ZnO, PLA-ZnO and many more have versatile applications in various areas due to its nanoscale shape and structure, nano-porosity, biocompatible nature, tuned optical property and scaffold morphology [44-47]. These biosynthesized Zn-based NMs are now used in a variety of fields to benefit people. They have a long history of usage as an antimicrobial agent including against bacteria, fungi, viruses, parasites, and prokaryotes (Fig. 4) [31, 48-51]. Using the Agar well method, the antibacterial activity of *Bauhinia racemosa*-mediated ZnO NMs was evaluated against bacterial species including *S. aureus, E. coli, P. vulgaris*, and *K. pneumoniae*. 20 μL of a solution containing 1 μg/mL of *Bauhinia racemosa*-mediated ZnO NMs was poured into the hole. The plates were sealed and then incubated for 24 hrs at 37 °C to observed the inhibition [19]. The antifungal impact of ZnO NMs on *Colletotrichum* sp. was examined in a study by Mosquera-Sanchez and co-workers. [52]. The behavior of the fungus was compared to that of a fungicide and a control sample after it was exposed to varying doses of ZnO NMs. At a dose of 15 mmol L^{-1}, the ZnO NMs demonstrated a noteworthy 96% reduction of fungal growth after 6 days, resulting in a loss of some hyphae's continuity and the development of clusters with distinct hyphal-structures. ZnO NMs derived from *Aquilegia pubiflora* were investigated for antimicrobial and anti-leishmanial properties against various bacterial and fungal strains. The findings indicate that ZnO NMs were more efficient against *F. solani* and *P. aeruginosa*, with inhibition zones that measured 13. ± 1.4 mm and 10.3 ± 0.19 mm, respectively. The susceptibility of *Leishmania tropica* (KWH23) to a dose-dependent cytotoxic effect was demonstrated, revealing substantial IC50 values for both the promastigote (48 μg/mL) and amastigote (51 μg/mL) forms of the parasite [53].

Using seed extract from *Ocimum tenuiflorum*, Panchal et al. [54] described a one-step synthesis of Ag/ZnO NCs. The results showed significantly improved antibacterial activity of 1% Ag/ZnO against *Escherichia coli*, 1 x 108 cfu in 15 min. Even against *S. aureus* and Methicillin-Resistant *S. haemolyticus* (MRSH), *Hibiscus sabdariffa*-assisted Au@ZnO NCs have demonstrated superior antibacterial and antibiofilm action than pure Au and ZnO NMs, even at 100 μg/mL concentrations [55]. Compared to ZnO NMs, Au@ZnO NCs are reported to be less hazardous to mouse fibroblast cells. Based on experimental results, it is anticipated that produced NCs will prevent the growth of *S. aureus* bacteria by producing reactive oxygen species (ROS) that damage the bacterial membrane and other parts of the cell. The study proposes that the ROS-dependent antimicrobial efficacy of synthesized Au@ZnO NCs against S. aureus strains is significant. The assessment of intracellular ROS production was conducted through fluorescence microscopy using DCF-DA, which facilitated the identification of hydrogen peroxide. The generation of ROS is believed to be the primary reason for the enhanced antimicrobial activity observed in ZnO NMs. DNA damage studies have revealed that ZnO NMs and Au@ZnO NCs infiltrated bacterial cells, resulting in notable oxidative stress and the production of reactive oxygen species (ROS). Consequently, this led to the degradation of DNA and ultimately the demise of the cells [55].

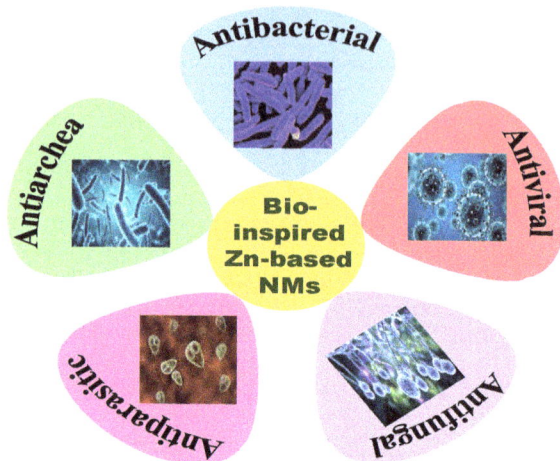

Figure 4. Antimicrobial applications of Bioinspired Zinc-based NMs

With zones of inhibition measuring 22.5 ± 0.5 mm and 22 ± 0.3 mm, respectively, the *Sida acuta*-mediated chitosan/ZnO NCs demonstrated impressive growth inhibition ability against *E. coli* and *B. subtilis* demonstrated by Madhan and his co-workers [56]. As reported by Nagaraj et al. [57], fabrication of graphene oxide sheet with *Dalbergia latifolia* derived-ZnO NPs, ZnO/GO NCs and evaluated against human pathogens, specifically *S. aureus*, *B. substills*, and *E. coli*. In contrast to ZnO NPs, and GO, ZnO/GO NCs exhibited strong antibacterial activity against *S. aureus*. Its large surface area and generation of reactive oxygen species were ascribed to graphene's interaction with lipid molecules. ZnO NPs, on the other hand, compromises the integrity of cells by releasing Zn^{2+} ions, which increases the production of ROS in positive and negative bacterium cells. Gram positive bacterial strains of *S. aureus* and *B. subtilis*, as well as gram negative strains of *E. coli* and *K. pneumoniae*, were used to test the antibacterial efficacy of GEA-ZnO hydrogel NCs [29]. At an optimal dosage of 100 µg/mL, gram positive bacteria exhibited greater antibacterial activity against gram negative bacteria as it came to polymer NCs. Table 1 shows some of the antimicrobial studies of bioinspired Zn-based NMs.

Table1. Antimicrobial applications of biosynthesized ZnO-based NMs

Sr. No	Bio-sources	ZnO-based NMs	Size (nm)	Shaped/ Morphology	Antimicrobial Activity Against	Refs.
1.	*Bauhinia racemosa*	ZnO	11-25	Spherically shaped	*P. vulgarus, E. coli, S. aureus and K. pneumoniae*	[19]
2.	Ethylene Glycol	ZnO	20-70	Spheroidal shaped	*Colletotrichum* sp.	[52]
3.	*Aquilegia pubiflora*	ZnO	19.58	Spherically shaped	*P. aeruginosa, B. Subtilis, E. coli, S. aureus, K. pneumoniae, A. niger, M. racemosus, F. solani, A. flavus, A. fumigatus, L. tropica*	[53]
4.	*Ocimum tenuiflorum*	Ag/ZnO (0.5%, 1%, and 2%)	28, 19 and 54	Agglomerated clusters	*E. coli*	[54]
5.	*Hibiscus sabdariffa*	Au/ZnO	31	Nearly spherical shaped	*S. aureus* and Methicillin Resistant *S. haemolyticus* (MRSH)	[55]
6.	*Sida acuta*	Chitosan/ZnO	24	Rod-shaped	*B. Subtilis* and *E. coli*	[56]
7.	*Dalbergia latifolia*	ZnO/GO	50-95	Spherically shaped	*B. Subtilis, E. coli* and *S. aureus*	[57]
9.	*Sterculia foetida*	GEA-ZnO	30-35	-	*B. Subtilis, E. coli, S. aureus* and *K. pneumoniae*	[29]
10.	*Azadirachta indica*	ZnO@TiO$_2$	~60-70	-	*P. aeruginosa, B. Subtilis, E. coli, S. aureus, S. pneumonia, K. pneumonia, P. vulgaris, V. cholera*	[58]
11.	Oxalic acid	CuO/ZnO	30	Spherical and clustered shaped	*P. aeruginosa, E. coli* and *S. aureus*	[59]
12.	*Serratia ureilytica bacteria*	ZnO	170-250	Nano flowers	*E. coli* and *S. aureus*	[31]
13.	*Moringa oleiferan*	CNTs/ZnO	35	Platelike structure	*P. aeruginosa, B. cereus* and *E. coli*	[60]
14.	*Ziziphus mauritiana* Lam	Cu doped ZnO (1% and 5%)	67 and 73 respectively	Spherically shaped	*E. coli* and *S. aureus*	[61]

3.1.2 Anticancer activity

Cancer is one of the pandemic diseases caused by unregulated cell proliferation in humans. This disease can be cured using advanced medicology. As with the convectional approaches, there are a number of medicines available for the diagnosis of this cancer, including immunotherapy, phototherapy, radiation, and chemotherapy, but all of which have adverse consequences. One of the most amazing developments in cancer treatment is the use of NMs. Because of their special qualities, NMs are currently attracting a lot of attention as an area to investigate as well as a promising alternative for conventional cancer therapy methods. Among all these, Zn-based NMs exhibits great potential for use in cancer treatment because of their strong potency and extreme selectivity towards cancerous cells. The primary cause of cytotoxicity in Zn-based NMs is the intracellular release of dissolved zinc ions, which triggers the production of ROS. This leads to oxidative stress and disequilibrium in zinc-mediated protein activity, which eventually cause cell death. Changes in the concentration of zinc inside the cell can lead to serious issues in numerous cellular activities, as zinc functions as a co-factor for over 300 human enzymes. The disruption in protein activity arises from the introduction of Zn-based NMs and the subsequent release of zinc ions within cells. This leads to an elevated concentration of zinc, causing an imbalance in protein activity known as zinc-mediated protein activity disequilibrium. This is harmful to the cell because it interferes with several essential cellular functions, such as oxidative stress, apoptosis, the electron transport chain, replication of DNA, and cellular homeostasis (Fig 5.) [62]. During a variety of cellular functions, such as mitochondrial respiration, inflammatory responses, microsomal activity, peroxisome activity, etc., the cell produces ROS. It functions as a biomolecule and contributes significantly to homeostasis and cell signaling [63].

The *Monsonia burkeana*-derived ZnO NPs exhibited concentration and time-dependent activity against A549 cell lines. The A549 cells' viability was decreased after 48 hrs of exposure to ZnO NPs at several doses. There was a small drop in cell viability at lower doses (62.5 and 125 µg/mL), but at higher concentrations (250 and 500 µg/mL), and there was a significant fall to 40% and 29%, respectively [64]. According to Senthilkumar and co-workers, *Tectona grandis* (L.) derived-ZnO NPs had excellent anticancer activity when evaluated against cancer osteoblast MC3t3-E1 cell line [65]. The viability of cancerous MC3t3-E1 osteoblast cells decreased to approximately 13% when treated with 50 µg/mL of ZnO NPs. Notably, no cytotoxic effects were observed in normal untreated MC3t3-E1 cells, even at a concentration as high as 1 mg/mL. The MTT experiment provided further insights, revealing an IC50 value of 41.29 µg/mL. *Gracilaria edulis* mediated-ZnO NPs produced have antitumor potential against the SiHa cervical cancer cell line. In the cytotoxicity experiment, it was observed that ZnO NPs exhibited a concentration-dependent increase in cytotoxicity at 24 hours (IC$_{50}$ value = 35 ± 0.03 µg/ml), surpassing the positive control, cisplatin (IC$_{50}$ = 10 µg/ml) [66]. As an alternative to traditional animal tests, ZnO NPs demonstrated the biocompatible nature of the medicine by demonstrating neither cytotoxic impact against human peripheral blood mononuclear cells nor hemolytic activity against erythrocytes, the *in-vitro* model systems utilized to assess the toxicity of the drug.

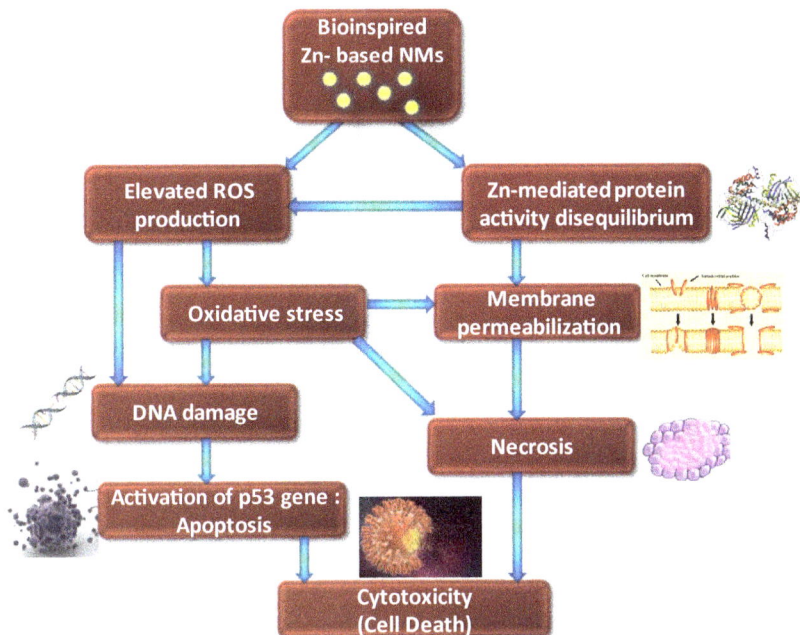

Figure 5. Stepwise action of bioinspired Zn-based NMs for anticancer activity

Similarly, bimetallic Ag-ZnO NPs were tested at various doses on Vero cells to determine their cytotoxicity. The produced NPs were safe to use, as confirmed by the results, which revealed an IC_{50} of 155.1 µg/mL. An IC_{50} more than 90 µg/mL implies non-cytotoxicity. Likewise, Hashem and El-Sayyad [67] biosynthesized bimetallic Ag-ZnO NPs from pomegranate peel extract were tested for their anticancer potential against the cancer cell lines MCF7 and Caco2. The results showed that bimetallic Ag-ZnO nanoparticles have anticancer activity with IC_{50} values of 104.9 and 52.4 µg/mL against cancer cell lines MCF7 and Caco2, respectively. Pandiyan et al. [25] described Ag-Au/ZnO NPs derived from *Justicia adhatoda* shows a dose-dependent cytotoxic action against HeLa cell lines. The results demonstrated that Ag-Au/ZnO NPs, as compared to ZnO, Ag-doped ZnO, and Au doped ZnO NPs, more successfully promoted cell mortality in HeLa cells. Also, with concentration of 6.25, 12.5, 25, 50, and 100 µg/mL, Ag-Au/ZnO NPs caused cell death at 19.66, 30.75, 49.07, 58.42, and 73.07%, respectively of cancerous cell. Similarly, Ag-Au/ZnO NPs have a stronger anticancer impact because they emit Ag^+, Au^+, and Zn^{2+} ions, which cause ROS to enter the cancer cell membrane and cause cancer cell death. Elimike et al. [68] used the MTT assay technique to assess the cytotoxicity *of Alchornea cordifolia* derived Cu_2O/CuO, ZnO, and Cu_2O/CuO–ZnO NPs against Hela cells. The outcomes demonstrated that as nanoparticle concentration increased, cell viability reduced. The cytotoxic potential of all the produced nanoparticles against Hela cells was exhibited; the maximum activity was observed by

Cu_2O/CuO–ZnO NCs. In contrast to CuO, ZnO NPs had greater anticancer efficacy, however they were not more effective than the reference drug.

Nagaraj et al. [57] investigated the cytotoxicity of ZnO, GO, and ZnO/GO NCs against MCF-7 and A549 cell lines using the MTT test. When compared to ZnO/GO NCs, the anticancer activities of the nanocomposite appear to complement those of ZnO, GO, despite both having lesser activity. ZnO/GO toxicity to MCF-7 cells may work in harmony *via* taking into account interactions between cells. In MCF-7 cells, it can lead to formation of ROS and disruptions in the cell membrane. Vijaykumar and his colleagues [69] conducted an in vitro cytotoxicity assessment using human colon cancer (HT-29) and human dermal fibroblasts (HDF) cells. In their study, they discovered that the presence of laminarin embedded ZnO NPs (Lm-ZnO) and chitosan capped ZnO NCs (Ch-Lm-ZnO) significantly reduced cell viability in a dose-dependent manner. The results imply that Ch-Lm-ZnO and Lm-ZnO may be employed as potential nanotherapeutic agents to treat cancer. Batool et al. [70] investigated the loading capacities (LC) and drug loading efficiencies (LE) of polyethylene glycolated (PEGylated) and unstabilized ZnO NPs were measured using the drugs gemcitabine (GEM) and doxorubicin (DOX). On *Aloe barbadensis* derived-ZnO NPs, DOX exhibited better LE 65% and greater LC 32% than GEM LE 30.5% and LC 16.25%. PEG-ZnO-NPs revealed increased LC of 35% and LE of 68%, respectively. The triple-negative breast cancer (TNBC) cell line was used to test the nanoparticles' antiproliferative capability after DOX was selected to encapsulate them. PEGylated biogenically produced ZnO NPs may enhance triple breast cancer therapy approaches, based on such drug delivery mechanism.

3.1.3 Bioimaging and biosensing

Zinc in the form of ZnS or ZnO NMs obtained chemically or biosynthetically [71, 72], are efficiently implemented in Sensing [73], bio-imaging, antitumor activity in human body and also as optoelectronics. ZnO NPs are inherently resistant to microbes, this property of ZnO provides it extensive scope in biomedical application. It is safe to consume zinc (Zn^{2+}), a component of zinc oxide, because it is widely recognized as a crucial micronutrient for the human body. In addition to use as nanocarriers in drug delivery with minimal toxic effects, ZnO NPs are implemented in bioimaging being highly stable and photo luminescent [74]. The band gap of ZnO can be effectively tuned by doping it with other elements or forming complexes with inorganic/organic compounds. This manipulation significantly enhances the capabilities of nano-scaled ZnO materials, establishing them as prominent agents in the realms of bio-imaging and drug delivery. The utilization of fluorescence imaging has become prevalent in preclinical research due to its affordability and practicality. By harnessing the inherent fluorescence of ZnO NPs, the penetration of these particles in human skin was observed in both *in vitro* and *in vivo* settings.

Due to their numerous desirable optical features, quantum dots (QDs) are the nanoparticles for optical imaging that have undergone the most research. However, the CdSe or CdTe cores of QDs, which are frequently used, have the potential to be harmful to biological systems. As a result, substantial work has gone into designing fluorescent nanoparticles that are less harmful, including ZnO-based QDs. Sudhagar and coworkers reported transferrin-conjugated ZnO QDs as nanoprobes for in-vitro targeting the cancerous cell imaging with minimal cytotoxicity [75]. In another intriguing investigation, Gd-doped ZnO QDs were created for optical and magnetic resonance imaging (MRI) with diameters of around 6 nm. It was discovered that the Gd-doped ZnO QDs' emission intensity increased with rising Gd^{3+} concentrations, reaching a peak at 550 nm. The Gd-doped ZnO QDs that were created after the surface was coated with N-(2-

aminoethyl)aminopropyltrimethoxysilane (AEAPS) showed little toxicity to HeLa cells and could be viewed *in-vitro* using confocal microscopy and MRI [76]. Wanas et.al. reported graphene doped with ZnO NSs in bio-imaging of cancer tumor. The synthesized G/FA-ZnO NSs with luminescence, wane toxicity and great photo stabilities were found to be highly efficient in selectivity in tumor bio-imaging [77]. In order to carry out mouse cancer immunotherapy, Fe_3O_4@ZnO core-shell NPs were created to transfer carcinoembryonic antigen into dendritic cells (DCs). These Fe_3O_4@ZnO core-shell NPs served as imaging agents for in vitro detection using confocal laser scanning microscopy (CLSM) and in vivo detection using MRI in addition to acting as nanocarriers to distribute antigen efficiently [78]. Lei and co-workers reported a new approach for designing amphibious ZnO QDs which is blue fluorescent within hyper-branched poly(ethylenimine) (HPEI). By varying $[Zn^{2+}]$:$[OH^-]$ ratio and heating time, ZnO QDs with a quantum yields of 30% were achieved in ethanol. In fact HPEI facilitated amphibious characteristic, thus it become possible to implement these water-soluble ZnO/HPEI NCs in bioimaging and optical electronic devices with great efficiency [79].

The remarkable affinity of the enzyme to adhere to the surface of ZnO NPs was achieved due to the minute dimensions of the ZnO NPs. This facilitated the modification of the enzyme, leading to an enhanced performance of the biosensor system. Due to high surface area of NMs they can facilitate direct electron transport between the electrode and the active sites of the biomolecules. In addition to their semiconducting properties, ZnO NPs exhibit several advantageous qualities for biosensing. These include high catalytic efficiency, robust adsorption capability, and a high isoelectric point. These attributes make them well-suited for the electrostatic adsorption of certain proteins.

Amongst various attempts in developing biosensors, recently Sawakar and co-workers fabricated the nanocomposite electrode biosensor of graphene doped-ZnO NCs to observe the electrochemical oxidation behavioral study of cetirizine, an anti-inflammatory drug and found this behavior diffusion-controlled involving two protons and two electrons. Furthermore, the efficiency and stability of obtained biosensors were validated by analyzing pathological samples [80]. To carry out sensitive detection of *M. tuberculosis* electrochemical DNA (TB DNA), biosensor was fabricated from nanocomposite of ZnO and Au nanoparticles and was immobilized with thiolated TB DNA. The detection level of biosensor probe was found to be highly selective, precise, stable with reproducible results [81]. Field Effect Transistor (FET) based potassium sensor was developed by Valinomycin immobilization on Fe_2O_3 nanoparticles decorated ZnO nanorods for detection of potassium level. In order to maintain device stability while detecting potassium, the surface of ZnO nanorods can be modified with Fe_2O_3 NPs, which may help stabilize the ZnO surface against acidic or basic conditions. [82].

Due to effective surface area, biocompatibility, ease of synthesis, controllable morphologies, and high electron communication, the biosensors created utilizing ZnO-based nanostructures offer enormous potential for biomedical applications. Biosensors built on ZnO nanostructures modified by metal oxides offer great benefits for creating multifunctional and structural nanomaterial-based sensors. These nanohybrid-based biosensors have the potential to revolutionize a number of industries, including biomedical diagnostics, environmental remediation/monitoring, and food analysis.

3.2   Environmental applications

Nanomaterials are being produced for dealing with sustainability issues in the environment. Through design, synthesis, and modification, improved knowledge of controlled synthesis and structure-activity connections has resulted from advances in materials science and catalysis research. This has improved performance in environmental applications. Zn-based NMs also exhibits a captivating potential for photocatalysis applications due to its remarkable charge carrier recombination activity, scaffold structures, mesoporous nature, convenient accessibility of raw materials, cost-effectiveness, and non-toxic characteristics.

3.2.1   Photocatalytical activity

Many approaches have been used to breakdown various organic pollutants, but one of the most promising technologies seems to be advanced heterogeneous photocatalysis using different NMs. Many potential photocatalysts have been actively used in environmental waste management systems including TiO_2, ZnO, Fe_2O_3, ZrO_2, V_2O_5, Bi_2O_3, Co_3O_4, ThO_4, WO_3 and many more based photocatalysts [83-86]. Today, bioinspired Zn-based NMs is the most popular and promising option for green environmental management systems in the field of photocatalysis due to its distinctive features, which include a direct and wide bandgap in the near UV spectral region, potent oxidation capacity, good photocatalytic behavior, and a large free exciton binding energy [87, 88]. Kaliraj et al. [89] conducted an evaluation of the quaker ladies shaped ZnO (ZnO/QNF) flower type derived from panos extract, for its potential in photocatalytic degradation of various dyes. Moreover, the degradation of methylene blue, eosin, and malachite green were carried out under UV light irradiation. The ZnO/QNF photocatalyst exhibited exceptional performance in reducing the concentration of methylene blue, eosin, and malachite green to over 99% after 80, 90, and 110 minutes of contact time respectively. This significant reduction in the presence of these dyes, each at a concentration of 15 mg/L, underscores the efficacy of the ZnO/QNF photocatalyst as a potent solution for water treatment. Furthermore, after 30, 35, and 40 minutes of contact time respectively, the low concentration of 5 mg/L of methylene blue, eosin, and malachite green were eliminated by the ZnO/QNF photocatalyst. When compared to the numerous other benchmark materials previously described, the pedal structure stands out by offering unobstructed access to all active sites. This unique characteristic enables a seamless interaction with the dye molecule under UV light, resulting in accelerated rates of dye degradation. The *Leucaena leucocephala* synthesized ZnO NPs photocatalytic activity was evaluated on crystal violet dye solution under ultraviolet exposure [90]. In the presence of ZnO nanocatalyst, being exposed to UV light causes the intensity of absorption to diminish with increasing exposure duration. After 90 minutes, the intensity nearly disappears, demonstrating the total reduction of crystal violet dye under ultraviolet light. Yu and Yu, photocatalytically degraded a rhodamine-B at room temperature by glucose-assisted ZnO hollow nanosphere [91]. ZnO hollow nanosphere photocatalytic activity can possibly be altered by changing the molar ratio (R) of glucose to zinc ions. The photocatalytic activity rises with increasing R and reaches a maximum value at R = 15. The durability and resistance to photocorrosion of ZnO hollow spheres is evident as the catalyst maintains its activity level consistently even after multiple cycles of photodegradation of rhodamine B.

Similarly, Tian et al. [92] used rhodamine-B in the to assess Ag/ZnO photocatalytic efficacy. Under UV light, rhodamine-B absorbance progressively drops and vanishes in 50 minutes in presence of Ag/ZnO NPs. When it comes to photodegradation of the dye, the Ag/ZnO NCs outperform Degussa P25. Ag/ZnO NCs photocatalytic performance is influenced by Ag content and shows higher activity compared to pure ZnO NPs. The mixture containing 5% Ag/ZnO exhibits the greatest catalytic activity, outperforming pure ZnO by 11 times and Degussa P25 by 2 times. Likewise, Vishwanathan and Das [93] reported that the utilization of MgO-ZnO catalysts, produced under natural sun irradiation, exhibited remarkable photocatalytic activity in the degradation of phenol, methylene blue, and rhodamine-B. The results revealed that although pure ZnO samples only demonstrated 65% and 79% degradation, methylene blue and rhodamine-B solutions were completely broken down by 0.6 MgO-ZnO in 100 and 150 minutes respectively. Additionally, 0.6 MgO-ZnO demonstrated improved degradation efficiency of 72% in 240 min for the phenol solution, compared to ZnO, 58% degradation. Also, Zelekew et al. [94] reported photocatalytic methylene blue decomposition using the *Eichhornia crassipes* asissted-Cr_2O_3/ZnO photocatalyst. Within 90 minutes of irradiation, 85% of methylene blue degradation was achieved using the molar ratio of 0.08/0.1 of Cr/Zn in the photocatalyst compared to another molar ratio. The increased efficiency in photocatalysis can be ascribed to the electron/hole pairs separation and the porous structure derived from *Eichhornia crassipes*.

Similarly, Chaudhary et al. [24], observed that the utilization of *Sesbania bispinosa* derived rGO-ZnO NCs resulted in remarkable photocatalytic activity. The nanocomposites exhibited an impressive 85% degradation efficiency towards methylene blue when subjected to UV-Visible light irradiation for a duration of 70 minutes. The rapid deterioration of methylene blue by rGO-ZnO NCs signifies the enhanced phenomena of charge transfer, light absorption, and separation. Under the influence of visible light, the valence band (VB) in rGO-ZnO NCs experiences the electrons excitation (e^-) to the conduction band (CB), which possesses higher energy. Simultaneously, the corresponding holes (h^+) are also present. The generation of h^+ in the conduction band is initiated through this excitation process, which is made possible by the suitable bandgap energy of rGO-ZnO NCs. These nanocomposites receive electrons from the valence band to facilitate the creation of h^+. Termed as e^-h^+ phenomenon, this phenomenon occurs because of absorption of visible light by the rGO-ZnO NCs. The active sites of the rGO-ZnO photocatalyst exhibit pairs of electron-hole (e^-h^+) charges, which are located on the catalyst's surface. These charges are subsequently captured by water and oxygen molecules, along with the dye particles present in the surrounding environment. Consequently, the surface-active nature of rGO-ZnO NCs allows for the transfer of photo-induced charge carriers from the bulk to the catalyst's surface, effectively preventing the recombination of holes and electrons. This phenomenon plays a crucial role in the photocatalytic activity of the rGO-ZnO NCs (Fig. 6). Correspondingly, Atchudan et al. [95] examined the photocatalytic efficiency of ZnO@GO composite in breaking down methylene blue. Their findings indicate that the addition of GO leads to a significant enhancement in the effectiveness of photocatalysis. The introduction of GO in the synthesized ZnO@GO NCs has resulted in a remarkable degradation efficiency of 98.5% in a neutral solution under UV-light exposure for 15 minutes. In contrast, pure ZnO exhibits a degradation efficiency of only 49% after 60 minutes of irradiation. This improvement can be attributed to the enhanced light absorption and minimized charge recombination facilitated by the incorporation of GO. Likewise, Bhuvaneswari et al. [96] has proposed rGO/LTH/ZnO/g-C_3N_4 nanohybrid as photocatalyst, which exhibited a crucial part in the breakdown of methylene blue. Significantly, rGO and g-C_3N_4, which served as visible light harvesters and electron mediators, respectively, have an impact on the degradation activity. The vertical arrangement of LTH on r-GO and g-C3N4, combined with the 2D/2D/2D layered heterojunction arrangement, has the potential to greatly enhance electron transportation and synergetic effects among them. More efficient separation and effective recombination suppression are the results of this structure, which also lowers the rate of charge carrier recombination and increases visible light absorption.

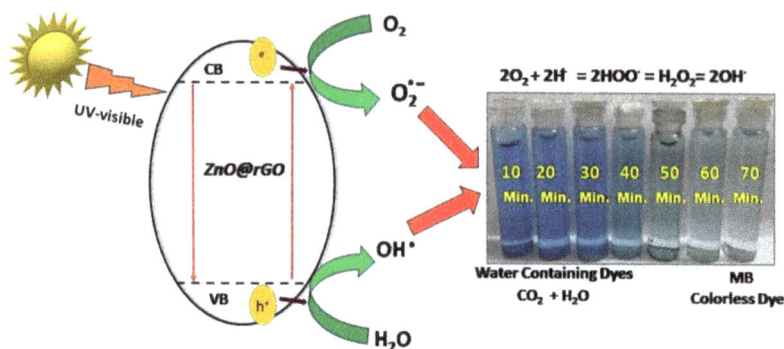

Figure 6. Photochemical degradation with plausible mechanism of methylene blue dye by rGO-ZnO nanocomposites.[24] Reproduce with permission of Elsevier.

3.2.2 Water treatment

Water is a fundamental necessity for all living organisms. As the human population continues to grow, the presence of pollutants in water bodies due to human activities also escalates. Consequently, the need for clean water rises in tandem with the population increase. The primary objective of water treatment is typically centered around eliminating or breaking down organic pollutants, heavy metals, and microorganisms [97-99]. Numerous scientists have devised various Zn-based NMs materials specifically for water treatment in order to obtain purified water for practical purposes. ZnO nanoparticles, synthesized from zerumbone as a green reducing agent, were utilized by Azizi et al. [100] to investigate the adsorption of Pb (II) ions from an aqueous solution. In the aqueous solution, the maximum adsorption capacity of Pb (II) ions was attained at a pH of 5 and a temperature of 70°C, reaching an impressive 93% [101]. In addition, the utilization of ZnO nanostructure resulted in a noteworthy enhancement in the adsorption efficiency of As (III) ions, reaching an impressive 96% at a pH of 7. This remarkable result was achieved by using a dosage of 0.4 g of adsorbent, a contact period of 105 minutes, and a constant temperature of 50°C [102].

Sharma and colleagues [103] explored the potential of mesoporous ZnO and TiO_2@ZnO NPs in adsorbing heavy toxic metals viz. Pb (II) and Cd (II) ions in an aqueous solution with a pH of 6, for a contact time of 80 minutes at 30 °C. The study revealed that TiO_2@ZnO exhibited a higher maximum adsorption efficiency for Pb (II) ions at 978 mg/L compared to ZnO at 790 mg/L. Similarly, TiO_2@ZnO also demonstrated a higher maximum adsorption efficiency for Cd (II) ions at 786 mg/L compared to ZnO at 643 mg/L. Similarly, Cruz et al. [104] reported biochar that has been impregnated with corncob-derived ZnO NPs shows promise as an adsorbent for eliminating Pb (II) and As (V) from polluted water. The most favorable results were obtained using ZnO NPs impregnated biochar made from corncobs, which showed a minimum of 25.8 mg of Pb (II)/g and

Green Synthesis and Emerging Applications of Frontier Nanomaterials   Materials Research Forum LLC
Materials Research Foundations 169 (2024) 81-112   https://doi.org/10.21741/9781644903278-4

for As (V)/g, the maximum equilibrium adsorption capacity was 25.9 mg. Also, Yu et al. [105] found in their investigation that biochar loaded with 30 weight percent ZnO can remove Cr (VI) with an efficiency of above 95%. The sorbent exhibits adsorption kinetics that are in line with the pseudo-second-order kinetic model, and the adsorption isotherm is described by the Langmuir model, with an adsorption capacity of 43.48 mg/g. ZnO NPs are function as a photo-catalyst, producing photogenerated electrons that help reduce Cr (VI) to Cr (III). Even after the fifth cycle, the ZnO-biochar composite shows good recyclability, with a removal ratio of almost 70%. This shows that efficient management and effective removal of pollutants are both achievable for aqueous system.

An assortment of organic pollutants, including dyes and insecticides, are frequently found in aquatic systems. These materials are the result of runoff from agriculture, industrial wastewater, and chemical deposition. The toxicity, stability, and lengthy half-lives of these substances pose a significant risk to human health. The results of Dehghani et al. [106] experiments showed that the UV/ZnO action performed optimally if alkaline pH and 0.2 g/L of ZnO NPs were used to remove dyes. Moreover, the removal effectiveness fell with increasing Acid–32–Cyanine 5R dye concentration while increasing the radiation period enhanced the removal efficiency. It should be emphasized that in order to completely remove the dye in the presence of UV light, a longer exposure time and higher radiation intensity are needed. Moghaddas et al. [100] investigated the elimination of methylene blue utilizing Quince seed mucilage derived ZnO NPs. The maximum removal effectiveness of 80% was observed at an adsorbent dosage of 50 mg/L and a contact duration of two hours. Similarly, Beevi et al. [107] successfully produced ZnO NPs using *Simarouba Glauca* and utilized them for the degradation of Congo red dye under sunlight rays. The results demonstrated an impressive degradation efficiency of 85% when the dye concentration was at 10 ppm, the catalyst dose was 0.1g, the pH was 6, and the contact time was 150 minutes.

Further, Maleki et al. [108] showcased a remarkable achievement by achieving almost complete reduction in the concentration of diazinon, an organophosphorus pesticide, in aqueous solutions. The greatest amount of diazinon degradation, which reached an amazing 99%, was achieved by using 2% WO_3 doped ZnO. Under particular circumstances, such as a pH of 7, a diazinon concentration of 10 mgL^{-1}, and a contact period of 180 minutes, this degradation was accomplished. Utilizing *Azadirachta indica*, Rani et al. [109] fabricated a composite of ZnO and CdS (ZnO@CdS NCs) that was activated by sunlight for the elimination of chlorpyrifos and atrazine pesticides from aqueous solutions. The maximum removal capacity of chlorpyrifos and atrazine pesticides was achieved at neutral pH with an adsorbent dose of 25 mg for chlorpyrifos and 20 mg for atrazine, and a contact time of less than 6 hours in the presence of sunlight, resulting in 91% and 89% removal capacity, respectively. Zhu et al. [110] demonstrated a study on the photodegradation properties of dimethoate using the ZnO/rGO NCs. Their results revealed that the rate and efficiency of photodegradation of dimethoate were significantly higher when using the ZnO/rGO NCs compared to bare ZnO. Specifically, the photodegradation rate was four times higher, while the photodegradation efficiency was 1.5 times higher. The ZnO/rGO NCs exhibited favorable characteristics, including a high surface area of 41.0 m^2/g and a pore volume of 4.72 cm^3/g. The adsorption and mass transfer of pesticides and oxygen were greatly enhanced by the unique properties of the nanocomposite, ultimately leading to its improved overall performance. The improved photocatalytic performance of the ZnO/rGO NCs can be ascribed to two main factors. Firstly, the presence of rGO reduced the electron-hole recombination rate, preventing the loss of photo-generated charge carriers. Secondly, the nanocomposite facilitated effective carrier transport, ensuring efficient utilization of the charge carriers for the degradation process. Huang and co-authors [111] conducted a study on the application of $ZnSnO_3/g-C_3N_4$ NCs with varying mass fractions for the photocatalytic degradation of tetracycline under UV-Visible light irradiation. The outcomes demonstrated that the most effective breakdown of tetracycline was achieved by 10% $ZnSnO_3/g-C_3N_4$ NCs, with a recorded efficiency of 85% after a contact time of 120 minutes. Moreover, Chaba et al. [112] reported the degradation of amoxicillin by employing a composite material composed of carbon nanofibers coated with ZnO. The most effective removal of amoxicillin, with an efficiency of 98.7%, was obtained by using an adsorbent mass of 75 mg, keeping the pH at 6.5, and allowing a contact duration of 60 minutes. Moreover, the fabricated composites exhibited exceptional recyclability, maintaining a removal efficiency of 90.3% even after undergoing 15 cycles of regeneration. These results emphasize the potential of the zinc oxide coated carbon nanofibers composite for effective elimination of amoxicillin from contaminated water sources.

3.3 Agriculture application

The foundation of third world economies lies in agricultural production; however, this sector is currently confronted with various global challenges. These challenges encompass climate change, urbanization, the need for sustainable resource utilization, as well as environmental concerns like overflow and the accumulation of pesticides and fertilizers. The application of nanotechnology is bringing about a remarkable transformation in the field of agriculture. By effectively addressing issues such as leaching, drifting, hydrolysis, photolysis, and microbial degradation, it is successfully reducing the loss of agrochemicals. This advancement is proving to be a game-changer in ensuring sustainable and efficient agricultural practices. By utilizing nanoparticles and nanomaterials, the application of pesticides and fertilizers can be precisely and selectively administered, effectively reducing the risk of unintended damage [113, 114]. Zn-based NMs are flexible material with a widespread variety of applications due to their outstanding features, which include wide band gap, optical and semiconducting nature, piezoelectricity, and biocompatibility. In agriculture, ZnO nanoparticles applied in a colloidal solution work well as fertilizer. This nanofertilizer not only provides essential nutrients to plants but also rejuvenates the soil naturally, eliminating the harmful effects associated with chemical fertilizers. A notable advantage of nanofertilizers is their ability to be utilized in minimal quantities. In comparison, a mature tree necessitates only 40-50 kg of nanofertilizer, whereas conventional fertilizers would require approximately 150 kg. Furthermore, nanopowders exhibit promising potential as both fertilizers and pesticides, as evidenced by the average 20-25% increase in wheat plant yield when seeds are treated with metal nanoparticles [115-117].

The potential of ZnO NPs to enhance the productivity and growth of food crops is noteworthy. Varying amounts of zinc oxide NPs were administered to peanut seeds. The utilization of zinc oxide NPs, with an average particle size of 25 nm and a concentration of 1000 ppm, was implemented to enhance seed germination, boost seedling vigor, and stimulate plant growth. Additionally, these nanoparticles exhibited efficacy in promoting the development of stems and roots in peanuts [118]. Keerthana et al. [119] stated that the yield of *Abelmoschus esculentus* (Ladies finger) pods and the quality of the seeds have been supported by the biogenesis of ZnO NPs employing *Citrus medica* peel extract. *Abelmoschus esculentus* was subjected to foliar applications of ZnO nano-fertilizer with varying concentrations derived from biogenesis. In comparison to the control group, plants treated with a 20 mg/L concentration of the nano-fertilizer exhibited the highest levels of biochemical contents. The application of this nano-fertilizer resulted in significant improvements in various growth parameters of the plants. These improvements included enhanced seed germination, overall plant height, shoot height, root height, total fresh weight, shoot fresh weight, root fresh weight, total dry weight, shoot dry weight, root dry weight, number of branches, leaf area, number of pods, and length of pods. Notably, these improvements were considerably higher when compared to plants that were solely hydro-primed prior to planting. Pandey et al. [120] reported that ZnO NPs enhanced *Cicer arietinum* (chick pea) root development and seed germination via a process that is assumed to include higher synthesis of the growth regulator indole acetic acid.

The study conducted by Durgude et al. [121] delved into the effects of iron and zinc NCs, synthesized from mesoporous nanosilica and reduced graphene oxide, on rice grain fortification, productivity, and micronutrient use efficiency. The results indicated that the foliar application of these nanocomposites resulted in a substantial increase in the uptake of iron and zinc, a 53% enhancement in grain yield, and a significant improvement in zinc and iron use efficiency by 527% and 380%, respectively. Dispat et al. [122] investigated how ZnO/SiO_2-modified starch-graft-polyacrylate superabsorbent polymer (MS-gPA SAP) affected the germination and growth traits of mung beans in their investigation. The indirect effects of MS-g-PA SAP on the development of mung bean seeds were especially examined by the researchers in order to determine the phytotoxicity of the substance. The length of mung bean shoots gradually shrank after germination, despite the fact that neither MS-g-PA SAP nor control SAP (cSAP) significantly affected mung bean germination. The atmosphere became more humid due to the SAPs' ability to store water, which prevented mung bean shoots from growing. In comparison to the control group, the average height of mung bean plants was marginally lower in both the SAP and cSAP treatments. Remarkably, compared to the MS-g-PA SAP treatment, a larger percentage of plants survived dry conditions. Martins et al. [123] have highlighted the significance of precise fertilization strategies in effectively fulfilling plant micronutrient requirements while minimizing adverse consequences such as waste generation, water contamination, metal leaching, and detrimental effects on soil organisms. By immobilizing ZnO nanoparticles on natural polymers, it becomes possible to create materials that facilitate controlled release of micronutrients in soils. The study suggests that ZnO NPs immobilized on alginate beads can be utilized for controlled release in agricultural soils, specifically catering to the Zn needs of maize plants in acidic soil LUFA 2.1. This long-term release mechanism allows for the reuse of these materials across multiple crop cycles [123].

Conclusion and future aspects

In the present chapter, we present an overview of recent research conducted on bioinspired Zn-based NMs in the fields of biomedical, environmental, and agriculture sectors. Bioinspired Zn-based NMs have shown remarkable properties, making them highly versatile for applications in these fields. The preparation of bioinspired Zn-based NMs can be easily achieved using biogenic sources such as plants, microbes, and biomolecules. However, the plants-mediate has proven to be more efficient in generating ZnO compared to nanocomposites. The biocompatible and non-toxic nature of bioinspired Zn-based NMs has led to their utilization in various biomedical applications. Additionally, their wide bandgap, high surface area, and unique scaffold morphology make them suitable for environmental applications. Furthermore, Zn-based NMs serve as micronutrients and nanofertilizers, providing zinc ions that can alter the growth of crops and enhance soil fertility. In the future, these bioinspired Zn-based NMs hold potential for utilization in electrochemical and energy storage applications due to their excellent optical and semiconducting properties. They can be employed in sensing, luminescence, imaging, anode material, and pseudocapacitance.

References

[1] S.C.C. Arruda, A.L.D. Silva, R.M. Galazzi, R.A. Azevedo, M.A.Z. Arruda, Nanoparticles applied to plant science: a review, Talanta, 131 (2015) 693-705. https://doi.org/10.1016/j.talanta.2014.08.050

[2] H. Mirzaei, M. Darroudi, Zinc oxide nanoparticles: Biological synthesis and biomedical applications, Ceramics International, 43 (2017) 907-914. https://doi.org/10.1016/j.ceramint.2016.10.051

[3] S. Mondal, M.S. Nagmote, S.V. Kombe, B.K. Dutta, T.L. Lambat, P.B. Chouke, A. Mondal, Ecofriendly microorganism assisted fabrication of metal nanoparticles and their applications, Biogenic Sustainable Nanotechnology, Elsevier, 2022, pp. 77-105. https://doi.org/10.1016/B978-0-323-88535-5.00002-0

[4] A. Haldar, K.M. Dadure, D. Mahapatra, R.G. Chaudhary, Natural extracts-mediated biosynthesis of Zinc Oxide nanoparticles and their multiple pharmacotherapeutic perspectives, Jordan Journal of Physics, 15 (2022) 67-79. https://doi.org/10.47011/15.1.10

[5] Y. Sun, J.H. Seo, C.J. Takacs, J. Seifter, A.J. Heeger, Inverted polymer solar cells integrated with a low-temperature-annealed sol-gel-derived ZnO film as an electron transport layer, Advanced Materials, 23 (2011) 1679-1683. https://doi.org/10.1002/adma.201004301

[6] Z. Fan, J.G. Lu, Zinc oxide nanostructures: synthesis and properties, Journal of nanoscience and nanotechnology, 5 (2005) 1561-1573. https://doi.org/10.1166/jnn.2005.182

[7] A. Janotti, C.G. Van de Walle, Fundamentals of zinc oxide as a semiconductor, Reports on Progress in Physics, 72 (2009) 126501. https://doi.org/10.1088/0034-4885/72/12/126501

[8] N. Padmavathy, R. Vijayaraghavan, Enhanced bioactivity of ZnO nanoparticles-an antimicrobial study, Science and Technology of Advanced Materials, (2008). https://doi.org/10.1088/1468-6996/9/3/035004

[9] W. Maret, Metals on the move: zinc ions in cellular regulation and in the coordination dynamics of zinc proteins, Biometals, 24 (2011) 411-418. https://doi.org/10.1007/s10534-010-9406-1

[10] Z. Song, T.A. Kelf, W.H. Sanchez, M.S. Roberts, J. Rička, M. Frenz, A.V. Zvyagin, Characterization of optical properties of ZnO nanoparticles for quantitative imaging of transdermal transport, Biomedical Optics Express, 2 (2011) 3321-3333. https://doi.org/10.1364/BOE.2.003321

[11] P.K. Mishra, H. Mishra, A. Ekielski, S. Talegaonkar, B. Vaidya, Zinc oxide nanoparticles: a promising nanomaterial for biomedical applications, Drug discovery Today, 22 (2017) 1825-1834. https://doi.org/10.1016/j.drudis.2017.08.006

[12] Z.-Y. Zhang, H.-M. Xiong, Photoluminescent ZnO nanoparticles and their biological applications, Materials, 8 (2015) 3101-3127. https://doi.org/10.3390/ma8063101

[13] S. Kim, S.Y. Lee, H.-J. Cho, Doxorubicin-wrapped zinc oxide nanoclusters for the therapy of colorectal adenocarcinoma, Nanomaterials, 7 (2017) 354. https://doi.org/10.3390/nano7110354

[14] H.M. Xiong, ZnO nanoparticles applied to bioimaging and drug delivery, Advanced Materials, 25 (2013) 5329-5335. https://doi.org/10.1002/adma.201301732

[15] I. Udom, M.K. Ram, E.K. Stefanakos, A.F. Hepp, D.Y. Goswami, One dimensional-ZnO nanostructures: synthesis, properties and environmental applications, Materials Science in Semiconductor Processing, 16 (2013) 2070-2083. https://doi.org/10.1016/j.mssp.2013.06.017

[16] P. Bhilkar, A. Bodhne, S. Yerpude, R. Madankar, S. Somkuwar, A. Chaudhary, A. Lambat, M. Desimone, R. Sharma, R. Chaudhary, Phyto-derived Metal Nanoparticles: Prominent Tool for Biomedical Applications, OpenNano, (2023) 100192. https://doi.org/10.1016/j.onano.2023.100192

[17] P.B. Chouke, T. Shrirame, A.K. Potbhare, A. Mondal, A.R. Chaudhary, S. Mondal, S.R. Thakare, E. Nepovimova, M. Valis, K. Kuca, Bioinspired metal/metal oxide nanoparticles: A road map to potential applications, Materials Today Advances, 16 (2022) 100314. https://doi.org/10.1016/j.mtadv.2022.100314

[18] P. B Chouke, A. K Potbhare, G. S Bhusari, S. Somkuwar, D. PMD Shaik, R. K Mishra, R. Gomaji Chaudhary, Green fabrication of zinc oxide nanospheres by Aspidopterys cordata for effective antioxidant and antibacterial activity, Advanced Materials Letters, 10 (2019) 355-360. https://doi.org/10.5185/amlett.2019.2235

[19] P. Chouke, A. Potbhare, K. Dadure, A. Mungole, N. Meshram, R. Chaudhary, A. Rai, R. Chaudhary, An antibacterial activity of Bauhinia racemosa assisted ZnO nanoparticles during lunar eclipse and docking assay, Materials Today: Proceedings, 29 (2020) 815-821. https://doi.org/10.1016/j.matpr.2020.04.758

[20] T. Sutharappa Kaliyamoorthy, V. Subramaniyan, S. Renganathan, J. Elavarasan, J. Ravi, P. Prabhakaran Kala, P. Subramaniyan, S. Vijayakumar, Sustainable Environmental-Based ZnO Nanoparticles Derived from Pisonia grandis for Future Biological and Environmental Applications, Sustainability, 14 (2022) 17009. https://doi.org/10.3390/su142417009

[21] M. Tariq, A.U. Khan, A.U. Rehman, S. Ullah, A.U. Jan, Z.U.H. Khan, N. Muhammad, Z.U. Islam, Q. Yuan, Green synthesis of Zno@ GO nanocomposite and its' efficient antibacterial activity, Photodiagnosis and Photodynamic Therapy, 35 (2021) 102471. https://doi.org/10.1016/j.pdpdt.2021.102471

[22] R. Sánchez-Albores, F.J. Cano, P. Sebastian, O. Reyes-Vallejo, Microwave-assisted biosynthesis of ZnO-GO particles using orange peel extract for photocatalytic degradation of methylene blue, Journal of Environmental Chemical Engineering, 10 (2022) 108924. https://doi.org/10.1016/j.jece.2022.108924

[23] M. Umekar, R. Chaudhary, G. Bhusari, A. Potbhare, Fabrication of zinc oxide-decorated phytoreduced graphene oxide nanohybrid via Clerodendrum infortunatum, Emerging Materials Research, 10 (2021) 75-84. https://doi.org/10.1680/jemmr.19.00175

[24] R.G. Chaudhary, A.K. Potbhare, S.T. Aziz, M.S. Umekar, S.S. Bhuyar, A. Mondal, Phytochemically fabricated reduced graphene Oxide-ZnO NCs by Sesbania bispinosa for

photocatalytic performances, Materials Today: Proceedings, 36 (2021) 756-762. https://doi.org/10.1016/j.matpr.2020.05.821

[25] N. Pandiyan, B. Murugesan, M. Arumugam, J. Sonamuthu, S. Samayanan, S. Mahalingam, Ionic liquid-a greener templating agent with Justicia adhatoda plant extract assisted green synthesis of morphologically improved Ag-Au/ZnO nanostructure and it's antibacterial and anticancer activities, Journal of Photochemistry and Photobiology B: Biology, 198 (2019) 111559. https://doi.org/10.1016/j.jphotobiol.2019.111559

[26] F.A. Jan, R. Ullah, N. Ullah, M. Usman, Exploring the environmental and potential therapeutic applications of Myrtus communis L. assisted synthesized zinc oxide (ZnO) and iron doped zinc oxide (Fe-ZnO) nanoparticles, Journal of Saudi Chemical Society, 25 (2021) 101278. https://doi.org/10.1016/j.jscs.2021.101278

[27] R. Subbiah, S. Muthukumaran, V. Raja, Phyto-assisted synthesis of Mn and Mg Co-doped ZnO nanostructures using Carica papaya leaf extract for photocatalytic applications, BioNanoScience, 11 (2021) 1127-1141. https://doi.org/10.1007/s12668-021-00890-x

[28] K.D. Dejen, E.A. Zereffa, H.A. Murthy, A. Merga, Synthesis of ZnO and ZnO/PVA nanocomposite using aqueous Moringa oleifera leaf extract template: antibacterial and electrochemical activities, Reviews on Advanced Materials Science, 59 (2020) 464-476. https://doi.org/10.1515/rams-2020-0021

[29] K. Subashini, S. Prakash, V. Sujatha, Biological applications of green synthesized zinc oxide and nickel oxide nanoparticles mediated poly (glutaric acid-co-ethylene glycol-co-acrylic acid) polymer nanocomposites, Inorganic Chemistry Communications, 139 (2022) 109314. https://doi.org/10.1016/j.inoche.2022.109314

[30] M. Barani, M. Masoudi, M. Mashreghi, A. Makhdoumi, H. Eshghi, Cell-free extract assisted synthesis of ZnO nanoparticles using aquatic bacterial strains: Biological activities and toxicological evaluation, International Journal of Pharmaceutics, 606 (2021) 120878. https://doi.org/10.1016/j.ijpharm.2021.120878

[31] P. Dhandapani, A.S. Siddarth, S. Kamalasekaran, S. Maruthamuthu, G. Rajagopal, Bio-approach: ureolytic bacteria mediated synthesis of ZnO nanocrystals on cotton fabric and evaluation of their antibacterial properties, Carbohydrate polymers, 103 (2014) 448-455. https://doi.org/10.1016/j.carbpol.2013.12.074

[32] S. Azizi, M.B. Ahmad, F. Namvar, R. Mohamad, Green biosynthesis and characterization of zinc oxide nanoparticles using brown marine macroalga Sargassum muticum aqueous extract, Materials Letters, 116 (2014) 275-277. https://doi.org/10.1016/j.matlet.2013.11.038

[33] E. Selvarajan, V. Mohanasrinivasan, Biosynthesis and characterization of ZnO nanoparticles using Lactobacillus plantarum VITES07, Materials Letters, 112 (2013) 180-182. https://doi.org/10.1016/j.matlet.2013.09.020

[34] V. Ganesan, M. Hariram, S. Vivekanandhan, S. Muthuramkumar, Periconium sp.(endophytic fungi) extract mediated sol-gel synthesis of ZnO nanoparticles for antimicrobial and antioxidant applications, Materials Science in Semiconductor Processing, 105 (2020) 104739. https://doi.org/10.1016/j.mssp.2019.104739

[35] C. Jayaseelan, A.A. Rahuman, A.V. Kirthi, S. Marimuthu, T. Santhoshkumar, A. Bagavan, K. Gaurav, L. Karthik, K.B. Rao, Novel microbial route to synthesize ZnO nanoparticles using Aeromonas hydrophila and their activity against pathogenic bacteria and fungi, Spectrochimica Acta Part A: Molecular and Biomolecular Spectroscopy, 90 (2012) 78-84. https://doi.org/10.1016/j.saa.2012.01.006

[36] J.A. Tanna, R.G. Chaudhary, H.D. Juneja, N.V. Gandhare, A.R. Rai, Histidine-capped ZnO nanoparticles: an efficient synthesis, spectral characterization and effective antibacterial activity, BioNanoScience, 5 (2015) 123-134. https://doi.org/10.1007/s12668-015-0170-0

[37] M. Ramani, S. Ponnusamy, C. Muthamizhchelvan, E. Marsili, Amino acid-mediated synthesis of zinc oxide nanostructures and evaluation of their facet-dependent antimicrobial activity, Colloids and Surfaces B: Biointerfaces, 117 (2014) 233-239. https://doi.org/10.1016/j.colsurfb.2014.02.017

[38] A. Taşdemir, N. Akman, A. Akkaya, R. Aydın, B. Şahin, Green and cost-effective synthesis of zinc oxide thin films by L-ascorbic acid (AA) and their potential for electronics and antibacterial applications, Ceramics International, 48 (2022) 10164-10173. https://doi.org/10.1016/j.ceramint.2021.12.228

[39] J.Z. Marinho, F.d.C. Romeiro, S.C.S. Lemos, F.V.d. Motta, C. Riccardi, M.S. Li, E. Longo, R.C.d. Lima, Urea-based synthesis of zinc oxide nanostructures at low temperature, Journal of Nanomaterials, 2012 (2012) 3-3. https://doi.org/10.1155/2012/427172

[40] S.-J. Kim, C.W. Na, I.-S. Hwang, J.-H. Lee, One-pot hydrothermal synthesis of CuO-ZnO composite hollow spheres for selective H2S detection, Sensors and Actuators B: Chemical, 168 (2012) 83-89. https://doi.org/10.1016/j.snb.2012.01.045

[41] B. Peychev, P. Vasileva, Novel starch-mediated synthesis of Au/ZnO nanocrystals and their photocatalytic properties, Heliyon, 7 (2021) e07402 . https://doi.org/10.1016/j.heliyon.2021.e07402

[42] M.T. Klug, N.m.-M. Dorval Courchesne, Y.E. Lee, D.S. Yun, J. Qi, N.C. Heldman, P.T. Hammond, N.X. Fang, A.M. Belcher, Mediated growth of zinc chalcogen shells on gold nanoparticles by free-base amino acids, Chemistry of Materials, 29 (2017) 6993-7001. https://doi.org/10.1021/acs.chemmater.7b02571

[43] S.T. Yerpude, A.K. Potbhare, P. Bhilkar, A.R. Rai, R.P. Singh, A.A. Abdala, R. Adhikari, R. Sharma, R.G. Chaudhary, Biomedical and clinical applications of platinum-based nanohybrids: An update review, Environmental Research, 231 (2023) 116148. https://doi.org/10.1016/j.envres.2023.116148

[44] S. Pal, S. Mondal, J. Maity, R. Mukherjee, Synthesis and characterization of ZnO nanoparticles using Moringa oleifera leaf extract: investigation of photocatalytic and antibacterial activity, International Journal of Nanoscience and Nanotechnology, 14 (2018) 111-119.

[45] A. Sirelkhatim, S. Mahmud, A. Seeni, N.H.M. Kaus, L.C. Ann, S.K.M. Bakhori, H. Hasan, D. Mohamad, Review on zinc oxide nanoparticles: antibacterial activity and toxicity mechanism, Nano-Micro Letters, 7 (2015) 219-242. https://doi.org/10.1007/s40820-015-0040-x

[46] H. Agarwal, S.V. Kumar, S. Rajeshkumar, A review on green synthesis of zinc oxide nanoparticles-An eco-friendly approach, Resource-Efficient Technologies, 3 (2017) 406-413. https://doi.org/10.1016/j.reffit.2017.03.002

[47] J.M. Galdopórpora, S. Municoy, F. Ibarra, V. Puente, P.E. Antezana, M.I.A. Echazú, M.F. Desimone, A green synthesis method to tune the morphology of CuO and ZnO nanostructures, Current Nanoscience, 19 (2023) 186-193. https://doi.org/10.2174/1573413717666210921152709

[48] S.J.M. Rosid, S.M. Rosid, N.A. Nasir, W.N.W. Abdullah, N.N. Mohamed, Antimicrobial potentialities: special emphasis on metal and metal oxide-based bionanomaterials, Bionanomaterials for Environmental and Agricultural Applications, IOP Publishing Bristol, UK2021, pp. 2-1-2-32. https://doi.org/10.1088/978-0-7503-3863-9ch2

[49] T. Shrirame, P. Bhilkar, A. Chaudhary, A. Rai, R. Singh, P. Dhongle, S. Thakare, A. Abdala, R. Chaudhary, Magnetic Nanoparticles: Fabrications and applications in cancer therapy and diagnosis, magnetic nanoparticles for biomedical applications, 143 (2023) 199-232. https://doi.org/10.21741/9781644902335-7

[50] S. Faisal, H. Jan, S.A. Shah, S. Shah, A. Khan, M.T. Akbar, M. Rizwan, F. Jan, Wajidullah, N. Akhtar, Green synthesis of zinc oxide (ZnO) nanoparticles using aqueous fruit extracts of Myristica fragrans: their characterizations and biological and environmental applications, ACS Omega, 6 (2021) 9709-9722. https://doi.org/10.1021/acsomega.1c00310

[51] D. Kalaimurgan, K. Lalitha, R.K. Govindarajan, K. Unban, M.S. Shivakumar, S. Venkatesan, C. Khanongnuch, F.M. Husain, F.A. Qais, I. Hasan, Biogenic synthesis of zinc oxide nanoparticles using Drynaria Quercifolia tuber extract for antioxidant, antibiofilm, larvicidal, and photocatalytic applications, Biomass Conversion and Biorefinery, (2023) 1-17. https://doi.org/10.1007/s13399-023-04751-3 https://doi.org/10.1007/s13399-023-05133-5

[52] L. Mosquera-Sánchez, P. Arciniegas-Grijalba, M. Patiño-Portela, B.E. Guerra-Sierra, J. Muñoz-Florez, J. Rodríguez-Páez, Antifungal effect of zinc oxide nanoparticles (ZnO-NPs) on Colletotrichum sp., causal agent of anthracnose in coffee crops, Biocatalysis and Agricultural Biotechnology, 25 (2020) 101579. https://doi.org/10.1016/j.bcab.2020.101579

[53] H. Jan, M. Shah, H. Usman, M.A. Khan, M. Zia, C. Hano, B.H. Abbasi, Biogenic synthesis and characterization of antimicrobial and antiparasitic zinc oxide (ZnO) nanoparticles using aqueous extracts of the Himalayan Columbine (Aquilegia pubiflora), Frontiers in Materials, 7 (2020) 249. https://doi.org/10.3389/fmats.2020.00249

[54] P. Panchal, D.R. Paul, A. Sharma, P. Choudhary, P. Meena, S. Nehra, Biogenic mediated Ag/ZnO nanocomposites for photocatalytic and antibacterial activities towards disinfection of water, Journal of Colloid and Interface Science, 563 (2020) 370-380. https://doi.org/10.1016/j.jcis.2019.12.079

[55] M.I. Khan, S.K. Behera, P. Paul, B. Das, M. Suar, R. Jayabalan, D. Fawcett, G.E.J. Poinern, S.K. Tripathy, A. Mishra, Biogenic Au@ ZnO core-shell nanocomposites kill Staphylococcus aureus without provoking nuclear damage and cytotoxicity in mouse fibroblasts cells under hyperglycemic condition with enhanced wound healing proficiency, Medical Microbiology and Immunology, 208 (2019) 609-629. https://doi.org/10.1007/s00430-018-0564-z

[56] G. Madhan, A.A. Begam, L.V. Varsha, R. Ranjithkumar, D. Bharathi, Facile synthesis and characterization of chitosan/zinc oxide nanocomposite for enhanced antibacterial and photocatalytic activity, International Journal of Biological Macromolecules, 190 (2021) 259-269. https://doi.org/10.1016/j.ijbiomac.2021.08.100

[57] E. Nagaraj, P. Shanmugam, K. Karuppannan, T. Chinnasamy, S. Venugopal, The biosynthesis of a graphene oxide-based zinc oxide nanocomposite using Dalbergia latifolia leaf extract and its biological applications, New Journal of Chemistry, 44 (2020) 2166-2179. https://doi.org/10.1039/C9NJ04961D

[58] K. Karthikeyan, M. Chandraprabha, R.H. Krishna, K. Samrat, A. Sakunthala, M. Sasikumar, Optical and antibacterial activity of biogenic core-shell ZnO@ TiO2 nanoparticles, Journal of the Indian Chemical Society, 99 (2022) 100361. https://doi.org/10.1016/j.jics.2022.100361

[59] A. Jafari, M. Ghane, M. Sarabi, F. Siyavoshifar, Synthesis and antibacterial properties of zinc oxide combined with copper oxide nanocrystals, Oriental Journal of Chemistry, 27 (2011) 811.

[60] S. Hussain, N. Khakwani, Y. Faiz, S. Zulfiqar, Z. Shafiq, F. Faiz, A. Elhakem, R. Sami, N. Aljuraide, T. Farid, Green production and interaction of carboxylated CNTs/Biogenic ZnO Composite for Antibacterial activity, Bioengineering, 9 (2022) 437. https://doi.org/10.3390/bioengineering9090437

[61] A. Rahman, A.L. Tan, M.H. Harunsani, N. Ahmad, M. Hojamberdiev, M.M. Khan, Visible light induced antibacterial and antioxidant studies of ZnO and Cu-doped ZnO fabricated using aqueous leaf extract of Ziziphus mauritiana Lam, Journal of Environmental Chemical Engineering, 9 (2021) 105481. https://doi.org/10.1016/j.jece.2021.105481

[62] C. Shen, S.A. James, M.D. de Jonge, T.W. Turney, P.F. Wright, B.N. Feltis, Relating cytotoxicity, zinc ions, and reactive oxygen in ZnO nanoparticle-exposed human immune cells, Toxicological Sciences, 136 (2013) 120-130. https://doi.org/10.1093/toxsci/kft187

[63] A. Manke, L. Wang, Y. Rojanasakul, Mechanisms of nanoparticle-induced oxidative stress and toxicity, BioMed Research International, 2013 (2013) 942916. https://doi.org/10.1155/2013/942916

[64] N. Ngoepe, Z. Mbita, M. Mathipa, N. Mketo, B. Ntsendwana, N. Hintsho-Mbita, Biogenic synthesis of ZnO nanoparticles using Monsonia burkeana for use in photocatalytic, antibacterial and anticancer applications, Ceramics International, 44 (2018) 16999-17006. https://doi.org/10.1016/j.ceramint.2018.06.142

[65] N. Senthilkumar, E. Nandhakumar, P. Priya, D. Soni, M. Vimalan, I.V. Potheher, Synthesis of ZnO nanoparticles using leaf extract of Tectona grandis (L.) and their anti-bacterial, anti-arthritic, anti-oxidant and in vitro cytotoxicity activities, New Journal of Chemistry, 41 (2017) 10347-10356. https://doi.org/10.1039/C7NJ02664A

[66] B. Gowdhami, M. Jaabir, G. Archunan, N. Suganthy, Anticancer potential of zinc oxide nanoparticles against cervical carcinoma cells synthesized via biogenic route using aqueous extract of Gracilaria edulis, Materials Science and Engineering: C, 103 (2019) 109840. https://doi.org/10.1016/j.msec.2019.109840

[67] A.H. Hashem, G.S. El-Sayyad, Antimicrobial and anticancer activities of biosynthesized bimetallic silver-zinc oxide nanoparticles (Ag-ZnO NPs) using pomegranate peel extract, Biomass Conversion and Biorefinery, (2023) 1-13. https://doi.org/10.1007/s13399-023-04126-8 https://doi.org/10.1007/s13399-023-04126-8

[68] E.E. Elemike, D.C. Onwudiwe, M. Singh, Eco-friendly synthesis of copper oxide, zinc oxide and copper oxide-zinc oxide nanocomposites, and their anticancer applications, Journal of Inorganic and Organometallic Polymers and Materials, 30 (2020) 400-409. https://doi.org/10.1007/s10904-019-01198-w

[69] S. Vijayakumar, J. Chen, V. Kalaiselvi, K. Tungare, M. Bhori, Z.I. González-Sánchez, E.F. Durán-Lara, Marine polysaccharide laminarin embedded ZnO nanoparticles and their based chitosan capped ZnO nanocomposites: Synthesis, characterization and in vitro and in vivo toxicity assessment, Environmental Research, 213 (2022) 113655. https://doi.org/10.1016/j.envres.2022.113655

[70] M. Batool, S. Khurshid, W.M. Daoush, S.A. Siddique, T. Nadeem, Green synthesis and biomedical applications of ZnO nanoparticles: role of PEGylated-ZnO nanoparticles as doxorubicin drug carrier against MDA-MB-231 (TNBC) cells line, Crystals, 11 (2021) 344. https://doi.org/10.3390/cryst11040344

[71] A. Regmi, B.R. Bhattarai, S.K. Gautam, Synthesis and microscopic study of zinc sulfide nanoparticles, 2019 International Conference on Computer, Communication, Chemical, Materials and Electronic Engineering (IC4ME2), IEEE, 2019, pp. 1-4. https://doi.org/10.1109/IC4ME247184.2019.9036683

[72] S.M. Husseiny, T.A. Salah, H.A. Anter, Biosynthesis of size controlled silver nanoparticles by Fusarium oxysporum, their antibacterial and antitumor activities, Beni-Suef University Journal of Basic and Applied Sciences, 4 (2015) 225-231. https://doi.org/10.1016/j.bjbas.2015.07.004

[73] A. Żaba, S. Sovinska, W. Kasprzyk, D. Bogdał, K. Matras-Postołek, Zinc sulphide (ZNS) nanparticles for advanced application, Czasopismo Techniczne, 2016 (2016) 125-134.

[74] R.K. Sahoo, S. Rani, V. Kumar, U. Gupta, Zinc oxide nanoparticles for bioimaging and drug delivery, Nanostructured Zinc Oxide, Elsevier, 2021, pp. 483-509. https://doi.org/10.1016/B978-0-12-818900-9.00021-8

[75] S. Sudhagar, S. Sathya, K. Pandian, B.S. Lakshmi, Targeting and sensing cancer cells with ZnO nanoprobes in vitro, Biotechnology Letters, 33 (2011) 1891-1896. https://doi.org/10.1007/s10529-011-0641-5

[76] Y. Liu, K. Ai, Q. Yuan, L. Lu, Fluorescence-enhanced gadolinium-doped zinc oxide quantum dots for magnetic resonance and fluorescence imaging, Biomaterials, 32 (2011) 1185-1192. https://doi.org/10.1016/j.biomaterials.2010.10.022

[77] W. Wanas, S.A. Abd El-Kaream, S. Ebrahim, M. Soliman, M. Karim, Cancer bioimaging using dual mode luminescence of graphene/FA-ZnO nanocomposite based on novel green technique, Scientific Reports, 13 (2023) 27. https://doi.org/10.1038/s41598-022-27111-z

[78] N.-H. Cho, T.-C. Cheong, J.H. Min, J.H. Wu, S.J. Lee, D. Kim, J.-S. Yang, S. Kim, Y.K. Kim, S.-Y. Seong, A multifunctional core-shell nanoparticle for dendritic cell-based cancer

immunotherapy, Nature Nanotechnology, 6 (2011) 675-682.
https://doi.org/10.1038/nnano.2011.149

[79] G. Lei, S. Yang, R. Cao, P. Zhou, H. Peng, R. Peng, X. Zhang, Y. Yang, Y. Li, M. Wang, In situ preparation of amphibious ZnO quantum dots with blue fluorescence based on hyperbranched polymers and their application in bio-imaging, Polymers, 12 (2020) 144. https://doi.org/10.3390/polym12010144

[80] R.R. Sawkar, M.M. Shanbhag, S.M. Tuwar, K. Mondal, N.P. Shetti, Zinc Oxide-Graphene nanocomposite-based sensor for the electrochemical determination of cetirizine, Catalysts, 12 (2022) 1166. https://doi.org/10.3390/catal12101166

[81] Z. Hatami, E. Ragheb, F. Jalali, M.A. Tabrizi, M. Shamsipur, Zinc oxide-gold nanocomposite as a proper platform for label-free DNA biosensor, Bioelectrochemistry, 133 (2020) 107458. https://doi.org/10.1016/j.bioelechem.2020.107458

[82] M.-S. Ahn, R. Ahmad, K.S. Bhat, J.-Y. Yoo, T. Mahmoudi, Y.-B. Hahn, Fabrication of a solution-gated transistor based on valinomycin modified iron oxide nanoparticles decorated zinc oxide nanorods for potassium detection, Journal of Colloid and Interface science, 518 (2018) 277-283. https://doi.org/10.1016/j.jcis.2018.02.041

[83] R. Tomar, A. A. Abdala, R.G. Chaudhary, N.B. Singh, Photocatalytic degradation of dyes by nanomaterials, Materials Today: Proceedings, 29, (2020), 967-973. https://doi.org/10.1016/j.matpr.2020.04.144

[84] V.N. Sonkusare, R.G. Chaudhary, G.S. Bhusari, A.R. Rai, H.D. Juneja, Microwave-mediated synthesis, photocatalytic degradation and antibacterial activity of α-Bi2O3 microflowers/novel γ -Bi2O3 microspindles, Nano-Structures & Nano-Objects, 13 (2018), 121-131 https://doi.org/10.1016/j.nanoso.2018.01.002

[85] V.N. Sonkusare, R.G. Chaudhary, G.S. Bhusari, A. Mondal, A.K. Potbhare, R.K. Mishra, H.D. Juneja A.A. Abdala Mesoporous Octahedron-Shaped Tricobalt Tetroxide Nanoparticles for Photocatalytic Degradation of Toxic Dyes, ACS Omega, 5 (2020), 7823-7835. https://doi.org/10.1021/acsomega.9b03998

[86] R.G. Chaudhary, V. Sonkusare, G. Bhusari, A. Mondal, A.K. Potbhare, R. Sharma, H.D. Juneja, A.A. Abdala, Preparation of mesoporous ThO2 nanoparticles: influence of calcination on morphology and visible-Light-driven photocatalytic degradation of indigo carmine and methylene blue, Enviornmental Research, 222 (2023) 115363. https://doi.org/10.1016/j.envres.2023.115363

[87] R. Vinu, G. Madras, Environmental remediation by photocatalysis, Journal of the Indian Institute of Science, 90 (2010) 189-230.

[88] A. Potbhare, P. Bhilkar, S. Yerpude, R. Madankar, S. Shingda, R. Adhikari, R. Chaudhary, Nanomaterials as Photocatalyst, Appl. Emerg. Nanomater. Nanotechnol, 148 (2023) 304-333. https://doi.org/10.21741/9781644902554-11

[89] L. Kaliraj, J.C. Ahn, E.J. Rupa, S. Abid, J. Lu, D.C. Yang, Synthesis of panos extract mediated ZnO nano-flowers as photocatalyst for industrial dye degradation by UV illumination, Journal of Photochemistry and Photobiology B: Biology, 199 (2019) 111588. https://doi.org/10.1016/j.jphotobiol.2019.111588

[90] K. Kanagamani, P. Muthukrishnan, K. Saravanakumar, K. Shankar, A. Kathiresan, Photocatalytic degradation of environmental perilous gentian violet dye using leucaena-mediated zinc oxide nanoparticle and its anticancer activity, Rare Metals, 38 (2019) 277-286. https://doi.org/10.1007/s12598-018-1189-5

[91] J. Yu, X. Yu, Hydrothermal synthesis and photocatalytic activity of zinc oxide hollow spheres, Environmental Science & Technology, 42 (2008) 4902-4907. https://doi.org/10.1021/es800036n

[92] C. Tian, Q. Zhang, B. Jiang, G. Tian, H. Fu, Glucose-mediated solution-solid route for easy synthesis of Ag/ZnO particles with superior photocatalytic activity and photostability, Journal of Alloys and Compounds, 509 (2011) 6935-6941. https://doi.org/10.1016/j.jallcom.2011.04.005

[93] S. Vishwanathan, S. Das, Glucose-mediated one-pot hydrothermal synthesis of hollow magnesium oxide-zinc oxide (MgO-ZnO) microspheres with enhanced natural sunlight photocatalytic activity, Environmental Science and Pollution Research, 30 (2023) 8512-8525. https://doi.org/10.1007/s11356-022-20283-1

[94] O.A. Zelekew, P.A. Fufa, F.K. Sabir, A.D. Duma, Water hyacinth plant extract mediated green synthesis of Cr_2O_3/ZnO composite photocatalyst for the degradation of organic dye, Heliyon, 7 (2021) e07652. https://doi.org/10.1016/j.heliyon.2021.e07652

[95] R. Atchudan, T.N.J.I. Edison, S. Perumal, D. Karthikeyan, Y.R. Lee, Facile synthesis of zinc oxide nanoparticles decorated graphene oxide composite via simple solvothermal route and their photocatalytic activity on methylene blue degradation, Journal of Photochemistry and Photobiology B: Biology, 162 (2016) 500-510. https://doi.org/10.1016/j.jphotobiol.2016.07.019

[96] K. Bhuvaneswari, G. Palanisamy, G. Bharathi, T. Pazhanivel, I.R. Upadhyaya, M.A. Kumari, R. Rajesh, M. Govindasamy, A. Ghfar, N.H. Al-Shaalan, Visible light driven reduced graphene oxide supported ZnMgAl LTH/ZnO/g-C3N4 nanohybrid photocatalyst with notable two-dimension formation for enhanced photocatalytic activity towards organic dye degradation, Environmental Research, 197 (2021) 111079. https://doi.org/10.1016/j.envres.2021.111079

[97] E.Y. Shaba, J.O. Jacob, J.O. Tijani, M.A.T. Suleiman, A critical review of synthesis parameters affecting the properties of zinc oxide nanoparticle and its application in wastewater treatment, Applied Water Science, 11 (2021) 1-41. https://doi.org/10.1007/s13201-021-01370-z

[98] Mayuri S. Umekar, Ganesh S. Bhusari, Toshali Bhoyar, Vidyasagar Devthade, Bharat P. Kapgate, Ajay P. Potbhare, Ratiram G. Chaudhary and Ahmed A. Abdala, Graphitic carbon nitride-based photocatalysts for environmental remediation of organic pollutants, Current Nanoscience, 19 (2) 2023, 148-169. https://doi.org/10.2174/1573413718666220127123935

[99] M.S. Nagmote, A.R. Rai, R. Sharma, M.F. Desimone, R.G. Chaudhary, NB Singh, Bioremediation of heavy metals using microorganisms, Genetically engineered organisms in bioremediation, CRC Press, 2024, pp.168-190. https://doi.org/10.1201/9781003188568-11

[100] S. Azizi, M. Mahdavi Shahri, R. Mohamad, Green synthesis of zinc oxide nanoparticles for enhanced adsorption of lead ions from aqueous solutions: equilibrium, kinetic and

thermodynamic studies, Molecules, 22 (2017) 831.
https://doi.org/10.3390/molecules22060831

[101] G. Yuvaraja, C. Prasad, Y. Vijaya, M.V. Subbaiah, Application of ZnO nanorods as an adsorbent material for the removal of As (III) from aqueous solution: kinetics, isotherms and thermodynamic studies, International Journal of Industrial Chemistry, 9 (2018) 17-25. https://doi.org/10.1007/s40090-018-0136-5

[102] M. Sharma, J. Singh, S. Hazra, S. Basu, Adsorption of heavy metal ions by mesoporous ZnO and TiO2@ ZnO monoliths: adsorption and kinetic studies, Microchemical Journal, 145 (2019) 105-112. https://doi.org/10.1016/j.microc.2018.10.026

[103] G.J. Cruz, D. Mondal, J. Rimaycuna, K. Soukup, M.M. Gómez, J.L. Solis, J. Lang, Agrowaste derived biochars impregnated with ZnO for removal of arsenic and lead in water, Journal of Environmental Chemical Engineering, 8 (2020) 103800. https://doi.org/10.1016/j.jece.2020.103800

[104] J. Yu, C. Jiang, Q. Guan, P. Ning, J. Gu, Q. Chen, J. Zhang, R. Miao, Enhanced removal of Cr (VI) from aqueous solution by supported ZnO nanoparticles on biochar derived from waste water hyacinth, Chemosphere, 195 (2018) 632-640. https://doi.org/10.1016/j.chemosphere.2017.12.128

[105] M.H. Dehghani, P. Mahdavi, Z. Heidarinejad, The experimental data of investigating the efficiency of zinc oxide nanoparticles technology under ultraviolet radiation (UV/ZnO) to remove Acid-32-Cyanine 5R from aqueous solutions, Data in Brief, 21 (2018) 767-774. https://doi.org/10.1016/j.dib.2018.10.037

[106] S.M.T.H. Moghaddas, B. Elahi, V. Javanbakht, Biosynthesis of pure zinc oxide nanoparticles using Quince seed mucilage for photocatalytic dye degradation, Journal of Alloys and Compounds, 821 (2020) 153519. https://doi.org/10.1016/j.jallcom.2019.153519

[107] A.F. Beevi, G. Sreekala, B. Beena, Synthesis, characterization and photocatalytic activity of SnO2, ZnO nanoparticles against congo red: A comparative study, Materials Today: Proceedings, 45 (2021) 4045-4051. https://doi.org/10.1016/j.matpr.2020.10.755

[108] A. Maleki, F. Moradi, B. Shahmoradi, R. Rezaee, S.-M. Lee, The photocatalytic removal of diazinon from aqueous solutions using tungsten oxide doped zinc oxide nanoparticles immobilized on glass substrate, Journal of Molecular Liquids, 297 (2020) 111918. https://doi.org/10.1016/j.molliq.2019.111918

[109] M. Rani, J. Yadav, U. Shanker, Green synthesis of sunlight responsive zinc oxide coupled cadmium sulfide nanostructures for efficient photodegradation of pesticides, Journal of Colloid and Interface Science, 601 (2021) 689-703. https://doi.org/10.1016/j.jcis.2021.05.152

[110] Z. Zhu, F. Guo, Z. Xu, X. Di, Q. Zhang, Photocatalytic degradation of an organophosphorus pesticide using a ZnO/rGO composite, RSC Advances, 10 (2020) 11929-11938. https://doi.org/10.1039/D0RA01741H

[111] X. Huang, F. Guo, M. Li, H. Ren, Y. Shi, L. Chen, Hydrothermal synthesis of ZnSnO3 nanoparticles decorated on g-C3N4 nanosheets for accelerated photocatalytic degradation of tetracycline under the visible-light irradiation, Separation and Purification Technology, 230 (2020) 115854. https://doi.org/10.1016/j.seppur.2019.115854

[112] J.M. Chaba, P.N. Nomngongo, Effective adsorptive removal of amoxicillin from aqueous solutions and wastewater samples using zinc oxide coated carbon nanofiber composite, Emerging Contaminants, 5 (2019) 143-149. https://doi.org/10.1016/j.emcon.2019.04.001

[113] H. Chen, R. Yada, Nanotechnologies in agriculture: new tools for sustainable development, Trends in Food Science & Technology, 22 (2011) 585-594. https://doi.org/10.1016/j.tifs.2011.09.004

[114] N.B. Singh, R.G. Chaudhary, M.F. Desimone, A. Agrawal, S.K. Shukla, Green synthesized nanomaterials for safe technology in sustainable agriculture, Current Pharmaceutical Biotechnology, 24 (2023) 61-85. https://doi.org/10.2174/1389201023666220608113924

[115] V. Selivanov, E. Zorin, Sustained Action of ultrafine metal powders on seeds of grain crops, Perspekt. Materialy, 4 (2001) 66-69.

[116] Al-Qudah Tamara; M.H. Sami; Abu-Zurayk Rund; Shibli Rida; Khalaf Aya;T.L. Lambat R.G. Chaudhary, Nanotechnology applications in plant tissue culture and molecular genetics: a holistic approach, Current Nanoscience, 18 (4) (2022) 442-464. https://doi.org/10.2174/1573413717666211118111333

[117] L.M. Batsmanova, L. Gonchar, N.Y. Taran, A. Okanenko, Using a colloidal solution of metal nanoparticles as micronutrient fertiliser for cereals, Sumy State University, 2013.

[118] T. Prasad, P. Sudhakar, Y. Sreenivasulu, P. Latha, V. Munaswamy, K.R. Reddy, T. Sreeprasad, P. Sajanlal, T. Pradeep, Effect of nanoscale zinc oxide particles on the germination, growth and yield of peanut, Journal of Plant Nutrition, 35 (2012) 905-927. https://doi.org/10.1080/01904167.2012.663443

[119] P. Keerthana, S. Vijayakumar, E. Vidhya, V. Punitha, M. Nilavukkarasi, P. Praseetha, Biogenesis of ZnO nanoparticles for revolutionizing agriculture: A step towards anti-infection and growth promotion in plants, Industrial Crops and Products, 170 (2021) 113762. https://doi.org/10.1016/j.indcrop.2021.113762

[120] A.C. Pandey, S. S. Sanjay, R. S. Yadav, Application of ZnO nanoparticles in influencing the growth rate of Cicer arietinum, Journal of Experimental Nanoscience, 5 (2010) 488-497. https://doi.org/10.1080/17458081003649648

[121] S.A. Durgude, S. Ram, R. Kumar, S.V. Singh, V. Singh, A.G. Durgude, B. Pramanick, S. Maitra, A. Gaber, A. Hossain, Synthesis of mesoporous silica and graphene-based FeO and ZnO nanocomposites for nutritional biofortification and sustained productivity of rice (Oryza sativa L.), Journal of Nanomaterials, (2022). https://doi.org/10.1155/2022/5120307. https://doi.org/10.1155/2022/5120307

[122] N. Dispat, S. Poompradub, S. Kiatkamjornwong, Synthesis of ZnO/SiO2-modified starch-graft-polyacrylate superabsorbent polymer for agricultural application, Carbohydrate Polymers, 249 (2020) 116862. https://doi.org/10.1016/j.carbpol.2020.116862

[123] N.C. Martins, A. Avellan, S. Rodrigues, D. Salvador, S.M. Rodrigues, T. Trindade, Composites of biopolymers and ZnO NPs for controlled release of zinc in agricultural soils and timed delivery for maize, ACS Applied Nano Materials, 3 (2020) 2134-2148. https://doi.org/10.1021/acsanm.9b01492

Green Synthesis and Emerging Applications of Frontier Nanomaterials   Materials Research Forum LLC
Materials Research Foundations 169 (2024) 113-138   https://doi.org/10.21741/9781644903278-5

Chapter 5

Bioinspired fabrication of copper nanoparticles and their potential applications

Minakshi Y. Deshmukh[1], Sudip Mondal[1]*, Rohit S. Madankar[1], Trimurty Lambat[2], Ratiram G. Chaudhary[1] and Anirudh Mondal[3]

[1]Post Graduate Department of Chemistry, Seth Kesarimal Porwal College of Arts and Science and Commerce, RTM Nagpur University, Kamptee-441001, Nagpur, Maharashtra, India

[2]Department of Chemistry, Manoharbhai Patel College of Arts, Commerce & Science, Deori-441901, India

[3]Division of Materials Science, Department of Engineering Sciences and Mathematics, Lulea University of Technology, SE-97187 Lulea, Sweden

sudipmondal5555@gmail.com

Abstract

A new era of materials with a wide range of uses in pharmacology, agriculture, and medicine has been brought about by the development of metallic nanoparticles (NPs). Nanoparticles with remarkable qualities are created using physical, chemical, and biological processes. The common element copper is essential to an organism's ability to operate normally. As compared to traditional antibiotics, Copper NPs have better antibacterial potential. Copper NPs are also known for better antiviral, antifungal, and anticancer effects. Moreover, bioinspired Copper NPs exhibit potential capacity against bacterial strains that are resistant to many drugs, even though the precise mechanisms of action are yet unclear. With this interest, the present chapter focuses on effects, modes of action and potential toxicity of bioinspired copper NPs.

Keywords

Bioinspired Copper NPs, Biogenic Synthesis, Antimicrobial Assay, Toxicity

Contents

1.   Introduction

Advancements in nanotechnology in current years have heightened our enthusiasm for developing NPs with precise shapes and dimensions. Nanomaterials with scaffolds morphology are very useful for various sectors like environmental remediation, medicinal and agricultural [1–4]. The utilization of plants in the green synthesis of NPs offers multiple advantages compared to conventional synthesis. Moreover, conventionally synthesized NPs involved poisonous chemicals and unwanted byproducts that could contribute to environmental pollution. The disposal of those byproducts poses a threat to human health. In addition to this, it required high priced chemicals. Therefore, plants, fungi, and microorganisms are alternatives for green fabrication of NPs (Fig. 1 and 2) [5]. Moreover, in instances, the synthesized NMs are capped *via* way of means of biomolecules consisting of phenols, tannins, alkaloids, and organic acids of the plant substances. These biomolecules decorate the NPs. Green techniques however utilize various waste materials such as peels of banana, lemon, dried leaves of medicinal plants, and microorganisms among others. Some NPs including novel and non-novel metal salts and metal oxides have been effectively synthesized by many researchers in recent times [6–12].

Copper NPs hold great attention due to their affordability, ready availability, and excellent properties [13–15]. Similarly, copper oxide (CuO) acted as a cost-effective material, particularly when it compared to silver or gold. Several workers have introduced the synthesis and characterization of CuO NPs, highlighting their potential roles as antioxidants, antibacterial agents, and antifungal agents. Cu and copper-based NPs are commonly utilized as antimicrobial, antioxidant, antidiabetic, anti-inflammatory, pesticides, fungicides, and fertilizers [16-17]. Moreover, this material could be employed in various applications like heat transfer fluids, solar cells, lithium-ion batteries, fuel sensors, heterogeneous catalyst, adsorbent, and anticancer agents. Keeping this interest, this chapter focuses on plant-derived fabrication of copper-based NPs and its applications in catalytic and antimicrobial.

2.   Biogenic synthesis of Cu/CuO NPs

There are several routes for biogenic synthesis of copper NPs (Figure1). In general, most approaches for the synthesis of copper NPs are reduction processes. In this approach, copper ions are transformed into copper NPs using different reducing and surfactants [12]. Nevertheless, it can pose protection risks because of the usage of poisonous chemicals. Template-assisted synthesis approach makes use of templates, along with surfactant micelles or polymeric matrices, to manual the formation of NPs. The green synthesis of copper NPs using plant extracts, microorganisms, and fungi is environment friendly. The biogenic fabrication of copper and copper oxide NPs using bioinspired resources reasserts presents an environmentally pleasant and sustainable approach. This technique capitalizes the herbal decreasing and stabilizing homes found in diverse plant extracts described in table 1. By utilizing this method, dangerous chemical compounds are avoided, and the inherent bioactive compounds observed in plant life are hired to facilitate the reduction metal salt solution into NPs. Plant extracts, enriched with compounds like polydiols and phenols, alkaloids, and terpenoids, play a vital function as powerful source for the transformation of copper ions into NPs [17]. The interplay among those bioactive compounds and copper ions triggers an advantageous method that leads to NPs with regular sizes and greater balance are achieved [18]. The biogenic synthesis method is straightforward and involves blending the plant extract with a copper salt solution [19]. The resultant copper NPs show off awesome traits inspired via way of means of elements together with the selected plant source, composition of the extract, and synthesis parameters. The biogenic synthesis method extends to each copper and copper oxide NPs, broadening its programs [20]. Diverse plant parts, along with leaves, stems, fruits, and seeds, were explored as capacity reasserts for NPs synthesis. The choice of plant species contributes to shaping the homes and capacity programs of the synthesized NPs. The programs of biogenically synthesized Cu/CuO NPs through plant reasserts span a huge variety, encompassing catalysis, antimicrobial programs, electronics, and environmental remediation [21]. The inherent sustainability and cost-effectiveness of this technique make it an attractive desire for NPs synthesis, aligning properly with the standards of inexperienced chemistry.

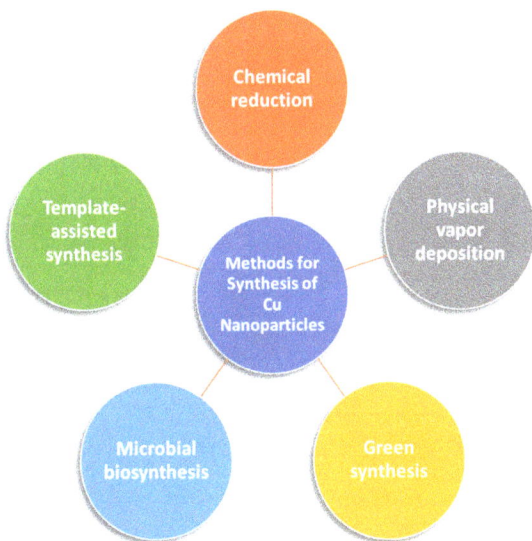

Figure 1: Various methods of synthesis of copper NPs

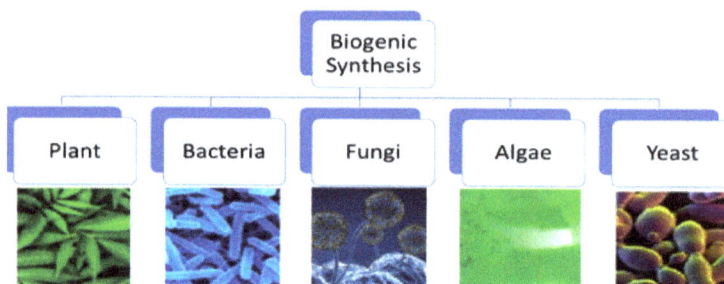

Figure 2: Biogenic synthesis of Cu NPs

In the present chapter we will focus mainly on the plant mediate synthesis of Cu NPs. The plant sources generally comprise leaf, stem, root, fruit, seed, and peels. Different parts of NPs are useful and have their own chemical composition which causes the reduction and leads to the formation of NPs (Fig. 3).

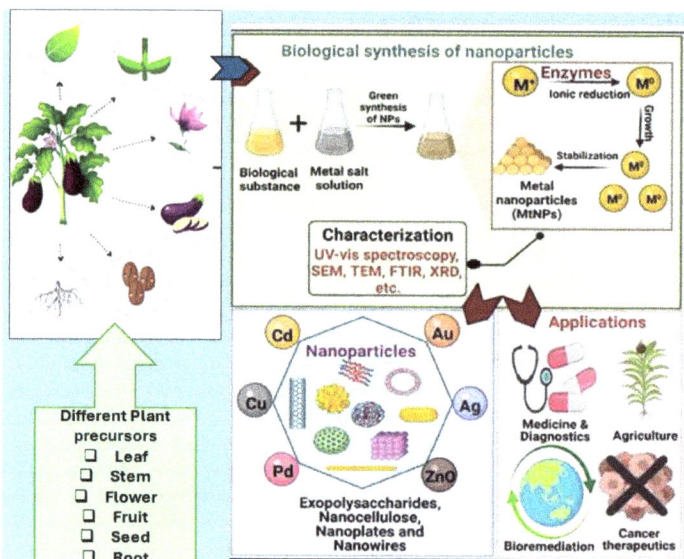

Figure 3. Schematic diagram of various types of NPs biosynthesis procedures, characterization, applications. Reproduced from ref. (134),

2.1   Synthesis of Cu/CuO NPs using flower:

Thiruvengadam et al. using *Millettia pinnata* plant flower extract manages to fabricate Cu NPs of round shape having length 24 nm approximately. The spectroscopic techniques including UV-Vis, SEM, TEM, XRD, FTIR, and SAED were used to confirm Cu NPs formation [21]. FTIR indicated the presence of coyun7ytmpounds like acid proteins, carboxylic acids, flavonoids, alkaloids, and polyphenols involved in reducing Cu^{2+} ions to Cu-NPs. *Flower extract of Tecoma stans (T. stans) was deployed to prepare copper nanoparticles. The formation of NPs was found to be spherical shape. The role of the NPs as anticancer agent was tested specifically for lung cancer by fluorescence microscopy [39].* Cu NPs were also synthesized by using flower extract of Q. Indica from copper acetate. As-synthesized Cu NPs were employed in vivo therapeutic efficacy studies in mice bearing B16F10 melanoma tumor. The study observed a decrease in tumor growth in nanoparticles treated animal model [40]. Neem flower was also used to produce Cu NPs [41].

2.2   Synthesis of Cu/CuO NPs using peels:

Cu-NPs with diameters of 34-45 nm were successfully fabricated using *lemon peel aqueous extract* at room temperature [29]. The characteristic peak at 560nm confirms the formation of Cu-NPs which also intensify with increased concentration. Higher temperatures enhanced Cu-NP yield, with complete transformation achieved at 90–95°C. Increasing the concentration of *lemon peel aqueous extract* to 20% resulted in the production of smaller and round Cu-NPs. Natural molecules or metabolites can act as capping agents, causing the aggregation of nanoparticles (NPs). This aggregation may result from impurities in the NPs colloidal solution and aqueous leaf extract. *Magnolia leaf* extract was employed to fabricate Cu NPs which is stable up to 40days due to its role as caping and protecting agent which makes him intact from arial oxidations. Whereas chemical process uses oxidation method with harmful chemicals & Stability is also very less [30]. Bio precursors such as *Eucalyptus sp.* leaf extract were investigated by *Kulkarni et al.* for the fabrication of Cu-NPs using copper salt at room temperature. The resultant UV-Vis spectrum at 572 nm confirms the formation of Cu NPs with the influence of pH change whereas pH of the leaf extract nearly 7. Ghidan et al. synthesized CuO-NPs using *pomegranate* peel extract, which contains compounds capable of reducing copper ions to form CuO NPs. The NPs were found to be crystalline and round, with an approximate length of 40 nm, but tended to aggregate into large clusters [67].

A plant-based method from peel extract of *Punica granatum* was employed to prepare Cu-NPs with diameters of 18 nm which also serves as reducing and capping agent [51]. The UV-Vis studies of the Cu-NPs showed minimal absorption within the 500-700 nm range, with a broad peak at 585 nm that cannot definitively be attributed to Cu-NPs [52]. The formation of composite material with carbon nanofiber coated Cu NPs were achieved by calcining the material at 350°C. [53].

2.3   Synthesis of Cu/CuO NPs using leaf extract:

The leaf extract acts as both reducing and capping agents in the synthesis process of NPs. In the *Eucalyptus* leaf extract the presence of phenols, amines, amino acids, and flavonoids was established by FTIR spectrum of the Cu-NPs [31]. The NPs exhibited a crystalline structure with a face-focused cubic (FCC) arrangement and size less than 40nm. The concentration and pH of the reducing agent were identified as the primary factors influencing the production of these NPs.

Similar observations were reported for the plant mediated synthesis of Cu NPs by using *Azadirachta indica* leaf broth. It is also observed that production of the NPs is influenced by leaf concentration, pH and nature of the salt solutions. These factors are known to play crucial roles in the synthesis and properties of NPs [31,32].

The study reports the successful synthesis of ultrasmall copper nanoparticles (Cu-NPs) with a diameter of 3 nm using a one-pot technique involving lemongrass tea [33]. The as-synthesized Cu-NPs are ultrasmall as the solution of both the mixture looks yellow and it also lacks the characteristic pick of Cu NPs. This is so because Cu-NPs prevent the display of surface plasmon resonance pertaining to the very small size of Cu NPs [34, 35] although large Cu NPs naturally exhibit a sharp absorption peak around 560-570 nm. Green-synthesis of CuO NPs carried out using aqueous leaf extract of the plant *Artemisia deserti* [36]. The NPs are found to be of 9.8 nm and spherical shaped. Importantly, these NPs were resistant to oxidation, likely by the presence of reducing agents that inhibit the oxidation process. The study also showed that apoptosis could be persuaded in A2780-CP cells in a significant level using the synthesized CuO NPs. *Centratherum punctatum leaf extract* has been used to synthesis CuO NPs. Spherical and hemispherical aggregated particles were formed with size between 20-30 nm. These bio-derived NPs were found effective in inhabitation in biofilm formation by Pseudomonas aeruginosa [37]. *Macroptilium lathyroides plant was employed to synthesize CuO NPs in an innovative manner. The particle size was found to be 18.9 nm with potential applications in Staphylococcus aureus and Escherichia coli via disc diffusion method as antibacterial agent [38].*

Novel green synthesis of monoclinic type Cu NPs was executed by *Jasminium sambac* leaf extract having particle size 13.4 nm confirms by band at 590 nm in FTIR for NPs. The as-synthesized CuO NPs is useful for dye degradation experiment with Methylene Blue & eventually it shows 97% dye degradation in 210 min [40]. *Cupressocyparis leylandii* leaf extracts is used to prepare green copper oxide nanoparticles (G-CuO NPs) via co-precipitation. The NPs were found to have bandgap energy of 3.24 eV with spherical structure and crystallite size of 18 nm [41]. In recent time greener synthesis attracts researchers due to its low toxicity. In line to the idea fabrication of copper oxide (CuO) nanoparticles (NPs) using Betel leaf (Piper betel) extracts as reducing, capping, and stabilizing agents. The NPs were used in photodegradation of Congo red dye (CR) and as a potential antibacterial performance against Bacillus subtilis and Pseudomonas aeruginosa [42]. *Allahabad Safeda* and *Hisar Safeda* leaf extracts were employed to synthesis copper oxide nanoparticles (CuO-NPs) having particle size of 2 nm and 15 nm, respectively they were found effective for antibacterial, antioxidant, antidiabetic, and photocatalytic studies [43].

The study documented the plant-based synthesis of copper nanoparticles (Cu-NPs) encompassing *Cissus arnotiana*. The method is green and economically viable. The UV-Vis spectrum reflects an absorption peak at 634 nm for Cu-NPs formation. Aluminum vessels are used to conduct the reduction as partial dissolution of aluminum into the citron juice takes place. As a result, the initial colour change of the solution of CuSO₄ happens on the inner surface of the aluminum vessel suggesting that more reactive metal can displace a less reactive metal [46-47].

Table1: *Various plant mediated method for the synthesis of Cu/CuO NPs*

Sr. No.	Plants Name	Part Used	Phytochemicals	Shape NPs	References
01	*Millettia pinnata*	*Flower extract*	Proteins, acids, favonoids, polyphenols, carboxylic acid and alkaloids	Spherical – shaped	[21]
02	*Tecoma stans (T. stans)*	*Flower extract*	Flavonoids, terpenoids and polyphenols	Spherical	[39]
03	*Quisqualis indica*	*Flower extract*	Alcohols and phenols	Spherical – shaped	[40]
04	*Azadirachta indica*	*Flower extract*	Lawsone and phenols	Spherical- shaped	[41]
05	*Lemon peel aqueous extract*	Peel extract	Flavonoids, terpenoids and polyphenols	Sphere-like	[29]
06	*Pomegranate*	Peel extract	Flavonoids, terpenoids and polyphenols	Large cluster	[67]
07	*Punica granatum*	Peel extract	Flavonoids, terpenoids and polyphenols	Spherical- shaped	[51]
08	*Azadirachta indica*	Leaf extract	Flavonoids, terpenoids and polyphenols	Cubic-shaped	[32]
09	*Lemongrass tea*	Leaf extract	Flavonoids, terpenoids and polyphenols	Spherical – shaped	[33]
10	*Artemisia deserti*	leaf extract	Flavonoids, terpenoids and polyphenols	Spherical – shaped	[36]
11	*Centratherum punctatum*	leaf extract	Flavonoids, terpenoids and polyphenols	Spherical and hemispherical aggregated	[37]
12	*Macroptilium lathyroides*	leaf extract	Flavonoids, terpenoids and polyphenols	Spherical – shaped	[38]
13	*Jasminium sambac*	leaf extract	Flavonoids, terpenoids and polyphenols	Spherical – shaped	[40]
14	*Cupressocyparis leylandii*	leaf extracts	Flavonoids, terpenoids and polyphenols	spherical structure	[41]

15	Betel leaf (Piper betle)	leaf extracts	Flavonoids, terpenoids and polyphenols	spherical structure	[42]
16	Allahabad Safeda and Hisar Safeda	leaf extracts	Flavonoids, terpenoids and polyphenols	spherical structure	[43]
17	Cissus arnotiana	leaf extracts	Biomolecules	Spherical-shaped	[46-47]
18	Pterospermum acerifolium	leaf extracts	Flavonoids and phenols	Spherical – shaped	[71]
19	Carica papaya	leaf extracts	Flavonoids, terpenoids and polyphenols	Spherical – shaped	[78]
20	R. cordifolia	bark extracts	Terpenoids and polyphenols	Spherical – shaped	[44]
21	Ziziphus spina-christi	Fruit	Alcohols and phenols	Cubical-shaped	[96]
22	Tribulus terrestris	Fruit	Alcohols and phenols	Cubical-shaped	[68-69]
23	Perse Americana	Seed	Carboxylic acid acid alkanes	Spherical – shaped	[86]
24	Theobroma cacao	Seed	Terpenoids and polyphenols	Spherical and hemispherical aggregated	[87]

The UV-Vis studies of CuO-NPs showed peaks at 261 nm, attributed to Cu_2O-NPs. The conversion of Cu^{2+} to CuO-NPs is facilitated by terpenoids in the algal extract [45,59-60]. The sample contains a combination of Cu_2O and CuO-NPs, indicated by the reddish-black hue and confirmed by TEM images showing spherical and elongated CuO-NPs with diameters of range from 4.9 to 41 nm. The primary additives found in algae, diterpenoids, act as stabilizers and inhibitors for CuO-NPs during their formation [61]. CuO-NPs were synthesized using *Aloe vera* leaf extract, resulting in the formation of monoclinic CuO-NPs with an average particle length of approximately 20 nanometers [62]. *Ixoro coccinea* leaf extract was employed to prepare CuO-NPs of length 301 nm due NP clusters formation [63].

The Fourier-Transform Infrared Spectroscopy (FTIR) analysis of the CuO-NPs revealed distinct peaks at specific wavenumbers, including 452, 612, 794, 893, 1120, 1652, and 3184 cm^{-1} within the range of 500–4000 cm^{-1}. These peaks correspond to various functional groups and vibrations present in the synthesized CuO-NPs. The bands at 3184 cm^{-1} are associated with O−H stretching, indicating the presence of carboxylic acids, while the bands at 1652 cm^{-1} are linked to −C=C− bending, suggesting the existence of alkenes. The prominent bands observed at 1120 cm^{-1} indicate C−O stretching vibrations, characteristic of alcohols and esters in the sample.

Furthermore, for CuO NPs peaks at 451 cm^{-1}, 611 cm^{-1}, and 793 cm^{-1} find in FTIR for Cu−O vibrations. These findings provide valuable insights into the chemical composition and functional groups present in the CuO-NPs fabrication. The FTIR analysis of biofabricated CuO-NPs revealed the presence of diverse functional groups including alkenes, carboxylic acids, alcohols, esters, and aromatic compounds. This offers comprehensive insight into the chemical composition and functional groups associated with the CuO-NPs formation. The preparation of CuO-NPs from *Psidium guajava* leaf solution was detailed by Singh et al., utilizing natural plant extracts as reducing agents for NPs synthesis. The study shows nanorods to nanospheres shape NPs was formed.

Additionally, in water medium *Eichhornia crassipes* derived CuO-NPs was formed in laboratory conditions, demonstrating the adaptability of CuO-NP synthesis methods and their potential applications across diverse fields. The UV-Vis studies give peak at 311 nm, indicating the presence of CuO-NPs in the solution along with weed solution. The analysis of SEM and TEM images revealed the presence of CuO-NPs with a round shape and an approximate diameter of 28 nm, free from impurities. The purity was confirmed through EDX pattern showing distinct peaks for copper and oxygen elements.

The study demonstrates *Pterospermum acerifolium* leaf extract and hydrated copper nitrate as a precursor was used to prepare CuO-NPs, showcasing the potential of herbal extracts for green synthesis of NPs [71]. The oval-shaped NPs exhibited greater stability compared to engineered NPs. The plant mediated synthesis can increase the hydrodynamic diameter while that of the engineered NPs particle size. It is suggested that the green-synthesized NPs may be capped by phytochemicals present in the leaf extract, leading to reduced aggregation due to faster release of Cu than that from plant mediated process. Additionally, the release of copper ions depends on may dependent factors to some extent [72,73].

Additionally, *Carica papaya* extract employed to form CuO-NPs have been reported, demonstrating the use of natural sources for NP synthesis [78]. The FTIR showed a sharp peak at 473 cm^{-1}, conforming their formation. The presence of Cu_2O-NPs has been ruled out due to the absence of characteristic peaks in the range of 605–660cm^{-1}. Rod-shaped CuO-NPs, 140 nanometers in length, were found to be crystalline with a Face-Centered Cubic (FCC) structure, providing insights into their structural characteristics [79]. The observation of larger particles (614 nanometers) that are not monodispersed is for the capping effect of plant chemicals of papaya leaf extract, which likely interact with the CuO-NPs, affecting their size and dispersion. Furthermore, the leaf solution of *Calotropis gigantea* was used to prepare CuO-NPs, with plant chemicals acting as reducers for Cu^{2+} ions and stabilizers for the NPs. This demonstrates the potential of various herbal extracts and compounds for green synthesis of NPs.

The biosynthesis Cu NPs contradiction in the argument may arise due to various factors, including differences in the behavior and interactions of different plant species with CuO-NPs and dissolved copper, as well as variations in experimental conditions. This complexity highlights the need for further research to comprehensively understand these interactions [76]. The acid exuded from growing shoots is suggested to solubilize copper from CuO NPs. This can be represented by the chemical reactions:

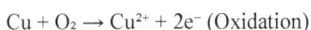

$CuO + H_2 \rightarrow Cu + H_2O$ (Reduction)

$Cu + O_2 \rightarrow Cu^{2+} + 2e^-$ (Oxidation)

By osmosis process transportation process becomes easy for Cu^{+2} ions as they remain ionized in water and the can decimated in any parts of the plant. Whereas due to large size compared to ions NPs remain accumulated [77].

It's important to clarify that statements regarding the formation of CuO-NPs and their relationship with phytochemicals and metal chelates are contradictory and scientifically problematic. Metal chelates typically involve the coordination of a metal ion with ligands, while the formation of CuO-NPs does not necessarily involve metal chelation reactions. The oxidation of copper to CuO typically occurs at higher temperatures, around 900°C, rather than the suggested 400°C. This raises concerns about the scientific basis for the formation of CuO-NPs through phytochemicals in leaf extract and subsequent generation of "pure" NPs after heating at 400°C. It's important for scientific research to be based on sound principles and empirical evidence, and inconsistencies or inaccuracies should be rigorously tested and clarified through further experimentation and analysis.

Absence of a peak at 610 cm^{-1} in the IR spectra rules out the presence of Cu$_2$O, indicating dominance of CuO. The synthesis of CuO-NPs using natural extracts from vegetable peels offers valuable insights [85]. Cu2O is distinguishable from CuO due to its deep purple color. Different vegetable waste sources can lead to variations in CuO-NPs morphology affecting catalytic activity.

2.3 Synthesis of Cu/CuO NPs using beark extract:

Chinese medicine and Indian Ayurveda originated R. cordifolia bark extracts employed to fabricate copper oxide nanoparticles (CuO NPs) having particle size of 51 nm and spherical-shaped. Photocatalytic dye degradation activity was studied as application [44,45].

2.4 Synthesis of Cu/CuO NPs using fruit extract:

The fruit extract composition was characterized based on IR spectral peaks, indicating ployphenols, amines, and proteins. The Cu NPs obtained were approximately 48±4nm in size and appeared to be in a natural state, as confirmed by EDX analysis showing sharp signals only for copper and oxygen elements, suggesting the absence of impurities [68-69]. NPs derived from *Tribulus terrestris* fruit extract show significant effect on E. coli even at a low concentration compared to tetracycline. They were found to contain carbonyl and hydroxyl groups in the extract. The IR spectrum indicated the presence of an unpaired main component in the extract, and the CuO-NPs, ranging from 5.5 to 22.1 nm in size, were well-dispersed, with organic molecules from the extract depositing onto them.

2.5 Synthesis of Cu/CuO NPs using Seed:

Bio sorce such as *Theobroma cacao* seed extract employed for Pd/CuO-NPs catalysts formation [86]. The stability of Pd/CuO-NPs over 30 days is attributed to antioxidants present in cocoa seed extract. The FCC structure of Pd combined with CuO-NPs and prevention of agglomeration are notable findings, likely due to capping by natural compounds. The discrepancy in the presence of oxygen in the EDS spectrum and its association with the oxidation of Cu NPs to CuO-NPs needs further clarification. The slow process of metal copper oxidation in the presence of water leading to the production of CuSO$_4$ suggests a need for further investigation into the proposed mechanism for the formation of NPs. This emphasizes the importance of comprehensive scientific evaluation and experimentation in NPs synthesis via *Perse Americana* to address the noted inconsistencies and ensure precision [87].

3.   Application of copper NPs in diverse field

Potential uses of Cu-NPs/CuO-NPs have been displayed in Figure 4.

3.1   Antimicrobial interest

Assessment compared the antibacterial efficacy of froth-covered and uncoated Cu NPs. Results showed that Cu-NPs synthesized from Magnoliako bus aqueous leaf extract were more effective against Escherichia coli than chemically synthesized NPs. This suggests that green synthesis using leaf extract produces Cu-NPs with superior antibacterial properties. Another study analyzed the antibacterial effects of Cu-NPs derived by *Citron juice* (Citrus medica) to inhibit bacterial proliferation [87-90].

The research findings indicate that Cu-NPs synthesized from Magnolia kobus leaf extract showed higher antibacterial potentiality for Escherichia coli compared to chemically synthesized NPs. This highlights the superior antibacterial properties of Cu-NPs produced through green synthesis using plant extracts. Another study focused on Cu-NPs from Citron juice demonstrated effectiveness in inhibiting bacterial growth, particularly against Escherichia coli and other bacterial strains. Additionally, in vitro screening against various microorganisms confirmed the strong antimicrobial efficacy of Cu-NPs, making them effective against a range of pathogens. These results suggest the potential of Cu-NPs as powerful antibacterial agents with broad-spectrum activity [91-94].

The study found that Fusarium culmorum was the most sensitive to Cu-NPs, successively by F. oxysporum and F. graminearum. This suggests that Cu-NPs derived by pomegranate (punica grantum) could be effective in controlling or inhibiting the growth of these plant pathogenic fungi, which is significant for agricultural and plant protection purposes [96].

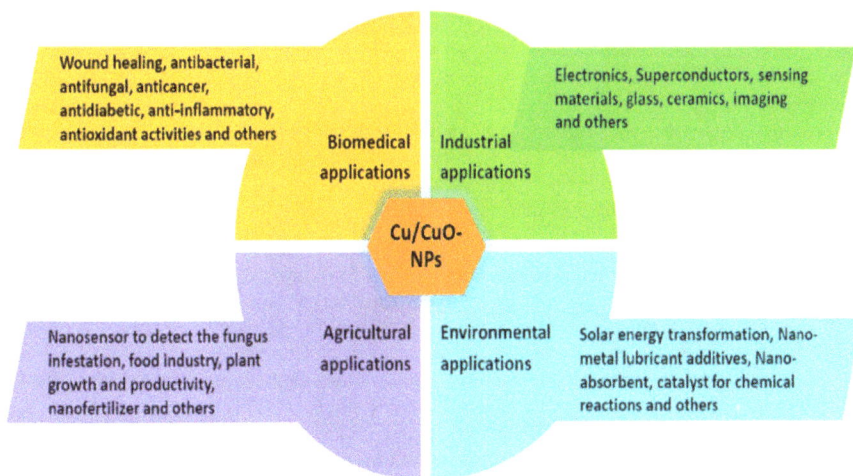

Figure 4: Applications of Cu/ CuO NPs

The peel extract has strong bactericidal properties against various microorganisms, including Micrococcus luteus, Pseudomonas aeruginosa, Salmonella enterica, and Enterobacter aerogenes. It can effectively inhibit the growth or kill these bacterial strains, making it a promising candidate for antimicrobial applications. The antibacterial efficacy of the peel extract is even better than that exhibited by the standard antibiotic streptomycin. Additionally, bio-synthesized Cu-NPs from Millettia pinnata flower extract have been found to be more effective against gram-positive bacteria such as Staphylococcus aureus and Bacillus subtilis than against gram-negative bacteria like Pseudomonas aeruginosa and E. coli. This difference in effectiveness against different types of microorganisms is not uncommon between gram-positive and gram-negative bacteria [97-100].

For antibacterial activity of CuO-NPs derived by gum karaya and Bifurcaria bifurcate has been demonstrated against E. coli, S. aureus, Enterobacter aerogenes, and other bacterial strains. Factors influencing the antibacterial efficacy of Cu-NPs include size, density of microorganism cell and contact period. Shorter NPs are greater effective despite not breaching the cell wall, highlighting the importance of NP size in antimicrobial activity. The interaction between microbial cells and the release of copper ions is responsible for the bactericidal effect. Cu-NPs from natural sources show promise for healthcare and antimicrobial applications [101, 102]. The passage discusses the potential of copper oxide nanoparticles (CuO-NPs) as antibacterial and antifungal supplemnts. It highlights their effectiveness in inhibiting bacterial and fungal growth, as well as their impact on cell viability. The study also mentions the complete rupture of bacterial cell walls when exposed to CuO-NPs, indicating their efficacy. [10-105].

3.2   Photocatalytic degradation:

The use of nanoscale CuO particles in degrading harmful dyes has been documented. A Cu/rGO-Fe_3O_4 was developed using barberry fruit juice as a reducing agent and studied as a catalyst in the ortho arylation of phenols with aryl halides. The nanocomposite is readily retrievable and can be reused without decline in catalytic effectiveness due to the presence of vitamin C and polyphenols in barberry fruit juice, which solve the criteria for preparing the Cu-NPs. CuO-NPs were used in the conversion of phenyl cyanamide to urea, initially yielding low results, but successfully addressed by raising the reaction mixture's temperature. Additionally, CuO-NPs were used in the reduction of 4-nitrophenol to 4-aminophenol in just 70 seconds when NaBH4 is present, while the same reaction failed to initiate without CuO-NPs even after an extended 3-hour duration [106-110]. The synthesis and catalytic performance of Pd/CuO-NPs, usingobroma cacao as a substrate, have been explored, including their utilization in the Heck coupling reaction to convert nitrophenol to aminophenol [86].

The passage discusses various applications of CuO-NPs synthesized from different bio extracts and methods. These applications include catalytic reduction of 4-nitrophenol, Ullmann-coupling reaction, and degradation of industrial dyes such as Nile blue and reactive yellow. Additionally, CuO-NPs were used in the photocatalytic degradation of Coomassie brilliant blue dye and Rhodamine B dye under visible light conditions. The synthesis methods involved the use of glycine and urea as fuel chember through the solution combustion process. The average crystallite size was found to be 9 nm and 17 nm for the two synthesis methods. However, the precise mechanistic pathway responsible for the catalytic activity of the CuO-NPs has not been fully elucidated [111-113].

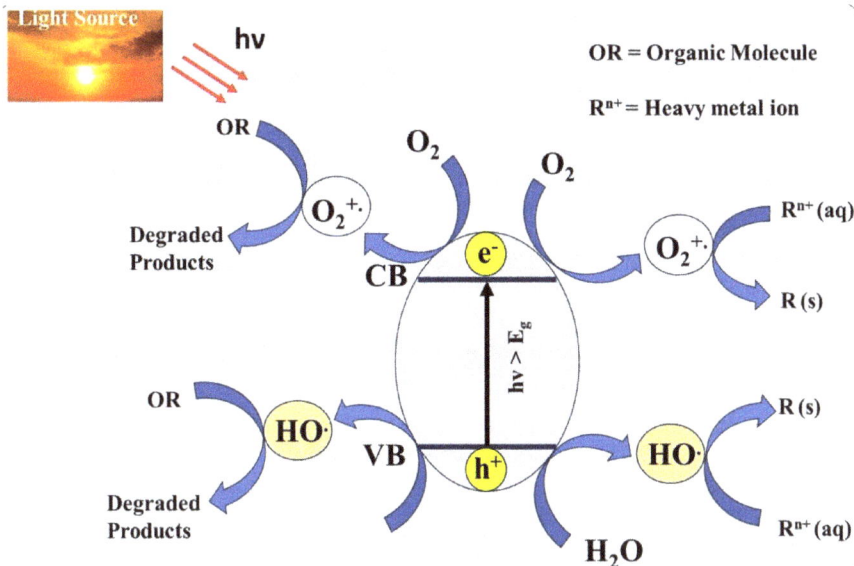

Figure 5: Cu/ CuO NPs for photocatalytic degradation

The text discusses the use of copper nanoparticles (Cu-NPs) and their impact on plant growth and yield. It mentions that Cu-NPs had a significant effect on growth of mung beans, while in wheat, they led to increased root growth but reduced shoot growth. Additionally, the study conducted by Yasmeen et al. fabricated Cu and Fe NPs using an onion extract and investigated their effects on wheat growth and yield. The results showed that 25 ppm Cu-NPs increased spike length and the number of grains per spike, but higher concentrations led to a decrease in grain number and overall weight. Furthermore, the combined application of Cu and Fe NPs at 25 ppm had a more favorable effect on wheat production and yield compared to using either Cu or Fe NPs alone. Overall, Cu-NPs were found to enhance grain production and tolerance in wheat plants, primarily through starch dilapidation [114-118].

The study involved the transformation of activated carbon fiber (ACF) covered with copper into Cu-CNF/ACF composites, which were used to treat chickpea plants and investigate the effects of engineered copper oxide nanoparticles (CuO-NPs) on Arabidopsis thaliana. The treatment with Cu-CNF showed improved growth in chickpea plants. On the other hand, Arabidopsis thaliana seedlings treated with CuO-NPs exhibited various changes, including reduced biomass, chlorophyll content, and root elongation, along with increased anthocyanin levels, lipid peroxidation, proline content, and lignin deposition in roots. The seedlings also showed elevated levels of superoxide and hydrogen peroxide, leading to oxidative stress that inhibited primary root growth but promoted lateral root formation due to metabolic imbalances and the plant's response to CuO-NPs-induced resistance [119-121].

The pressure on the plant machine triggers the production of antioxidants as a defense against reactive oxygen species (ROS), leading to changes in metabolic functions. A study by Chung et al. found that CuO-NPs reduced stem lengths, chlorophyll, carotenoids, and sugar content in Brassica rapa seedlings, while increasing proline and anthocyanin levels. Additionally, CuO-NP treatment led to elevated malondialdehyde and hydrogen peroxide production, linked to DNA damage. A Cu/rGO-Fe_3O_4 nanocomposite was formed using barberry fruit juice as a precursor and used as a catalyst. The nanocomposite was retrievable and could be reused without loss of effectiveness due to the presence of vitamin C and polyphenols in the fruit juice. CuO-NPs have been used in the conversion of phenyl cyanamide to urea, with low initial production, but this issue was resolved by increasing the reaction mixture's temperature. Additionally, CuO-NPs were utilized in the reduction of 4-nitrophenol to 4-aminophenol in the presence of NaBH4.

- CuO-NPs have shown efficient catalytic performance in various reactions such as Heck coupling and reduction of nitrophenol to aminophenol.
- Pd/CuO-NPs catalyst demonstrated durability through six cycles of reuse.
- Cu NPs derived from different plant extracts like Euphorbia esula, Psidium guajava, and Carica papaya have been effective in catalytic applications.
- CuO-NPs synthesized from Psidium guajava leaf extract showed impressive degradation capabilities against industrial dyes with high degradation percentages and apparent rate constants.
- CuO NPs synthesized using the solution combustion method have been utilized for photocatalytic degradation of dyes like Rhodamine B and Coomassie brilliant blue under visible light conditions.

The research article focuses on the synthesis of CuO nanoparticles (NPs) using the solution combustion method. The average crystallite sizes obtained were 19 nm and 16 nm for the two synthesis methods. The study includes characterization using different analytical techniques and evaluates the photocatalytic activity of the CuO NPs in degrading Rose Bengal dye [122-126].

3.3   Plant growth responses

The introduction of Copper (Cu) nanoparticles (NPs) had varying effects on different plant species. In mung beans, Cu-NPs positively affected stem growth, while in wheat, it increased root growth but reduced shoot growth. A study by Yasmeen et al. fabricated Cu and Iron (Fe) NPs using an onion extract, with irregular shapes and diameters ranging from 15 to 30 nanometers. When wheat varieties were treated with 25 ppm Cu-NPs, spike length either increased or remained unchanged, effectively boosted the number of grains per spike. More Cu concentrations led to a reduction in the number of grains per spike and an overall decrease in the 1000-grain weight. The combined application of Cu and Fe NPs at a concentration of 25 ppm had a more favorable impact on wheat production and yield compared to using either Cu or Fe NPs alone [127-129].

Many researchers focus on the carrier for copper transportation by the study utilized activated carbon fiber (ACF) coated with copper sodium dodecyl sulfate (SDS) which is obtained by calcination at 350^0C with the precursor which was then transformed in a Cu ACF carbon nanofibers under hydrogen atmosphere. These nanofibers were developed on Cu ACF to form Cu-CNF/ACF composites. Chickpea plants treated with Cu-CNF showed improved growth [130-132].

Another aspect of the study investigated the effects of engineered CuO nanoparticles (CuO-NPs) on Arabidopsis thaliana seedlings. Treatment with CuO-NPs led to decreased plant biomass, total chlorophyll content, and root elongation whereas, increasing anthocyanin levels, lipid peroxidation, proline content, and lignin deposition in roots. Elevated levels of superoxide and hydrogen peroxide were observed in leaves and roots, correlating with CuO-NP dose. Oxidative stress inhibited primary root growth but stimulated lateral root formation due to metabolic imbalances and plant response to CuO-NP-induced resistance.

In response to oxidative stress, plants activate antioxidant production as a defense mechanism against reactive oxygen species (ROS), leading to metabolic alterations. A recent study by Chung et al. investigated the impact of CuO-NPs on Brassica rapa seedlings

The CuO-NP treatment resulted in decreased root and shoot lengths, reduced chlorophyll, carotenoids, and sugar gratified levels, while increasing proline and anthocyanin levels in seedlings. Additionally, the treatment led to elevated malondialdehyde and hydrogen peroxide production, associated with DNA damage [133,134].

Conclusion

Green synthesis has many advantages, including a straightforward process for generating nanoparticles, cost-effectiveness, and environmental friendliness. The progress in the eco-friendly production of nanoparticles (NPs) has introduced a novel method in several domains such as antibacterial research, biomedical uses, and plant development. The bioinspired synthesis of Cu NPs has several desirable features, including antibacterial, antifungal, anticancer, anti-inflammatory, and wound healing effects. This is achieved by a biological technique that allows for precise control over the form and size of the synthesized nanoparticles. The papers listed in this chapter demonstrate various shapes and sizes of Cu/CuO NPs, including spherical, rod-like, and cylindrical morphologies. The sizes of these particles range from 2 to 200 nm. Although there are potential uses for Cu/CuO NPs, many research areas remain unexplored in order to reach a significant breakthrough.

References

[1] Husen A, Siddiqi KS. Phytosynthesis of nanoparticles: concept, controversy and application. Nano Res Lett. 2014;9:229. https://doi.org/10.1186/1556-276X-9-229

[2] Siddiqi KS, Husen A, Rao RAK. A review on biosynthesis of silver nanoparticles and their biocidal properties. J Nanobiotechnol. 2018;16:14. https://doi.org/10.1186/s12951-018-0334-5

[3] Husen A. Introduction and techniques in nanomaterials formulation: An overview. In: Husen A, Jawaid M, editors. Nanomaterials for Agriculture andForestry Applications. Cambridge: Elsevier Inc; 2020. p. 1-14. https://doi.org/10.1016/B978-0-12-817852-2.00001-9

[4] Siddiqi KS, Husen A. Fabrication of metal and metal oxide nanoparticles by algae and their toxic effects. Nano Res Lett. 2016;11:363. https://doi.org/10.1186/s11671-016-1580-9

[5] Siddiqi KS, Husen A. Fabrication of metal nanoparticles from fungi and metal salts: scope and application. Nano Res Lett. 2016;11:98. https://doi.org/10.1186/s11671-016-1311-2

[6] Philip D, Unni C, Aromal SA, Vidhu VK. Murrayakoenigii leaf-assisted rapid green synthesis of silver and gold nanoparticles. Spectrochem Acta A Mol BiomolSpectrosc. 2011;78:899-904. https://doi.org/10.1016/j.saa.2010.12.060

[7] Chouke, P.B., Potbhare, A.K., Dadure, K.M., Mungole, A.J., Meshram, N.P., Chaudhary, R.R., Rai, A.R., Chaudhary, R.G., An antibacterial activity of Bauhinia racemosa assisted ZnO nanoparticles during lunar eclipse and docking assay, Mater. Today Proc., 2020, 29, 815-821. https://doi.org/10.1016/j.matpr.2020.04.758

[8] Nagar N, Jain S, Kachhawah P, Devra V. Synthesis and characterization ofsilver nanoparticles via green route. Korean J Chem Eng. 2016;33:2990-7. https://doi.org/10.1007/s11814-016-0156-9

[9] Husen A. Gold nanoparticles from plant system: synthesis, characterizationand their application. In: Ghorbanpour M, Manika K, Varma A, editors. Nanoscience and Plant-Soil Systems. Soil Biology. Cham: Springer, 2017;48: 455-479. https://doi.org/10.1007/978-3-319-46835-8_17

[10] Chouke, P.B., Potbhare, A.K., Meshram, N.P., Rai, M.M., Dadure, K.M., Chaudhary, K., Rai, A.R., Desimone, M.F., Chaudhary, R.G., Masram, D.T., Bioinspired NiO nanospheres: Exploring in-vitro toxicity using Bm-17 and L. rohita liver cells, DNA degradation, docking and proposed vacuolization mechanism, ACS Omega, 2022,7, 6869−6884. https://doi.org/10.1021/acsomega.1c06544

[11] Siddiqi KS, Rashid M, Rahman A, Tajuddin HA, Rehman S. Biogenic fabrication and characterization of silver nanoparticles using aqueousethanolic extract of lichen (Usnea longissima) and their antimicrobial activity.Biomat Res. 2018;22:23. https://doi.org/10.1186/s40824-018-0135-9

[12] Mondal, A., Umekar, M.S., Bhusari, G.S., Chouke, P.B., Lambat, T., Mondal, S., Chaudhary, R.G. and Mahmood, S.H., Biogenic synthesis of metal/metal oxide nanostructured materials. Curr. Pharm. Biotech. 2021, 22, 1782-1793. https://doi.org/10.2174/1389201022666210111122911

[13] Umer A, Naveed S, Ramzan N, Rafiqui MS. Selection of a suitable method for the synthesis of copper nanoparticles. Nano. 2012;7:1230005. https://doi.org/10.1142/S1793292012300058

[14] Jain S, Jain A, Kachhawah P, Devra V. Synthesis and size control of copper nanoparticles and their catalytic application. Trans Nonferrous Met SocChina. 2015;25:3995-4000. https://doi.org/10.1016/S1003-6326(15)64048-1

[15] K. Tharani, L. Nehru, Synthesis and chareterization of copper oxide nanoparticles by solution combustion method: photocatalytic activity under visible light irradiation, Rom. J. Biophys. 30 (2) (2020).

[16] Potbhare, A.K., Chaudhary, R.G., Chouke, P.B., Yerpude, S., Mondal, A., Sonkusare, V.N., Rai, A.R., Juneja, H.D., Phytosynthesis of nearly monodisperse CuO nanospheres using Phyllanthus reticulatus/Conyza bonariensis and its antioxidant/antibacterial assays. Mater. Sci. Eng. C., 2019, 99, 783-793. https://doi.org/10.1016/j.msec.2019.02.010

[17] Tiwari M, Jain P, Hariharapura RC, Narayanan K, Udaya BK, Udupa N, Rao JV. Biosynthesis of copper nanoparticles using copper-resistant Bacillus cereus,a soil isolate. Process Biochem. 2016;51:1348-56. https://doi.org/10.1016/j.procbio.2016.08.008

[18] Borkow G, Gabbay J. Copper, an ancient remedy returning to fight microbial, fungal and viral infections. Cur Chem Biol. 2009;3:272-8. https://doi.org/10.2174/2212796810903030272

[19] Zheng XG, Xu CN, Tomokiyo Y, Tanaka E, Yamada H, Soejima Y. Observationof charge stripes in cupric oxide. Phys Rev Lett. 2000;85:5170-3. https://doi.org/10.1103/PhysRevLett.85.5170

[20] Ren G, Hu D, Cheng EW, Vargas-Reus MA, Reip P, AllakerRP.Characterisation of copper oxide nanoparticles for antimicrobial applications. Int J Antimicrob Agent. 2009;33:587-90. https://doi.org/10.1016/j.ijantimicag.2008.12.004

[21] Din MI, Arshad F, Hussain Z, Mukhtar M. Green adeptness in the synthesisand stabilization of copper nanoparticles: catalytic, antibacterial, cytotoxicity,and antioxidant activities. Nano Res Lett. 2017;12:638. https://doi.org/10.1186/s11671-017-2399-8

[22] Apostolov AT, Apostolova IN, Wesselinowa JM. Dielectric constant of multiferroic pure and doped CuO nanoparticles. Solid State Commun. 2014;192:71-4. https://doi.org/10.1016/j.ssc.2014.05.014

[23] Thiruvengadam M, Chung IM, Gomathi T, Ansari MA, Khanna VG, Babu V, Rajakumar G. Synthesis, characterization and pharmacological potential of green synthesized copper nanoparticles. Bioprocess Biosyst Eng. 2019;42: 1769-77. https://doi.org/10.1007/s00449-019-02173-y

[24] Pariona N, Mtz-Enriquez AI, Sanchez-Rangel D, Carrion G, Paraguay-DelgadoF, Rosas-Saito G. Green-synthesized copper nanoparticles as a potentialantifungal against plant pathogens. RSC Adv. 2019;9:18835-43. https://doi.org/10.1039/C9RA03110C

[25] Lee Y, Choi JR, Lee KJ, Stott NE, Kim D. Large-scale synthesis of coppernanoparticles by chemically controlled reduction for applications of inkjetprinted electronics. Nanotechnology. 2008;19:598-604. https://doi.org/10.1088/0957-4484/19/41/415604

[26] Rubilar O, Rai M, Tortella G, Diez MC, Seabra AB, Durán N. Biogenic nanoparticles: copper, copper oxides, copper sulphides, complex coppernanostructures and their applications. Biotechnol Lett. 2013;35:1365-75. https://doi.org/10.1007/s10529-013-1239-x

[27] Waser O, Hess M, Güntner A, Novák P, Pratsinis SE. Size controlled CuO nanoparticles for Li-ion batteries. J Power Sour. 2013;241:415-22. https://doi.org/10.1016/j.jpowsour.2013.04.147

[28] Sharma JK, Akhtar MS, Ameen S, Srivastava P, Singh G. Green synthesis ofCuO nanoparticles with leaf extract of Calotropis gigantea and its dyesensitized solar cells applications. J All Comp. 2015;632:321-5. https://doi.org/10.1016/j.jallcom.2015.01.172

[29] Kir I, Mohammed HA, Laouini SE, Souhaila M, Hasan GG, Abdullah JA, Mokni S, Naseef A, Alsalme A, Barhoum A. Plant extract-mediated synthesis of CuO nanoparticles from lemon peel extract and their modification with polyethylene glycol for enhancing photocatalytic and antioxidant activities. Journal of Polymers and the Environment. 2024 Feb;32(2):718-34. https://doi.org/10.1007/s10924-023-02976-x

[30] Joshi A, Sharma A, Bachheti RK, Husen A, Mishra VK. Plant-mediated synthesis of copper oxide nanoparticles and their biological applications. In:Husen A, Iqbal M, editors. Nanomaterials and Plant Potential. Cham: Springer International Publishing AG; 2019. p. 221-37. https://doi.org/10.1007/978-3-030-05569-1_8

[31] Lee HJ, Song JY, Kim BS. Biological synthesis of copper nanoparticles using Magnoliakobus leaf extract and their antibacterial activity. J Chem TechnolBiotechnol. 2013;88:1971-7. https://doi.org/10.1002/jctb.4052

[32] Song JY, Jang HK, Kim BS. Biological synthesis of gold nanoparticles usingMagnoliakobus and Diopyros kaki leaf extracts. Process Biochem. 2009;44:1133-8. https://doi.org/10.1016/j.procbio.2009.06.005

[33] Kulkarni V, Suryawanshi S, Kulkarni P. Biosynthesis of copper nanoparticles using aqueous extract of Eucalyptus sp. plant leaves. Curr Sci. 2015;109:255-27.

[34] Nagar N, Devra V. Green synthesis and characterization of copper nanoparticles using Azadirachta indica leaves. Mat Chem Phys. 2018;213: 44-51. https://doi.org/10.1016/j.matchemphys.2018.04.007

[35] Brumbaugh AD, Cohen KA, Angelo SKS. Ultrasmall copper nanoparticles synthesized with a plant tea reducing agent. ACS Sustain Chem Eng. 2014;2:1933-9. https://doi.org/10.1021/sc500393t

[36] Umekar, M., Chaudhary, R., Bhusari, G., Potbhare, A., Fabrication of zinc oxide- decorated phytoreduced graphene oxide nanohybrid via Clerodendrum infortunatum, Emerg. Mater. Res., 2021, 10, 75-84. https://doi.org/10.1680/jemmr.19.00175

[37] Chouke, P.B Chouke, P.B., Bhusari, G. S., Somkuwar, S., Shaik, PMD., Mishra, R. K., Chaudhary, R.G., Green fabrication of zinc oxide nanospheres by Aspidopterys cordata for effective antioxidant and antibacterial activity, Adv. Mater. Lett., 2019, 10, 355-360. https://doi.org/10.5185/amlett.2019.2235

[38] Prabu P, Losetty V. Green synthesis of copper oxide nanoparticles using Macroptilium Lathyroides (L) leaf extract and their spectroscopic characterization, biological activity and photocatalytic dye degradation study. Journal of Molecular Structure. 2024 Apr 5;1301:137404. https://doi.org/10.1016/j.molstruc.2023.137404

[39] Sonkusare, V.N., Chaudhary, R.G., Bhusari, G.S., Mondal, A., Potbhare, A.K., Mishra, R.K., Juneja, H.D., Abdala, A.A., Mesoporous octahedron-shaped tricobalt tetraoxide nanoparticles for photocatalytic degradation of toxic dyes, ACS Omega, 2020, 5, 7823-7835. https://doi.org/10.1021/acsomega.9b03998

[40] Nouren S, Bibi I, Kausar A, Sultan M, Bhatti HN, Safa Y, Sadaf S, Alwadai N, Iqbal M. Green synthesis of CuO nanoparticles using Jasmin sambac extract: Conditions optimization and photocatalytic degradation of Methylene Blue dye. Journal of King Saud University-Science. 2024 Mar 1;36(3):103089. https://doi.org/10.1016/j.jksus.2024.103089

[41] Halfadji A, Naous M, Rajendrachari S, Ceylan Y, Ceylan KB, Shekar PR. Effective investigation of electro-catalytic, photocatalytic, and antimicrobial properties of porous CuO nanoparticles green synthesized using leaves of Cupressocyparis leylandii. Journal of Molecular Structure. 2024 Apr 5;1301:137318. https://doi.org/10.1016/j.molstruc.2023.137318

[42] Ahmad A, Khan M, Osman SM, Haassan AM, Javed MH, Ahmad A, Rauf A, Luque R. Benign-by-design plant extract-mediated preparation of copper oxide nanoparticles for environmentally related applications. Environmental Research. 2024 Apr 15;247:118048. https://doi.org/10.1016/j.envres.2023.118048

[43] Relhan A, Guleria S, Bhasin A, Mirza A, Zhou JL. Biosynthesized copper oxide nanoparticles by Psidium guajava plants with antibacterial, antidiabetic, antioxidant, and photocatalytic capacity. Biomass Conversion and Biorefinery. 2024 Apr 12:1-8. https://doi.org/10.1007/s13399-024-05544-y

[44] Vinothkanna A, Mathivanan K, Ananth S, Ma Y, Sekar S. Biosynthesis of copper oxide nanoparticles using Rubia cordifolia bark extract: characterization, antibacterial, antioxidant, larvicidal and photocatalytic activities. Environmental Science and Pollution Research. 2023 Mar;30(15):42563-74. https://doi.org/10.1007/s11356-022-18996-4

[45] Wang Y, Biradar AV, Wang G, Sharma KK, Duncan CT, Rangan S, Asefa T.Controlled synthesis of water-dispersible faceted crystalline coppernanoparticles and their catalytic properties. Chemistry. 2010;16:10735-43. https://doi.org/10.1002/chem.201000354

[46] Demirskyi D, Agrawal D, Ragulya A. Neck formation between copperspherical particles under single-mode and multimode microwave sintering.Mat Sci Eng: A. 2010;A527:2142-5. https://doi.org/10.1016/j.msea.2009.12.032

[47] Swarnkar RK, Singh SC, Gopal R. Effect of aging on copper nanoparticlessynthesized by pulsed laser ablation in water: structural and opticalcharacterizations. Bull Mater Sci. 2011;34:1363-9. https://doi.org/10.1007/s12034-011-0329-4

[48] Shende S, Ingle AP, Gade A, Rai M. Green synthesis of copper nanoparticlesby Citrus medica Linn. (Idilimbu) juice and its antimicrobial activity. World JMicrobiolBiotechnol. 2015;31:865-73. https://doi.org/10.1007/s11274-015-1840-3

[49] Sastry ABS, Aamanchi RBK, Rama Linga Prasad CS, Murty BS. Large-scalegreen synthesis of cu nanoparticles. Environ Chem Lett. 2013;11:183-7. https://doi.org/10.1007/s10311-012-0395-x

[50] Hirai H, Wakabayashi H, Komiyama M. Preparation of polymer-protectedcolloidal dispersions of copper. Bull Chem Soc Japan. 1986;59:367-72. https://doi.org/10.1246/bcsj.59.367

[51] Zhu YJ, Qian YT, Zhang MW, Chen ZY, Xu DF. Preparation andcharacterization of nanocrystalline powders of cuprous oxide by using Cradiation. Mater Res Bull. 1994;29:377-83. https://doi.org/10.1016/0025-5408(94)90070-1

[52] Nasrollahzadeh M, Sajadi SM, Khalaj M. Green synthesis of copper nanoparticles using aqueous extract of the leaves of Euphorbia esula L andtheir catalytic activity for ligand-free Ullmanncoupling reaction and reduction of 4-nitrophenol. RSC Adv. 2014;4:47313-8. https://doi.org/10.1039/C4RA08863H

[53] Kaur P, Thakur R, Chaudhury A. Biogenesis of copper nanoparticles using peel extract of Punica granatum and their antimicrobial activity against opportunistic pathogens. Green Chem Lett Rev. 2016;9:33-8. https://doi.org/10.1080/17518253.2016.1141238

[54] Hashemi pour H, Zadeh ME, Pourakbari R, Rahimi P. Investigation on synthesis and size control of copper nanoparticle via electrochemical andchemical reduction method. Int J Phys Sci. 2011;6:4331-6.

[55] Ashfaq M, Verma N, Khan S. Carbon nanofibers as a micronutrient carrier in plants: efficient translocation and controlled release of cu nanoparticles.Environ Sci: Nano. 2017;4:138-48. https://doi.org/10.1039/C6EN00385K

[56] Padil VVT, Černík M. Green synthesis of copper oxide nanoparticles usinggum karaya as a biotemplate and their antibacterial application. Int JNanomedicine. 2013;8:889-98. https://doi.org/10.2147/IJN.S40599

[57] Das D, Nath BC, Phukon P, Dolui SK. Synthesis and evaluation of antioxidantand antibacterial behavior of CuO nanoparticles. Coll Surf B Biointerf. 2013;101:430-3. https://doi.org/10.1016/j.colsurfb.2012.07.002

[58] Abboud Y, Saffaj T, Chagraoui A, El Bouari A, Brouzi K, Tanane O, IhssaneB.Biosynthesis, characterization and antimicrobial activity of copper oxidenanoparticles (CONPs) produced using brown alga extract (Bifurcariabifurcata). Appl Nanosci. 2014;4:571-6. https://doi.org/10.1007/s13204-013-0233-x

[59] Krithiga N, Jayachitra A, Rajalakshmi A. Synthesis, characterization andanalysis of the effect of copper oxide nanoparticles in biological systems.Ind J Nano Sci. 2013;1:6-15.

[60] Borgohain K, Murase N, Mahamuni S. Synthesis and properties of Cu2Oquantum particles. J Appl Phys. 2002;92:1292-7. https://doi.org/10.1063/1.1491020

[61] Yin M, Wu CK, Lou Y, Burda C, Koberstein JT, Zhu Y, O'Brien S. Copper oxidenanocrystals. J Am Chem Soc. 2005;127:9506-11. https://doi.org/10.1021/ja050006u

[62] Rahman A, Ismail A, Jumbianti D, Magdalena S, Sudrajat H. Synthesis ofcopper oxide nanoparticles by using Phormidium cyanobacterium. Indo JChem. 2009;9:355-60. https://doi.org/10.22146/ijc.21498

[63] Kiruba Daniel SCG, Nehru K, Sivakumar M. Rapid biosynthesis of silvernanoparticles using Eichornia crassipes and its antibacterial activity. CurrNanosci. 2012;8:1-5. https://doi.org/10.2174/1573413711208010125

[64] Vijay Kumar PPN, Shameem U, Kollu P, Kalyani RL, Pammi SVN. Greensynthesis of copper oxide nanoparticles using Aloe vera leaf extract and itsantibacterial activity against fish bacterial pathogens. BioNanoSci. 2015;5:135-9. https://doi.org/10.1007/s12668-015-0171-z

[65] Vishveshvar K, Aravind Krishnan MV, Haribabu K, Vishnuprasad S. Greensynthesis of copper oxide nanoparticles using Ixiro coccinea plant leavesand its characterization. BioNanoSci. 2018;8:554-8. https://doi.org/10.1007/s12668-018-0508-5

[66] Vasantharaj S, Sathiyavimal S, Saravanan M, Senthilkumar P, Kavitha G,Shanmugavel M, Manikandan E, Pugazhendhi A. Synthesis of ecofriendlycopper oxide nanoparticles for fabrication over textile fabrics:characterization of antibacterial activity and dye degradation potential. JPhotochemPhotobiol B Biol. 2018;191:149. https://doi.org/10.1016/j.jphotobiol.2018.12.026

[67] Chouke, P.B., Dadure, K.M., Potbhare, A.K., Bhusari, G.S., Mondal, A., Chaudhary, K., Singh, V., Desimone, M.F., Chaudhary, R.G. and Masram, D.T., Biosynthesized δ-Bi2O3 nanoparticles from Crinum viviparum flower extract for photocatalytic dye degradation and molecular docking, ACS Omega, 2022,7, 20983-20993. https://doi.org/10.1021/acsomega.2c01745

[68] Vanathi P, Rajiv P, Sivaraj R. Synthesis and characterization of Eichhorniamediated copper oxide nanoparticles and assessing their antifungal activityagainst plant pathogens. Bull Mater Sci. 2016;39:1165-70. https://doi.org/10.1007/s12034-016-1276-x

[69] SAziz, S.T., Ummekar, M., Karajagi, I., Riyajuddin, S.K., Siddhartha, K.V.R., Saini, A., Potbhare, A., Chaudhary, R.G., Vishal, V., Ghosh, P.C. and Dutta, A., A Janus cerium-doped bismuth oxide electrocatalyst for complete water splitting, Cell. Rep. Phys. Sci., 2022, 3(11) 101106. https://doi.org/10.1016/j.xcrp.2022.101106

[70] Sivaraj R, Rahman PK, Rajiv P, Salam HA, Venckatesh R. Biogenic copperoxide nanoparticles synthesis using Tabernaemontana divaricate leaf extractand its antibacterial activity against urinary tract pathogen. SpectrochimActa A Mol BiomolSpectrosc. 2014;133:178-81. https://doi.org/10.1016/j.saa.2014.05.048

[71] Potbhare, A. K., Chouke, P. B., Mondal, A., Thakare, R. U., Mondal, S., Chaudhary, R. G., Rai, A. R., Rhizoctonia solani assisted biosynthesis of silver nanoparticles for antibacterial assay. Mater. Today Proc., 2020, 29, 939-945. https://doi.org/10.1016/j.matpr.2020.05.419

[72] Gopinath V, Priyadarshini S, Al-Maleki AR, Alagiri M, Yahya R, Saravanan S,Vadivelu J. In vitro toxicity, apoptosis and antimicrobial effects of phytomediated copper oxide nanoparticles. RSC Adv. 2016;6:110986-95. https://doi.org/10.1039/C6RA13871C

[73] Saif S, Tahir A, Asim T, Chen Y. Plant mediated green synthesis of CuOnanoparticles: comparison of toxicity of engineered and plant mediatedCuO nanoparticles towards Daphnia magna. Nanomaterials. 2016;6:205. https://doi.org/10.3390/nano6110205

[74] Odzak N, Kistler D, Behra R, Sigg L. Dissolution of metal and metal oxidenanoparticles in aqueous media. Environ Pollut. 2014;191:132-8. https://doi.org/10.1016/j.envpol.2014.04.010

[75] Adam N, Leroux F, Knapen D, Bals S, Blust R. The uptake of ZnO and CuOnanoparticles in the water-flea Daphnia magna under acute exposurescenarios. Environ Pollut. 2014;194:130-7. https://doi.org/10.1016/j.envpol.2014.06.037

[76] Regier N, Cosio C, von Moos N, Slaveykova VI. Effects of copper-oxidenanoparticles, dissolved copper and ultraviolet radiation on copperbioaccumulation, photosynthesis and oxidative stress in the aquaticmacrophyte Elodea nuttallii. Chemosphere. 2015;128:56-61. https://doi.org/10.1016/j.chemosphere.2014.12.078

[77] Perreault F, Oukarroum A, Melegari SP, Matias WG, Popovic R. Polymercoating of copper oxide nanoparticles increases nanoparticles uptake andtoxicity in the green alga Chlamydomonas reinhardtii. Chemosphere. 2012;87:1388-94. https://doi.org/10.1016/j.chemosphere.2012.02.046

[78] Shi J, Abid AD, Kennedy IM, Hristova KR, Silk WK. To duckweeds (Landoltiapunctata), nanoparticulate copper oxide is more inhibitory than the solublecopper in the bulk solution. Environ Pollut. 2011;159:1277-82. https://doi.org/10.1016/j.envpol.2011.01.028

[79] Raja Naika H, Lingaraju K, Manjunath K, Kumar D, Nagaraju G, Suresh D,Nagabhushana H. Green synthesis of CuO nanoparticles using Gloriosa superbaL. extract and their antibacterial activity. J Taibah Univ Sci. 2015;9:7-12. https://doi.org/10.1016/j.jtusci.2014.04.006

[80] Sankar R, Manikandan P, Malarvizhi V, Fathima T, Shivashangari KS,Ravikumar V. Green synthesis of colloidal copper oxide nanoparticles usingCarica papaya and its application in photocatalytic dye degradation.Spectrochim Acta A Mol BiomolSpectrosc. 2014;121:746-50. https://doi.org/10.1016/j.saa.2013.12.020

[81] Ethiraj AS, Kang DJ. Synthesis and characterization of CuO nanowires by asimplewetchemical method. Nano Res Lett. 2012;7:70. https://doi.org/10.1186/1556-276X-7-70

[82] Nasrollahzadeh M, Maham M, Sajadi SM. Green synthesis of CuOnanoparticles by aqueous extract of Gundeliatournefortii and evaluation of their catalytic activity for the synthesis of N-monosubstituted ureasandreduction of 4-nitrophenol. J Colloid Interface Sci. 2015; 455:245-53. https://doi.org/10.1016/j.jcis.2015.05.045

[83] Adzet T, Puigmacia M. High-performance liquid chromatography ofcaffeoylquinic acid derivatives of Cynara scolymus L. leaves. J Chromatograph A. 1985; 348:447-53. https://doi.org/10.1016/S0021-9673(01)92486-0

[84] Haghi G, Hatami A, Arshi R. Distribution of caffeic acid derivatives inGundeliatournefortii L. Food Chem. 2011;124:1029-35. https://doi.org/10.1016/j.foodchem.2010.07.069

[85] Nasrollahzadeh M, Mohammad Sajadi S, Rostami-Vartooni A. Greensynthesis of CuO nanoparticles by aqueous extract of Anthemis nobilisflowers and their catalytic activity for the A3 coupling reaction. J Colloid Interface Sci. 2015; 459:183-8. https://doi.org/10.1016/j.jcis.2015.08.020

[86] Ullah H, Ullah Z, Fazal A, Irfan M. Use of vegetable waste extracts forcontrolling microstructure of CuO nanoparticles: green synthesis,characterization, and photocatalytic applications. J Chem. 2017; 2721798:5. https://doi.org/10.1155/2017/2721798

[87] Umekar, M. S., Bhusari, G.S., Potbhare, A.K., Mondal, A., Kapgate, B.P., Desimone, M.F. and Chaudhary, R.G., Bioinspired reduced graphene oxide based nanohybrids for photocatalysis and antibacterial applications. Curr. Pharm. Biotech. 2021, 22, 1759-1781. https://doi.org/10.2174/1389201022666201231115826

[88] Nasrollahzadeh M, Sajadi SM, Rostami-Vartooni A, Bagherzadeh M. Greensynthesis of Pd/CuO nanoparticles by Theobroma cacao L. seeds extract andtheir catalytic performance for the reduction of 4-nitrophenol andphosphine-free heck coupling reaction under aerobic conditions. J ColloidInterface Sci. 2015;448:106-13. https://doi.org/10.1016/j.jcis.2015.02.009

[89] Geil P, Anderson J. Nutrition and health implications of dry beans: a review.J Am Coll Nutr. 1994;13:549-58. https://doi.org/10.1080/07315724.1994.10718446

[90] Potbhare, A.K., Tarik Aziz, S.K., Ayyub, M.M., Kahate, A., Madankar, R., Wankar, S., Dutta A., Abdala A.A., Sami H.M., Adhikari, R., Chaudhury, R.G., Bioinspired Graphene-based metal oxide nanocomposites for photocatalytic and electrochemical performances: an updated review, Nanoscale Adv., 2024, DOI: 10.1039/D3NA01071F. https://doi.org/10.1039/D3NA01071F

[91] Bachheti A, Sharma A, Bachheti RK, Husen A, Pandey DP. Plantallelochemicals and their various application. In: Mérillon JM, RamawatKG,editors. Co-Evolution of Secondary Metabolites, Reference Series inPhytochemistry. Cham: Springer International Publishing AG. https://doi.org/10.1007/978-3-319-76887-8_14-1 (2019). https://doi.org/10.1007/978-3-319-76887-8_14-1

[92] Nagajyothi PC, Muthuraman P, Sreekanth TVM, Kim DH, Shim J. Greensynthesis: in-vitro anticancer activity of copper oxide nanoparticles againsthuman cervical carcinoma cells. Arab J Chem. 2017;10:215-25. https://doi.org/10.1016/j.arabjc.2016.01.011

[93] Bawadi HA. Inhibition of Caco-2 colon, MCF-7, and Hs578T breast, and DU145 prostatic cancer cell proliferation by water soluble black bean condensed tannins. Can Lett. 2005;218:153-62. https://doi.org/10.1016/j.canlet.2004.06.021

[94] Bobe G, Barret KG, Mentor-Marcel RA, Saffiotti U, Young MR, Colburn Chaudhary, R.G., Bhusari, G.S., Tiple, A.D., Rai, A.R., Somkuvar, S.R., Potbhare, A.K., Lambat, T.L., Ingle, P.P., Abdala, A.A., Metal/metal oxide nanoparticles: toxicity, applications, and prospects, Curr. Pharm. Des., 2019, 25, 4013-4029. https://doi.org/10.2174/1381612825666191111091326

[95] Hangen L, Bennik MR. Consumption of black beans and navy beans (Phaseolus vulgaris) reduced azoxymethane-induced colon cancer in rats.Nutr Cancer. 2002;44:60-5. https://doi.org/10.1207/S15327914NC441_8

[96] Thompson MD, Mensack MM, Jiang W, Zhu Z, Lewis MR, McGinley JN, BrickMA, Thompson HJ. Cell signaling pathways associated with a reduction inmammary cancer burden by dietary common bean (Phaseolus vulgaris L.).Carcinogenesis. 2012;33:226-32. https://doi.org/10.1093/carcin/bgr247

[97] Mukhopadhyay R, Kazi J, Debnath MC. Synthesis and characterization ofcopper nanoparticles stabilized with Quisqualis indica extract: evaluation ofits cytotoxicity and apoptosis in B16F10 melanoma cells. BiomedPharmacother. 2018;97:1373-85. https://doi.org/10.1016/j.biopha.2017.10.167

[98] Khani R, Roostaei B, Bagherzade G, Moudi M. Green synthesis of coppernanoparticles by fruit extract of Ziziphus spina-christi (L.) Willd.: applicationfor adsorption of triphenylmethane dye and antibacterial assay. J Mol Liq.2018;255:541-9. https://doi.org/10.1016/j.molliq.2018.02.010

[99] Machado TDB, Leal ICR, Amaral ACF, Dos Santos KRN, Da Silva MG, KusterRM. Antimicrobial Ellagitannin of Punica granatum Fruits. J Braz Chem Soc.2002;13:606-10. https://doi.org/10.1590/S0103-50532002000500010

[100] Voravuthikunchai SP, Sririrak T, Limsuwan S, Supawita T, Iida T, Honda T.Inhibitory effects of active compounds from Punica granatum pericarp onVerocytotoxin production by Enterohemorrhagic Escherichia coli O157: H7. JHealth Sci. 2005;51:590-6. https://doi.org/10.1248/jhs.51.590

[101] Yerpude, S.T., Potbhare, A.K., Bhilkar, P., Rai, A.R., Singh, R.P., Abdala, A.A., Adhikari, R., Sharma, R., Chaudhary, R.G., Biomedical and clinical applications of platinum- based nanohybrids: An update review, Environ. Res., 2023, 231, 116148. https://doi.org/10.1016/j.envres.2023.116148

[102] Azam A, Ahmed AS, Oves M, Khan MS, Memic A. Size-dependentantimicrobial properties of CuO nanoparticles against gram-positive and-negative bacterial strains. Int J Nanomedicine. 2012;7:3527-35. https://doi.org/10.2147/IJN.S29020

[103] Nasrollahzadeh M, Maham M, Rostami-Vartooni A, Bagherzadeh M,Sajadi SM. Barberry fruit extract assisted in situ green synthesis of cunanoparticles supported on a reduced graphene oxide-Fe3O4nanocomposite as a magnetically separable and reusable catalyst forthe O-arylation of phenols with aryl halides under ligand-freeconditions. RSC Adv. 2015;5:64769-80. https://doi.org/10.1039/C5RA10037B

[104] Umekar, M.S., Chaudhary, R.G., Bhusari, G.S., Mondal, A., Potbhare, A.K., Sami, M., Phytoreduced graphene oxide-titanium dioxide nanocomposites using Moringa oleifera stick extract, Mater. Today Proc., 2020, 29, 709-714. https://doi.org/10.1016/j.matpr.2020.04.169

[105] Embiale A, Hussein M, Husen A, Sahile S, Mohammed K. Differentialsensitivity of Pisum sativum L. cultivars to water-deficit stress: changes ingrowth, water status, chlorophyll fluorescence and gas exchange attributes.J Agron. 2016;15:45-57. https://doi.org/10.3923/ja.2016.45.57

[106] Siddiqi KS, Husen A. Engineered gold nanoparticles and plant adaptationpotential. Nano Res Lett. 2016;11:400. https://doi.org/10.1186/s11671-016-1607-2

[107] Siddiqi KS, Husen A. Plant response to engineered metal oxidenanoparticles. Nano Res Lett. 2017;12:92. https://doi.org/10.1186/s11671-017-1861-y

[108] Chaudhary, R.G., Potbhare, A.K., Chouke, P.B., Rai, A.R., Mishra, R.K., Desimone, M.F., Abdala, A.A. Graphene-based materials and their nanocomposites with metal oxides: biosynthesis, electrochemical, photocatalytic and antimicrobial applications, Magnetic Oxides and Composites II, MRF , 2020, 83, 79-116. https://doi.org/10.21741/9781644900970-4

[109] Husen A, Iqbal M, Sohrab SS, Ansari MKA. Salicylic acid alleviates salinitycaused damage to foliar functions, plant growth and antioxidant system inEthiopian mustard (Brassica carinata A. Br.). Agri Food Sec. 2018;7:44. https://doi.org/10.1186/s40066-018-0194-0

[110] Husen A, Iqbal M, Khanum N, Aref IM, Sohrab SS, Meshresa G. Modulationof salt-stress tolerance of Niger (Guizotiaabyssinica), an oilseed plant, byapplication of salicylic acid. J Environ Biol. 2019;40:94-104. https://doi.org/10.22438/jeb/40/1/MRN-808

[111] Dimkpa CO, McLean JE, Latta DE, Manangon E, Britt DW, Johnson WP,Boyanov MI, Anderson AJ. CuO and ZnO nanoparticles: phtotoxicity, metalspeciation, and induction of oxidative stress in sand-grown wheat. JNanopart Res. 2012;14:1125-9. https://doi.org/10.1007/s11051-012-1125-9

[112] Chaudhary, R.G., Potbhare, A.K., Aziz, S.T., Umekar, M.S., Bhuyar, S.S., Mondal, A., Phytochemically fabricated reduced graphene Oxide-ZnO NCs by Sesbania bispinosa for photocatalytic performances. Mater. Today Proc., 36 (2021) 756-762. https://doi.org/10.1016/j.matpr.2020.05.821

[113] Yasmeen F, Raja NI, Razzaq A, Komatsu S. Proteomic and physiologicalanalyses of wheat seeds exposed to copper and iron nanoparticles. BiochimBiophys Acta. 1865;2017:28-42. https://doi.org/10.1016/j.bbapap.2016.10.001

[114] Ashfaq M, Singh S, Sharma A, Verma N. Cytotoxic evaluation of thehierarchical web of carbon micronanofibers. Ind Eng Chem Res. 2013;52:4672-82. https://doi.org/10.1021/ie303273s

[115] Nair PMG, Chung IM. Impact of copper oxide nanoparticles exposure onArabidopsis thaliana growth, root system development, root lignificaion, andmolecular level changes. Environ Sci Pollut Res. 2014;21:12709-22. https://doi.org/10.1007/s11356-014-3210-3

[116] Lequeux H, Hermans C, Lutts S, Nathalie V. Response to copper excess inArabidopsis thaliana: impact on the root system architecture, hormonedistribution, lignin accumulation and mineral profile. Plant Physiol Biochem.2010;48:673-82. https://doi.org/10.1016/j.plaphy.2010.05.005

[117] Potbhare, A.K., Bhilkar, P.R., Yerpude, S.T., Madankar, R.S., Shingda, S.R., Rameshwar Adhikari, R., Chaudhary, R.G., Applications of Emerging Nanomaterials and Nanotechnology, Applications of Emerging Nanomaterials and Nanotechnology, MRF, 2023, 148, 304-333. https://doi.org/10.21741/9781644902554-11

[118] Chung IM, Rekha K, Venkidasamy B, Thiruvengadam M. Effect of copperoxide nanoparticles on the physiology, bioactive molecules, andtranscriptional changes in Brassica rapa ssp. rapa seedlings. Water Air SoilPollut. 2019;230:48. https://doi.org/10.1007/s11270-019-4084-2

[119] Chouke, P.B., Shrirame, T., Potbhare, A.K., Mondal, A., Chaudhary, A.R., Mondal, S., Thakare, S.R., Nepovimova, E., Valis, M., Kuca, K., Sharma, R., Bioinspired metal/metal oxide nanoparticles: A road map to potential applications, Mater. Today Adv. 2022, 16, 100314. https://doi.org/10.1016/j.mtadv.2022.100314

[120] Heinlaan M, Ivask A, Blinova I, Dubourguier HC, Kahru A. Toxicity ofnanosized and bulk ZnO, CuO and TiO2 to bacteria Vibrio fischeriandcrustaceans Daphnia magna and Thamnocephalusplatyurus. Chemosphere.2008;71:1308-16. https://doi.org/10.1016/j.chemosphere.2007.11.047

[121] Tavares KP, Caloto-Oliveira Á, Vicentini DS, Melegari SP, Matias WG, BarbosaS, Kummrow F. Acute toxicity of copper and chromium oxide nanoparticlesto Daphnia similis. Ecotoxicol Environ Contam. 2014; 9:43-50. https://doi.org/10.5132/eec.2014.01.006

[122] Adam N, Vakurov A, Knapen D, Blust R. The chronic toxicity of CuOnanoparticles and copper salt to Daphnia magna. J Hazard Mater. 2015; 283:416-22. https://doi.org/10.1016/j.jhazmat.2014.09.037

[123] Chaudhary, R. G., Sonkusare, V., Bhusari, G., Mondal, A., Potbhare, A., Juneja, H., Abdala, A., Sharma, R., Preparation of mesoporous ThO2 nanoparticles: influence of calcination on morphology and visible-Light-driven photocatalytic degradation of indigo carmine and methylene blue, Environ. Res., 2023, 222, 115363. https://doi.org/10.1016/j.envres.2023.115363

[124] Asghar MA, Zahir E, Shahid SM, Khan MN, Asghar MA, Iqbal J, Walker G. Iron,copper and silver nanoparticles: green synthesis using green and black tealeaves extracts and evaluation of antibacterial, antifungal and aflatoxin B1adsorption activity. LWT. 2018; 90:98-107. https://doi.org/10.1016/j.lwt.2017.12.009

[125] Roy K, Ghosh CK, Sarkar CK. Rapid detection of hazardous H2O2 by biogeniccopper nanoparticles synthesized using Eichhornia crassipes extract.Microsyst Technol. 2019; 25:1699-703. https://doi.org/10.1007/s00542-017-3480-z

[126] Zangeneh MM, Ghaneialvar H, Akbaribazm H, Ghanimatdan M, Abbasi N,Goorani S, Pirabbasi E, Zangeneh A. Novel synthesis of Falcaria vulgaris leafextract conjugated copper nanoparticles with potent cytotoxicity,antioxidant, antifungal, antibacterial, and cutaneous wound healingactivities under in vitro and in vivo condition. J PhotochemPhotobiolBBiol. 2019; 197:111556. https://doi.org/10.1016/j.jphotobiol.2019.111556

[127] Rajeshkumar S, Rinith G. Nanostructural characterization of antimicrobial andantioxidant copper nanoparticles synthesized using novel Perseaamericanaseeds. OpenNano. 2018; 3:18-27. https://doi.org/10.1016/j.onano.2018.03.001

[128] Kerour A, Boudjadar S, Bourzami R, Allouche B. Eco-friendly synthesis ofcuprous oxide (Cu2O) nanoparticles and improvement of their solarphotocatalytic activities. J Solid State Chem. 2018; 263:79-83. https://doi.org/10.1016/j.jssc.2018.04.010

[129] Mehr ES, Sorbiun M, Ramazani A, Fardood ST. Plant-mediated synthesis ofzinc oxide and copper oxide nanoparticles by using Ferulago angulate (Schlecht) Boiss extract and comparison of their photocatalytic degradationof Rhodamine B (RhB) under visible light irradiation. J Mater Sci MaterElectron. 2018; 29:1333-40. https://doi.org/10.1007/s10854-017-8039-3

[130] Potbhare, A., Bhilkar, P., Yerpude, S., Madankar, R. S., Shingda, S. Adhikari, R., Chaudhary, R., Nanomaterials as Photocatalyst, Appl. Emerg. Nanomater. Nanotechnol, 2023, 148, 304-333. https://doi.org/10.21741/9781644902554-11

[131] Jadhav MS, Kulkarni S, Raikar P, Barretto DA, Vootla SK, Raikar US. Greenbiosynthesis of CuO & Ag-CuO nanoparticles from Malus domestica leafextract and evaluation of antibacterial, antioxidant and DNA cleavageactivities. New J Chem. 2018; 42:204-13. https://doi.org/10.1039/C7NJ02977B

[132] Khatami M, Varma RS, Heydari M, Peydayesh M, Sedighi A, Askari HA, RohaniM, Baniasadi M, Arkia S, Seyedi F, Khatami S. Copper oxide nanoparticlesgreener synthesis using tea and its antifungal efficiency on Fusarium solani.Geomicrobiol J. 2019;36:777-81. https://doi.org/10.1080/01490451.2019.1621963

[133] Akhter SMH, Mohammad F, Ahmad S. Terminalia belerica mediated greensynthesis of nanoparticles of copper, iron and zinc metal oxides as thealternate antibacterial agents against some common pathogens.BioNanoSci. 2019; 9:365-72 https://doi.org/10.1007/s12668-019-0601-4

[134] Koul, B., Poonia, A. K., Yadav, D., & Jin, J. O. (2021). Microbe-mediated biosynthesis of nanoparticles: Applications and prospects. Biomolecules, 11(6), 886. https://doi.org/10.3390/biom11060886

Green Synthesis and Emerging Applications of Frontier Nanomaterials   Materials Research Forum LLC
Materials Research Foundations 169 (2024) 139-172   https://doi.org/10.21741/9781644903278-6

Chapter 6

Synthesis and applications of nickel-based nanomaterials

Chiranjibi Dhakal[1,2], Samjhana Dahal[1,2], Prakash Lamichhane[3], Ratiram Chaudhary[4], Rameshwar Adhikari[1,5,6*]

[1] Research Centre for Applied Science and Technology (RECAST), Tribhuvan University, Kirtipur 44618, Kathmandu, Nepal

[2] Central Department of Physics, Tribhuvan University, Kirtipur 44618, Kathmandu, Nepal

[3] Department of Chemical Science and Engineering, Kathmandu University, Dhulikhel, Nepal

[4] P. G. Department of Chemistry, Porwal College, Kamptee, Maharashtra 441001, India

[5] Central Department of Chemistry, Tribhuvan University, Kirtipur 44618, Kathmandu, Nepal

[5] Nepal Polymer Institute (NPI), P. O. Box 24411, Kathmandu, Nepal

* rameshwar.adhikari@cdc.tu.edu.np, nepalpolymer@yahoo.com

Abstract

Nanoparticles (NPs), due to their small size, exhibit unique and enhanced properties compared to bulk materials possessing potential applications in energy conversion technology, catalysis, environmental remediation, technological advancements, medicine, etc. In this chapter, we discuss the synthesis process of Nickel NPs (Ni-NPs) and nickel-based nanomaterials. The basic synthetic technique considering the bottom-up approach and top-down approach are discussed. 'Green' synthesis of Ni-NPs production is the central focus of the study because it promotes sustainability, enhances biocompatibility, and supports safer and more cost-effective methods. The extracts obtained from different parts of plants (such as *Ocimum sanctum, Medicago sativa, Azadirachta Psidium guajava,* etc.) are used for the biosynthesis of nickel-based NMs. It has been found that biosynthesized Ni-NPs constitute a broad spectrum of applications in antimicrobial, antileishmanial, anti-cancer, anti-diabetic activity, drug delivery, battery electrodes, wastewater management, and biosensors. Strategies are still to be developed to overcome the issues of disparity in particle size, stability, shapes, and nanoparticle yield.

Keywords

Nickel-based NMs, Green Synthesis, Plant Extract, Drug Delivery, Anti-Cancer

Contents

Green Synthesis and Emerging Applications of Frontier Nanomaterials   Materials Research Forum LLC
Materials Research Foundations 169 (2024) 139-172   https://doi.org/10.21741/9781644903278-6

1. Introduction and overview

The technologies emerging in different eras have led to the discovery of new materials in search of a wide range of electric, thermal, and magnetic properties. Our forefathers were in search of materials like iron, copper, gold, diamond, and silver for daily purposes such as home appliances, jewelry, armor, and agricultural tools. In the past, we were aware that armor and tools used on the battlefield were made up of iron. Similarly, gold was used for ornaments, copper for pottery, and silver especially for jewelry, coins, food vessels, and utensils. Among these elements, nickel was an element often used by our forefathers [1]. It is known that the metal from iron/nickel meteorites was used by our ancestors as a superior form of iron. It was employed in tools and swords believing it like the stainless steel of today. Natives of Peru regarded it as silver because nickel didn't have rusting properties. Besides these applications, nickel alloyed with copper in 1857 was used in coins in the USA, while Switzerland in 1881 used pure nickel for coin making [2]. With progress in time, chemists started searching different areas where nickel could be employed. The consistent curiosity and desire to discover findings for nickel use persisted for a longer time and also opened a window for its use in numerous areas such as electrical, electronic as well as home appliances, medicines, environmental protection, etc. [2]. On the other hand, nickel has become today an imperative resource for the development of cleaner technologies such as electric vehicle (EV) batteries, and is used in harsh environments such as in jet engines, offshore installations, and various power generation facilities [3].

As nickel rarely occurs in the pure metallic state, it can be obtained through a series of mixing, smelting, and refining processes. Generally, mineral salts and rocks embeds nickel [4]. Several types of nickel ores exist from which one can mine the element in bulk amounts. Nickel ore can be found in several forms, such as laterite, sulfide, nickel-cobalt laterite, and magmatic sulfide deposits [5]. The most common type of nickel ore deposits are called laterite deposits, and they are generally found in tropical and subtropical areas. They are formed when ultramafic rocks weather and leach. This accumulation of nickel-rich limonite and saprolite ores results due to weathering and leaching. They represent 73% of the continental world's nickel resources [6].

Due to strong leaching of magnesium and silica, Limonite-type laterites incorporate high amount of iron [7]. As a volcanic or intrusive rock cools and solidifies, nickel and other sulfide minerals separate from the magma to create sulfide deposits, which are associated with ultramafic or mafic rocks [8]. Nickel-Cobalt laterite deposits contain significant amounts of Cobalt and Nickel, and they are especially found in tropical and subtropical regions [8]. In this context, due to their extraordinary properties, the nanometric Ni- and Ni-based NPs (Ni-NPs), with diameters in the range of 10-40 nm, have occupied a special place in the advancement of novel technologies and processes [9]. The so-called green synthetic routes for the fabrication of the Ni-NPs have demonstrated particularly special promises towards the application of these materials [10].

Studying the properties of various forms of nickel and its potential applications is crucial for technological advancements. By 2040, it is expected that the demand for nickel would increase 40 times due to the growth of the EV industry and energy storage [11]. To meet such an increasing demand, it is paramount to think about the advanced synthesis and mixing processes [12].

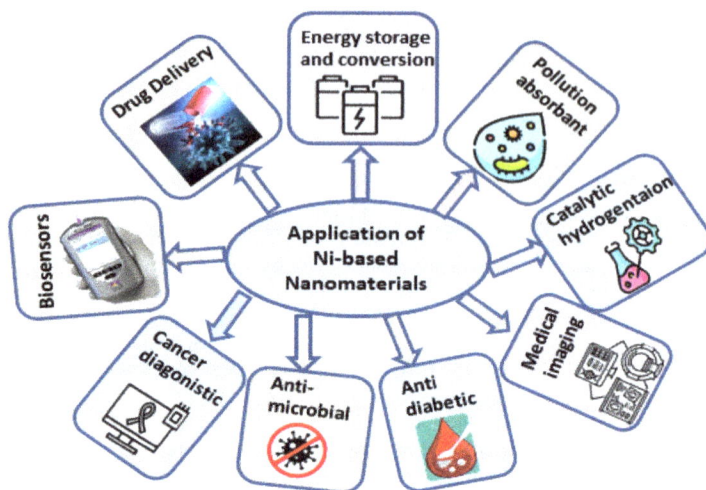

Figure 1. General applications of nickel-based nanomaterials in energy storage, pollution management, biomedical devices, and medicines.

As the size of nickel is reduced to nanosized, its chemical, electrical, thermal, and magnetic properties also differ from the bulk nickel. Ni-NPs are black silvery lustrous powders that are hard and ductile [13]. Due to their superior ferromagnetic properties such as magneto-crystalline anisotropy, high coercive forces, and chemical stability, they are the emerging ground for various potential applications. General applications of nickel-based NMs are shown in **Figure 1**. They are used in biomedical devices, catalysis, supercapacitors, dye-sensitized solar cells, propellant, and sintering additives, coating, nanofibers, textiles, anode of solid oxide fuel cells, and battery manufacturing. It is economical to use nickel as a catalyst because of its abundance in the earth's crust [14]. A large surface-to-volume ratio and active energy carriers provide these NPs surface-controlled properties. Furthermore, a great deal of research has been done on a variety of inorganic NPs because of their unique uses in biomedicine, photocatalysis, magnetic devices, and sensors. Due to their inexpensive and non-toxic nature, nickel oxide NPs (NiO-NPs) are among the many NPs that make suitable candidates for use in medical applications such as medication delivery, imaging, biological detection, and antibacterial [15]. Also, NiO-NPs play a crucial role in removing organic and organic pollutants and aiding environmental protection. The leaf extract has been a great idea for the green synthesis of Ni and Ni-based NMs. The green synthesis of Ni-based NMs from plant extract along with their potential applications are shown in **Table 1**.

The table shows various plants such as Ocimum Sanctum, Azadirachta, Psidium gaujava, Aegle marmelous correa, Furraria Officinalis, Aloe vera, Arabic gum, Limonia acidissima, Nigella sativa, Calotropis gigantea, Rhamnus Triquetra, Cres. Nudicaulis, Banana peels, Asparagus racemosus, Urtica, Mukia maderaspatana, and Calendula officinalis employed in green synthesis of Ni and Ni-based NMs such as NiO, NiS, $Ni(OH)_2$, Ni-TiO_2, and $NiFe_2O_4$. When making plant extracts, various plant parts are taken into consideration, including seeds of *Nigella sativa*, fruits of *Limonia acidissima*, gum of *Arabic gum*, and leaves of various plants mentioned in **Table 1**. It can be seen that Ni and Ni-based NMs prepared from these plant extracts show anticancer, antibacterial, and anti-microbial properties. Some other plants such as Phoenix dactylifera (Dates), Zingiber Officinale (roots), allium sativum (roots), Berberis balochistani are also used to prepare NiO-NPs.

Table 1. List of previous works on green synthesis of Ni and Ni-based NMs with their potential applications

S.N	Plant Name	Precursor	NMs type	Applications	Refs.
1	*Ocimum Sanctum*	$Ni(NO_3)_2$	Ni	Dye and pollutant-absorbent	[16]
2	*Azadirachta*	$NiCl_2.6H_2O$	Ni	Photocatalytic, anti-cancer activity	[17]
3	*Psidium gaujava*	$NiCl_2.6H_2O$	NiO	Decolorization of reactive dyes	[18]
4	*Aegle marmelous correa*	$NiCl_2$, $Ni(NO_3)_2$	Ni, NiO	Anti-inflammatory, anti-microbial	[19]
5	*Furraria Officinalis*	$NiSO_4.6H_2O$	Ni	Chemotherapeutic drugs	[20]
6	*Raphanus sativus*	$Ni(CH_3COO)_2.6H_2O$	NiO, Ni	Antioxidant, antimicrobial	[21]
7	*Aloe vera*	$Ni(NO_3)_2.6H_2O$	NiO	Anti-microbial activity	[22]
8	*Arabic gum(gum)*	$Ni(NO_3)_2.6H_2O$	NiO	Anti-cancer activity, photocatalytic	[23]
9	*Limonia acidissima (fruit)*	$Ni(NO_3)_2.6H_2O$	NiO	Anti-angiogenic, Photocatalytic activities	[24]
10	*Nigella sativa (seed)*	$Ni(NO_3)_2.6H_2O$	NiO	Electro-catalytic activity	[25]
11	*Calotropis gigantea*	$Ni(NO_3)_2.6H_2O$	NiO	Anti-microbial activity	[26]
12	Rhamnus Triquetra	$Ni(NO_3)_2$	NiO	Anti- Leishmanial, pharmaceutical industries	[27]
13	*Cres. nudicaulis*	$Ni(NO_3)_2$	NiO	Drug delivery activity	[28]
14	*Calendula officinalis*	$NiSO_4.6H_2O$	NiO	Anti-oxidant, anti-esophageal carcinoma	[29]
15	*Banana peels(Capping agent)*	$Ni(NO_3)_2 \cdot 6H_2$, Na_2S	NiS	Battery devices, solar cells, Li-ion batteries	[30, 31]
16	*Urtica Plant*	$NiCl_2 \cdot 6H_2O$	$NiFe_2O_4$	Cell culture and treatment of cells	[32]
17	*Mukia maderaspatana leaf*	Titanium tetraisopropoxide(TTIP), $NiCl_2$	Ni-TiO_2 Nanoflakes	Photocatalytic degradation of congo red dye	[33]
18	*Asparagus racemosus leaf*	$NiSO_4$	$Ni(OH)_2$	Anti-microbial, anti-oxidant	[34]

Green Synthesis and Emerging Applications of Frontier Nanomaterials   Materials Research Forum LLC
Materials Research Foundations 169 (2024) 139-172   https://doi.org/10.21741/9781644903278-6

In this chapter, we will first discuss on general method of synthesis of Ni and Ni-based NMs. We focus the discussion on physical and chemical methods such as thermal decomposition, sputtering, ball milling, chemical vapor deposition, sol-gel method, etc. Then the central focus of the paper incorporates the 'green' approach to synthesizing NMs. The structure and properties of the NMs using various plant extracts will be discussed. Then the potential application of biosynthesized Ni and Ni-based NMs will be studied.

2.   General Methods of Synthesis

The challenging work is the preparation of Ni-NPs of the desired size and property. Different methods and techniques are used for its preparation. Several references on the synthetic procedures employed in the synthesis of Ni-based NMs are available [35]. **Figure 2** shows the general procedures for creating Ni-based NMs using both top-down and bottom-up approaches. There are typically two methods used to prepare Ni-NPs [36]. These approaches are "bottom-up" and "top-down." Top-down synthesis techniques convert bulk resources into materials at the nanoscale. Mechanical milling, laser ablation, nanolithography, thermal decomposition, sputtering, and chemical etching are widely used in top-down nanoparticle synthesis (plasma processing, robust catalyst reactions) [37]. Chemical vapor deposition, sol-gel, and micro-emulsion are some methods employed in bottom-up approaches.

Apart from those synthetic methods, the biosynthesis of nanomaterials is central to our discussion. We discuss various thermal top-down and bottom-up approaches, and then 'green' synthesis techniques over these methods.

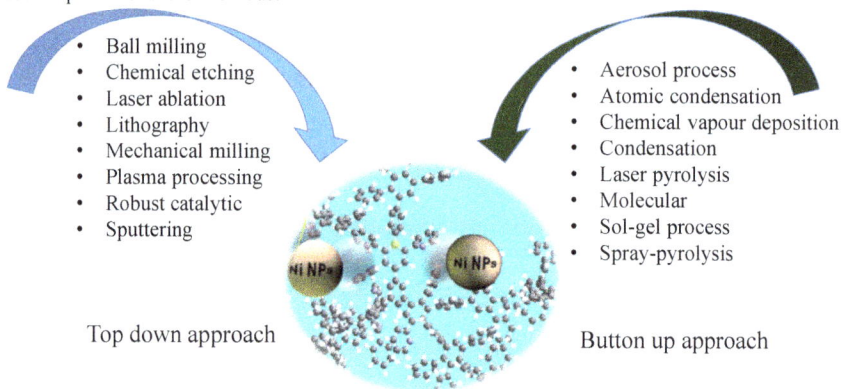

- Ball milling
- Chemical etching
- Laser ablation
- Lithography
- Mechanical milling
- Plasma processing
- Robust catalytic
- Sputtering

- Aerosol process
- Atomic condensation
- Chemical vapour deposition
- Condensation
- Laser pyrolysis
- Molecular
- Sol-gel process
- Spray-pyrolysis

Top down approach   Button up approach

Figure 2. Schematic representation of the synthesis of nickel-based nanomaterials using different physical and chemical methods

2.1   Top-down synthesis

2.1.1   Thermal decomposition

This technique breaks the compound's binding by using heat produced by endothermic chemical breakdown. There is a decomposition temperature for each element. When nickel (II) acetylacetonate is thermally decomposed in alkylamines, nickel NPs can be produced. The crystalline phase is mostly controlled by the reaction temperature, heating rate, and kind of solvent. Synthesis of 9 nm sized Ni-NPs using the thermal decomposition method involves the preparation of Ni-Oleylamine. Ni-oleylamine is prepared at 100 °C by reacting 1.55 g of oleylamine with 0.257 g of Nickel (II) acetylacetonate under an inert atmosphere [38].

The oleylamine complex and dioctyl ether (6 mL) solution is vacuum-degassed for one hour at 100 °C. Next, under an argon atm, tributyl phosphine (TBP) is injected into the nickel complex solution. After heating the mixture to 250 °C, the color of the mixture begins to change at 200 °C, suggesting the creation of Ni-NPs. **Figure 3** displays the Ni-NPs TEM image obtained at precursor concentrations of 167 and 200 milli mole, respectively. A wide and broad (III) peak at 44° and a minor (220) peak at 77° are visible in the crystal structure of the Ni-NPs that were characterized by XRD. These patterns indicate that the particles were made up of face-centered cubic (fcc) Ni, as seen in **Figure 4**.

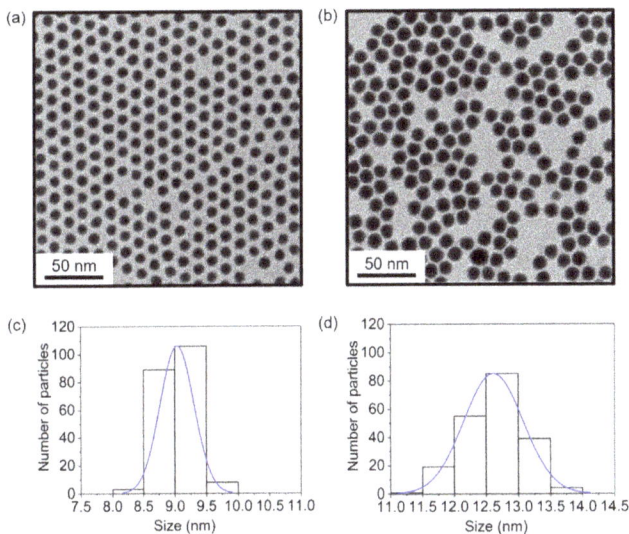

*Figure 3. Transmission Electron Microscopy TEM pictures of (a) 9 nm and (b) 13 nm Ni-NPs produced at 167 and 200 mM precursor concentrations, respectively, and **(c), (d)** the corresponding size distribution synthesized using the thermal decomposition method [38]. Reproduced from open-source article (https://creativecommons.org/licenses/by/4.0/).*

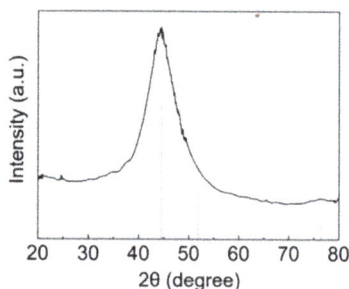

Figure 4. The XRD spectra of Ni-NPs produced at 200 mM precursor concentration, using TBP as the surfactant and dioctyl ether as the solvent, display broad, big (111) peaks at 44° [38]. Reproduced from open-source article (https://creativecommons.org/licenses/by/4.0/).

Though this method is employed for NMs synthesis, it possesses some disadvantages like limited control over the size and shape and polydispersity. Also, NMs synthesized using thermal decomposition tend to agglomerate due to high surface energy. Moreover, the choice of precursors for thermal decomposition is critical, and not all materials are suitable for this method [39].

2.1.2　Laser ablation

NPs are produced using several solvents in this synthesis technique. This technique involves irradiating a metal immersed in a liquid solution with a laser beam [40]. This irradiation of a metal condenses the plasma plume to produce the NPs. It allows NPs to be synthesized steadily in organic solvent with the help of a water-stabilizing or capping agent. **Figure 5** depicts the experimental arrangement for the synthesis of Ni-NPs utilizing the pulse laser ablation approach along with the associated UV-Vis absorption spectra.

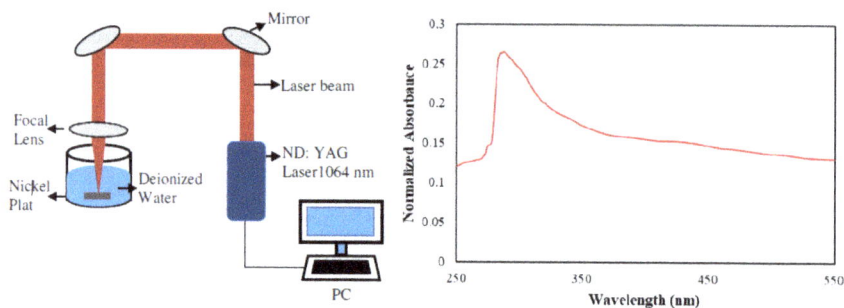

Figure 5. The experimental setup for Ni-NPs synthesis using the pulse laser ablation method is shown on the left, while the UV-Vis absorption spectra of the Ni-NPs colloid with plasma resonance centered at 290 nm is shown on the right [40]. Reproduced from open-source article (https://creativecommons.org/licenses/by/3.0/)

In ambient settings, pulsed laser ablation (PLA) in liquid is a straightforward and environmentally friendly technological technique that often works in water or other organic liquids. A laser beam is focused and directed onto a high-purity nickel plate (99.95%) in order to create Ni -NPs using the laser ablation process. It is important to wash nickel plates with alcohol 70% to kill bacteria and other contamination before user bombardment. In **Figure 5**, it can be seen that the Neodymium yttrium aluminum garnet (NdiYAG) laser beam is focused on a nickel plate immersed in 10 mL of deionized water in a Petri dish for 20 min, by a 15 Hz laser pulse spherical Ni-NPs with an average diameter of 25 nm are produced. The creation of Ni-NPs is confirmed by the plasma resonance observed at 290 nm in the UV-Vis absorption peak[40]. Although it offers several benefits, including the capacity to create precisely defined NMs and compositional control, there are some drawbacks as well. Expensive equipment, such as powerful lasers and specialized chambers, is needed for laser ablation installations.

2.1.3　Nanolithography

This method can be used to produce nanoparticles (NPs) with at least one dimension between one and one hundred nanometers. It's a technique that prints the required structure or form on a light-sensitive substrate while selectively extracting some of the materials to obtain the desired structure and shape [41]. This method enables the creation of single nanoparticles to clusters of nanoparticles with the required size and form. Due to the potential to manufacture homogeneous 1D, 2D, or 3D nanostructures and compatibility with water-scale processes, the nanosphere lithography technique is appealing [42]. The schematic representation of Ni-NPs synthesis using the Lithography process is shown in **Figure 6**. In the general lithography process, colloidal suspension is prepared at the beginning following dip coating. After dip coating, Ni deposition is carried. Then lift- off of the colloids is performed which leaves only the nano/micro-patterned material in between the particles [43].

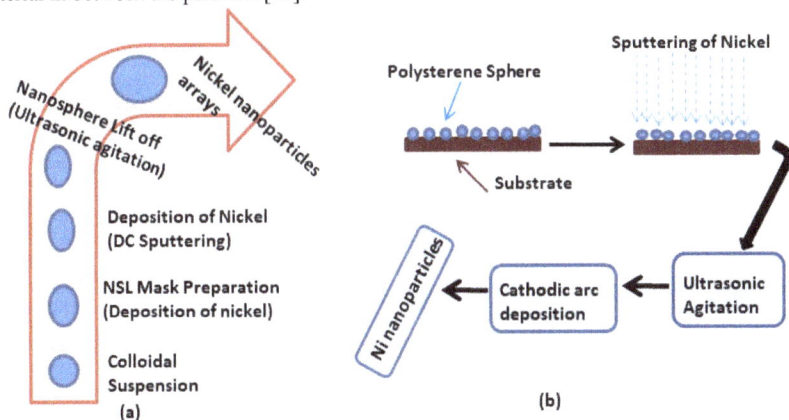

Figure 6. Diagram illustrating the synthesis process used to produce nickel nanoparticles via lithography [44].

A suspension of monodisperse spherical colloids (polystyrene) is applied to the flat substrate in **Figure 6b**. A colloidal crystal mask, also known as a hexagonal close-packed (HCP) monolayer or bilayer, forms. After these steps, the metallization is carried out using DC sputtering of nickel, and the polystyrene balls are removed by ultrasonic agitation. Ultimately, uniform distribution of Ni-NPs can be achieved. There are some drawbacks to using this technique. The rate at which nanomaterials or nanostructures are produced is somewhat slow. Compared to certain alternative synthesis processes, this constraint renders it less appropriate for large-scale production.

2.1.4 Sputtering

This method involves ejecting particles which collide with ions to deposit nanoparticles (NPs) on a surface. The process of sputtering involves annealing once a thin coating of NPs is deposited. Annealing is the process of heating a metal to a specific temperature, holding it there for a predetermined amount of time, and then allowing it to drop to room temperature. The size and form of NMs in the sputtering process are determined by the kind of substrate, layer thickness, annealing time, and operating temperature [45]. The practical parameters that determine a sputtering process are working pressure (p), target-to-substrate distance (d), discharge voltage (V), and discharge current (i). The energy of spewed atoms (E) is related to V and p. λ= p.d gives the relationship between d and p to the mean free path of the sputtered atom [45]. Low deposition rate, limited control over NPs size and distribution, and risk of contamination make this process less effective in NMs synthesis.

2.1.5 Mechanical ball milling

Of all the top-down approaches available, this one is most frequently used to generate different NPs. It is used to mill many elements in an inert atmosphere during synthesis and after the annealing of NPs [46]. Mechanical milling can produce an NMs composite of copper and nickel. Copper (Cu) (particle size < 63 μm), nickel (particle size < 150 μm), and other elemental metal powder blends can be ball milled at 1425 rpm in a SPEX 8000M mill employing 20:1 ball-to-powder weight ratio and hardened steel vials. The ball milling continues for up to 20 hr. Cu_xNi_{1-x} can thus be generated in series [47]. **Figure 7** displays the transmission electron micrograph for the mechanically produced Cu–Ni–NPs of $Cu_{27.5}Ni_{72.5}$. According to these micrographs, the average particle size of NPs is 10 nm, which is similar to the size distribution calculated from magnetic measurements, X-ray diffraction broadening, the Scherrer equation, and other sources. In **Figure 7 (right)**, it can be seen that the particles are partially agglomerated, and in some locations larger grains in the form of platelets 200 nm long and 5 nm thick.

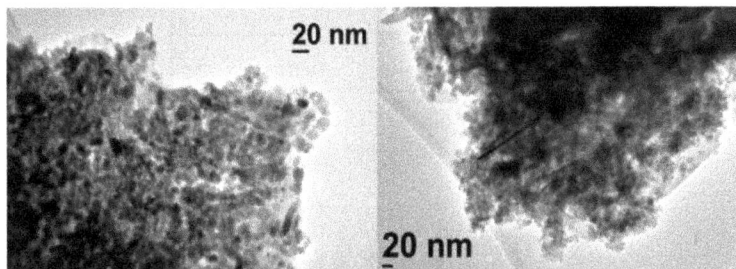

Figure 7. TEM micrographs of mechanically alloyed Cu $_{27.5}$ Ni $_{72.5}$ particles [47].

This approach has several benefits, such as simplicity and versatility, but it also has some drawbacks. For instance, impurities may enter the product due to wear and tear on the container or the milling media, lowering hence the purity of the finished product.

2.2 Bottom-up approach [48]

The bottom-up approach includes the use of metallic oxides and soluble metallic salts (such as chloride, sulphate, nitrate, and acetate) as precursors for the reaction. These metallic salts and oxides are reduced to metallic nanoparticles (NPs) with the aid of a suitable reducing agent and solvent. Precursor and reducing agent concentration, pH level, temperature, heating time, and stabilizing agent type can all be adjusted to produce the appropriate dispersal mode, shape, and dimensions of NPs. Solvent-gel, gas evaporation, and microemulsion are a few techniques included in the bottom-up methodology. Molecular condensation, laser/spray pyrolysis, and biological synthesis.

2.2.1 Chemical vapor deposition (CVD)

It is a vacuum deposition method. It is employed in the production of high-performing and high-quality solid materials. By using this method, a thin layer of gaseous reactants is placed onto a substrate [49]. At ambient temperature, gas molecules combine in the reaction chamber. There is a chemical reaction going on between the combined gas and the heated substrate. Finally, products are placed as thin films on the substrate surface. The NPs produced by this method are strong, rigid, homogeneous, and extremely pure. Because CVD can yield excellent, conformal coatings and films, it is a commonly used technology. Unintended byproducts, the use of dangerous gasses or high temperatures, and excessive energy consumption are some of the disadvantages of this approach.

2.2.2 Sol-gel method

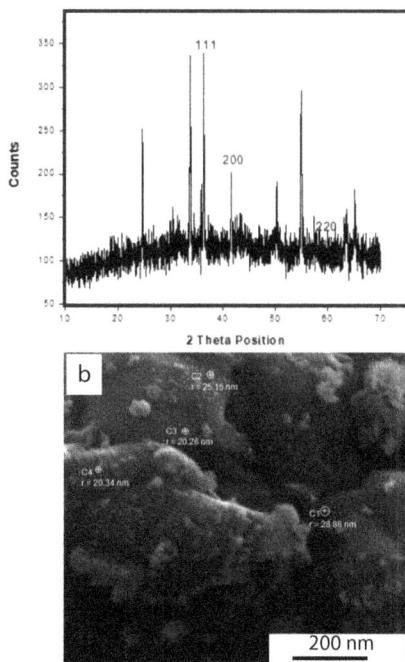

Figure 8. (a) XRD Spectrum of NiO-NPs and (b) SEM micrograph of NiO-NPs [51].

The sol-gel method produces NPs with homogeneous mixing, good crystallinity, and a uniform and sharp size distribution. An aqueous solution of nickel nitrate hexahydrate is made by dissolving 3–4 grams of salt in 1000 milliliters of deionized water, which is then used to synthesize nickel oxide nanoparticles. Subsequently, the 0.5M NaOH solution is added to the salt solution dropwise while stirring continuously in the burette. Nickel precipitates have a "green" tint, and to eliminate moisture, they are washed with deionized water and dried at 95 °C. Ultimately, precipitates are crushed and calcined at 55 °C to produce NiO-NPs.

The chemical process for NiO-NPs synthesis follows [51].

$$Ni^{2+} + 20\ H^{1-} + x\ H_2O \longrightarrow Ni(OH)_2.xH_2O_{(S)} \tag{1}$$

$$Ni(OH)_2.xH_2O_{(S)} \longrightarrow Ni(OH)_{2\ (S)} + xH_2O_{(g)} \tag{2}$$

$$Ni(OH)_{2\ (S)} \longrightarrow NiO_{\ (S)} + H_2O_{(g)} \tag{3}$$

Figure 8a displays the diffraction peak for NiO-NPs, and the average size of NiO NMs is 45 nm. The crystallographic planes 111, 200, and 220 that show the NiO-NP crystallization are represented by the peaks in Figure 8b. SEM study of NiO-NPs presented in Figure 6 (right) shows the successful formation of spherical black uniform-sized NPs because of the selection of suitable calcination temperature. Thus, utilizing an ideal calcination temperature of 550 °C, NiO-NPs with an average diameter of 40 nm, very close to its XRD size of 45 nm, were produced.

Despite its ease of usage, it has a lot of disadvantages when it comes to the synthesis of NMs. Longer processing periods could arise from the need for more time to form a stable sol and gel. The NPs can be hard to distribute uniformly across the gel matrix, which can lead to problems with inhomogeneity.

2.2.3 Microemulsion technique

The size of the NPs in microemulsion techniques can be adjusted based on the microemulsion parameters [52]. Ni (II) in a microemulsion droplet may be reduced using sodium borohydride or hydrazine during the Ni-NPs manufacturing process. 6.5 g (dioctylsodium sulfosuccinate 13.5 g of n-heptane) and 4.5 g ionic liquid contain 1 M Ni (II) and 3 M sodium borohydride [53]. Reduction to Ni is completed when a dark black color appears. Finally, centrifugation is performed to separate the product and then it is cleared in ethanol and dried at 60 °C. In Figure 9a, a microemulsion process for the Ni-NPs synthesis is depicted which shows the oil as a continuous phase, surfactant molecule, and Ni-NPs suspended in an ionic liquid pool. Generally, microemulsions are formed spontaneously by surfactants, additives, oil, water, or saliva water under an appropriate mixing ratio [54]. It constitutes water in oil type (w/O) in water type (O/w) and successive type. High-resolution TEM images for the morphology of Ni-NPs are displayed in **Figure 9b**.

It can be seen that the spherical NiNPs along with some agglomerated and elongated particles. TEM images of NiNPs exhibit fringes with 0.21 nm spacing corresponding to the plane (103). Microemulsion synthesis is valued for its ability to produce NPs with controlled properties and narrow size distributions; however, selecting appropriate surfactants that stabilize the microemulsion without obstructing the intended chemical reactions might be difficult. It has been found that numerous techniques employed in nanomaterials synthesis produce substantial waste and involve the use of toxic reagents. The synthesized nanomaterials with the traditional approach are less biocompatible, therefore, the alternative approach has become central for the researcher. 'green' synthesis can be an alternative technique to promote biocompatibility, homogeneity production, reduced agglomeration, precise shape and size, hazardous waste reduction, biodegradability [55], etc. Thus, plant extract-assisted synthesis of NPs, so-called 'green' synthesis, is discussed.

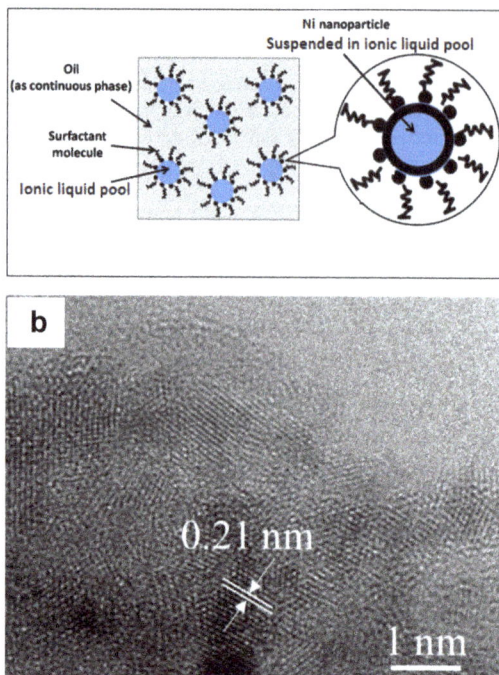

Figure 9. *High-resolution TEM images of Ni-NPs (right) and the microemulsion synthetic approach of NiNPs synthesis (left) [53]. Reproduced from open-source article (https://creativecommons.org/licenses/by/4.0/).*

2.3   Plants extracts mediated synthesis

This method produces NPs by combining the precursors with plant extracts, bacteria, and fungi. Leaf extracts of Ocimum sanctum, Azadirachta and psidium guajava, Medicago sativa, Annona squamosa, Aegle marmelos corroa, etc. are studied and prepared nickel nanoparticles owing to 'green' synthesis [16-34]. Preparing metal NPs using plant extracts involves a 'green' synthesis approach, which is considered environmentally friendly compared to conventional methods. Numerous phytochemicals, including polyphenols, flavonoids, and alkaloids, are found in plant extracts and can function as reducing, stabilizing, and capping agents during the synthesis process [56], **Figure 10**. Briefly, the leaf, root, or bark of certain plants possessing rich amounts of bioactive compounds with reducing properties, such as green tea, neem, aloe Vera, or others are chosen for the sample [57]. The fresh plant material is cleaned thoroughly using distilled water and then dried followed by crushed in a grinder and dispersed in water.

Figure 10. Scheme showing the processes involved in 'green' synthesis of Ni-based nanomaterials

2.3.1 Ocimum sanctum

Ocimum sanctum is an erect, many-branched subshrub in tropical climates, often grown as an annual elsewhere. In Nepal, it has been embraced into spiritual ritual and lifestyle practices, considered a cred plant by Hindus. Pandian and coworkers have produced Ni-NPs using aqueous leaf extract of Ocimum sanctum. After collecting fresh leaves, they are cleaned using distilled water and running tap water. After washing, the leaves are dried for three days at (25 ± 1) °C. As leaves are dried [58], they are mechanically ground into a powder and then sieved through a sieve. After the powder has been sieved, it is cleaned with a 2% HCL solution. Next, 1 g of leaf powder is dissolved in 50 mL of double-distilled water to create the leaf extract [58]. After two hours of shaking the contents, precipitates are removed by filtering. The filtrate can now be utilized to create Ni nanoparticles. Aqueous 1 mol/L $Ni(NO_3)_2$ is added with 10 mL of *Ocimum Sanctum* leaf extract, and the mixture is forcefully agitated. The reactor can last for 3 h at 60 °C. After that, the solution is freeze-dried for 24 hours to produce a powdered form of NPs.

2.3.2 Azadirachta and Psidium guajava

NiO and Ni-NPs are produced by boiling *Azadirachta* and *Psidium guajava* leaves [59]. During the preparation of NiO and Ni NMs, collected leaves are washed with distilled water, and leaves are crushed after leaf broth (5 gm leaves + 15 mL milli Q water). The resulting crushed leaves are centrifuged for 10 min at 4 to 5 °C. When combined with *A. indica* and *P. guajava* to provide NiO and Ni, the suggested mechanism is supported by the presence of Na^+, Cl^-, and BO^{-2} ions, as revealed by the analysis of the solution left over after the solid removal. The precursor, hydrated salt ($NiCl_2.6H_2O$), supplied the water.

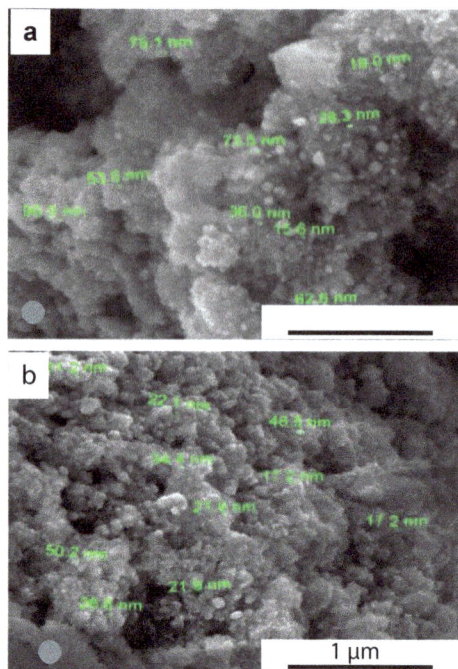

Figure 11. SEM examination of biosynthesized NPs made from the leaves of (a) guava (NiO) and (b) azadirachta indica plants (Ni) [59].

NiO and Ni-NPs are derived from the filtered outcomes. Next, the extract is combined with 7 mL of Nickel chloride solution (1% Ni chloride $NiCl_2$. $6H_2O$) and added to 25 mL of boiling mini Q water. After that, it takes on a brownish-red hue and undergoes centrifugation purification. The general reactions that are suggested for the mechanisms that result in the creation of NiO and Ni-NPs can be illustrated as follows [59]:

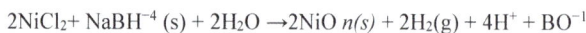

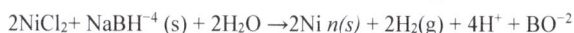

$$2NiCl_2 + NaBH^{-4} (s) + 2H_2O \rightarrow 2NiO \, n(s) + 2H_2(g) + 4H^+ + BO^{-1}$$

$$2NiCl_2 + NaBH^{-4} (s) + 2H_2O \rightarrow 2Ni \, n(s) + 2H_2(g) + 4H^+ + BO^{-2}$$

The morphology of synthesized NiO and Ni-NPs for Guavas and *Azadirachta indica* respectively is shown in **Figure 11**. The particles, which range in size from 17 to 70 nm, are spherical and aggregate because of the bioactive substances found in *P. Guajava* and *A. indica*. By applying the Tauc relation, the optical band gap value for NiO is found to be 2.69eV.

2.3.3 Medicago sativa (Alfalfa)

Ni-NPs were synthesized by Chen and colleagues using an extract from the alfalfa plant. Alfalfa leaf plant extract is made according to the method described in Ocimum sanctum leaf extract making. In order to prepare Ni-NPs, alfalfa extract (10 mL, 3–15 g/L) is mixed with aqueous Ni $(NO_3)_2$ (1 mm) while being vigorously stirred. The reaction is allowed to continue for 4 hours at 60 °C [60]. Consequently, after the solution is dried in vacuum freezing for 24 hours, Ni-NPs are produced. Furthermore, it has been found that varying the concentration of extract can alter the size of Ni-NPs. It turned out that when the content of alfalfa extract was raised, the diameters of Ni-NPs rose and the size distribution widened [60]. The size of the Ni-NPs production can be varied by changing the extraction concentration as shown in **Figure 12**. The histogram in **Figure 12 (d, e, f)** shows the strong dependence of size distribution on extract concentration. A greater concentration of *alfalfa* extract produced an abundant Ni(0) supply, which simultaneously promoted the growth of the Ni-NPs, given the same $Ni(NO_3)_2$ concentration. Large NPs formed as a result of this. Size-controlled Ni-NPs are thus the result of the bioreduction of Ni (II), which can be intensified further by increasing the extract concentration.

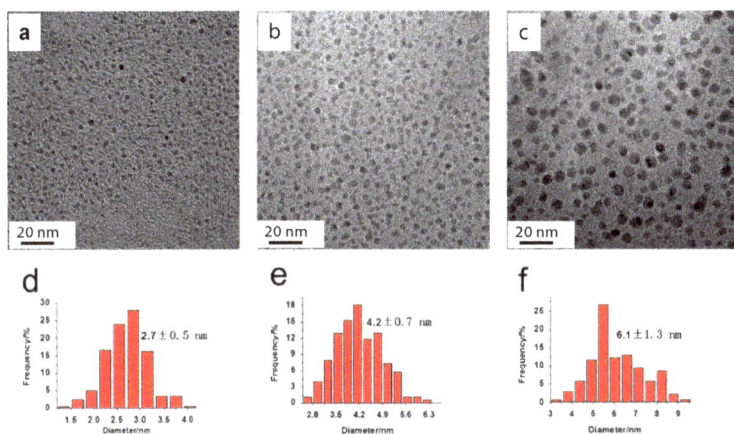

Figure 12. TEM images of the NiNPs produced at three distinct alfalfa extract concentrations: 3 g/L (a), 8 g/L (b), and 15 g/L (c); the corresponding Ni-NP size distributions were (d), €, and (f), respectively [60].

2.3.4 Annona squamosa

The *Annona squamosa* plant extract has been shown by Mamuru and colleagues to be useful for the synthesis of Ni and Cr NPs [61]. The phytochemicals in the leaf extract include an amino acid called arylamine, and this reduces Ni oxide to Ni-NPs. Leaf extract is produced using the previously described procedure, with the synthesis method being taken from Gopinath et al (2013). Plant extract (10 mL) is added dropwise to 50 mL of each solution (0.001 M nickel oxide and chromium (III) chloride) while stirring continuously for one hour or until the color changes. Following this stage, the products are dried at room temperature, rinsed with ethanol and distilled water, then centrifuged for ten minutes at 3000 rpm. Besides Ni-NPs, UV-visible and FTIR spectra are observed by the Mamuru groups as shown in **Figure 13**. The spectrum illustrates the typical surface plasmon resonance (SPR) for nickel and chromium NPs, with absorbance at roughly 280–

295 nm and peak maxima at 285 nm. Symmetrical plasma bands demonstrate the uniform and well-dispersed nature of the NPs.

According to the FTIR spectrum, chromium chloride (Cr) and nickel oxide (Ni) may have been reduced by an aryl amine group.

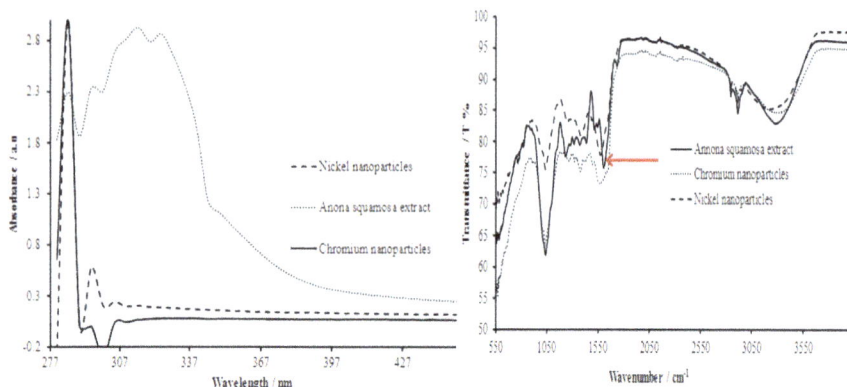

Figure 13. The extract of Annona squamosa, chromium, and nickel nanoparticles' (a) UV-visible absorption and (b) FTIR spectra [61].

2.3.5 Aegle marmelos correa

Angajala and colleagues have shown that *Aegle marmelos Correa* (AMC) extract can be used for the making of Ni-NPs [62]. It has been demonstrated that Ni-NPs have stronger anti-inflammatory properties. To prepare the plant extract, *A. Marmelos correa* is properly cleaned with distilled water and let to air dry for approximately four days. subsequently, 10g of the powdered medication is extracted using 100 mL of double-distilled water. After the filtrate has evaporated, ethyl acetate and methanol are added to it. At this point, 80 mL of a 1 mM $NiCl_2$ solution is combined with 10 mL of aqueous extract. After that, the temperature is maintained between 25 and 60 °C. After that, the solid made of the solution is filtered, and Ni-NPs are collected. The reaction is stopped and the precipitation of colloidal solution is permitted after heating at 60 °C for five hours. It has been noted that a layer of organic biomaterial derived from the leaf extract of *A. marmelos Correa* encompasses the nickel nanoparticles and has functional OH- groups on its surface that actively aid in the reduction of nickel to small particles [62]. **Figure 14** shows the size distribution and color change during the formation of Ni-NPs. The bar graph shows that almost 50 % of Ni-NPs of diameter 80 nm are found to be formed.

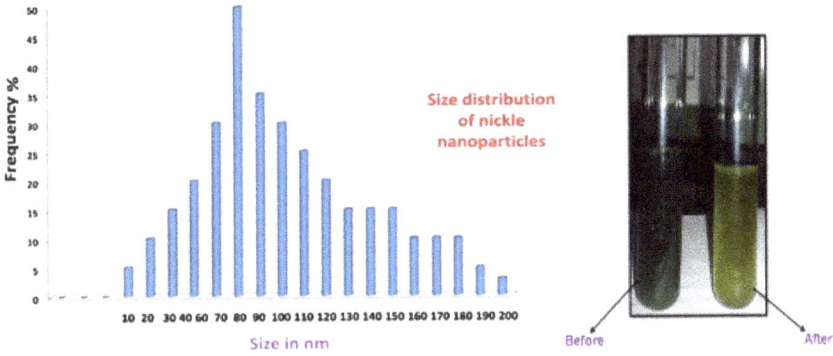

Figure 14. Size distribution (Left) and color change from dark green to light green during the formation of Ni-NPs with a maximum number of Ni-NPs having a diameter as size 80 nm in diameter [62].

2.3.6 Furraria officinalis

This herb is used to cure rheumatism, hypertension, and skin diseases in various countries. It has been found that the metal nanoparticles (NPs) created and combined with these plants have a higher potential to exhibit anti-cancer properties when applied to ovarian cancer cells [63]. The preparation of the Fumaria officinalis extract comes first. Next, 30 mL of 15 mM $NiSO_4.6H_2O$ is mixed with 10 mL of plant extract (2 g in 20 mL of water). For NiO and Ni-NPs synthesis, the filtered results are used. Now, add 25 mL of boiling mini Q water to 7 mL of nickel chloride solution (1% Ni chloride $NiCl_2.6H_2O$) that has been combined with the extract. After that, centrifugation is used to purify it and give it a brown-red tint. The following could be a sketch of the general reactions suggested for the processes producing NiO and Ni-NPs [59]:

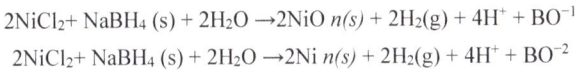

$$2NiCl_2 + NaBH_4 \text{ (s)} + 2H_2O \rightarrow 2NiO \, n(s) + 2H_2(g) + 4H^+ + BO^{-1}$$

$$2NiCl_2 + NaBH_4 \text{ (s)} + 2H_2O \rightarrow 2Ni \, n(s) + 2H_2(g) + 4H^+ + BO^{-2}$$

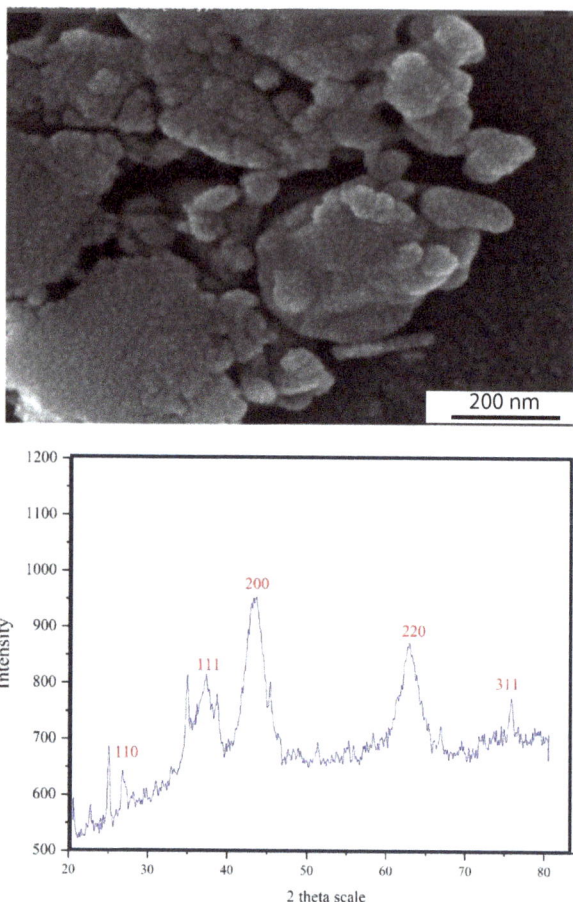

Figure 15. XRD patterns and FE-SEM pictures of Ni-NPs made using Fumaria officinalis extract [63].

2.3.7 Raphanus sativas

The first step is to make the plant extract from *Raphanus sativas*. To make nickel NPs, mix 10 mL of extract with 1 mmol/L of aqueous $Ni(NO_3)_2$. Ni^{+2} cations and negative nitrate anions are the products of nickel nitrate dissociation in an aqueous solution. Ni^{2+} from aqueous leaf extract is reduced by hydrated electrons to zero-valent nickel Ni (0) [64]. For UV-Vis, the absorption peak was measured at 395 nm. The FTIR spectra of the produced NPs show peaks that are characteristic of functional groups, such N-H, C-N, C-H, and O-H. The XRD pattern confirms the FCC structure of the nickel NPs. The generated NiNPs' almost spherical form and size range of 12 to 36 nm are seen using transmission electron microscopy (TEM). Additionally, the abstract of *Raphanus sativas* can be used to obtain NiO-NPs [65]. The 40 mM stock solution aqueous solution is made by dissolving 9.95 g of $Ni(CH_3COO)_2.4H_2O$ in 1000 mL. NiO-NPs were produced by mixing 80 milliliters of the stock solution with 20 milliliters of concentrated root extract and agitating the mixture for 10 minutes. After heating and stirring the reaction mixture for thirty minutes, 0.1 M NaOH solution was added to raise the pH of the mixture to 10. The reaction mixture was allowed to cool for 12 hours at room temperature before being washed three times with deionized water and then ethanol. The solid product was oven-dried at 100 °C, then calcined at 300 °C, 600 °C, and 900 °C. It was then crushed in a muffle furnace to a fine powder. The XRD examination verifies the synthesis of highly crystalline NiO-NPs, and when the calcination temperature rises, so does the crystallinity along with the size of the crystallites [65].

3.   Applications of Nickel-based NPs

Antibacterial properties of nickel and $Ni(OH)_2$ NPs have been shown against multidrug-resistant E. Coli and K. Preumeria [66]. These nanoparticles have their uses in food packaging, pharmaceuticals, textiles, and water disinfection. Since surface modification techniques have been developed, Ni-based nanomaterials have demonstrated antibacterial capabilities [67]. Ni-NPs' antibacterial qualities can be enhanced by adding polymers to their surface. For instance, when compared to pure Ni-NPs, chitin nanogels loaded with Ni-NPs (120–150 nm) exhibit greater potency against bacterial strains at lower doses. In vitro compatibility tests using Ni nanoparticle-loaded chitin nanogels on A549 and 1929 cells demonstrate their nontoxicity [68].

3.1   Antimicrobial activity

Treatments based on nanotechnology have been widely used to identify and treat illnesses as well as create new medications. Among other NP types, NiO-NPs have been examined against a range of human pathogenic bacteria. The strong bactericidal potential is exhibited by NiO-NPs produced from *moringa oleifera* leaves against gram-negative and multidrug-resistant gram-positive bacterial extract [69]. Furthermore, NiO-NPs prepared from leaf extract of *Aegle marmelos* have comparable bactericidal effects on the same bacterial strains [70]. It is widely known that "green" produced nickel oxide nanoparticles have bactericidal and antibiofilm properties against *S. aureus*, *P. aeruginosa*, and E. coli. NiO-NPs made from *Calotropis gigantea* have demonstrated bactericidal action against *P. neurinoma* when combined with the common medication chloramphenicol. Five pathogenic fungal strains are tested in the manufacture of NiO-NPs using *sagoretia* leaf extract [71].

3.2   Antileishmanial activity

Commonly used antileishmanial drugs have been shown to have several issues, such as toxicity, side effects, and diminished efficacy as a result of drug resistance [72]. On the other hand, scientists are looking for a way to treat it. Using the MTT cytotoxic assay, the antileishmanial activity of bio-inspired NiO-NPs was evaluated against leishmania tropica amastigote and promastigote cultures. It is demonstrated that both cultures are effectively repressed in a dose-dependent manner by the IC_{50} values against the promastigotes and amastigotes cultures, which are found to be 24.13 µg/mL and 26 µg/mL, respectively [71]. NiO-NPs produced from *Rhamnus virgate* have also demonstrated significant antileishmanial activity against promastigotes and amastigotes, with IC_{50} values of 27.58 µg/mL and 10.6 µg/mL, respectively [73]. Because of their concentration-dependent and lower IC_{50} values, NPs may be used in future medicine as a powerful drug delivery system to treat leishmaniasis.

3.3   Anti-diabetic activity

NiO-NPs made using Averrhoa bilimbi have been shown to exhibit α-α-amylase inhibitory effectiveness and anti-diabetic action, with an IC_{50} of 311.26 µg/mL [74]. Furthermore, it has been discovered that, in comparison to metformin, the as-synthesized NiO-NPs exhibit strong anti-diabetic efficacy.

3.4   Anticancer activity

The application of NiO-NPs has demonstrated anticancer potential, particularly against cancers of the colon, breast, liver, and lung. NiO-NPs made with *Geranium Wallichianum* are being investigated utilizing the 3-(4,5 dimethylthiazol-2-yl)-2, 5-diphenyltetrazolium bromide (MTT) assay against liver cancer cells [75]. Prior to treatment, the cells are seeded on 96 plates and kept at 37 °C with a constant supply of 5% CO_2 in their respective media. Concentration-dependent suppression of the cell lines has been seen when human hepatocarcinoma (HepG2) cells are exposed to different doses of NiO-NPs (500-3.9 µg/mL) during a 24-hour period [76]. Additionally, nickel oxide nanoparticles (NPs) made with leaf extract from Andrographis panicalata have demonstrated dose-dependent suppression of human breast cancer (MCF-7) cell lines.

3.5   Drug loading and delivery

Metal oxide nanoparticles (NPs) have received significant attention due to their exceptional potential for use in medication delivery [77]. The success of the loading process is revealed by the ability of NPs to load the specific medicament molecules. Reports on drug loading under various conditions are provided, including pH, NiO-NP concentration, and reaction time. Nickel oxide NPs derived from Cressa nudicaulis plant extract were used to deliver anti-cancer and anti-depressive doxepin medicines, according to research done by Yan Lu et al. To investigate loading capacity, his team treated the NPs with doxepin. When pH is raised from 4 to 6, drug loading is observed to be inclined, with the maximum drug loading (about 68%) occurring at pH 6. Drug loading decreases at acidic circumstances (pH=4) because there are less H-bonds between NiO and drug molecules. The interaction of drug molecules with NPs improves with time (maximum drug loading is approximately 68% in 12 hours); however, drug loading may decrease over an extended period because of potential NiO aggregation [78]. Furthermore, when the quantity of NiO is increased from 0.002 to 0.04 g, the drug loading increases from 13 to 74%. Conversely, when a larger quantity of NPs (> 0.04 g) is present, the loading of drug molecules falls. About 73% of loaded pharmaceuticals can be released after 80 hours, according to NiO-NPs' drug release characteristics at pH 6.5 and 37 °C [78].

3.6   Cytotoxicity activities

The property of being harmful to cells is known as cytotoxicity. A growing body of research is being done on the use of NPs in cancer detection and treatment. By half of the highest inhibitory concentration (IC_{50}), or as low as 16 µg/mL, NiO-NPs have been shown to contribute to a 50% reduction in cancer cells (Sabouri [79] et al.). They used gum Arabic polymer to examine the cytotoxic effects of NiO-NPs on cancer U87MG cell lines and normal CN cell lines. There is also evidence of a dose-dependent cytotoxicity. NiO-NPs produced using "green" synthesis processes have a more fatal effect on cancer cells than on normal cells, which supports the crucial role that NiO-NPs play in the treatment of cancer. Using the Cydonia oblonga extract, Ghazal et al. [80] carried out an environmentally friendly and cost-effective biosynthesis of NiO-NPs, measuring cell viability and bioavailability with the MTT assay. The results of in vitro cell viability experiments on L929 cells following treatment with NiO-NPs at doses ranging from 0 to 400 mg/mL are shown in **Figure 16**. The study's findings demonstrate that at dosages up to 400 mg/mL, no appreciable cytotoxicity was seen. Put more simply, it indicates that there are no deleterious consequences that the studied cells (L929 cells) can't resist or tolerate while the produced nickel oxide NPs are present. The results indicate that the NPs are not a major threat to the viability of the L929 cells, suggesting that they are biocompatible.

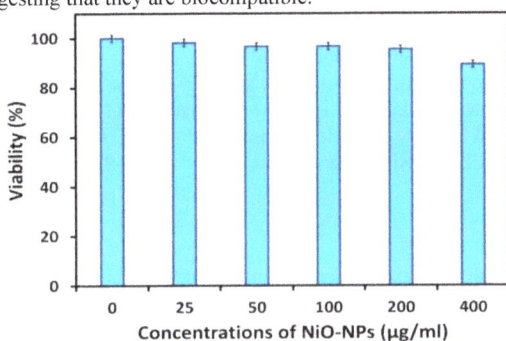

Figure 16. Studies on the cytotoxicity of NiO-NPs at different concentrations (up to 400 µg/mL) using the MTT assay [80].

3.7   Wastewater treatment and pollution absorbent

The chemical and magnetic characteristics of NiO-NPs make them an excellent absorbent. Aqueous solutions containing dyes such as crystal violet (CV), eosin Y (EY), orange II (OR), and anionic pollutants like nitrate (NO_3^-) and sulfate (SO_4^{2-}) can be effectively removed using Ni-NPs, which are synthesized from *Ocimum sanctum*, as an adsorbent. For CV, EY, OR, NO_3^-, and SO_4^{2-}, the maximum percentage removal of pollutants and dyes was measured at 40, 20, 30, 10, and 10 mg•L^{-1} initial concentration, respectively [81].

The co-precipitation approach was used by Uzaira Rafique et al. (2012) to create nano NiO, and the resulting particles were examined to determine whether they absorbed sulfate and nitrate [82]. According to reports, the absorbent can effectively remove both anions from water at a pH of 7. Similar to this, Fereshten Motahari et al. (2015) used nickel nitrate, NaOH, and H_2 acacen as precursors for hydrothermally synthesizing β-Ni(OH)$_2$, then calcined to yield nano NiO. In the end, it is discovered that Rhodamine B may be effectively removed from contaminated waterways using nano NiO. The nano Ni was created by Amira M et al. (2015) by mixing an ethanolic solution of oxalic acid with an ethanolic solution of nickel nitrate. The nano Ni was found to be useful in eliminating Pb^{+2}—a possible pollutant—from contaminated waters [82]. Additionally, commercially available NiO was employed by Roya Nateghi et al. (2012) as an absorbent for the successful removal of mona azo orange II dye from wastewater.

3.8   Biosensors

It is crucial to keep an eye out for chemical and antibiotic residues in food that comes from both plant and animal sources. These residues can include amoxicillin, penicillin, tetracycline, and vancomycin and can shield people from antibiotic-induced allergic reactions [83]. Therefore, electrochemical sensors composed of Ni-NPs can detect the presence of these compounds. For the detection of biomolecules in liquid samples, electrochemically based sensory techniques have shown to be viable and affordable options [84]. Electrochemical sensors are capable of yielding rapid results and can be sized down to a portable size to monitor even extremely low concentrations of penicillin G. Lead, arsenic, cadmium, and mercury are not good choices for electrode materials due to their inherent toxicity. It's interesting to note that Ni-NPs retain their oxidoreductive capabilities and are inexpensive (as opposed to Pt, Pd, or Ag); they are also unstable (having characteristics similar to those of noble metals); and they are not as toxic as Hg^{+2}, As^{+3}, Pb^{+2}, or Cd^{+2}. It is discovered that the minimum detection and quantitation limits for the Ni-NP composite for the electrochemical sensory detection of penicillin G are 0.00031 µM and 0.0010 µM, respectively. As a result, the Ni-NPs composite offers good adaptability, reproducibility, and repeatability.

3.9   Battery

Ni-based sulfides, which have excellent electrical conductivity and large capacity, have become more and more important in electrochemical energy storage applications [86]. Ball milling was used by Han et al. to make NiS powders, and it was found that the Ni powders had excellent cycling properties and a specific discharge capacity of 580 mAhg^{-1} at 1.4 V versus Li/Li. In the application of lithium-ion batteries (LIBs), nickel sulfides will perform incredibly well when paired with other materials. After 500 cycles, the NiS-C composite powder made by Son et al. showed outstanding discharge performance values of 472 and 273 mAhg^{-1}, respectively, at a steady, fantastic current density of 1000 mAg^{-1}.

Additionally, alkaline, sodium, and aluminum-ion batteries all use nickel-based sulfide negative electrodes. Most nickel-based sulfide NMs electrodes have low conductivity, but by combining them with highly conductive materials, such as carbon nanotubes or reduced graphene oxides, the conductivity can be increased. After conducting research, Matsumura et al. [87] discovered that Li_2S-GeS_2-P_2S_5 served as a solid electrolyte and that Ni_3S_2 materials might function as a cathode in all-solid-state LIBs using the LISICON. The cycle capacity of the NiS or Ni_9S_8 NPs-based all-solid-state cells at 25°C and a constant current density of 0.13 mAcm^{-2} is shown in **Figure 17**.

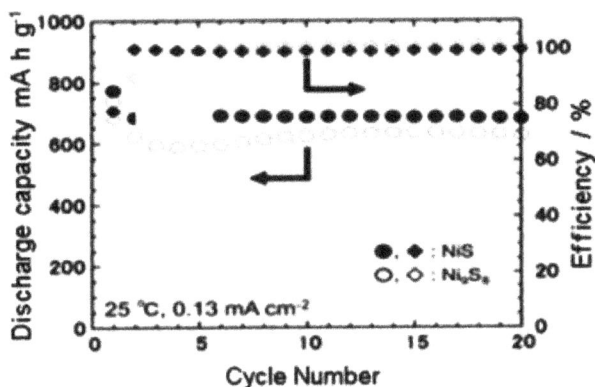

Figure 17. Cycle performance of the NiS or Ni₂S₈ all-solid-state cells [87].

Over the course of two to twenty cycles, the NiS NPs cell showed a discharge capacity of 680 mAhg⁻¹ and a charge-discharge efficiency of about 100%. The results show that in the all-solid form, NiS NPs have a greater capacity to operate as an active material.

3.10 Catalytic hydrogenation

Catalytic hydrogenation and dehydrogenation processes frequently involve nickel [88]. Du et al. found that when P-nitrophenol was hydrogenated to P-aminophenol, nanosized nickel had more activity, selectivity, and stability than traditional Raneyenickel (RNi) catalyst [89]. According to Zuo et al., nickel nanocatalysts have superior activity, selectivity, and thermal stability when it comes to hydrogenating nitrobenzene when compared to RNi. An essential procedure for adding amine functionality to intermediates in the polyurethane, pharmaceutical, and agrochemical industries is the catalytic hydrogenation of aromatic nitro compounds [90]. Hydrogenation is commonly used to reduce or saturate organic compounds. Nickel (0) NPs are found to catalyze the transfer hydrogenation of olefins and carbonyl compounds as well as the reductive amination of aldehydes, with 2-propanol as the hydrogen donor [91]. Additionally, Ni-NPs have the ability to activate primary alcohols for the reductive aza Wittig reaction and α-alkylation of ketones.

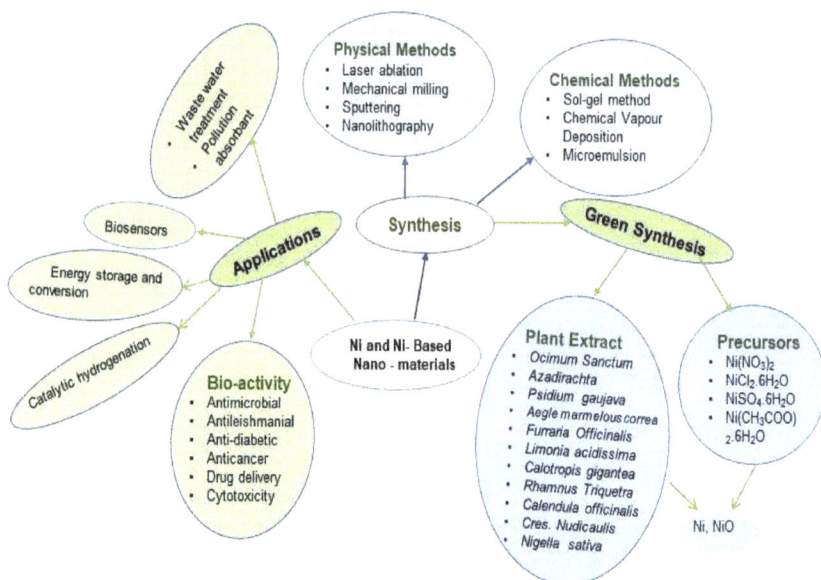

Figure 18. Summary of Ni and Ni-based nanomaterials' synthesis with applications

Ocimum sanctum, Azadirachta, Psidium gaujava, Aegle marmelous correa, Furraria officinalis, Aloe vera, Arabic gum, Limonia acidissima, Nigella sativa, Calotropis gigantea, Rhamnus Triquetra, and Cres. Nudicaulis are some plants found to be used in Ni-based NMs synthesis. It has been found that Ni-based NMs have potential in anticancer, antimicrobial, drug delivery, water pollution management, catalytic hydrogenation, and energy storage and conversion applications. It is found that Ni powders have exceptional cycling characteristics and a 580 mAhg^{-1} specific discharge capacity at 1.4 V vs Li/Li. It comes to light that the minimum detection and quantitation limits for the Ni-NPs composite for the electrochemical sensory detection of penicillin G are 0.00031 µM and 0.0010 µM, respectively. NiO-NPs' drug release characteristics at pH 6.5 and 37 °C have demonstrated the ability to release around 73% of loaded pharmaceuticals after 80 hours. These different biological activities of 'green' synthesized Ni-based NMs show promising applications in biomedicine, energy technology, and technological advancements.

Conclusions and prospects

In summary, the 'green' approach of NPs and NMs synthesis is an appealing research field for the synthesis of novel nanoparticulate materials (see **Figure 18**). Biosynthesis assures a non-toxic, sustainable eco-friendly, clean, and safe approach for regenerative medicine, tissue engineering, removal of pollutants from air and soil, enhancement of fertilizers, and promotion of sustainable practice. Different physical approaches for Ni and Ni-based NMs can be employed such as laser ablation, ball milling, sputtering, and lithography. Similarly, numerous chemical approaches such as sol-gel, thermal decomposition, chemical vapor deposition, and microemulsion reduction for nanomaterials preparation have been used. These physical and chemical methods are considered to be more environmentally unfriendly and less compatible in several applications. Thus, 'green' synthesis is replacing these methods of preparations and is observed to be feasible for NPs research in terms of their promised biocompatibility, homogeneity in their production, and hazardous waste reduction. Despite several advantages and prospects, the yield of the NPs synthesized by 'green' methods has been reported much lower than that synthesized by chemical and physical methods. There is hence a need to optimize the greener approaches, particularly looking at the yield and homogeneity in NPs dispersion, as well as control of their sizes, and shapes.

Acknowledgments

RA acknowledges the Alexander von Humboldt (AvH) Foundation for supporting his research stay in Germany from Aug. 2023 - Oct. 2023.

References

[1] Forbes, R. J. (1965). Studies in ancient technology. 5 (Vol. 2). Brill Archive.

[2] Carter, B. (2021). Boom, Bust, Boom: a Story about Copper, the Metal that Runs the World. Simon and Schuster.

[3] Li, Z., Khajepour, A., & Song, J. (2019). A comprehensive review of the key technologies for pure electric vehicles. Energy, 182, 824-839. https://doi.org/10.1016/j.energy.2019.06.077

[4] Howard-White, F. B. (2024). Nickel: a historical review. Taylor & Francis. https://doi.org/10.4324/9781032638836

[5] Bide, T., Hetherington, L., & Gunn, G. (2008). Nickel.

[6] Golightly, J. P. (2010). Progress in understanding the evolution of nickel laterites. https://doi.org/10.5382/SP.15.2.07

[7] Teitler, Y., Cathelineau, M., Ulrich, M., Ambrosi, J. P., Munoz, M., & Sevin, B. (2019). Petrology and geochemistry of scandium in New Caledonian Ni-Co laterites. Journal of Geochemical Exploration, 196, 131-155. https://doi.org/10.1016/j.gexplo.2018.10.009

[8] Naldrett, A. J. (2004). Magmatic sulfide deposits: geology, geochemistry and exploration. Springer Science & Business Media. https://doi.org/10.1007/978-3-662-08444-1

[9] Tanaka, T., Montanari, G. C., & Mulhaupt, R. (2004). Polymer nanocomposites as dielectrics and electrical insulation: perspectives for processing technologies, material characterization, and future applications. IEEE Transactions on Dielectrics and Electrical Insulation, 11(5), 763-784. https://doi.org/10.1109/TDEI.2004.1349782

[10] Yáñez-Sedeño, P., Campuzano, S., & Pingarrón, J. M. (2017). Electrochemical sensors based on magnetic molecularly imprinted polymers: A review. Analytica Chimica Acta, 960, 1-17. https://doi.org/10.1016/j.aca.2017.01.003

[11] Tang, C., Sprecher, B., Tukker, A., & Mogollón, J. M. (2021). The impact of climate policy implementation on lithium, cobalt and nickel demand: The case of the Dutch automotive sector up to 2040. Resources Policy, 74, 102351. https://doi.org/10.1016/j.resourpol.2021.102351

[12] Anand, U., Carpena, M., Kowalska-Góralska, M., Garcia-Perez, P., Sunita, K., Bontempi, E. & Simal-Gandara, J. (2022). Safer plant-based NPs for combating antibiotic resistance in bacteria: A comprehensive review on its potential applications, recent advances, and future perspective. Science of The Total Environment, 821, 153472. https://doi.org/10.1016/j.scitotenv.2022.153472

[13] Barhoum, A., García-Betancourt, M. L., Jeevanandam, J., Hussien, E. A., Mekkawy, S. A., Mostafa, M., ... & Bechelany, M. (2022). Review on natural, incidental, bioinspired, and engineered nanomaterials: history, definitions, classifications, synthesis, properties, market, toxicities, risks, and regulations. Nanomaterials, 12(2), 177. https://doi.org/10.3390/nano12020177

[14] Kaur, M., & Pal, K. (2019). Review on hydrogen storage materials and methods from an electrochemical viewpoint. Journal of Energy Storage, 23, 234-249. https://doi.org/10.1016/j.est.2019.03.020

[15] Chouke, PB., Potbhare, A.K., Meshram, N.P., Rai, M.M., Dadure, K.M., Chaudhary, K., Rai A.R., Desimone, M.F. Chaudhary, R.G. Masram, D.T. (2022) Bioinspired NiO nanospheres: Exploring in vitro toxicity using Bm-17 and L. rohita liver cells, DNA degradation, docking, and proposed vacuolization mechanism. ACS Omega, 7(8) 6869-6884. https://doi.org/10.1021/acsomega.1c06544

[16] Pandian, C. J., Palanivel, R., & Dhananasekaran, S. (2015). Green synthesis of nickel NPs using Ocimum sanctum and their application in dye and pollutant adsorption. Chinese Journal of Chemical Engineering, 23(8), 1307-1315. https://doi.org/10.1016/j.cjche.2015.05.012

[17] Patil, S. P., Chaudhari, R. Y., & Nemade, M. S. (2022). Azadirachta indica leaves mediated green synthesis of metal oxide NPs: A review. Talanta Open, 5, 100083. https://doi.org/10.1016/j.talo.2022.100083

[18] Rafique, M. A., Kiran, S., Jamal, A., Abrar, S., Jalal, F., & Rahman, N. (2023). Nickel NPs synthesized from Psidium guajava peels mediated degradation of Orange E3 dye reactive dye: a sustainable approach. International Journal of Environmental Science and Technology, 20(3), 2733-2744. https://doi.org/10.1007/s13762-022-04509-w

[19] Kudlur, D. S., Meghashree, A. M., Vinutha, S. A., Kumar, K. S., Karthik, G., Venkatesh, P. A., ... & Mallikarjunaswamy, C. (2023). One pot synthesis of CuO-NiO-NPs using Aegle marmelos fruit extract and their antimicrobial activity. Materials Today: Proceedings. https://doi.org/10.1016/j.matpr.2023.03.256

[20] Huang, Y., Zhu, C., Xie, R., & Ni, M. (2021). Green synthesis of nickel NPs using Fumaria officinalis as a novel chemotherapeutic drug for the treatment of ovarian cancer. Journal of Experimental Nanoscience, 16(1), 368-381. https://doi.org/10.1080/17458080.2021.1975037

[21] Haq, S., Dildar, S., Ali, M. B., Mezni, A., Hedfi, A., Shahzad, M. I., ... & Shah, A. (2021). Antimicrobial and antioxidant properties of biosynthesized of NiO-NPs using Raphanus

sativus (R. sativus) extract. Materials Research Express, 8(5), 055006.
https://doi.org/10.1088/2053-1591/abfc7c

[22] Ahmad, B., Khan, M. I., Naeem, M. A., Alhodaib, A., Fatima, M., Amami, M., ... & Iqbal, M. (2022). Green synthesis of NiO-NPs using Aloe vera gel extract and evaluation of antimicrobial activity. Materials Chemistry and Physics, 288, 126363. https://doi.org/10.1016/j.matchemphys.2022.126363

[23] Sabouri, Z., Akbari, A., Hosseini, H. A., Khatami, M., & Darroudi, M. (2021). Green-based bio-synthesis of nickel oxide NPs in Arabic gum and examination of their cytotoxicity, photocatalytic and antibacterial effects. Green Chemistry Letters and Reviews, 14(2), 404-414. https://doi.org/10.1080/17518253.2021.1923824

[24] Kumar, M. S., Soundarya, T. L., Nagaraju, G., Raghu, G. K., Rekha, N. D., Alharthi, F. A., & Nirmala, B. (2021). Multifunctional applications of Nickel oxide (NiO) NPs synthesized by facile green combustion method using Limonia acidissima natural fruit juice. Inorganica Chimica Acta, 515, 120059. https://doi.org/10.1016/j.ica.2020.120059

[25] Boudiaf, M., Messai, Y., Bentouhami, E., Schmutz, M., Blanck, C., Ruhlmann, L., ... & Mekki, D. E. (2021). Green synthesis of NiO-NPs using Nigella sativa extract and their enhanced electro-catalytic activity for the 4-nitrophenol degradation. Journal of Physics and Chemistry of Solids, 153, 110020. https://doi.org/10.1016/j.jpcs.2021.110020

[26] Din, M. I., Nabi, A. G., Rani, A., Aihetasham, A., & Mukhtar, M. (2018). Single step green synthesis of stable nickel and nickel oxide NPs from Calotropis gigantea: Catalytic and antimicrobial potentials. Environmental Nanotechnology, Monitoring & Management, 9, 29-36. https://doi.org/10.1016/j.enmm.2017.11.005

[27] Iqbal, J., Abbasi, B. A., Ahmad, R., Mahmoodi, M., Munir, A., Zahra, S. A., ... & Capasso, R. (2020). Phytogenic synthesis of nickel oxide NPs (NiO) using fresh leaves extract of Rhamnus triquetra (wall.) and investigation of its multiple in vitro biological potentials. Biomedicines, 8(5), 117. https://doi.org/10.3390/biomedicines8050117

[28] Lu, Y., Han, M., Shahri, E. E., Abbaspour, S., & Tayebee, R. (2023). Delivery of anti-cancer and anti-depression doxepin drug by nickel oxide NPs originated from the Cressa nudicaulis plant extract. RSC Advances, 13(18), 12133-12140. https://doi.org/10.1039/D2RA07545H

[29] Zhang, Y., Mahdavi, B., Mohammadhosseini, M., Rezaei-Seresht, E., Paydarfard, S., Qorbani, M., Karimian, M., Abbasi N., Ghaneialva, H., & Karimi, E. (2021). Green synthesis of NiO-NPs using Calendula officinalis extract: Chemical characterization, antioxidant, cytotoxicity, and anti-esophageal carcinoma properties. Arabian Journal of Chemistry, 14(5), 103105. https://doi.org/10.1016/j.arabjc.2021.103105

[30] Reddy, P. L., Deshmukh, K., Kovářík, T., Reiger, D., Nambiraj, N. A., Lakshmipathy, R., & SK, K. P. (2020). Enhanced dielectric properties of green synthesized Nickel Sulphide (NiS) NPs integrated polyvinylalcohol nanocomposites. Materials Research Express, 7(6), 064007. https://doi.org/10.1088/2053-1591/ab955f

[31] Wang, Y., & Pang, H. (2018). Nickel-Based Sulfide Materials for Batteries. Chemistryselect, 3(45), 12967-12986. https://doi.org/10.1002/slct.201802348

[32] Amiri, M., Pardakhti, A., Ahmadi-Zeidabadi, M., Akbari, A., & Salavati-Niasari, M. (2018). Magnetic nickel ferrite NPs: Green synthesis by Urtica and therapeutic effect of frequency magnetic field on creating cytotoxic response in neural cell lines. Colloids and Surfaces B: Biointerfaces, 172, 244-253. https://doi.org/10.1016/j.colsurfb.2018.08.049

[33] Indira, K., Shanmugam, S., Hari, A., Vasantharaj, S., Sathiyavimal, S., Brindhadevi, K., ... & Pugazhendhi, A. (2021). Photocatalytic degradation of congo red dye using nickel-titanium dioxide nanoflakes synthesized by Mukia madrasapatna leaf extract. Environmental Research, 202, 111647. https://doi.org/10.1016/j.envres.2021.111647

[34] Parveen, A., Sonkar, S., Yadav, T. P., Sarangi, P. K., Singh, A. K., Singh, S. P., & Gupta, R. (2022). Asparagus racemosus leaf extract mediated bioconversion of nickel sulfate into nickel/nickel hydroxide NPs: in vitro catalytic, antibacterial, and antioxidant activities. Biomass Conversion and Biorefinery, 1-21. https://doi.org/10.1007/s13399-022-02843-0

[35] Chaudhary, R.G., Tanna, J.A., Gandhare, N.V., Rai, A.R., Juneja. H.D., (2015) Synthesis of nickel NPs: microscopic characterization, an efficient catalyst and effective antibacterial activity. Advanced Materials Letters, 6(11) 990-998. https://doi.org/10.5185/amlett.2015.5901

[36] Chen, Y., Peng, D. L., Lin, D., & Luo, X. (2007). Preparation and magnetic properties of nickel NPs via the thermal decomposition of nickel organometallic precursor in alkylamines. Nanotechnology, 18(50), 505703. https://doi.org/10.1088/0957-4484/18/50/505703

[37] Shamsi, J., Urban, A. S., Imran, M., De Trizio, L., & Manna, L. (2019). Metal halide perovskite nanocrystals: synthesis, post-synthesis modifications, and their optical properties. Chemical reviews, 119(5), 3296-3348. https://doi.org/10.1021/acs.chemrev.8b00644

[38] Krishnia, L., Thakur, P., & Thakur, A. (2022). Synthesis of NPs by physical route. In Synthesis and Applications of NPs (pp. 45-59). Singapore: Springer Nature Singapore. https://doi.org/10.1007/978-981-16-6819-7_3

[39] Witkowski, A., Stec, A. A., & Hull, T. R. (2016). Thermal decomposition of polymeric materials. SFPE handbook of Fire Protection Engineering, 167-254. https://doi.org/10.1007/978-1-4939-2565-0_7

[40] Shalichah, C., & Khumaeni, A. (2018, May). Synthesis of nickel NPs by pulse laser ablation method using Nd: YAG laser. In Journal of Physics: Conference Series (Vol. 1025, No. 1, p. 012002). IOP Publishing. https://doi.org/10.1088/1742-6596/1025/1/012002

[41] Lan, H., & Ding, Y. (2010). Nanoimprint lithography (pp. 457-494). Croatia: InTech. https://doi.org/10.5772/8189

[42] Ye, X., & Qi, L. (2011). Two-dimensionally patterned nanostructures based on monolayer colloidal crystals: Controllable fabrication, assembly, and applications. Nano Today, 6(6), 608-631. https://doi.org/10.1016/j.nantod.2011.10.002

[43] Colson, P., Henrist, C., & Cloots, R. (2013). Nanosphere lithography: a powerful method for the controlled manufacturing of nanomaterials. Journal of Nanomaterials, 2013, 21-21. https://doi.org/10.1155/2013/948510

[44] Nirupama, M. P., Bheemaraju, A., Panwar, O. S., & Satyanarayana, B. S. (2018). Nickel Nanoparticle Arrays Prepared Using Nanosphere Lithography. IJRASET, 6(II), 75-78. https://doi.org/10.22214/ijraset.2018.2013

[45] Wender, H., Migowski, P., Feil, A. F., Teixeira, S. R., & Dupont, J. (2013). Sputtering deposition of NPs onto liquid substrates: Recent advances and future trends. Coordination Chemistry Reviews, 257(17-18), 2468-2483. https://doi.org/10.1016/j.ccr.2013.01.013

[46] Yadav, T. P., Yadav, R. M., & Singh, D. P. (2012). Mechanical milling: a top down approach for the synthesis of nanomaterials and nanocomposites. Nanoscience and Nanotechnology, 2(3), 22-48. https://doi.org/10.5923/j.nn.20120203.01

[47] Ban, I., Stergar, J., Drofenik, M., Ferk, G., & Makovec, D. (2011). Synthesis of copper-nickel NPs prepared by mechanical milling for use in magnetic hyperthermia. Journal of Magnetism and Magnetic Materials, 323(17), 2254-2258. https://doi.org/10.1016/j.jmmm.2011.04.004

[48] Chouke, P.B., Shrirame, T., Potbhare, A.K., Mondal, A., Chaudhary, A.R., Mondal, S., Thakare, S.R., Nepovimova, E., Valis, M., Kuca, K. (2022) Bioinspired metal/metal oxide NPs: A road map to potential applications, Materials Today Advances, 16, 100314. https://doi.org/10.1016/j.mtadv.2022.100314

[49] Morosanu, C. E. (2016). Thin films by chemical vapour deposition (Vol. 7). Elsevier.

[50] Ealia, S. A. M., & Saravanakumar, M. P. (2017, November). A review on the classification, characterisation, synthesis of NPs and their application. In IOP conference series: materials science and engineering (Vol. 263, No. 3, p. 032019). IOP Publishing. https://doi.org/10.1088/1757-899X/263/3/032019

[51] Shamim, A., Ahmad, Z., Mahmood, S., Ali, U., Mahmood, T., & Nizami, Z. A. (2019). Synthesis of nickel NPs by sol-gel method and their characterization. Open Journal of Chemistry, 2(1), 16-20. https://doi.org/10.30538/psrp-ojc2019.0009

[52] Ganguli, A. K., Ganguly, A., & Vaidya, S. (2010). Microemulsion-based synthesis of nanocrystalline materials. Chemical Society Reviews, 39(2), 474-485. https://doi.org/10.1039/B814613F

[53] Rehman, Z. U., Nawaz, M., Ullah, H., Uddin, I., Shad, S., Eldin, E., ... & Javed, M. S. (2022). Synthesis and Characterization of Ni-NPs via the Microemulsion Technique and Its Applications for Energy Storage Devices. Materials, 16(1), 325. https://doi.org/10.3390/ma16010325

[54] Chime, S. A., Kenechukwu, F. C., & Attama, A. A. (2014). Nanoemulsions-advances in formulation, characterization and applications in drug delivery. Application of Nanotechnology in Drug Delivery, 3, 77-126. https://doi.org/10.5772/58673

[55] Ingale, A. G., & Chaudhari, A. N. (2013). Biogenic synthesis of NPs and potential applications: an eco-friendly approach. Journal of Nanomedicine Nanotechology, 4(165), 1-7.

[56] Mohamad, N. A. N., Arham, N. A., Jai, J., & Hadi, A. (2014). Plant extract as reducing agent in synthesis of metallic NPs: a review. Current Pharmaceutical Biotechnology, 22 (13), 1782-1793..

[57] Mondal, A., Umekar, M.S., Bhusari, G.S., Chouke, P.B., Lambat, T. Mondal, S., Chaudhary, R.G., Mahmood, S.H. (2021). Biogenic synthesis of metal/metal oxide nanostructured materials, Current Pharmaceutical Biotechnology, 22 (13), 1782-1793. https://doi.org/10.2174/1389201022666210111122911

[58] Imran Din, M., & Rani, A. (2016). Recent advances in the synthesis and stabilization of nickel and nickel oxide NPs: a green adeptness. International Journal of Analytical Chemistry, 2016. https://doi.org/10.1155/2016/3512145

[59] Mariam, A. A., Kashif, M., Arokiyaraj, S., Bououdina, M., Sankaracharyulu, M., Jayachandran, M., & Hashim, U. (2014). Bio-synthesis of NiO and Ni-NPs and their characterization. Digest Journal of Nanomaterials and Biostructures, 9(3), 1007-1019.

[60] Chen, H., Wang, J., Huang, D., Chen, X., Zhu, J., Sun, D., ... & Li, Q. (2014). Plant-mediated synthesis of size-controllable Ni-NPs with alfalfa extract. Materials Letters, 122, 166-169. https://doi.org/10.1016/j.matlet.2014.02.028

[61] Mamuru, S. A., Bello, A. S., & Hamman, S. B. (2015). Annona squamosa leaf extract as an efficient bioreducing agent in the synthesis of chromium and nickel NPs. International Journal of Applied Sciences and Biotechnology, 3(2), 167-169. https://doi.org/10.3126/ijasbt.v3i2.11651

[62] Angajala, G., Ramya, R., & Subashini, R. (2014). In-vitro anti-inflammatory and mosquito larvicidal efficacy of nickel NPs phytofabricated from aqueous leaf extracts of Aegle marmelos Correa. Acta Tropica, 135, 19-26. https://doi.org/10.1016/j.actatropica.2014.03.012

[63] Huang, Y., Zhu, C., Xie, R., & Ni, M. (2021). Green synthesis of nickel NPs using Fumaria officinalis as a novel chemotherapeutic drug for the treatment of ovarian cancer. Journal of Experimental Nanoscience, 16(1), 368-381. https://doi.org/10.1080/17458080.2021.1975037

[64] Adams, E. Q., & Rosenstein, L. (1914). The color and ionization of crystal violet. Journal of the American Chemical Society, 36(7), 1452-1473. https://doi.org/10.1021/ja02184a014

[65] Haq, S., Dildar, S., Ali, M. B., Mezni, A., Hedfi, A., Shahzad, M. I., ... & Shah, A. (2021). Antimicrobial and antioxidant properties of biosynthesized of NiO-NPs using Raphanus sativus (R. sativus) extract. Materials Research Express, 8(5), 055006. https://doi.org/10.1088/2053-1591/abfc7c

[66] Hafshejani, B. K., Mirhosseini, M., Dashtestani, F., Hakimian, F., & Haghirosadat, B. F. (2018). Antibacterial activity of nickel and nickel hydroxide NPs against multidrug resistance K. pneumonia and E. coli isolated urinary tract. Nanomedicine Journal, 5(1).

[67] Golkhatmi, F. M., Bahramian, B., & Mamarabadi, M. (2017). Application of surface-modified nano ferrite nickel in catalytic reaction (epoxidation of alkenes) and investigation on its antibacterial and antifungal activities. Materials Science and Engineering: C, 78, 1-11. https://doi.org/10.1016/j.msec.2017.04.025

[68] Kumar, N. A., Rejinold, N. S., Anjali, P., Balakrishnan, A., Biswas, R., & Jayakumar, R. (2013). Preparation of chitin nanogels containing nickel NPs. Carbohydrate polymers, 97(2), 469-474. https://doi.org/10.1016/j.carbpol.2013.05.009

[69] Ezhilarasi, A. A., Vijaya, J. J., Kaviyarasu, K., Maaza, M., Ayeshamariam, A., & Kennedy, L. J. (2016). Green synthesis of NiO-NPs using Moringa oleifera extract and their biomedical applications: Cytotoxicity effect of NPs against HT-29 cancer cells. Journal of Photochemistry and Photobiology. B, Biology, 164, 352-360. https://doi.org/10.1016/j.jphotobiol.2016.10.003

[70] Singh, A., Goyal, V., Singh, J., Kaur, H., Kumar, S., Batoo, K. M., ... & Hussain, S. (2022). Structurally and morphologically engineered single-pot biogenic synthesis of NiO-NPs with enhanced photocatalytic and antimicrobial activities. Journal of Cleaner Production, 343, 131026. https://doi.org/10.1016/j.jclepro.2022.131026

[71] Berhe, M. G., & Gebreslassie, Y. T. (2023). Biomedical applications of biosynthesized nickel oxide NPs. International Journal of Nanomedicine, 4229-4251. https://doi.org/10.2147/IJN.S410668

[72] Götte, M., Berghuis, A., Matlashewski, G., Wainberg, M. A., & Sheppard, D. (Eds.). (2017). Handbook of antimicrobial resistance. New York, NY, USA: Springer. https://doi.org/10.1007/978-1-4939-0694-9

[73] Iqbal, J., Abbasi, B. A., Mahmood, T., Hameed, S., Munir, A., & Kanwal, S. (2019). Green synthesis and characterizations of Nickel oxide NPs using leaf extract of Rhamnus virgata and their potential biological applications. Applied Organometallic Chemistry, 33(8), e4950. https://doi.org/10.1002/aoc.4950

[74] Haritha, V., Gowri, S., Janarthanan, B., Faiyazuddin, M., Karthikeyan, C., & Sharmila, S. (2022). Biogenic synthesis of Nickel Oxide NPs using Averrhoa bilimbi and Investigation of its Antibacterial, Antidiabetic and Cytotoxic properties. Inorganic Chemistry Communications, 144, 109930. https://doi.org/10.1016/j.inoche.2022.109930

[75] Abbasi, B. A., Iqbal, J., Mahmood, T., Ahmad, R., Kanwal, S., & Afridi, S. (2019). Plant-mediated synthesis of nickel oxide NPs (NiO) via Geranium wallichianum: Characterization and different biological applications. Materials Research Express, 6(8), 0850a7. https://doi.org/10.1088/2053-1591/ab23e1

[76] Karthik, K., Shashank, M., Revathi, V., & Tatarchuk, T. (2019). Facile microwave-assisted green synthesis of NiO-NPs from Andrographis paniculata leaf extract and evaluation of their photocatalytic and anticancer activities. Molecular Crystals and Liquid Crystals. https://doi.org/10.1080/15421406.2019.1578495

[77] Chaudhary, R.G., Bhusari, G.S., Tiple, A.D., Rai, A.R., Somkuvar, S.R., Potbhare, A.K., Lambat, T.L., Ingle, P.P., Abdala, A.A. (2019). Metal/metal oxide nanoparticles: toxicity, applications, and future prospects. Current Pharmaceutical Design, 25(37), 4013-4029. https://doi.org/10.2174/1381612825666191111091326

[78] Lu, Y., Han, M., Shahri, E. E., Abbaspour, S., & Tayebee, R. (2023). Delivery of anti-cancer and anti-depression doxepin drug by nickel oxide NPs originated from the Cressa nudicaulis plant extract. RSC Advances, 13(18), 12133-12140. https://doi.org/10.1039/D2RA07545H

[79] Sabouri, Z., Akbari, A., Hosseini, H. A., Khatami, M., & Darroudi, M. (2021). Green-based bio-synthesis of nickel oxide NPs in Arabic gum and examination of their cytotoxicity, photocatalytic and antibacterial effects. Green Chemistry Letters and Reviews, 14(2), 404-414. https://doi.org/10.1080/17518253.2021.1923824

[80] Ghazal, S., Akbari, A., Hosseini, H. A., Sabouri, Z., Forouzanfar, F., Khatami, M., & Darroudi, M. (2020). Sol-gel biosynthesis of nickel oxide NPs using Cydonia oblonga extract and evaluation of their cytotoxicity and photocatalytic activities. Journal of Molecular Structure, 1217, 128378. https://doi.org/10.1016/j.molstruc.2020.128378

[81] Pandian, C. J., Palanivel, R., & Dhananasekaran, S. (2015). Green synthesis of nickel NPs using Ocimum sanctum and their application in dye and pollutant adsorption. Chinese Journal of Chemical Engineering, 23(8), 1307-1315. https://doi.org/10.1016/j.cjche.2015.05.012

[82] Ravindhranath, K., & Ramamoorty, M. (2017). Nickel based nano particles as adsorbents in water purification methods-a review. Oriental Journal of Chemistry, 33(4), 1603. https://doi.org/10.13005/ojc/330403

[83] Arsène, M. M. J., Davares, A. K. L., Viktorovna, P. I., Andreevna, S. L., Sarra, S., Khelifi, I., & Sergueïevna, D. M. (2022). The public health issue of antibiotic residues in food and feed: Causes, consequences, and potential solutions. Veterinary World, 15(3), 662. https://doi.org/10.14202/vetworld.2022.662-671

[84] Pandit, S., Dasgupta, D., Dewan, N., & Prince, A. (2016). Nanotechnology-based biosensor and its application. The Pharma Innovation, 5(6, Part A), 18.

[85] Salihu, S., Yusof, N. A., Mohammad, F., Abdullah, J., & Al-Lohedan, H. A. (2019). Nickel nanoparticle-modified electrode for the electrochemical sensory detection of penicillin G in bovine milk samples. Journal of Nanomaterials, 2019, 1-11. https://doi.org/10.1155/2019/1784154

[86] Chen, X., Liu, Q., Bai, T., Wang, W., He, F., & Ye, M. (2021). Nickel and cobalt sulfide-based nanostructured materials for electrochemical energy storage devices. Chemical Engineering Journal, 409, 127237. https://doi.org/10.1016/j.cej.2020.127237

[87] Aso, K., Hirokazu, H., Akitoshi, T., & Tatsumisago, M. (2011). Synthesis of nanosized nickel sulfide in high-boiling solvent for all-solid-state lithium secondary batteries. Journal of Materials Chemistry, 21(9), 2987 -2990. doi:10.1039/c0jm02639e

[88] Liang, G., He, L., Cheng, H., Li, W., Li, X., Zhang, C., ... & Zhao, F. (2014). The hydrogenation/dehydrogenation activity of supported Ni catalysts and their effect on hexitols selectivity in hydrolytic hydrogenation of cellulose. Journal of Catalysis, 309, 468-476. https://doi.org/10.1016/j.jcat.2013.10.022

[89] Du, Y., & Chen, R. (2007). Effect of Nickel Particle Size on Alumina Supported Nickel Catalysts for p-Nitrophenol Hydrogenation. Chemical and Biochemical Engineering Quarterly, 21(3), 251-255.

[90] Orlandi, M., Brenna, D., Harms, R., Jost, S., & Benaglia, M. (2016). Recent developments in the reduction of aromatic and aliphatic nitro compounds to amines. Organic Process Research & Development, 22(4), 430-445. https://doi.org/10.1021/acs.oprd.6b00205

[91] Alonso, F., Riente, P., & Yus, M. (2011). Nickel NPs in hydrogen transfer reactions. Accounts of Chemical Research, 44(5), 379-391. https://doi.org/10.1021/ar1001582

Green Synthesis and Emerging Applications of Frontier Nanomaterials   Materials Research Forum LLC
Materials Research Foundations 169 (2024) 173-196   https://doi.org/10.21741/9781644903278-7

Chapter 7

Biomedical applications of green synthesized cerium oxide nanoparticles

Sarita Rai[1], Preeti Gupta[2] and N.B. Singh [3*]

[1] Department of Chemistry, Dr. Harisingh Gour Central University, Sagar, India

[2] Department of Chemistry, DDU Gorakhpur University, Gorakhpur, India

[3] Department of Chemistry and Biochemistry, Sharda University, Greater Noida, India

* nbsingh43@gmail.com)

Abstract

Nanotechnology has emerged as a pivotal domain in the realm of science and technology, finding diverse applications across electronics, imaging, industry, and healthcare. In the healthcare sector, nanotechnology has proven instrumental in disease diagnostics, treatment, drug delivery, and the formulation of innovative pharmaceuticals. Among the myriad nanoparticles (NPs), Cerium Oxide (CeO_2) NPs stand out due to their distinctive surface chemistry, robust stability, and biocompatibility. Recent advancements have witnessed the synthesis of CeO_2 NPs through various bio-directed methods, leveraging natural and organic matrices as stabilizing agents. This approach aims to create biocompatible CeO_2 NPs, thereby addressing safety concerns and establishing conducive conditions for their efficacious utilization in biomedicine. In the pursuit of green synthesis, CeO_2 NPs have been successfully generated using plant extracts, microbial organisms, and other biological derivatives. Plants have proven to be a highly efficient source owing to their abundance, inherent safety, and rich reservoir of reducing and stabilizing agents. Different parts of plants, including leaves, flowers, and stems, have been harnessed for synthesizing CeO_2 NPs, with a predominant focus on leaves in existing green synthesis studies. Characterization of the synthesized NPs has been accomplished through various techniques such as UV spectroscopy, Fourier-transform infrared spectroscopy (FT-IR), Field-emission scanning electron microscopy (FESEM), Thermal methods (TG/DTA), and powder X-ray diffraction (PXRD) techniques. The applications of CeO_2 NPs across diverse biological fields have been explored and discussed in this context.

Keywords

Nanomaterials, Biomedical, Green Method, Drug

Contents

1.   Introduction

In past few decades, nanotechnology and nanomaterials (NMs) has emerged as a potential field due to the enormous application in variety of disciplines [1-5]. Because of the small size and high surface area, these materials exhibit unique optical, mechanical, catalytic, electrical as well as biological properties [6-12]. Usage of the NPs are very much implemented in the development of industrial, optical, electrical tools. Along with that, it has wide application in healthcare sector which includes drugs, drug delivery tools, treatment of infection [13-15] or diagnosis [16–21] of pathogen-fighting systems [22,23] or improvement of the fuel quality [20,21].

Nanosized particles of metal oxide exhibit potential biological applications. This is because of the shape and size, chemical components and as well as their valance state [24]. Extensively used metal oxides with nano sized are FeO_2, TiO_2 and CeO_2 because they can intensively connect with the cells involved in the metabolic networks [25]. Nowadays, rare-earth oxides have become more apparent in the field of information and biotechnology [26]. This is because the availability of electrons in 4f orbitals which is shielded by 4d and 5p electrons. With these unusual properties, lanthanides have potential to explore the new applications.

The first element of lanthanide series is cerium, which has been attracted by various fields. NPs of CeO_2 are used in biological applications, solar cells, solid oxide fuel cells, catalytic materials [27,28] and good pharmacological agents. In the past few years, NPs of CeO_2 and CeO_2 containing materials have been intensively used as an active component in the catalytic reaction. Taking the account of the chemistry of cerium, it has been reported that cerium forms two oxides i.e. CeO_2 and Ce_2O_3 [29,30]. In the CeO_2 lattice cerium exist in both the forms on the surface of nanoscale material. There is charge deficiency because of Ce^{+3} and oxygen vacancies compensate its scarcity. This type of imperfection is known as intrinsic oxygen defect [31] which is an active for catalytic reaction. Literature reveals that as the particle size decrease the concentration of defect increases, thus CeO_2 have enhanced redox properties [32]. It has also been reported that these materials are massively used in biological fields including biomedicine [33] drug delivery [34,35], bioanalysis [36-41], and bio-scaffolding [42-43] etc. (Figs. 1 and 2)

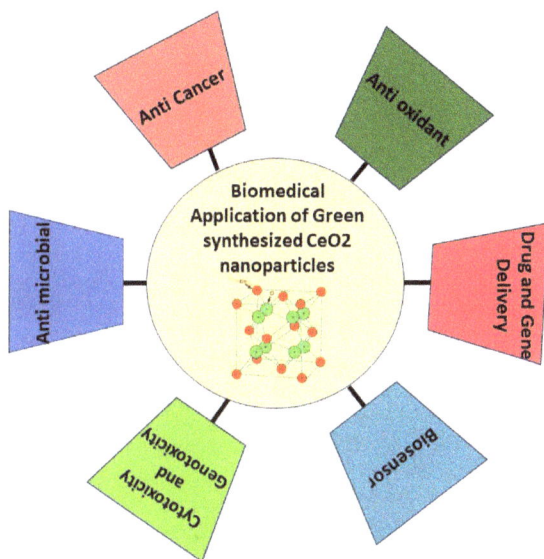

Figure 1. diagram showing different biomedical applications of green synthesized CeO₂ NPs.

In the past decades, green synthesis of CeO_2 NPs has been discussed at length [44-53]. The green synthesis methods have many advantages over the traditional methods such as sol–gel methods, hydrothermal, solvothermal, thermal decomposition, thermal hydrolysis, ball milling, solution precipitation, sonochemical, and spray pyrolysis etc.

Green Synthesis and Emerging Applications of Frontier Nanomaterials   Materials Research Forum LLC
Materials Research Foundations 169 (2024) 173-196   https://doi.org/10.21741/9781644903278-7

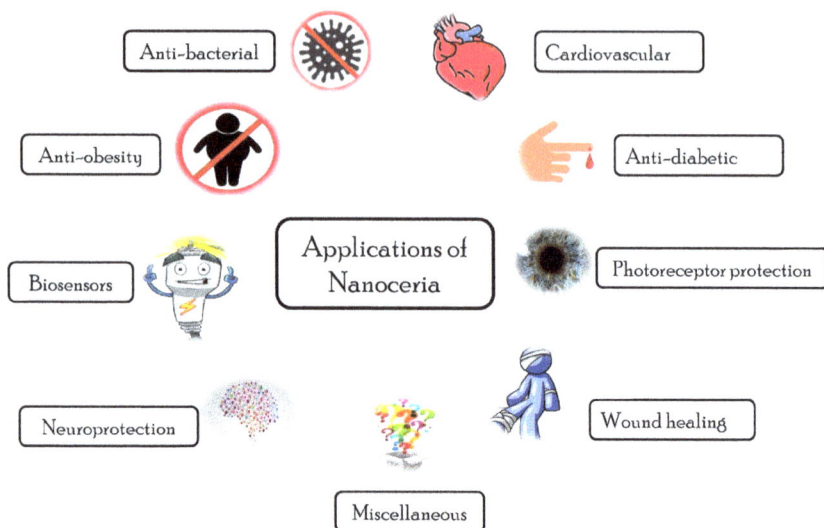

Figure 2. Biomedical applications of ceria NPs

2.   Synthesis

2.1   Chemical method

Many chemical methods are reported by researchers for the synthesis of nanoceria. Different have proved the synthesis of nanoceria by precipitation method [21–23] like co-precipitation [24] and chemical precipitation [25,26], hydrothermal [31–34], sonochemical [29,30], reverse-co-precipitation [35], microwave and hydrothermal methods [54-69, 36].

2.1.1   Precipitation method

Nanoparticles of CeO_2 has been synthesized by precipitation method by number of researchers [70-76]. In majority of studies, cerium nitrate hexahydrate has been used as a precursor and capping agent are dextran, sarium, PVP, MEECTES samarium etc. The capping agents are employed to alter the polydispersity of the NMs. With this precipitation method, produced NPs fall in the range of 3-18 nm. The size of the particles has also been controlled by making changes in pH, concentration of substance, temperature and stabilizing agents etc.

2.1.2   Hydrothermal method

Hydrothermal is another method to synthesize nanoparticles. Cerim oxide NPs have been prepared by using such method which is temperature dependent. It has been reported in the literature that cerium nitrate hexahydrate, cerium acetate, cerium chloride are used as precursor and different capping agent viz; citric acid, trisodium phosphate dodecahydrate, dithiodopamine etc are used as capping agent [77-82].

2.1.3 Solvothermal method

Among all the chemical methods, synthesis through Solvothermal is an important method for the preparation of cerium oxide NPs. In this method, NPs of various size have been prepared by using suitable organic solvents under high temperature and pressure. Concentration of the reagent and type of the solvent has also been used to control the properties [83-85].

2.1.4 Sol-Gel method

In sol-gel method, generally cerium nitrate hexahydrate, a precursor, formed cerium hydroxide by hydrolysis. This further undergo condensation to form a gel which gives a final product after drying it at high temperature. Number of works has been reported in which different surfactants as capping agents are used. The addition of these surfactants brought major changes in the morphology, surface area as well as change in the physiochemical properties. Different surfactant which are used in the synthesis of nanosized cerium oxide are diphenyl ether, olylamine, methanol, Lu seeds, Pullulan etc [86-90].

2.2 Green method

In past years, NPs of CeO_2 were prepared by using resources as stabilizing/ capping agent for the synthesis of biocompatible NPs. The initiatives of green synthesis route are an insight to work on the challenges to carefully and dynamically utilize this metal oxides in biomedical applications.

2.2.1 Synthesis from plant extract

Various parts of the plant such as leaf, stem and flowers are the good source of capping, stabilizing, shielding and surfactant agents [91-97]. Majority of research work has been done on leaves extract. This extract contains wide variety of metabolites such as phenols, carboxylic acid, ketones and ascorbic acid etc. The leaves of *Gloriosa superba, Acalypha indica, Aloe vera, Rheum turkestanicum, Aquilegia Pubiflora, Moriga Olifera, Olea europaea, Leucas asper, Petroselinumcrispum, dill odorous, Anethum graveolens, Lactuca sativa, Brassica oleracea, Pisum sativum, Agastache foeniculum, Prosopis juliflora, Ceratonia siliqua* [57], *Justicia adhatoda Jatropha curcus, Origanum majorana, Pedalium murex, Elaeagnus angustifolia, Piper betle, Azadirachta Indica* [98-109], *Orange Peel* [110], Stem of *Euphorbia tirucalli* [111], Fruit juice of *Watermelon* [112]. Lemon grass [113], seeds of *Cydonia oblonga, Salvia Macrosiphon Boiss* [114-115] are employed as stabilizing agents in the synthesis of CeO_2 NPs (Table 1). It is also found in the literature that the particle size of the cerium oxide is bit large and are not suitable for biomedical application. Employing the stabilizing agents extracted from the different parts of plants majorly effect the particles size and subsequently different biological effect and antibacterial activity has also been altered. In the preparation of NPs of Cerium oxide, adequate amount of leaf extract and the solution of cerium salt has been stirred for 2-6 hrs on 50 -80^{0}C, precipitate will be obtained.

Table 1. *Synthesis of Cerium oxide NPs using different parts of plants [9].*

Plants used	Part used	Morphology and particle size
Gloriosa superba	leaf	Spherical and 5 nm
Acalypha indica	leaf	Spherical and 36 nm
Aloe vera	leaf	Spherical and 63.6 nm
Rheum turkestanicum	leaf	Spherical and 30 nm
Aquilegia Pubiflora	leaf	Spherical and 28 nm
MorigaOlifera	leaf	-
Olea europaea	leaf	Spherical shape, 24 nm
Leucas aspera	leaf	
Petroselinum crispum	leaf	50 nm
Prosopis juliflora	leaf	Spherical ad 15 nm
Olive	Leaf]	Spherical and 6 nm
Rubia cordifolia	Leaf	Hexagonal and 26 nm
Prosopis farcta		Spherical and 30 nm
China rose	Petal	Nano-sheet and 7 nm
Salvia macrosiphonBoiss seeds Extract	leaf	Spherical and 47 nm
Cydonia oblonga	seeds	-
Ceratonia siliqua	Leaf	Spherical and 22 nm
Salvadora persica		Spherical and 10–15 nm
Morus nigra	Fruit	Irregular and 10–15 nm
Annona muricata	Fruit	7.5 nm
Justicia adhatoda	Leaf	Stick-like and 20–25 nm
Jatropha curcus	Leaf	Monodispersed, 3-5 nm
Origanum majorana	Leaf	Spherical and 20 nm
Pedalium murex	Leaf	Nano-rod and 5–55 nm
Elaeagnus angustifolia	Leaf	Spherical and 30–75 nm
Euphorbia amygdaloides		Spherical and 42 nm
Orange	Peel	Cubic structure, 20 nm
Piper betle	Leaf	
Azadirachta Indica]	Leaf	Spherical and 10–15 nm
Euphorbia tirucalli	Stem	Flaky and 37–40 nm
Lemon grass	Grass	10–40 nm
Watermelon	Fruit	Irregular and 36 nm

2.3   Synthesis from nutrients

In the green synthesis method for the preparation of CeO_2 NPs, natural nutrient materials have drawn significant attention. Reported literature has suggested that the carbohydrate sugar (Fructose), Sugar (Glucose, lactose), honey and protein of egg white (ovalbumin and lysozyme) microbial exopolysaccharides (levan, pullulan) are employed as green capping agents [117-120]. The synthesis of cerium oxide NPs has been prepared by ammonium cerium (IV) nitrate, sodium hydroxide solution and fructose, glucose and lactose as capping agent by microwave assisted method. The particle size is in the range of 5-20 nm at 400-600 ^{0}C respectively. In the egg white medium, there is an interaction between Ce^{3+} and oppositely charged protein leading to the isotropic growth of small sized CeO_2 NPs with high stability. The other material which has been used as capping agent is honey. It is full of vitamins, enzymes and carbohydrates which are containing number of hydroxyl and amine groups. The helps in the complexation of Ce^{3+} ion and are also capable of coating the metal species and inhibits the aggregation.

2.3.1   Synthesis from biopolymer

Naturally occurring polymers can also be used as base to synthesis of cerium oxide NPs. They act as efficient stabilizing agent as they are having hydroxyl moieties which controls the particle size for the desired synthesised material. Employing sol-gel method, cerium oxide NPs has been synthesized with the stabilizing agent agarose [121]. in recent year there are several biopolymers i.e. starch, dextran [122-131], polyacrylic acid, polyethylene glycol, folic acid and phosphonated polyethylene glycol, alginate-ascorbic acid biopolymer, Gum tragacanth, chitosan, Carboxymethyl cellulose and fructose are used as stabilizing agent to reduce the particle size of CeO_2 up to nano level.

Sol-gel method has been applied to synthesize spherical NPs with the particle size 10.5 nm using agarose biopolymer. It is dissolved in water at 90 ^{0}C and semisolid gel is obtained while reducing the temperature up to 35-40 ^{0}C over a pH range of 3 to 9. A nanochannel is formed by the interpenetrating Hydrogen bonding among the sugar molecules with the pore size 200nm. Another efficient capping agent is starch in which metal get attracted towards by the oxygen atom of -OH branches. It decreases the particle size up to 6 nm [121]. A complex, biocompatible and well water-soluble polysaccharide, Dextran is used to form spherical CeO_2 nanoparticle of 5nm. It is highly toxic for the cancer cells up to pH 6 [122]. Shegal et al. [123] obtained polyacrylic acid coated single nanoparticle of via precipitation-redispersion method. These NPs have ~ 2 nm and is stable over a broad range of pH. Karkoti et al. [124] has reported a synthesis of CeO_2 NPs in Polyethylene glycol solution and studied the increasing effect of Polyethylene glycol on the Superoxide dismutase mimetic properties which is shown by nanoceria. A report has been revealed the regeneration of Ce^{3+} and the also the function of polyethylene glycol in the redox reaction of CeO_2 NPs catalysed by H_2O_2.Folic acid is used as cell targeting ligand with nanoceria by Vassie et al [126]. Phosphonate polyethylene glycol has been used to functionalized CeO_2 NPs. Gum tragacanth (GT) has also been used in the synthesis of CeO_2 NPs by employing biological and chemical method. Gum has two fractions soluble and insoluble. Soluble fraction gives sol in water whereas insoluble part will remain as gel. Sol and gel fraction were heated at 40 ^{0}C where it become soluble and looses its semicrystalline nature. GT is a polysaccharide with -OH branches and when the $Ce(NO_3)_4$ is added to it , the metal gets attached to the oxygen of hydroxyl group thus formed $Ce(OH)_4$. This cerium hydroxide gets changed into CeO_2 NPs on dehydration followed by recrystallization. The role of GT attributes to form layer around cerium hydroxide and cerium oxide NPs [128]. This may be because the increase in suspension viscosity and electrostatic interaction. Chitosan has also been used as a good stabilizing agent. A.Kaushik et. al. [129] and H. Hassimiyad et al. [130] has discussed its nontoxicity, antibacterial activity, biodegradability. Table 2 discusses the morphology, particle size of Cerium oxide NPs based on nutrients and biopolymers

Table 2 Synthesis of Cerium oxide NPs using nutrients and biopolymers [9].

Name of nutrient	size	Morphology
Agarose	6	Spherical
Egg protein	8–17	Spherical
Honey	23	Spherical
Polyethylene glycol	~2	Spherical
Dextran	5–10 _	Spherical
Pectin	≤ 40	Spherical
Chitosan	~ 4	Spherical
Gum	10	Spherical
starch	6	Spherical

2.3.2 Synthesis from fungus

Synthesis of NPs of CeO_2 using microbes such as fungus has emerged as a different, innovative, low cost and environmental friendly method. This useful process helps in biomass handling and quiet easy downstream process due to the capability of extracellular synthesis of NPs. Moreover, microbes are expensive source of secondary metabolites such as heterocyclic moieties, proteins (amino acids) and enzymes which play a significant part in stabilizing the CeO_2 bulk matrices and reducing it into nanosize. In recent years, many researchers have focused on such type of synthesis by using different species of fungus viz; *Trichoderma viridae* [133], thermophilic fungus *Humicolasp.[*134], *aspergilusniger* [135], *Fusarium solani* [136], *Curvulunalunta* [137]. Nanopraticles of CeO_2 using *Trichoderma viridae* has been synthesize of *R. Tekupalli* [133]. Obtained NPs are spherical in shape with diameter ~18 nm. The FT-IR studies revealed that there is strong bond in the presence of -OH group. Its antifungal activity has also been studied. The biosynthesis of CeO_2 NPs by using a thermophilic *fungus Humicola* capping agent produced spherical shaped (12–20 nm) [134]. They are good candidate for treatment of such as Parkinson's and Alzheimer's disease. Extract of *Aspergillus niger* produced spherical NPs with ~ 5nm size [135]. *Fusarium solani* based CeO_2 NPs have spherical shaped particles and the particle size is from 20–30 nm [136]. *Curvularialunata* extract based cerium oxide NPs again yielded spherical shaped NPs and the size ranges from 5–20 nm [137].

Green Synthesis and Emerging Applications of Frontier Nanomaterials   Materials Research Forum LLC
Materials Research Foundations 169 (2024) 173-196   https://doi.org/10.21741/9781644903278-7

3.   Applications

3.1   Antibacterial activity

The antibacterial efficiency is decided by the establishment of an electrostatic attraction between the bacterial cell and CeO_2 NPs. The bacterial cell membrane is enriched with protein so as thiol group (-SH). The release of ions from CeO_2 and the reactive oxygen species is induced the intracellular functions such as cell division, cellular respiration and DNA replication The generation of Reactive oxygen species (ROS) and this causes the death of bacterial cell [10]. The antibacterial potential of the CeO_2-NPs is mostly dependent on morphological features, like shape, size and surface area too. Jan et al. [95] had explored the effect of CeO_2 NPs on five bacterial strains (2 gram-positive, 3 gram-negative) using different concentration of solution ranges from 5mg.mL^{-1} to 500 µg mL^{-1}. The measured zone of inhibition of E. coli, S. epidermis, *K.pneunomia* and *B.subtilis* is detailed in Table 2. Studies of different strain of fungus has also been studied and mentioned in the same table (Table 2). Another green synthesized CeO_2 nano particles using capping agent Acorus calamus is used to decrease the development of biofilms of S. aureus MTCC 3160 and P. aeruginosa (PA01). The dose of 1600 µg/mL of CeO_2 NPs are much more effective to present the growth of biofilms of these two bacterial strains. This is the first study on antibiofilm activity of CeO_2 NPs through green synthesis method [96].

3.2   Toxicity studies (Impact on human health)

Use of nanosized CeO_2 has major impact of the human health through two major routes that are ingestion and inhalation. Nanoceria gets absorbed in intestine and cerium excreted through faeces. Another major way that cerium oxide gets into the body through respiratory system. Once it gets inhaled, reached to lungs and lymph nodes and majorly affect the respiratory tract. When it gets absorbed via circulation, it may affect other organs like kidney, spleen, and liver.

3.3   Cytotoxicity

Cytotoxicity and oxidative stress have also been caused by CeO_2 NPs. A study has revealed that CeO_2 with particle size 20 nm was toxic for cultured huma. There was also a quantitative study of total malondialdehyde, glutathione, lactate dehydrogenase, ROS, α-tocopherol. These are the key indicators of cytotoxicity and oxidative stress in lung cancer cells. Balaji et al. [138] has used biosynthesized (using leaf extract of E. Globus) CeO_2 NPs for cytotoxic study to colon HCT116 carcinoma cell lines and lungA549 cell line. Studies has revealed that cell viability by IC_{50} and MTT assay value which are 58.2 µg/L for HCT116 and 45.5 µg/L for A549 cell line. Hence, it is reported that it is a potential anti-cancer agent. A green synthesis of CeO_2 NPs with *A. publiflora* has also shown a good cancer agent for hepatocarcinoma (HepG2 cell line). This cancer is yet not diagnosed by researchers. It is suggested that CeO_2 NPs enhanced to produced ROS in the cancerous cells and changed its own oxidation state from +3 to+4 and the reconversion takes place during the scavenging ROS in normal cell. In this way, the NPs induced apoptosis and oxidative stress to the cancer cell with the protection of normal cell.

3.4   Genotoxicity

Genotoxins can induce chromosomal aberration and recombination and it can also be defined as the destructive genetic alteration involving gene mutations. Metal NPs play a major role in the initiation and progression of different abnormalities. These genotoxins can damage the sequence of DNA and also disturbed the addition, duplication, ring formation and deletion of chromosomal structure. Recent studies have revealed that the genotoxicity of NPs of CeO_2 for the lung adenocarcinoma (A549) cells. The increased level ROS caused the damage of DNA and also cell cycle. It caused the apoptotic cell death in A549 because of nanoceria. The genotoxic and cytotoxic of nanoceria has been reported in human neuroblastoma cell line (IMR 32). Assessment of genotoxicity was verified by using the cytokines-block micronucleus and comet assays. The result concludes that ROS affects the toxicity of nanoceria.

4. Antioxidant activity

Cerium oxide NPs synthesize through bio-mediated sources is a potential Antioxidant for DPPH and ABTS assay. It has been explained that ROS and nitrogen is an important oxidizing agent. The disproportion between ROS and nitrogen species production raises the anti-oxidant level thus cause oxygen stress. Cerium oxide NPs could exhibit antioxidant properties [3] (Fig. 3)[7].

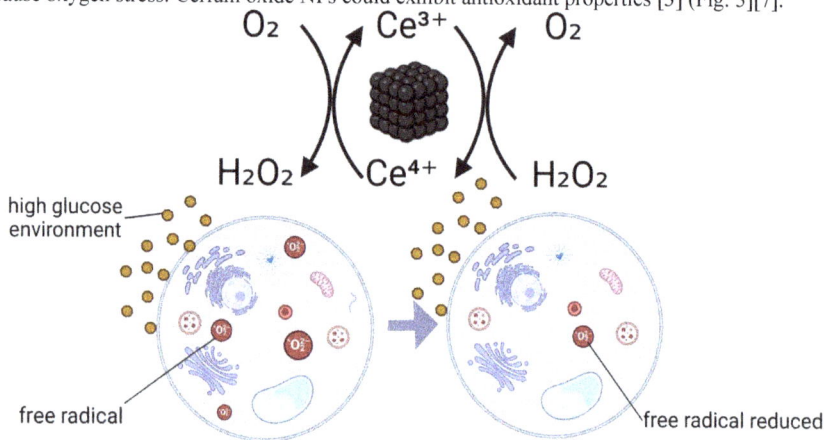

Figure 3. Mechanism of antioxidant activity of CeO_2 NPs

In a study, the antioxidant property of biosynthesized CeO_2 with leaf extract of *A. publiflora* has reported the total reducing power, total antioxidant capacity, ABTS assay and DPPH free radical scavenging in the concentration range of 12.5 -200 $\mu g.mL^{-1}$. Study revealed that NPs are concentration dependent (Table 3). Levan polysaccharide based CeO_2 NPs had also shown antioxidant property against H_2O_2 in NIH3T3 cells. Levan and its derivatives have antiflammatory and anti-tumour property along with antioxidant. An oxidative stress for $KBrO_3$ is possessed by these CeO_2 in the epithelial cell line. Moreover, it impedes H_2O_2 and also prevent the over production of ROS. This helps to inhibit cardiovascular disease.

Table 3 Antioxidant property of biosynthesized CeO$_2$ with leaf extract of A. publiflora

Conc. ($\mu g\ mL^{-1}$)	TRP (μg AAE per mg)	TAC (μg AAE per mg)	ABTS (TEAC)	DPPH (% FRSA)
12.5	27.64 ± 1.13	16.93 ±1.22	21.90 ± 1.91	9.61±1.8
25	31.87 ± 0.95	21.51 ± 0.71	39.71 ±2.21	12.32 ± 0.97
50	52.43 ±0.92	26.39 ± 0.84	54.48 ±2.25	21.18 ± 1.74
100	59.05 ± 0.98	34.64±1.32	63.31 ± 1.99	24.33 ± 1.2
200	87.06 ± 1.24	52.71 ± 1.14	74.21 ± 2.64	38.72 ± 1.38

5. Drug delivery

Biomediated-CeO$_2$ NPs show promise as effective anticancer agents and are crucial for drug delivery applications. Research indicates that CeO$_2$, particularly in nanocomposite serves as an efficient nanocarrier. Additionally, polyacrylic acid-coated CeO$_2$ NPs demonstrate potential for drug delivery, proving effective in treating non-small cell lung cancer. Dextran-coated CeO$_2$ NPs have been explored for the treatment of childhood neuroblastoma, and their application, when loaded with curcumin, induces cytotoxicity in neuroblastoma cells without posing a threat to normal cells. Consequently, CeO$_2$ NPs synthesized through biomediated processes emerge as highly suitable for targeted drug delivery [9].

6. Gene delivery

In 2016, the pioneering carrier for gene delivery employing CeO$_2$ NPs emerged as hybrid NPs combining CeO$_2$ with dimethyl dioctadecylammonium bromide (DODAB). This innovative approach replaced the conventional viral vector and successfully transfected plasma DNA (pEGFPN1) into various cell lines. The investigation focused on the tibialis muscle of mice, demonstrating that the nanovector exhibited approximately 17% lower transfection efficiency compared to the commercially available vivo-jetPEI transfection reagents. Consequently, the study concludes that CeO$_2$ NPs are efficacious and can serve as efficient carriers in gene delivery applications.

7. Treatment of diseases

Cerium metal has unique pharmacological property for the treatment of disease. Many neurogenerative diseases are caused by oxidative stress and free radical production. The NPs are CeO$_2$ damages the formation of free radicals by neutron shielding effect. The signal transduction pathways is also affected by this and the death of neuronal cell. So, the neurodegenerative diseases (Alzheimer's disease) can be treated with these potential NPs. The ischemic stroke can be protected by CeO$_2$ NPs encapsulated with phospholipid-polyethylene glycol. The treatment of a diabetic has also been treated by CeO$_2$ NPs.

8.   Cerium oxide-based biosensors

An analytical tool, biosensors, are used to analyses, recognizes, and converts the biological or physical or chemical signals into measurable electrochemical or an optical signal. Such tools are made up of elements for sensing, transducers, and signal processors. Among the potential metal oxide NPs, CeO_2 NPs are mostly applicable for the manufacturing of biosensors because of their distinct properties such as high storage capacity, biocompatibility, large surface area, high mechanical strength, oxygen ion conductivity, and nontoxic to the living cells. Detection of cholesterol has been done by the biosensor made up of CeO_2 NPs from sol–gel method. CeO_2 NPs based electrochemical biosensors is proved as a good tool for the detection of H_2O_2 detection in as low as 1 μM of water [2]. Oxygen-rich Pt doped CeO_2 NPs has created a biosensor to in vivo and in vitro monitor lactate in lactate oxidase monitoring during hypoxia [3].

9.   Enzyme mimetic applications

CeO_2 NPs have good catalytic stability in the severe condition. They form robust artificial oxidase enzymes which are capable of mimicking Super oxide dismutase (SOD), catalase and peroxidase enzymes. With the lower oxidation state (Ce^{3+}) in CeO_2 NPs, the surface showed peroxidase or catalase mimetic activity and it can break down H_2O_2 into water and oxygen. However, the higher oxidation state (Ce^{3+}) effectively scavenges superoxide (SOD mimetic activity) radicals and produce H_2O_2. Bhushan and Gopinath [138] have synthesize ceria-albumin nanoparticle (BCNP) which are biocompatible and sufficient enough to reduce intracellular ROS. Catalase is an enzyme that helps in the degradation of toxic oxidizing agent H_2O_2. Good catalyse mimetic activity is shown with low $3^+/4^+$ ratios of CeO_2 NPs [139]. An intracellular glutathione in cells is increased due to the catalase activity of CeO_2 -NPs and protected the cells from oxidative damage [12]. Enzymes like Phosphatases are used to eliminate phosphate groups from their substrates by hydrolysis reaction of esters into phosphate ions [140-142]. Phosphate mimic activity has been shown by low $3^+/4^+$ ratios of CeO_2 NPs. Cerium oxide NPs are also potential candidate in catalytic scavenging of ROS, in which the hydroxyl radical is known to be the active free radical in the biology. It has been investigated that dextran coated CeO_2 NPs have shown the anti-oxidant, auto-catalytic behaviour and biocompatibility for the treatment of neurological complications. Figure 4 [142] clearly shows the regeneration of CeO_2 NPs by utilizing H_2O_2. The scheme suggest that this oxide have auto-regenerative properties and can play a significant role in neuroprotective action as an antioxidant.

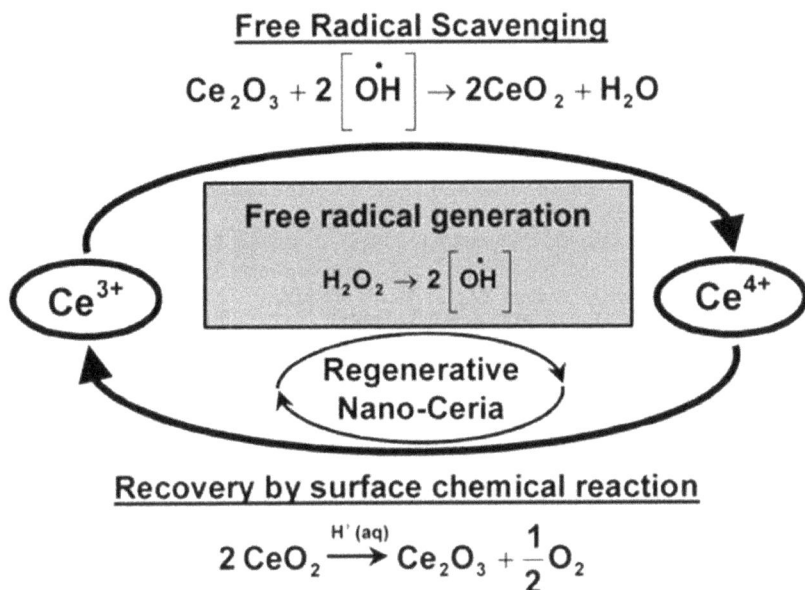

Figure 4 Reaction showing autoregenerative properties for CeO$_2$ NPs coated with dextran [142].

Conclusions

All the fields of science including biology, physics, material science, chemistry etc update by nanotechnology. A rare earth element, cerium, exhibits oxygen vacancies on the surface. This helps in the co-existence of two oxidation states: cerium^{3+} and cerium^{4+}, thus exhibit catalytic activity. Preparation of NPs through extracts of different parts of the plant, nutrients, biopolymer and microbes have drawn ample attention. Because this bio mediated process is non-toxic, low cost and biocompatible. In this chapter, detailed studies of the synthesis through chemical route and green synthesis route have been discussed. The particle obtained by the green synthesis route are in nano range and mostly are spherical in morphology. CeO$_2$ NPs are potential material for different biomedical applications such as antioxidant, antifungal activity, anticancer activity, anti-diabetic, toxicity, cytotoxicity, genotoxicity, drug delivery, gene delivery, biosensing and enzyme mimetic applications. From the detailed study, it is found that the green synthesized CeO$_2$ NPs are cost effective, and easy in production at large scale for biomedical use.

References

[1] A.P. F. Isabela, C. C.L. Santos, A. L. Xavier, T. M. Batista, Y. M. Nascimento, J.M.F.F. Nunes, P.M.F. Silva, R. A. Menezes-Ju'nior, J. M. Ferreira, E. O. Lima, J. F. Tavares, M. V. Sobral, D. Keyson, F. C. Sampaio, Synthesis, physicochemical characterization, antifungal

activity and toxicological features of cerium oxide nanoparticles, Arabian Journal of Chemistry, 14 (2021) 102888. https://doi.org/10.1016/j.arabjc.2020.10.035

[2] S Deshpande, S Patil, SV Kuchibhatla, S. Seal Size dependency variation in lattice parameter and valency states in nanocrystalline cerium oxide. Appl. Phys. Lett. 87(13) (2005)133113. https://doi.org/10.1063/1.2061873

[3] NP Sardesai, M Ganesana, A Karimi, JC Leiter, S Andreescu, Platinumdopedceria based biosensor for in vitro and in vivo monitoring of lactate during hypoxia. Anal. Chem. 87 (2015) 2996-3003. https://doi.org/10.1021/ac5047455

[4] A. Mondal, M.S. Umekar, G.S. Bhusari, P.B. Chouke, T. Lambat, S. Mondal, R.G. Chaudhary, S.H. Mahmood, Biogenic synthesis of metal/metal oxide nanostructured materials, Curr. Pharm. Biotechnol. 22 (13) 2021, 1782-1793. https://doi.org/10.2174/1389201022666210111122911

[5] N.B. Singh, R.G. Chaudhary, M.F. Desimone, A. Agrawal, S.K. Shukla, Green synthesized nanomaterials for safe technology in sustainable agriculture, Curr. Pharm. Biotechnol. 24 (2023) 61-85. https://doi.org/10.2174/1389201023666220608113924. https://doi.org/10.2174/1389201023666220608113924

[6] C. Jianrong, M Yuqing, H Nongyue et al. Nanotechnology and biosensors. Biotechnol Adv. 22(7) (2004) 505-518. doi:10.1016/j. biotechadv.2004.03.004 https://doi.org/10.1016/j.biotechadv.2004.03.004

[7] Haotian Xu, Shiqi Li, XiaoxuanMa , Tingting Xue, Fang Shen, Yi Ru, Jingsi Jiang , Le Kuai, Bin Li, Hang Zhao, Xin Ma, Cerium oxide nanoparticles in diabetic foot ulcer management: Advances, limitations, and future directions, Colloids and Surfaces B: Biointerfaces, 231 (2023) 113535. https://doi.org/10.1016/j.colsurfb.2023.113535

[8] R. Tomar, A. A. Abdala, R.G. Chaudhary, N.B. Singh, Photocatalytic degradation of dyes by nanomaterials, Materials Today: Proceedings, 29, (2020), 967-973. https://doi.org/10.1016/j.matpr.2020.04.144

[9] A.P. Farias, Carlos C.L. Santos a , Aline L. Xavier a , Tatianne M. Batista a , Yuri M. Nascimento a , Jocianelle M.F.F. Nunes a , Patrı'cia M.F. Silva b , Raimundo A. Menezes-Ju'nior a , Jailson M. Ferreira c , Edeltrudes O. Lima a , Josean F. Tavares a , Marianna V. Sobral a , Dawy Keyson a , Fa'bio C. Sampaio aVjvi, Arabian Journal of Chemistry 14, (2021) 102888 https://doi.org/10.1016/j.arabjc.2020.10.035

[10] V.N. Sonkusare, R.G. Chaudhary, G.S. Bhusari, A.R. Rai, H.D. Juneja, Microwave-mediated synthesis, photocatalytic degradation and antibacterial activity of α-Bi2O3 microflowers/novel γ-Bi2O3 microspindles, Nano-Structures & Nano-Objects, 13 (2018), 121-131. https://doi.org/10.1016/j.nanoso.2018.01.002

[11] J.A. Tanna, R.G. Chaudhary, H.D. Juneja, N.V. Gandhare, A.R. Rai, Histidine-capped ZnO nanoparticles: an efficient synthesis, spectral characterization and effective antibacterial activity, BioNanoScience, 5 (2015) 123-134. https://doi.org/10.1007/s12668-015-0170-0

[12] M.S. Nagmote, A.R. Rai, R. Sharma, M.F. Desimone, R.G. Chaudhary, NB Singh, Bioremediation of heavy metals using microorganisms, Genetically engineered organisms in bioremediation, CRC Press, 2024, pp.168-190. https://doi.org/10.1201/9781003188568-11

[13] N. Sanvicens, M. P. Marco, Multifunctional nanoparticles-properties and prospects for their use in human medicine, Trends Biotech., 26, (8),(2008) 425-433. https://doi.org/10.1016/j.tibtech.2008.04.005

[14] S.T. Yerpude, A.K. Potbhare, P.R. Bhilkar, P. Thakur, P. Khiratkar, M.F. Desimone, P.R. Dhongle, S.S. Sonawane, C. Goncalves and R.G. Chaudhary, Computational analysis of nanofluids-based drug delivery system: Preparation, current development and applications of nanofluids, Applications of Nanofluids in Chemical and Bio-medical Process Industry, Elsevier, 2022, pp. 335-364. https://doi.org/10.1016/B978-0-323-90564-0.00014-3

[15] S.T. Yerpude, A.K. Potbhare, P. Bhilkar, A.R. Rai, R.P. Singh, A.A. Abdala, R. Adhikari, R. Sharma, R.G. Chaudhary, Biomedical and clinical applications of platinum-based nanohybrids: An update review, Environmental Research, 231 (2023) 116148. https://doi.org/10.1016/j.envres.2023.116148

[16] H.Bridle, Nanotechnology for Detection of Waterborne Pathogens. In Waterborne Pathogens; Elsevier: Amsterdam, The Netherlands, 2014; pp. 291-318. Nanomaterials, 10, (2020)) 1614 13 of 15 https://doi.org/10.1016/B978-0-444-59543-0.00009-8

[17] S. Chigurupati, M.R. Mughal, E. Okun, S. Das, A. Kumar, M. McCaffery, S. Seal, M.P. Mattson, Effects of cerium oxide nanoparticles on the growth of keratinocytes, fibroblasts and vascular endothelial cells in cutaneous wound healing. Biomater. 34 (2013), 2194-2201. https://doi.org/10.1016/j.biomaterials.2012.11.061

[18] H. Choi, K. Lee, N. Hur, H. Lim, Cerium oxide-deposited mesoporous silica nanoparticles for the determination of carcinoembryonic antigen in serum using inductively coupled plasma-mass spectrometry, Anal. Chim. Acta, 847 (2014) 10-15. https://doi.org/10.1016/j.aca.2014.08.041

[19] C.Coman, L.F. Leopold, O.D. Rugină, L. Barbu-Tudoran, N. Leopold, M. Tofană, C. Socaciu, Green synthesis of gold nanoparticles by Allium sativum extract and their assessment as SERS substrate. J. Nanoparticle Res. 16 (2014) 2158. https://doi.org/10.1007/s11051-013-2158-4

[20] M.S.Wason, J.Colon, S. Das, S. Seal, J. Turkson, J. Zhao, C.H. Baker, Sensitization of pancreatic cancer cells to radiation by cerium oxide nanoparticle-induced ROS production. Nanomed. Nanotechnol. Biol. Med., 9 (2013) 558-569. https://doi.org/10.1016/j.nano.2012.10.010

[21] M. Hosseini, M. Mozafari, Cerium oxide nanoparticles: Recent advances in tissue engineering. Materials, 13, (2020) 3072. https://doi.org/10.3390/ma13143072

[22] T. Mori, D.R. Ou, J. Zou, J.Drennan, Present status and future prospect of design of Pt-cerium oxide electrodes for fuel cell applications, Prog. Nat. Sci. Mater. Int., 22 (2012) 561-571. https://doi.org/10.1016/j.pnsc.2012.11.010

[23] V.A.M. Selvan, R.Anand, M. Udayakumar, Effect of Cerium Oxide Nanoparticles and Carbon Nanotubes as fuel-borne additives in Diesterol blends on the performance, combustion and emission characteristics of a variable compression ratio engine, Fuel, 130 (2014) 160-167. https://doi.org/10.1016/j.fuel.2014.04.034

[24] K. Kandhasamy, K. Premkumar, Fabrication of Cerium Oxide Nanoparticles with Improved Antibacterial Potential and Antioxidant Activity, Biosci. Biotech. Res. Asia, 20(2) (2023), 487-497. https://doi.org/10.13005/bbra/3104

[25] L. Ghibelli, S. Mathu, Biological interactions of oxide nanoparticles: The good and the evil. Mrs Bulletin. 39(11) (2014) 949-54. https://doi.org/10.1557/mrs.2014.250

[26] C. Bouzigues, T.Gacoin A. Alexandrou, Biological applications of rare-earth based nanoparticles, Acs Nano, 5 (2011) 8488-8505. https://doi.org/10.1021/nn202378b

[27] S.K. Tarik Aziz, M. Awasthi, S. Guria, M. Umekar, I. Karajagi, S. K. Riyajuddin, K. V. R. Siddhartha, A. Saini, A.K. Potbhare, R.G. Chaudhary, V. Vishal, P. C. Ghosh, A. Dutta, Electrochemical water splitting by a bidirectional electrocatalyst, STAR Protocols, 2023, 4, 102448. https://doi.org/10.1016/j.xpro.2023.102448

[28] R.G. Chaudhary, P.B. Chouke, R.Bagade, A.K. Potbhare, Molecular docking and antioxidant activity of Cleome simplicifolia assisted synthesis of cerium oxide nanoparticles, Mater. Today: Procs, 29 (4), (2020) 1085-1090: doi.org/10.1016/j.matpr.2020.05.062. https://doi.org/10.1016/j.matpr.2020.05.062

[29] X. Beaudoux, M. Virot, Chave T, Durand G, Leturcq G, Nikitenko SI. Vitamin C boosts ceria-based catalyst recycling. Green Chem.18(2016) 3656-3668. https://doi.org/10.1039/C6GC00434B

[30] C. Xu, X. Qu Cerium oxide nanoparticle: a remarkably versatile rare earth nanomaterial for biological applications. NPG Asia Mater., 6, (2014) 90. https://doi.org/10.1038/am.2013.88

[31] S. Das, J.M. Dowding, K.E. Klump, J.F. McGinnis, W. Self, S. Seal Cerium oxide nanoparticles: applications and prospects in nanomedicine, Nanomedicine (Lond). 8(9), (2013) 1483-1508. https://doi.org/10.2217/nnm.13.133

[32] S. Deshpande, S. Patil, S.V. Kuchibhatla, S.Seal, Size dependency variation in lattice parameter and valency states in nanocrystalline cerium oxide. Appl Phys Lett. 87(13), (2005), 33113. https://doi.org/10.1063/1.2061873

[33] I.Celardo, J. Z.Pedersen, E.Traversa, L.Ghibelli, Pharmacological potential of cerium oxide nanoparticles. Nanoscale 3, (2011) 1411-1420. https://doi.org/10.1039/c0nr00875c

[34] Can Xu, Youhui Lin, Jiasi Wang, Li Wu, Weili Wei, Jinsong Ren, Xiaogang Qu,. Nanoceria-triggered synergetic drug release based on CeO2-capped mesoporous silica host-guest interactions and switchable enzymatic activity and cellular effects of CeO2. Adv. Healthcare Mater. 2(2013) 1591-1599. https://doi.org/10.1002/adhm.201200464

[35] M. Li, P. Shi, C. Xu, J. S. Ren, X. G. Qu, Cerium oxide caged metal chelator: anti-aggregation and anti-oxidation integrated H2O2-responsive controlled drug release for potential Alzheimer's disease treatment. Chem. Sci. 4, (2013), 2536-2542. https://doi.org/10.1039/c3sc50697e

[36] A. Asati, S. Santra, C. Kaittanis, S. Nath, J. M. Perez, Oxidase-like activity of polymer-coated cerium oxide nanoparticles. Angew. Chem. Int. Ed. 48, (2009) 2308-2312. https://doi.org/10.1002/anie.200805279

[37] A. Asati, C. Kaittanis, S. Santra, J. M. Perez, pH-tunable oxidase-like activity of cerium oxide nanoparticles achieving sensitive fluorigenic detection of cancer biomarkers at neutral pH. Anal. Chem. 83, (2011), 2547-2553. https://doi.org/10.1021/ac102826k

[38] X. Li, L. Sun, A. Ge, Y. Guo, Enhanced chemiluminescence detection of thrombin based on cerium oxide nanoparticles. Chem. Commun. 47, (2011), 947-949. https://doi.org/10.1039/C0CC03750H

[39] C. Kaittanis, S. Santra, A. Asati, J. M. Perez, A cerium oxide nanoparticle-based device for the detection of chronic inflammation via optical and magnetic resonance imaging. Nanoscale (2012), 2117-2123. https://doi.org/10.1039/c2nr11956k

[40] Umekar, M. S., Chaudhary, R. G., Bhusari, G. S., Mondal, A., Potbhare, A. K., & Sami, M. (2020). Phytoreduced graphene oxide-titanium dioxide nanocomposites using Moringa oleifera stick extract. Materials Today: Proceedings, 29, 709-714. https://doi.org/10.1016/j.matpr.2020.04.169

[41] Y. H. Lin, C. Xu, J. S. Ren, X. G. Qu, Using Thermally Regenerable Cerium Oxide Nanoparticles In Biocomputing To Perform Label-Free, Resettable, And Colorimetric Logic Operations. Angew. Chem. Int. Ed. 51, (2012) 12579-12583. https://doi.org/10.1002/anie.201207587

[42] A. S. Karakoti, O. Tsigkou, S. Yue, , P. D. Lee, M. M.Stevens, J. R. Jones, S.Seal, Rare earth oxides as nanoadditives in 3-D nanocomposite scaffolds for bone regeneration. J. Mater. Chem. 20, (2010) 8912-8919. https://doi.org/10.1039/c0jm01072c

[43] Mohd Aslam Saifi, Sudipta Seal, Chandraiah Godugu, Nanoceria, the versatile nanoparticles: Promising biomedical applications, Journal of Controlled Release 338 (2021)168-189 https://doi.org/10.1016/j.jconrel.2021.08.033

[44] A. Arumugam, C. Karthikeyan, A.S. Haja Hameed, K. Gopinath, S. Gowri, V. Karthika Synthesis of cerium oxide nanoparticles using Gloriosa superba L. leaf extract and their structural, optical and antibacterial properties. Mater. Sci. Eng. C Mater. Biol. Appl., 49: (2015) 408-415. 34. https://doi.org/10.1016/j.msec.2015.01.042

[45] S.K. Kannan, M. Sundrarajan, A green approach for the synthesis of a cerium oxide nanoparticle: characterization and antibacterial activity, Int. J. Nanosci. 13(03), (2014)1450018. https://doi.org/10.1142/S0219581X14500185

[46] G.S. Priya, A. Kanneganti, K.A. Kumar, K.V. Rao, S. Bykkam, Bio synthesis of cerium oxide nanoparticles using Aloe arbadensis Miller Gel. Int. J. Sci. Res. Publications.;4(6) (2014) 1.

[47] S.K. Tarik Aziz, M. Umekar, I. Karajagi, S.K. Riyajuddin, K.V.R. Siddhartha, A. Saini, A.K. Potbhare, R.G. Chaudhary, V. Vishal, P.C. Ghosh, A. Dutta. A Janus cerium-doped bismuth oxide electrocatalyst for complete water splitting, Cell Reports: Physical Science, 2022, 3(11) 101106. https://doi.org/10.1016/j.xcrp.2022.101106

[48] M. Darroudi, M. Sarani, R. Kazemi Oskuee, A. Khorsand Zak, H.A. Hosseini, L. Gholami, Green synthesis and evaluation of metabolic activity of starch mediated nanoceria. Ceramics Int. 40(1, Part B) (2014), 2041-2045. https://doi.org/10.1016/j.ceramint.2013.07.116

[49] H Kargar, H. Ghazavi, M.Darroudi Size-controlled and bio-directed synthesis of ceria nanopowders and their in vitro cytotoxicity effects. Ceramics Int.,41(3, Part A): (2015), 4123-4128. https://doi.org/10.1016/j.ceramint.2014.11.108

[50] M. Darroudi, Hoseini SJ, Kazemi Oskuee R, Hosseini HA, Gholami L, Gerayli S. Food-directed synthesis of cerium oxide nanoparticles and their neurotoxicity effects. Ceramics Int.,40(5): (2014), 7425-7430. https://doi.org/10.1016/j.ceramint.2013.12.089

[51] P.B. Chouke, T. Shrirame, A.K. Potbhare, A. Mondal, A.R. Chaudhary, S. Mondal, S.R. Thakare, E. Nepovimova, M. Valis, K. Kuca, R. Sharma, R.G. Chaudhary, Bioinspired

metal/metal oxide nanoparticles: A road map to potential applications, Mater. Today Adv. 16(2022) 100314. https://doi.org/10.1016/j.mtadv.2022.100314

[52] L. Qi, J. Fresnais, P. Mullera, O. Theodoly, F. Berretb, P. Chapel. Interfacial activity of phosphonated-polyethylene glycol functionalized cerium oxide nanoparticles. Langmuir; 28(31) (2012)11448-11456. https://doi.org/10.1021/la302173g

[53] A. Kaushik, P.R. Solanki, M.K. Pandey, S. Ahmad, B.D. Malhotra. Cerium oxide-chitosan based nanobiocomposite for food borne mycotoxin detection. Appl Phys Lett., 95(17), (2009), 173703. https://doi.org/10.1063/1.3249586

[54] Y.H. Liu, et al., Synthesis and character of cerium oxide (CeO₂) nanoparticles by the precipitation method, Metalurgija, 53 (4) (2014) 463-465.

[55] Q. L. Zhang, Y. M. Zhi, B. J. Ding, Synthesis of nanoceria by the precipitation method, Mater. Sci. Forum, Trans Tech Publications, vol. 610, (2009). https://doi.org/10.4028/www.scientific.net/MSF.610-613.233

[56] P. Kavitha, et al., Synthesis and characterization of nanoceria by using rapid precipitation method, PARIPEX-Indian J. Res. 4 (12) (2016).

[57] M. Farahmandjou, M. Zarinkamar, T.P. Firoozabadi, Synthesis of cerium oxide (CeO₂) nanoparticles using simple CO-precipitation method, Rev. Mex. Fis. 62,(2016) 496-499

[58] K.K. Babitha, et al., Structural characterization and optical studies of CeO₂ nanoparticles synthesized by chemical precipitation, Indian J. Pure Appl. Phys. (IJPAP) 53 (9) (2015) 596-603.

[59] J. Ketzial, A. Jasmine, S. Nesaraj, Synthesis of CeO₂ nanoparticles by chemical precipitation and the effect of a surfactant on the distribution of particle sizes, J. Ceram. Process. Res. 12 (1) (2011) 74-79.

[60] B.S. Shirke, et al., Synthesis of nanoceria by microwave technique using propylene glycol as a stabilizing agent, J. Mater. Sci. Mater. Electron. 22 (2) (2011) 200-203. https://doi.org/10.1007/s10854-010-0114-y

[61] Siba Soren, et al., Antioxidant potential and toxicity study of the nanoceria synthesized by microwave-mediated synthesis, Appl. Biochem. Biotechnol. 177 (1) (2015) 148-161. https://doi.org/10.1007/s12010-015-1734-8

[62] D. V. Pinjari, A. B. Pandit, Room temperature synthesis of crystalline CeO2 nanopowder: advantage of sonochemical method over conventional method, Ultrason. Sonochem. 18 (5) (2011) 1118-1123. https://doi.org/10.1016/j.ultsonch.2011.01.008

[63] Lunxiang Yin, et al., Sonochemical synthesis of nanoceria-effect of additives and quantum size effect, J. Colloid Interface Sci. 246 (1) (2002) 78-84. https://doi.org/10.1006/jcis.2001.8047

[64] S. I. Mutinda, Hydrothermal synthesis of shape/size-controlled cerium-based oxides, Dissertations, Youngstown State University, (2013).

[65] O. Kepenekci, Hydrothermal preparation of single crystalline CeO₂ nanoparticles and the influence of alkali hydroxides on their structure and optical behavior, MS Thesis, İzmir Institute of Technology (2009).

[66] T. Masui, et al., Synthesis of nanoceria by hydrothermal crystallization with citric acid, J. Mater. Sci. Lett. 21 (6) (2002) 489-491. https://doi.org/10.1023/A:1015342925372

[67] T. Masui, et al., Synthesis and characterization of nanoceria coated with turbostratic boron nitride, J. Mater. Chem. 13 (3) (2003) 622-627. https://doi.org/10.1039/b208109a

[68] M. Jalilpour, M. Fathalilou, Effect of aging time and calcination temperature on the nanoceria synthesis via reverse co-precipitation method, Int. J. Phys. Sci. 7 (6) (2012) 944-948. https://doi.org/10.5897/IJPS11.131

[69] F. Gao, Q. Lu, S. Komarneni, Fast synthesis of nanoceria and nanorods, J. Nanosci. Nanotechnol. 6 (12) (2006) 3812-3819. https://doi.org/10.1166/jnn.2006.609

[70] E.Alpaslan, M. B. Geilich; H. Yazici, T. Webster, pH-Controlled Cerium Oxide nanoparticle inhibition of both gram-positive and gram-negative bacteria growth., J. Sci. Rep. 7 (2017) 45859. https://doi.org/10.1038/srep45859

[71] P. L. Chen, I.W. Chen, Reactive Cerium (IV) Oxide Powders by the Homogeneous Precipitation Method. J. Am. Ceram. Soc., 76, (1993) 1577-1583. https://doi.org/10.1111/j.1151-2916.1993.tb03942.x

[72] H.I. Chen, H.-Y. Chang, Synthesis of nanocrystalline cerium oxide particles by the Precipitation method. Ceram. Int., 31 (2005), 795-802. https://doi.org/10.1016/j.ceramint.2004.09.006

[73] M. Ramachandran, R. Subadevi, M. Sivakumar, Role of pH on synthesis and characterization of cerium oxide (CeO2) nano particles by modified co-precipitation method. Vacuum, 161, (2019), 220-224. https://doi.org/10.1016/j.vacuum.2018.12.002

[74] H.S. Nanda, Preparation and Biocompatible Surface Modification of Redox Altered Cerium Oxide Nanoparticle Promising for Nanobiology and Medicine. Bioengineering ,3 ,28(2016) https://doi.org/10.3390/bioengineering3040028

[75] S.K. Nethi, H.S. Nanda, T.W. Steele, C.R. Patra, Functionalized nanoceria exhibit improved angiogenic properties. J. Mater. Chem. B, 5 (2017) 9371-9383. https://doi.org/10.1039/C7TB01957B

[76] F. Corsi, F. Caputo, E. Traversa, L. Ghibelli, Not Only Redox: The Multifaceted Activity of Cerium Oxide Nanoparticles in Cancer Prevention and Therapy. Front. Oncol., 8 (2018), 309. https://doi.org/10.3389/fonc.2018.00309

[77] A. Arya, N.K. Sethy, M. Das, S.K. Singh, A. Das, S.K. Ujjain, R.K. Sharma, M. Sharma, K. Bhargava, Cerium oxide nanoparticles prevent apoptosis in primary cortical culture by stabilizing mitochondrial membrane potential. Free Radic Res. 48, (2014, 784-793. https://doi.org/10.3109/10715762.2014.906593

[78] T. Masui, Characterization of Cerium (IV) Oxide Ultrafine Particles Prepared Using Reversed Micelles. Chem. Mater., 9 (1997), 2197-2204. https://doi.org/10.1021/cm970359v

[79] A. Cimini, B. D'Angelo, S. Das, R. Gentile, E. Benedetti, V.Singh, A.M. Monaco, S. Santucci, S. Seal, Antibody-conjugated PEGylated cerium oxide nanoparticles for specific targeting of Aβ aggregates modulate neuronal survival pathway, Acta Biomater, 8 (2012), 2056-2067. https://doi.org/10.1016/j.actbio.2012.01.035

[80] J.M. López, A.L. Gilbank, T. García, B. Solsona, S. Agouram, L. M. Torrente-, The prevalence of surface oxygen vacancies over the mobility of bulk oxygen in nanostructured ceria for the total toluene oxidation. Appl. Catal. B Environ., 174, (2015), 403-412. https://doi.org/10.1016/j.apcatb.2015.03.017

[81] A.I.Y. Tok, Hydrothermal synthesis of CeO2 nano-particles. J. Mater. Process. Technol., 190, (2007)217-222. https://doi.org/10.1016/j.jmatprotec.2007.02.042

[82] I. Trenque, G.C. Magnano, M.A. Bolzinger, L. Roiban, F. Chaput, I. Pitault, S. Briançon, T.Devers, K. Masenelli-Varlot, M. Bugnet, et al. Shape-selective synthesis of nanoceria for degradation of paraoxon as a chemical warfare simulant. Phys. Chem. Chem. Phys., 21, (2019), 5455-5465. https://doi.org/10.1039/C9CP00179D

[83] H. Zhang, X.He, Z. Zhang, P.Zhang, Y. Li, Y.Ma, Y.Kuang, Y. Zhao, Z. Chai, Nano-CeO2 exhibits adverse effects at environmental relevant concentrations. Env. Sci. Technol., 45, (2011) 3725-3730. https://doi.org/10.1021/es103309n

[84] S. Soren, S.R. Jena, L. Samanta, P. Parhi, Antioxidant potential and toxicity study of the cerium oxide nanoparticles synthesized by microwave-mediated synthesis. Appl. Biochem. Biotechnol., 177, (2015), 148-161. https://doi.org/10.1007/s12010-015-1734-8

[85] S. Machmudah, S. Winardi, H. Kanda, M. Goto, Synthesis of Ceria Zirconia Oxides using Solvothermal Treatment. In MATEC Web of Conferences; EDP Sciences: Les Ulis, France (2018). https://doi.org/10.1051/matecconf/201815605014

[86] T. Yu, J. Joo, Y.I. Park, T. Hyeon, Large-scale nonhydrolytic sol-gel synthesis of uniform-sized ceria nanocrystals with spherical, wire, and tadpole shapes. Angew. Chem. Int. Ed. Engl., 44, (2005), 7411-7414. https://doi.org/10.1002/anie.200500992

[87] M. Darroudi, Green synthesis and evaluation of metabolic activity of starch mediated nanoceria. Ceram. Int., 40, (2014) 2041-2045. https://doi.org/10.1016/j.ceramint.2013.07.116

[88] B.Elahi, M. Mirzaee, M.Darroudi, K. Sadri, R.K. Oskuee, Bio-based synthesis of Nano-Ceria and evaluation of its bio-distribution and biological properties, Colloids Surf. B Biointerfaces, 181, (2019),830-836. https://doi.org/10.1016/j.colsurfb.2019.06.045

[89] S.Gnanam, V. Rajendran, Synthesis of CeO2 or α-Mn2O3 nanoparticles via sol-gel process and their optical properties. J. Solgel Sci. Technol., 58, (2011) 62-69. https://doi.org/10.1007/s10971-010-2356-9

[90] E. Nourmohammadi, Cytotoxic activity of greener synthesis of cerium oxide nanoparticles using carrageenan towards a WEHI 164 cancer cell line. Ceram. Int., 44, (2018) 19570-19575. https://doi.org/10.1016/j.ceramint.2018.07.201

[91] P. Chouke, A. Potbhare, P. Dadure, A. Mungole, N. Meshram, R. Chaudhary, A. Rai, R. Chaudhary, An antibacterial activity of Bauhinia racemosa assisted ZnO nanoparticles during lunar eclipse and docking assay, Materials Today: Proceedings, 29 (2020) 815-821. https://doi.org/10.1016/j.matpr.2020.04.758

[92] P. B Chouke, A. K Potbhare, G. S Bhusari, S. Somkuwar, D. PMD Shaik, R. K Mishra, R. Gomaji Chaudhary, Green fabrication of zinc oxide nanospheres by Aspidopterys cordata for effective antioxidant and antibacterial activity, Advanced Materials Letters, 10 (2019) 355-360. https://doi.org/10.5185/amlett.2019.2235

[93] M. Umekar, R. Chaudhary, G. Bhusari, A. Potbhare, Fabrication of zinc oxide-decorated phytoreduced graphene oxide nanohybrid via Clerodendrum infortunatum, Emerging Materials Research, 10 (2021) 75-84. https://doi.org/10.1680/jemmr.19.00175

[94] Al-Qudah Tamara, M.H. Sami, Abu-Zurayk Rund, Shibli Rida, Khalaf Aya, T.L. Lambat, R.G. Chaudhary, Nanotechnology applications in plant tissue culture and molecular genetics:

a holistic approach, Current Nanoscience, 18 (4) (2022) 442-464.
https://doi.org/10.2174/1573413717666211118111333

[95] A. Haldar, K.M. Dadure, D. Mahapatra, R.G. Chaudhary, Natural extracts-mediated biosynthesis of Zinc Oxide nanoparticles and their multiple pharmacotherapeutic perspectives, Jordan Journal of Physics, 15 (2022) 67-79. https://doi.org/10.47011/15.1.10

[96] R. G. Chaudhary, N. B. Singh, A. R. Daddemal-Chaudhary and Rohit Sharma, Review on Agrobiowaste-mediated nanohybrids for removal of toxic heavy metals from wastewater, ChemistrySelect, 2024, 9(4) e202304230. https://doi.org/10.1002/slct.202304230.
https://doi.org/10.1002/slct.202304230

[97] A.K. Potbhare, R.G. Chaudhary, P.B. Chouke, S.Yerpude, A. Mondal, V.N. Sonkusare, A.R. Rai, H.D. Juneja. Phytosynthesis of nearly monodisperse CuO nanospheres using Phyllanthus reticulatus/Conyza bonariensis and its antioxidant/antibacterial assays. Materials Science and Engineering: C 99 (2019): 783-793. https://doi.org/10.1016/j.msec.2019.02.010

[98] Q. Maqbool, M. Nazar, S. Naz, T. Hussain, N. Jabeen, R. Kausar, S. Anwaar, F. Abbas, T.Jan, Antimicrobial potential of green synthesized CeO2 nanoparticles from Olea europaea leaf extract, Int. J. Nanomed. 11 (2016) 5015-5025. https://doi.org/10.2147/IJN.S113508

[99] J. Malleshappa, H. Nagabhushana, S. C. Sharma, Y.S. Vidya, K.S. Anantharaju, S. C. Prashantha, B. Daruka Prasad, H. Raja Naika, K. Lingaraju, B. S. Surendra, Leucas aspera mediated multifunctional CeO2 nanoparticles: Structural, Photoluminescent, Photocatalytic and Antibacterial properties DOI: http://dx.doi.org/10.1016/j.saa.(2015),04.073

[100] VV Makarov, SS Makarova, AJ Love, et al. Biosynthesis of stable iron oxide nanoparticles in aqueous extracts of Hordeum vulgare and Rumex acetosa plants. Langmuir; 30(20): (2014), 5982-8 https://doi.org/10.1021/la5011924

[101] A. M. Korotkova, P. O. Borisovna, G. I. Aleksandrovna et al., Green Synthesis of Cerium Oxide Particles in Water Extracts Petroselinum crispum, Current Nanomaterials, 4, (2019)176-190. https://doi.org/10.2174/2405461504666190911155421

[102] T. Arunachalam, M. Karpagasundaram, N. Rajarathinam, Ultrasound assisted green synthesis of cerium oxide nanoparticles using Prosopis juliflora leaf extract and their structural, optical and antibacterial properties, Materials Science-Poland, 35(4), (2017), 791-798. https://doi.org/10.1515/msp-2017-0104

[103] F. Avadi, ME Yazdi Taghavizadeh, M. Baghani et al. Biosynthesis, characterization of cerium oxide nanoparticles using Ceratonia siliqua and evaluation of antioxidant and cytotoxicity activities. Materials Research Express 6(6)(2019). https://doi.org/10.1088/2053-1591/ab08ff

[104] P. Nithya, B Murugesan, J Sonamuthu, S Samayanan, S.Mahalingam Facile biological synthetic strategy to morphologically aligned CeO 2/ZrO 2 core nanoparticles using Justicia adhatoda extract and ionic liquid: enhancement of its bio-medical properties. J. Photochem. Photobiol B. (2017);178. https://doi.org/10.1016/j.jphotobiol.2017.11.036

[105] S. Nezhad, A Haghi, M. Homayouni Green synthesis of cerium oxide nanoparticle using Origanum majorana L. leaf extract, its characterization and biological activities: green synthesis of nanoparticle. Appl. Organomet Chem. 34,(2019) 5314
https://doi.org/10.1002/aoc.5314

[106] R. Magudieshwaran, J. Ishii, KCN Raja, et al. Green and chemical synthesized CeO2 nanoparticles for photocatalytic indoor air pollutant degradation. Mater Lett. 239 (2019) 40-44. https://doi.org/10.1016/j.matlet.2018.11.172

[107] P Nithya, B Murugesan, J Sonamuthu, S Samayanan, S.Mahalingam [BMIM] PF6 ionic liquid mediated green synthesis of ceramic SrO/ CeO2 nanostructure using Pedalium murex leaf extract and their antioxidant and antibacterial activities. Ceramic Int. (2019).

[108] A Singh, I. Hussain, NB Singh, et al. Uptake, translocation and impact of green synthesized nanoceria on growth and antioxidant enzymes activity of Solanum lycopersicum L. Ecotoxicol Environ Saf.;182:109410. https://doi.org/10.1016/j.ecoenv.2019.109410

[109] D. Pinheiro, KR Sunaja Devi, A Jose, et al. Experimental design for optimization of 4-nitrophenol reduction by green synthesized CeO2/ g-C3N4/Ag catalyst using response surface methodology. J. Rare Earths. doi:10.1016/j.jre.(2019)10.001

[110] I. M. Sultan, MH Aziz, M Fatima et al. Green synthesis, cytotoxicity, antioxidant and photocatalytic activity of CeO2 nanoparticles mediated via orange peel extract (OPE). Mater Res Express. (2019)

[111] J. Malleshappa, H Nagabhushana, SC Prashantha, et al. Eco-friendly green synthesis, structural and photoluminescent studies of CeO2: eu3+nanophosphors using E. tirucalli plant latex. J. Alloys Compd. 612 (2014) 425-434. https://doi.org/10.1016/j.jallcom.2014.05.101

[112] LS Reddy Yadav, K Manjunath, B,Archana et al. Fruit juice extract mediated synthesis of CeO2 nanoparticles for antibacterial and photocatalytic activities. Eur Phys J Plus. 131(5):154. https://doi.org/10.1140/epjp/i2016-16154-y

[113] S Maensiri, S Labuayai, P Laokul, et al. Structure and optical properties of CeO2 nanoparticles prepared by using lemongrass plant extract solution. Jpn J Appl Phys. 53(6S):06JG14. https://doi.org/10.7567/JJAP.53.06JG14

[114] E. Behrouz, M. Mahdi, D. Majid, R. K. Oskuee, K. Sadri, L.eila Gholami Role of oxygen vacancies on photo-catalytic activities of green synthesized ceria nanoparticles in Cydonia oblonga miller seeds extract and evaluation of its cytotoxicity effects PII: S0925-8388(19)33799-5. https://doi.org/10.1016/j.jallcom.(2019)152553

[115] Behrouz Elahi, Mahdi Mirzaee, Majid Darroudi, Reza Kazemi Oskuee, Kayvan Sadri, Mohammad Sadegh Amiri Preparation of Cerium Oxide Nanoparticles in Salvia MacrosiphonBoiss seeds Extract and Investigation of Their Photo-catalytic Activities PII: S0272-8842(18)33291. https://doi.org/10.1016/j.ceramint.(2018)11.173

[116] F. Sadat Sangsefidi, M. Nejati, J. Verdi, M. Salavati-Niasari, Green synthesis and characterization of cerium oxide nanostructures in the presence carbohydrate sugars as a capping agent and investigation of their cytotoxicity on the mesenchymal stem cell. Journal of Cleaner Production, 156 (2017) 741e749. https://doi.org/10.1016/j.jclepro.2017.04.114

[117] M Darroudi, SJ Hoseini, R Kazemi Oskuee, et al. Food-directed synthesis of cerium oxide nanoparticles and their neurotoxicity effects. Ceram Int. 2014;40(5):7425-7430. doi:10.1016/j.ceramint.2013.12.089. https://doi.org/10.1016/j.ceramint.2013.12.089

[118] H Kargar, H Ghazavi, M Darroudi. Size-controlled and bio-directed synthesis of ceria nanopowders and their in vitro cytotoxicity effects. Ceram Int. 2015;41(3):4123-4128. doi:10.1016/j.ceramint.2014.11. https://doi.org/10.1016/j.ceramint.2014.11.108

[119] I. Milenković, K. Radotić, B. Matović, M. Prekajski, L. Živković, D. Jakovljević, G. Gojgić-Cvijović, V. Beškoski, Improving stability of cerium oxide nanoparticles by microbial polysaccharides coating J. Serb. Chem. Soc. 83 (6) (2018)745-757. https://doi.org/10.2298/JSC171205031M

[120] M Darroudi, MB Ahmad, AH Abdullah, NA. Ibrahim Green synthesis and characterization of gelatin-based and sugar-reduced silver nanoparticles. Int J Nanomedicine. 6:5(2011)69-574. https://doi.org/10.2147/IJN.S16867

[121] H Kargar, F Ghasemi, M.Darroudi Bioorganic polymer-based synthesis of cerium oxide nanoparticles and their cell viability assays. Ceramics Int. 41(1, Part B) (2015)1589-1594. https://doi.org/10.1016/j.ceramint.2014.09.095

[122] JM Perez, A Asati, S Nath, C Kaittanis Synthesis of biocompatible dextran-coated nanoceria with pH-dependent antioxidant properties. Small 4: (2008) 552-556. https://doi.org/10.1002/smll.200700824

[123] A Sehgal, Y Lalatonne, JF Berret, M Morvan, Precipitationre dispersion of cerium oxide nanoparticles with poly (acrylic acid): toward stable dispersions. Langmuir, 21(2005) 9359-9364. https://doi.org/10.1021/la0513757

[124] AS Karakoti, S Singh, A Kumar, M Malinska, SVNT Kuchibhatla, K Wozniak, WT Self, S Seal, PEGylated nanoceria as radical scavenger with tunable redox chemistry. J Am Chem Soc 131 (2009)14144-14145. https://doi.org/10.1021/ja9051087

[125] L Qi, J Fresnais, P Muller, O Theodoly, J-F Berret, J-P Chapel Interfacial activity of phosphonated-peg functionalized cerium oxide nanoparticles. Langmuir 28:(2012)11448-11456. https://doi.org/10.1021/la302173g

[126] JA Vassie, JM Whitelock, MS Lord Targeted delivery and redox activity of folic acid-functionalized nanoceria in tumor cells. Mol Pharm 15(2018), 994-1004. https://doi.org/10.1021/acs.molpharmaceut.7b00920

[127] A. Rana, I. Hasan, B. Heun Koo, R. Ahmad Khan, Green synthesized CeO2 nanowires immobilized with alginate-ascorbic acid biopolymer for advance oxidative degradation of crystal violet, Colloids and Surfaces A: Physicochemical and Engineering Aspects, 637, (2022), 128225. https://doi.org/10.1016/j.colsurfa.2021.128225

[128] M Darroudi, M Sarani, R Oskuee Kazemi A, Zak Khorsand, MS Amiri. Nanoceria: gum mediated synthesis and in vitro viability assay. Ceramics Int.40(2),(2014)2863-2868. https://doi.org/10.1016/j.ceramint.2013.10.026

[129] A Kaushik, PR Solanki, MK Pandey, S Ahmad, BD. Malhotra Cerium oxide-chitosan based nanobiocomposite for food borne mycotoxin detection. Appl Phys Lett. 95(17),(2009)173703. https://doi.org/10.1063/1.3249586

[130] H. Hassanneja, A.Nouri Synthesis and evaluation of self-healing cerium-doped chitosan nanocomposite coatings on AA5083-H321. Int J Electrochem Sci. 11(2016)2106-2118. https://doi.org/10.1016/S1452-3981(23)16086-X

[131] V. Shah, S.Shah , H. Shah, F. J. Rispoli, K. T. McDonnell, S. Workeneh , A. Karakoti, A. Kumar, Sudipta Seal, Antibacterial Activity of Polymer Coated Cerium Oxide Nanoparticles.7 10 (2012) e47827. https://doi.org/10.1371/journal.pone.0047827

[132] V.V. Spiridonov, A. V. Sybachin, V. A. Pigareva, M. I. Afanasov, S. A. Musoev, A. V. Knotko, S. B. Zezin, One-Step Low Temperature Synthesis of CeO2 Nanoparticles Stabilized

by carboxymethyl cellulose, Polymers 15,(2023) 1437.
https://doi.org/10.3390/polym15061437

[133] R. Tekupalli Biosynthesis of Cerium Oxide Nanoparticles by Fungus Trichoderma viridae, International Journal of Pharmacy and Biological Sciences-IJPBSTM (1) (2019)718-724

[134] SA Khan, A.Ahmad Fungus mediated synthesis of biomedically important cerium oxide nanoparticles. Mater Res Bull. 48 (10),(2013)4134-4138.
https://doi.org/10.1016/j.materresbull.2013.06.038

[135] K Gopinath, V Karthika, C Sundaravadivelan, et al. Mycogenesis of cerium oxide nanoparticles using Aspergillus niger culture filtrate and their applications for antibacterial and larvicidal activities. J Nanostructure Chem. 5(3)(2015):295-303.
https://doi.org/10.1007/s40097-015-0161-2

[136] V S. Kunga, K Gopinath, NS Palani et al. Plant pathogenic fungus F. solani mediated biosynthesis of Nanoceria: antibacterial and antibiofilm activity. RSC Adv. 6(2016) 42720-9.
https://doi.org/10.1039/C6RA05003D

[137] S Munusamy, K Bhakyaraj, L Vijayalakshmi, A Stephen, V Narayanan, Synthesis and characterization of cerium oxide nanoparticles using Curvularialunata and their antibacterial properties. Int J Innov Res Sci Eng. 2(1)(2014)318.

[138] S. Balaji, B.K. Mandal, L. V. Kumar Reddy, D. Sen, Biogenic Ceria Nanoparticles (CeO2 NPs) for Effective Photocatalytic and Cytotoxic Activity, Bioengineering, 7(2020) 26.
https://doi.org/10.3390/bioengineering7010026

[139] B. Bhushan, P. Gopinath. Antioxidant nanozyme: a facile synthesis and evaluation of the reactive oxygen species scavenging potential of nanoceria encapsulated albumin nanoparticles. J. Mater Chem B. 3(24) (2015)4843-4852.
https://doi.org/10.1039/C5TB00572H

[140] T.Pirmohamed, J.M.Dowding,; S.Singh, B. Wasserman, E. Heckert, A.S. Karakoti,; J.E.King,; S.Seal, W.T. Self, Nanoceria exhibit redox state-dependent catalase mimetic activity. Chem. Commun. 46, (2010)2736-2738. https://doi.org/10.1039/b922024k

[141] P. Cohen, The origins of protein phosphorylation. Nat. Cell Biol. 2002, 4, E127
https://doi.org/10.1038/ncb0502-e127

[142] M. Das, S. Patil, N. Bhargava, J.-F. Kang, L. M. Riedel, S. Seal and J. J. Hickman, Biomaterials, 28 (2007),1918-192. https://doi.org/10.1016/j.biomaterials.2006.11.036

Green Synthesis and Emerging Applications of Frontier Nanomaterials   Materials Research Forum LLC
Materials Research Foundations 169 (2024) 197-230   https://doi.org/10.21741/9781644903278-8

Chapter 8

Conventional and green synthesis techniques of carbon nanotubes and its environmental/biomedical applications

Rohit S. Madankar[1], Pavan R. Bhilkar[1], Ajay K. Potbhare[1], Ankita R. Daddemal-Chaudhary[2,3], Mayuri S. Umekar[1], Ashish P. Lambat[4], Sudip Mondal[1], Ratiram G. Chaudhary[1]* and Ahmed A. Abdala[5]*

[1]Post Graduate Department of Chemistry, Seth Kesarimal Porwal College of Arts and Science and Commerce, RTM Nagpur University, Kamptee-441001, Nagpur, Maharashtra, India

[2]L.A.D College for Women of Arts, Commerce and Science, Nagpur-440010, India

[3]Post Graduate Department of Botany, R.T.M. Nagpur University, Nagpur-440033, India

[4]Sevadal Mahila Mahavidyalaya, Umred Road, Nagpur-440024, India

[5]College of Science and Engineering, Hamada Bin Khalifa University, Qatar Foundation, Doha, Qatar

* chaudhary_rati@yahoo.com, ahabdalla@hbku.edu.qa

Abstract

Carbon nanotubes (CNTs) are popularly known for their incredible applications because of their outstanding physicochemical properties like high surface area, scaffold morphology, functional versatility, high electrical and thermal conductivity. The production of CNTs on a large scale is very costly and non-environmentally. Nonetheless, few methods of scalable fabrication are existed, besides an eco-friendly and cost-effective fabrication of CNTs. An eco-friendly fabrication includes microbes-mediated, plant extracts-mediated and agricultural bio-wastes-mediated. Moreover, research community ought to adopt some natural and renewable sources for CNTs production in large scale. Indeed, these approaches provides a cost-effectiveness, ecofriendly, straightforward, environmental benefits and so forth. Moreover, the properties of CNTs can be improved by doping with some precious transition metals for their advanced applications like environmental and biomedical. Therefore, with this perspective the present chapter is focuses on the conventional and eco-friendly synthesis, types of CNTs, environmental and biomedical applications.

Keywords

Carbon Nanotube (CNTs), Types of CNTs, Green Synthesis, Environmental Applications of CNTs, Biomedical Applications of CNTs

Contents

1.   Introduction

Over billions of years, biological systems were developed into complex structures with advanced functions. Consequently, drawing inspiration from these living organisms to design a novel material with exceptional properties has proven invaluable for addressing some challenging issue, for instance an environmental mitigation. To date, various synthetic materials have been developed with specific structures, properties, and functions. Recent examples of bioinspired materials often can be fabricated from natural resources or living organisms. Thus, the biogenic source (wheat straw, rice husk, sugarcane bagasse, coconut peal ash, etc.) has been utilized for the production of different types of carbon-based materials [1]. The biogenic carbon-based materials have revolutionized various scientific and technological fields because of their versatility and exceptional properties. For instance, from graphene to carbon nanotubes, fullerenes and activated-carbon exhibits unique properties viz. structure, electrical, and mechanical that make indispensable for wide range of applications in energy storage, environmental remediation, biomedical applications and beyond.

The carbon-based nanomaterials (CNMs) are not only widely used in domestic applications but also play a crucial role in industries such as metallurgy, medicine, optics, and environmental science. The rapid urbanization and industrialization have significantly increased the demand for new CNMs viz. hybrid materials (HMs). In the development of HMs, the individual properties of different materials are combined to create synergistic effects, resulting in advanced smart materials. These hybrid materials inherit certain properties and functions from their components, but their structures are distinct. The success of an HMs largely depends on its internal structure, and therefore, its physicochemical properties can be precisely controlled by manipulating this aspect. HMs are various forms including metal-metal oxide nanocomposites (NCs), mixed metal oxide NCs, polymer NCs, and chitin/chitosan NCs. The HMs exhibits an improved mechanical [2-5], electrical [6-8], thermal [9-11], sorptive [12-13], high surface area/porosity [14-17] and catalytic properties [18-20]. The various carbon composite materials including CNTs, graphene, graphite, carbon nanofibers, carbon fibrous materials and diamonds have been extensively developed in recent decades [21-28]. In particular, CNTs, offer numerous advantages, resulting in filamentous structures with high specific surface areas.

For instance, Radushkevich and Lukyanovich first reported CNTs in 1952, with Iijima providing a detailed scientific report in 1991. CNTs typically exhibit diameters ranging from a few nanometres to few millimetres. It is possible to have a single-layered CNTs (Single-wall carbon nanotubes: SWCNT's) or multi-layered CNTs (multi-wall carbon nanotubes: MWCNT's), depending on the number of layers (**Figure 1**).

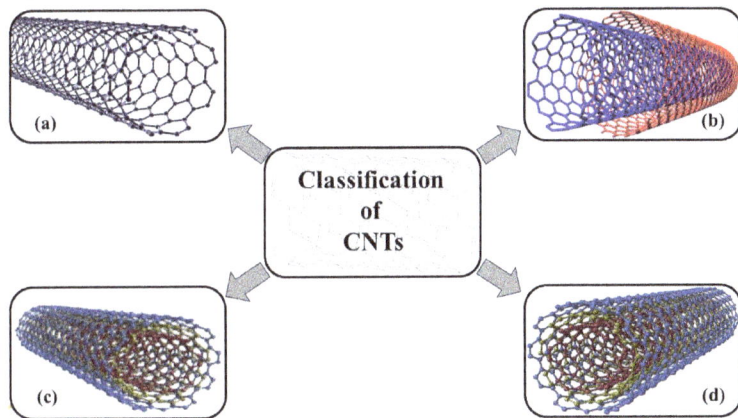

Figure 1: Classification of CNTs: (a) SWCNT, (b) DWCNT, (c) TWCNT and (d) MWCNT

A significant part of the appeal of MWCNTs is their cost-effectiveness compared to SWCNTs. MWCNTs consist of concentric pipes of graphene, while SWCNTs are individual graphene layers that can function as semiconductors or metals. The structure of CNTs is inherently porous. Similar to CNTs, carbon nanofibers (CNFs) also exhibit notable mechanical and electrical properties, though these can vary based on their structure, composition, and manufacturing process. CNFs are nano-sized carbon materials utilized in various applications due to their unique properties. However, there are key differences between CNTs and CNFs. CNFs are not hollow and can be characterized by the arrangement of their graphene layers, which may be perpendicular. Their structure, influenced by the growth mechanism, depends on the geometric properties of the metal catalyst particles and the carbon source gas. Typically, CNFs have diameters ranging from 5 to 200 nm, and lengths extending several μm. CNTs exhibit properties such as ballistic electron transport and tensile strength along their axis, similar to diamonds. For functionalization and electrochemical applications, CNFs are more reactive and offer greater electron transfer through their sidewalls. Considering the demand for nanostructure materials and their diverse applications, many aspects of CNTs are summarized in the present chapter, including their classification, properties, fabrication, and environmental and biomedical applications.

2.   Functionalization of CNTs

During the process of hybrid materials formation, intermediately CNTs functionalized and purified [29]. Purification is a crucial step in the formation of CNTs and can be achieved using various methods. The primary consideration when selecting a purification method should be its effectiveness. Factors such as the purity of the precursor, pH, choice of oxidizing agents, duration, and temperature of oxidation will all influence the final outcome. To eliminate impurities from CNTs, an air mixture containing H_2S and O_2 can be passed over them. However, combining ultrasonic treatment and high-temperature processing can increase the risk of oxidation and potential damage to the surface, including broken fibers or tubes. Non-oxidizing acids such as HCl and H_2SO_4 are commonly used to remove metal catalyst particles from carbon nanostructures without affecting their structure. Another purification method involves annealing in the presence of a vacuum, which offers the advantages of being faster and more efficient than previously mentioned methods. Process temperatures range from 600 °C to 2,000 °C depending on the purpose. Surface functionalization of nanotubes has gained significant attention in research, as it can produce new materials with distinctive features. This involves treating CNTs with different chemicals to introduce various functional groups on their surface. Functional groups may emerge on the sidewalls or ends of nanotubes and nanofibers. This type of functionalization can alter the electrical properties of single-walled CNTs, affecting their conductivity. In the case of MWCNTs, their electrical morphology remains intact while novel surface features are introduced, expanding their potential applications. Organic compounds that interact with carboxyl groups following surface oxidation are commonly used for covalent functionalization. [30-33]. Non-covalent functionalization, on the other hand, is achieved through van der Waals forces and hydrogen bonds, resulting in a different type of functionalization [29]. Non-covalent functionalization differs from covalent functionalization in that it does not introduce numerous surface flaws or alter the mechanical and conductive properties of the material. Instead, it uses various active ingredients and polymers to enhance the solubility of CNTs in hydrophilic solvents and improve their dispersion in a polymer or ceramic matrix, serving as modifying agents. Aromatic chemicals such as porphyrins and pyrenes can also be used, as they interact with the delocalized electron cloud of CNTs. The reactivity, influenced by the significant curvature of CNTs and related to orbital mismatch, is higher at the fullerene hemispheres at the nanotube ends compared to the sidewalls.

These properties can be leveraged for the selective functionalization of CNTs, and many studies specifically highlight the beneficial outcomes of surface modification of CNTs. [30-33]. During this process, the metal catalyst particles exit the nano system and enter the solution in the form of a salt [34]. Furthermore, the oxidation of carbon nanomaterials (CNMs) with substances such as $NaOCl$, HNO_3, and $KMnO_4$ can be used to introduce hydroxyl, carbonyl, and carboxyl groups onto their surfaces for surface modification [34-37]. The surface of CNFs can also be functionalized; however, unlike CNTs, the entire surface of CNFs can be modified. Oxygen-containing groups can be created through electrochemical oxidation or nitric acid activation of CNFs without degrading their structure [38].

3. Properties of CNTs

The CNTs exhibit astonishing mechanical properties attributed to their strong sp^2 C-C bonds. These materials possess an exceptional thermal, mechanical, and electronic properties that are unparalleled by any other material (**Figure 2**). The CNTs have incredibly low densities (1.31 g/cm^3), which is approximately $1/6^{th}$ of the density of stainless steel. Furthermore, there is a substantial disparity between CNTs and all carbon fibers in their Young's moduli, a measure of material stiffness, with CNTs exhibiting values that exceed 1 TPa, around 5 times greater than steel [39]. Indeed, it is their strength that truly sets CNTs apart. Humanity has never encountered a material with the exceptional strength exhibited by CNTs. The tensile strength of CNTs was measured at 63 GPa, more than 50 times that of steel at the time [39, 40]. The CNTs exhibit impressive environmental and chemical stability and with values reaching approximately 3000 W/m/K, surpassing even that of diamond. Their lightweight nature further enhances their appeal, making CNTs a highly promising material for various aerospace applications.

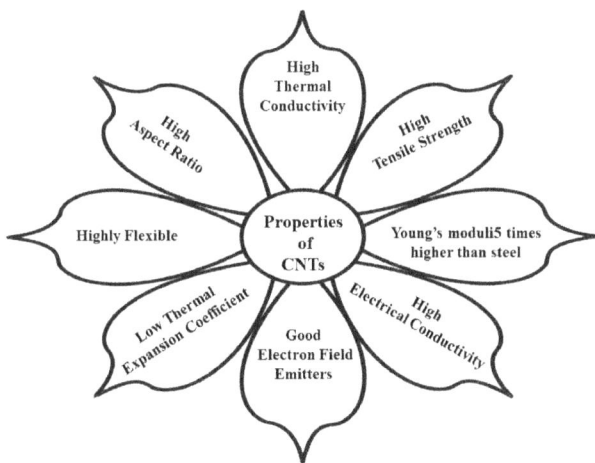

Figure 2: Properties of CNTs

Furthermore, CNTs have remarkable electronic property. Particularly, a remarkable observation found that nanotubes can exhibit either metallic or semiconducting properties. During the rolling action, a hexagonal lattice and an axial direction are established in a unique orientation, disrupting the symmetry. These nanotubes may be electrically a semiconductor or a metal, depending on relationship to the hexagonal lattice's unit vectors. The bandgap of a semiconducting nanotube varies inversely with its diameter, spanning from approximately 1.79 eV for the narrowest diameter tubes to 0.179 eV for the widest stable SWCNTs [41]. As a result, certain nanotubes exhibit conductivities surpassing that of copper, while others mimic the behaviour of silicon. Nanotubes have attracted considerable interest for its potential as a material for building nanoscale electronic devices. As a result of their properties, CNTs are already being used in a various application. Devices like sensors, scanning probe microscopes, flat-panel displays, and fuel cells are examples of these technologies.

4.   Synthesis methods of CNTs

Originally, the arc discharge process was used in a coincidental way to fabricate CNTs. Nonetheless, a variety of unique methods are currently accessible for the deliberate and regulated production of CNTs [42]. Numerous aspects of the manufacturing process can affect and define the properties of CNTs. These variables include the precursor material selected, the synthesis heat source, reaction time, temperature, the surrounding environment, and frequently the particular method utilised. These factors are essential in determining the features and attributes of the final CNTs [43]. CVD, laser ablation, and arc discharge are three of the most widely utilised methods for producing CNTs. These methods provide various ways to create CNTs, each having benefits and uses of their own [44-46]. There exist several alternative methods for the synthesis of CNTs, such as submerging graphite in cold water, employing solar energy, heat treating polymers, pyrolysis, liquid phase, and electrolysis, or synthesising them by breaking down silicon carbide (SiC). **Figure 3**, illustrates the most common methods for synthesizing carbon nanotubes CNTs [47-55].

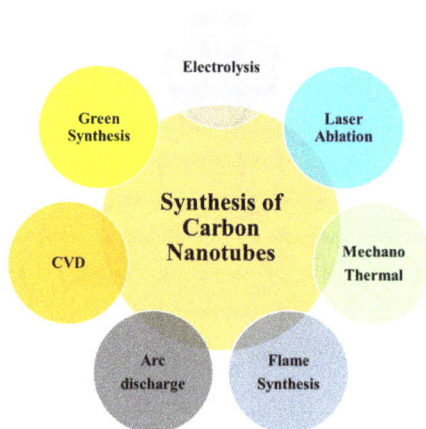

Figure 3: Synthesis methods of CNTs

4.1   Electrolysis

The technique was developed in 1995, utilizes electrowinning in order to separate alkali or alkaline-earth metals from their chloride salts. Following the electrowinning process, CNTs are deposited onto a substrate. This approach allows for the production of CNTs while simultaneously recovering valuable metals from their chloride salts, making it a potentially valuable method for both CNTs synthesis and metal recycling. The MWCNTs can be produced by applying a direct current (DC) voltage across two electrodes within a chamber filled with molten alkali-alkaline earth metals [56]. Below equation 1 illustrates the formation of lithium carbide [57].

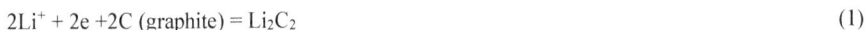

$$2Li^+ + 2e + 2C \text{ (graphite)} = Li_2C_2 \tag{1}$$

The synthesis of CNTs in a liquid phase can be initiated by the formation of lithium carbide (Li_2C_2). Typically, CNTs produced through this method exhibits diameters ranging from 2 to 10 nanometers (nm) and lengths of 0.5 micrometers (μm) or more. Additionally, this process yields byproducts such as amorphous carbon, carbon nanofibers, nanographites, and encapsulated CNTs [58]. The CNTs obtained through this method typically consist of MWCNTs; however, in certain research studies, the production of SWCNTs has also been reported [59]. Different salts, including NaCl, LiCl, KCl, and LiBr are used to produce CNTs [60]. The parameters that affect reaction include reaction time, molten salt, temperature, current density, and electrolysis regimes [61,62]. The reaction yields can be increased by 20–40% by optimising the process conditions to produce MWCNTs. Low temperatures, the ability to manage the synthesis process, the use of simple equipment, good quality, low energy consumption, and inability to produce large quantities are all advantages of electrolysis [61].

4.2   Chemical vapor deposition (CVD)

The CVD process is a straightforward approach that uses gas precursors containing carbon such as CO_2, C_2H_2, C_2H_4 and other hydrocarbons with a temperature range of 350–1000 °C [63,64]. The growth of CNTs is influenced by various factors, including reaction duration, temperature, catalyst size, growth rate, and the type of reactant gas used [65]. The CVD methods can be classified based on the energy source used, leading to various forms. For example, when the heat source is provided through infrared lamps, flame, or thermal resistance, it is referred to as thermal CVD. This categorization helps distinguish between different CVD techniques based on the energy input method [66].

General methods for the fabrication of CNTs include:

- Plasma Enhanced CVD (PE-CVD)

- Aerosol CVD (AA-CVD)

- Aerogel-assisted CVD

- Alcohol Catalyst Supported CVD (AC-CVD)

- Laser CVD (LCVD)

- Water Supported CVD (WA-CVD)

- Hot Filament CVD (HF-CVD)
- Oxygen Supported Plasma RF-PE-CVD
- Plasma Microwave (MPE-CVD)
- Catalytic CVD (CCVD)

These methods offer different approaches to synthesize CNTs, allowing for versatility in their production for various applications [67-83]. The plasma-enhanced method is considered more efficient due to the increased rate of reactions it offers. One of its key characteristics is that it operates at lower temperatures in comparison to some other CVD methods. This may be advantageous in few applications, especially when lower temperatures are required to prevent unwanted reactions or to work with temperature-sensitive materials. The CVD processes typically involve higher temperatures, and a minimum temperature of around 500 °C is often reported. However, it is possible to achieve synthesis temperatures for CNTs and carbon nanofibers as low as 120 °C in some specific methods. The actual synthesis temperature can vary depending on the CVD technique and the specific conditions used. In comparison with other low temperature methods, this type also has more vertical growth [84].

4.3   Mechanical thermal

A mechanical thermal method comprises two steps: first, the production of amorphous carbon, followed by annealing in a vacuum furnace. Carbon amorphization is achieved through high-energy ball milling. The duration of milling time for the fabrication of amorphous carbon can vary significantly, ranging up to 180 hrs. It is influenced by factors such as the atmospheric conditions, the cup speed, the ball-to-powder ratio, the number of balls used, and the purity of the powder. When milling time is increased, crystallite size decreases, and eventually amorphous structures are formed. During extended milling processes, a minimal amount of metal powders may infiltrate the graphite powder as a result of the friction between the milling cup and balls. This phenomenon can potentially useful in growth of CNTs and to the nucleation. A vacuum furnace was utilized to produce amorphous carbon, and CNTs are formed when the atoms of amorphous carbon are interconnected by heating to approximately 1400 °C for several hours. Mechanical heating offers several advantages, including its simplicity, suitability for mass production, cost-effectiveness, and the absence of a requirement for specialized equipment [85-87].

4.4   Laser ablation

In processes such as laser ablation or vaporization, the nucleation and growth of CNTs are initiated by either striking the graphite target with a pulsed laser or using a continuous wave laser. The first phase of the process is the evaporation, followed by a quick cooling. Van der Waals forces persist in binding the CNTs together as the sample cools. During this phase, minuscule molecules and carbon atoms start to condense, giving rise to larger clusters. In order to manufacture MWCNTs, pure graphite rods must be used, and composite blocks of graphite should be used for single-wall structures [88, 89]. In order to synthesize SWCNTs, graphite is combined with metal catalysts like Fe, Ni, and Co as well as ambient gases like He-H_2 and Ar. In comparison with continuous laser, pulsed laser requires more intense laser light. The most popular lasers used for laser ablation are Nd:YAG and CO_2. The CNTs fabricated using this technique typically have diameters in between 5 to 25 nm and a length of approximately few μm. Amorphous carbon, catalyst particles, fullerene, and other contaminants are byproducts which are formed during the synthesis. In addition to accelerating CNTs growth, catalysts also facilitate the process. In laser ablation, the most commonly used catalysts for quality production of CNTs include Ni, Pt, Co/Ni, Co/Pt, Co/Cu, Co, Cu, Nb, and Ni/Pt. The superiority of synthesis also relies on factors like composition of the target material, the power of the laser beam, laser properties, the type of catalyst employed, the ambient gas used, the reaction temperature, and the distance between the substrate and the target [90-94]. This process produces the most single wall CNTs with the highest purity and yield, however it is not appropriate for scale production because it also requires expensive and specialised equipment.

4.5   Flame synthesis

This process involves creating CNTs by burning a carbon source directly while it is being oxidised [59]. Flame synthesis typically involves three phases *viz.* hydrolyzing a hydrocarbon as a carbon source, the diffusion of carbon atoms into the metallic catalyst, and followed by growth of CNTs on the surface of catalyst. The superiority of CNTs and the quantity of amorphous carbon in the final product can change which is depend on the type of flame produced during the synthesis process. To achieve optimal conditions, it is crucial to closely control reaction catalysts, fuel gas composition, and temperature during the synthesis process [95, 96]. This method is economical, appropriate for mass manufacturing, and most effective at synthesising single-walled CNT, although its growth rate is modest.

4.6   Arc discharge

Arc discharge is indeed the initial methods used for the fabrication of CNTs [97]. A continuous arc is generated by employing two extremely pure graphite rods as the cathode and anode and applying either direct current or, at times, pulsed current. Due to the intensity of the arc, carbon from the anode separated and condensed onto the cathode, leading to the formation of soot. Arc discharge can be conducted in various environments, including liquid ones such as liquid nitrogen, toluene, and distilled water, as well as gas environments like argon, hydrogen-enriched argon, helium, or even plasma rotating arc discharge setups. A cost-effective method for large-scale production of CNTs is the plasma rotating arc discharge technique [98]. The rotation-induced centrifugal force accelerates the evaporation of the anode and ensures a uniform and stable dispersion of the arc, leading to increased volume and temperature of the discharge plasma. While the majority of CNTs produced using the arc method are multi-walled, it is possible to fabricate SWCNTs by penetrating graphite rods and filling them with graphite powder and a metal catalyst. Two essential variables for process control are steam chamber pressure and flow rate. Arc discharge has a high process rate and a regulated synthesis condition, but both the quality of the produced goods and the process efficiency are subpar [99].

4.7   Green Synthesis of CNTs

Greener synthesis can minimize the deployment of hazardous/toxic agents and expensive and rare materials / equipment to avoid adverse environmental impacts. Catalytic procedure has been broadly studied for increasing the yield of production and reducing the reaction temperature and time. The chemical vapor deposition technique with large-scale production potentials has been typically initiated by decomposition of hydrocarbon gases or organic solvents, when the nano-sized catalysts are generated in a heated furnace. However, this technique may suffer from high energy consumption and toxic/hazardous chemicals utilization; several techniques such as electrostatic spray-assisted or combustion chemical vapor deposition have been studied wherein safer reagents can be utilized. Via the optimization of the reaction conditions and applying renewable precursors/catalysts, different types of CNTs with unique properties can be achieved. Plant extracts as natural precursors can be deployed for sustainable synthesis of CNTs (**Figure 4**). They serve as greener catalysts for fabricating CNTs with the advantages of cost effectiveness, renewability, and abundancy, thereby avoiding the application of potentially hazardous metal catalysts. Besides, plant extracts as green catalysts can be converted into porous and activated carbons with considerable nucleation sites for growing CNT-based nanostructures. The activated carbons with porous surfaces illustrated larger surface area than transition metals. Thus, nucleation sites for growing CNTs became larger, and the activated carbon could decompose hydrocarbons at a lower temperature; the growth of CNTs using plant extracts has been reported below ~ 575 1C, whereas the typical reactions applying transition metal catalysts were performed at ~ 700-1200 °C. By applying a wall-nut extract catalyst, high yield of CNTs production could be obtained at chemical vapor deposition temperature of 576 °C. [100-101].

Qu et al. reported the greener formation of CNTs-Cu/ZnO nanocomposites utilizing Brassica juncea as a plant source of Cu, Zn, and C. The prepared CNTs had middle-hollow structures and were not at all crystalline, with a few defects in the walls. They provided an innovative greener plant-based strategy for the development of CNTs; however, future investigations should be focused on their surface functionalization and optimization process [102]. Besides, the nanostructures of MWCNTs were fabricated via chemical vapor deposition using Cocos nucifera Linn (coconut oil) where nitrogen gas was utilized as an inert atmosphere and a suitable carrier for the evaporated precursor [103].

Olive (*Olea europaea*) and coconut oils served as natural carbon precursors and $NiCl_2$ functioned as catalyst for the eco-friendly synthesis of CNTs through a simple and cost-effective pyrolysis technique at low temperatures. Consequently, uniformed SWCNTs from olive oil and the carbon rods from coconut oil were successfully obtained. The uniformity in SWCNTs was not reported by using coconut oil, because of high proportion of saturated fat content (81.9%) and less reactivity in comparison with the unsaturated hydrocarbons [104]. These CNTs should be further evaluated for agricultural, biotechnological and biomedical applications [105]. By applying a simple spray pyrolysis technique, SWCNTs were fabricated through catalytic decomposition of the eucalyptus oil on a high silica-zeolite support impregnated with Fe/Co catalyst at 850 °C [106]. Suriani et al. synthesized vertically aligned CNTs with a diameter of 0.6 1.2 nm on silicon substrates via thermal catalytic-chemical vapor reactor technique utilizing natural palm oil (as carbon source); these CNTs had high purity of B90% [107]. Additionally, the deployment of chemical vapor deposition technique was reported for synthesizing MWCNTs using bamboo charcoals derived from the heat-treated fresh bamboo culms at 1000–1500 °C. It was revealed that Mg_2SiO_4 and mostly calcium silicate was responsible for growing the MWCNTs [108]. The minerals in the bamboo charcoal served as catalysts similar to the transition metals for nucleating CNTs. Basta et al. illustrated the eco-friendly fabrication of MWCNTs from hydrochars of rice straw via a simple pyrolysis technique. The pyrolysis of rice straw-alkaline pulp and sulphite pulp hydrochars resulted in almost needles-like CNTs produced in between graphene nanosheets, with diameters ranging from 2.5–6.8 nm and 4–8 nm, respectively.[109] Besides, aligned CNTs were fabricated using turpentine oil and ferrocene mixture via a simple spray pyrolysis technique. These CNTs with outer diameters between 16 and 41 nm should be further evaluated for biomedical applications [110]. Sesame oil as a natural and low-cost hydrocarbon precursor was utilized for manufacturing aligned-stack of branched nitrogen doped CNTs via a simple spray pyrolysis-assisted chemical vapor deposition technique; ferrocene and acetonitrile were utilized as the catalyst and nitrogen-doping agent, respectively for growing the N-doped aligned CNTs [111]. Neem oil obtained from the seeds of *Azadirachta indica* was utilized as a carbon source to produce aligned CNTs nanostructures via a simple spray pyrolysis method. The main constituents of neem oil are hydrocarbons with less oxygen, offering the precursors in spray pyrolysis production of CNTs; the aligned CNT bundles were grown directly inside the quartz tubes. These CNTs were found to be multi-walled structures with an inner diameter of 16–29 nm [112].

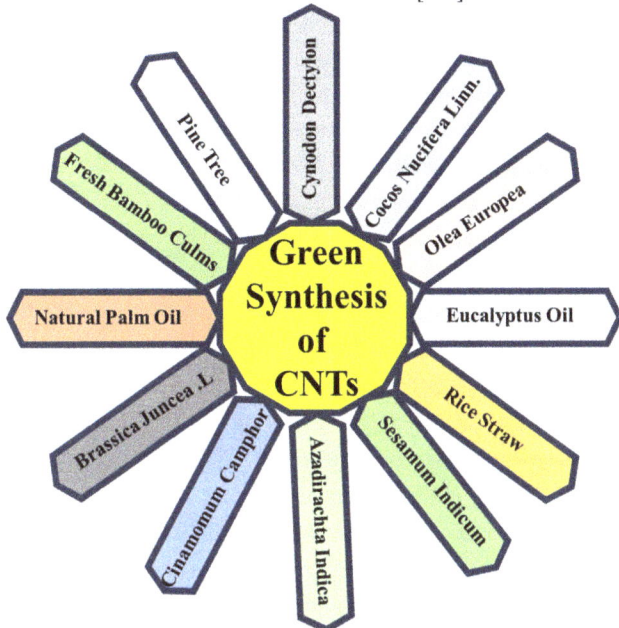

Figure 4: Synthesis of CNTs from plant extracts

5.   Potential Applications of CNTs in various fields:

The CNTs have a greater mechanical strength, thermal conductivity, improved electrical conductivity, little aggregation, high porosity and greater surface area. Additionally, CNT's highly interconnected structures produce an effective mass transmission reachable internal surface area, etc. These exceptional qualities, along with the materials' conductivity, stability, and hydrophobicity, give a broad platform for the potential applications. The various applications of CNTs are represented in **Table 2** and **Figure 5**.

Table 2: Potential applications of CNTs in different fields

Sr.No.	Material	Application	Roles	References
1	CNTs	Water decontamination	CNTs possess the capability to remove a wide range of pollutants.	[113]
2	CNTs	Electrical applications	carbon nanotubes provide the advantage of small size and remarkable flexibility.	[114]
3	CNTs	Electrode materials	CNTs were used to increase the capacity of the capacitor by providing a large surface area which enhanced its ability to store energy.	[115]
4	CNTs	Energy storage	The development of CNTs has spurred advancements in the field of lithium-ion batteries including solar cells, fuel cells, hydrogen storage etc.	[116]
5	CNTs	Electrochemical biosensors	CNTs enhance electron transfer, integrating electronic and electrochemical biosensors.	[117,118]
6	CNT	Drug development, gene therapy, and other biological uses	CNTs are suitable for a wide range of biomedical applications, making them ideal for drug delivery systems.	[119]
7	CNT-based filters	Air pollution	Using carbon nanotubes as sorbents for reducing pollution is a novel way of managing and controlling emissions.	[120-121]
	CNTs	Agricultural	CNTs are useful for delivering nutrients to plants. MWCNTs and SWCNTs are both useful for this purpose.	[121-126]
9	CNTs	Minimizing environmental pollution	CNTs are environmentally conscious materials that help to lessen the amount of carbon left behind in the environment.	[127,128]
10	CNTs	Removal of oil spills from water	CNT sponges are useful for successfully eliminating huge amounts of oil and work well with a variety of oils and solvents.	[129]
11	Hybrid SWCNTs ceramic filters	impressive thermal resistance merge seamlessly with the porous aggregate structure of SWNTs.	The field of nanomaterials is being improved and people's acceptance is sought after in an attempt to create a favourable atmosphere.	[130]
12	Green CNTs	CNTs fabricated from organic camphor	The CVD approach can be used to synthesise high purity CNTs from camphor hydrocarbons.	[131]

13	CNT based materials	Hybrid materials	Hybrid materials are generated through the fusion of diverse structural elements, leading to materials that inherit specific properties and functions from each constituent.	[132]
14	Nanotechnology	Food Science	Food preservation and protection can also be aided by nanotechnology through creative food packaging methods, among other things.	[133]
15	CNT based activated carbon	Converting waste biomass into goods with high levels of CNTs and activated carbon	These CNTs have a broad range of industrial applications such as wastewater treatment, composites, tissue engineering, microelectronics, and water adsorption.	[134]

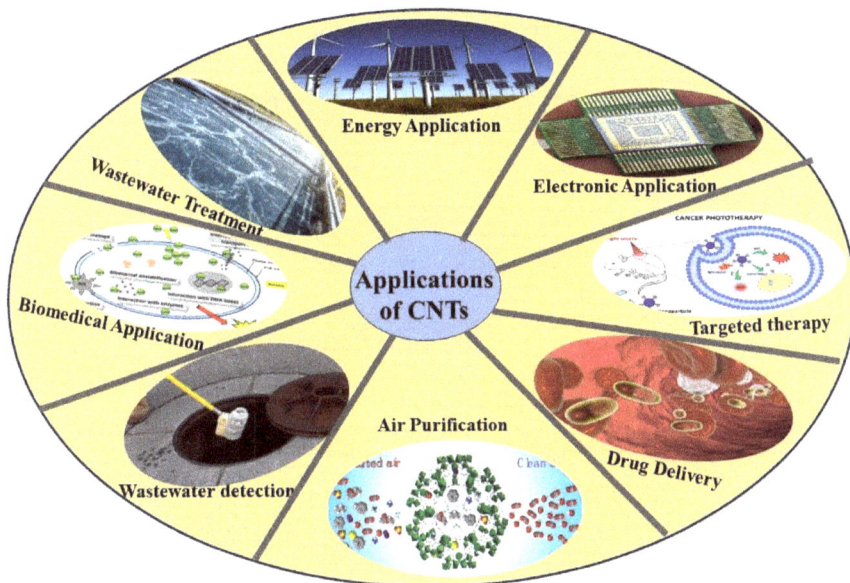

Figure 5: Applications of CNTs

5.1   Environmental applications

The CNTs have numerous environmental applications in various field such as removal of organic solvents and oil, adsorption of dyes, removal of heavy toxic metals, catalytic conversion of pollutants (**Figure 6**).

5.1.1 Removal of organic solvents and oils

During the extraction, storage, and transportation processes, the catastrophic leaking of oils and chemical compounds offers a variety of environmental pollution risks to marine and terrestrial ecosystems. Oil spills also lead to the depletion of energy resources. As a result, developing efficient remediation strategies is unavoidable. This crucial problem has been addressed by a plethora of methods, such as chemical dispersants, photodecomposition, biological separation, membrane filtration, adsorption, combustion, wet oxidation, and skimmers. It is strongly advised that adsorption be used as the most practical, economical, and environmentally friendly method due to its accessibility, ease of use, uncomplicated procedure, and low environmental effect. [135].

Carbon allotropes, such as CNTs, are hydrophobic by nature, which makes them potentially useful materials for reducing the effects of pollution and oil spills. Different sorbents use either absorption, adsorption, or both to remove or retrieve contaminants. As opposed to chemical processes, oil absorption occurs when oil seeps through the adsorbent's holes. This process is dependent on the viscosity, adhesion, and porous nature of the adsorbent. Oil builds up or is retained on the adsorbent as a result of adsorption. However, the process of adsorption occurs when oil sticks to the outside of sorbents. Numerous physicochemical interactions, including as hydrogen bonding, hydrophobicity, steric interactions, polarity, and van der Waals forces, have an impact on this attraction.

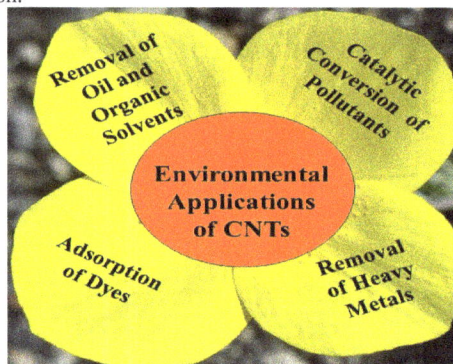

Figure 6: Environmental Applications of CNT

Kabiri *et al.* documented a one-step technique for producing three-dimensional graphene/CNTs networks. This method involved heating an aqueous blend of GO and CNTs in Fe^{2+} ions. After fabrication, aerogels demonstrated outstanding organic solvent adsorption properties. The whole quantity of oil rapidly adsorbed and completely absorbed in just 20 seconds by dipping it in the graphene-CNTs aerogels. The oil-saturated aerogels demonstrated the capability to remain buoyant on the water's surface, preventing oil discharge and water infiltration into their structure [136].

Similarly, Wan *et al.* introduced an environmentally friendly and straightforward method for synthesizing graphene/CNTs aerogels with remarkably low density. This process was achieved through a hydrothermal reaction. By optimizing the CNT/GO mass ratio, the adsorption was significantly enhanced, reaching up to 110-260 times their mass, contingent on the density of the adsorbed organic substances [137]. Likewise, Zhan *et al.* synthesized one-dimensional polydopamine functionalized MWCNTs into a graphene aerogel. This composite aerogel proved highly effective in absorbing organic pollutants. The addition of MWCNTs to the composite led to a substantial enhancement in both surface area and stability [138]. Further, Tao *et al.* have successfully produced feathery spheres composed of nickel/graphene/CNTs, utilizing natural carbon sources. Nickel doped carbon encapsulated NPs was synthesized by subjecting potato starch to hydrothermal carbonization. Subsequently, they were catalytically graphitized and splintered using biogas. During this process, the graphene shell was detached to synthesize graphene nanoplatelets, and were seamlessly integrated in situ with these platelets [139]. Moreover, Cai *el al.* described the fabrication of an aerogel, referred to as amino/rGO MWCNT, achieved over a chemical reaction among the carboxylic groups on graphene oxide and the amino groups on MWCNTs. The hybrid aerogel's hydrophobic assets and microstructure were significantly enhanced due to robust chemical bonds [140].

5.1.2   Adsorption of dyes

Due to the world's fast industrialisation, there is a sharp rise in water contamination. Highly poisonous organic dyes and chemicals, organometallic compounds, domestic trash, industrial effluents, agricultural pesticides, etc. are among the contaminants that have contaminated the water. In particular, because organic dyes are coloured and have high chemical oxygen requirements, they are extremely dangerous for the environment, aquatic habitats, and human health. The discharge of highly coloured garbage disrupts natural cycles and jeopardises aquatic environments by obstructing the passage of light. There is a problem with many organic colours' biodegradability. These compounds' high toxicity and carcinogenic properties are further exacerbated by their resistance to light, chemicals, and other therapies. Therefore, the development of novel techniques for the effective removal of dyestuffs is required [141]. The most common dissolving types for dyes are anionic or cationic. Dyestuff molecules may diffuse and potentially be stored in the nanopores due to their existence and the CNTs' larger surface area. Furthermore, due to their electrostatic properties and effective p–p interactions, these materials can be employed as efficient adsorbents for dyes, particularly those having aromatic structures. In this context, the CNT hybrid aerogels that were produced by drying their aqueous gel precursors in supercritical CO_2 were examined [142]. The ultralight hybrid aerogels' strong conductivity, generous pore volume, and big surface area made them suitable for use as sorbents for simple dyes, but they showed no appreciable scavenging ability. With Rhodamine B, it demonstrated an impressive adsorption capability of 146.8 mg per gramme, surpassing that of other compounds utilised for the same dye. In the instance of methylene blue, the acid-treated MWCNT/graphene hybrid aerogel registered at 189.9 mg g-1, demonstrating a substantially higher adsorption capacity than conventional adsorbents under identical conditions [143]. Acid fuchsin adsorption was measured at 66.7 mg g^{-1} for graphene/MWCNT hybrid aerogels and at 35.8 mg g^{-1} for MWCNT/graphene hybrid aerogels treated with acid respectively.

Similarly, Ai and Jiang's graphene-CNTs aerogel demonstrated an 81.97 mg g^{-1} adsorption capacity against methylene blue in an aqueous solution, with MB concentrations of 10 mg L^{-1}, resulting in a 97% removal efficiency [144]. Likewise, Kotal and Bhowmick's study revealed that HMs of CNTs and rGO displayed a highest adsorption capacity for crystal violet, surpassing the performance of magnetically modified MWCNTs. Additionally, the same hybrid displayed an adsorption capacity of 219 mg/g for Rhodamine 6G, outperforming cane sugar reduced graphene [145].

As advancements continue in the field of adsorption research focused on graphene/CNTs hybrid structures, ongoing efforts are directed towards the development of all-carbon nanoarchitecture membranes. These membranes aim to enhance selectivity and increase water permeability, thereby offering superior performance. Goh *et al.* achieved the development of exceptionally stable membranes of surface functionalized MWCNTs with a small diameter into reduced graphene oxide sheets. These membranes exhibited remarkable performance, capable of rejecting nearly 100% of organic dyes with varying charges, including cationic dyes like methylene blue and Rhodamine B, as well as anionic dyes like Acid Orange 7. The outstanding stability observed in these materials can be attributed to the van der Waals forces and pi-pi interactions occurring among the rGO sheets and multi-walled carbon nanotubes (MWCNTs) [146].

5.1.3 Removal of heavy metal ions

A significant number of toxic metal ions are discharged into the environment as a result of urbanisation and fast industrialisation. These manufacturing processes include the use of fertilisers, mining, battery storage, processing metals, plate metal, and alloy production-all without following the correct treatment protocols. These non-biodegradable contaminants can accumulate quickly, endangering both human health and ecosystems. Different heavy hazardous metals can be removed using a variety of techniques, including oxidation, chemical precipitation, reverse osmosis, electrochemical treatment, and adsorption [147,148,149]. One of them, adsorption, is widely used to remove toxic metals ions due to its higher productivity, ease of usage, and ability to recover metal. Carbon-based nanocomposites have shown encouraging adsorption properties for the elimination of several contaminants. Electrostatic attractions, ion exchange processes, and surface complexation involving metal ions drive the adsorption mechanism [150-154]. Metal cations replace the H_3O^+ groups in the adsorption sites during ion exchange. Increasing pH can help both cationic and non-cation dyes adsorb more effectively. Conversely, it is imperative to prevent an overly high pH as this may cause metal hydroxides to precipitate. It's also critical to remember that the adsorption of metals on graphene is an endothermic process, meaning that increasing the temperature will increase the adsorption capacity. The field of metal ion removal has given hybrid polymer-GO sponges a lot of attention. Chitosan-gelatin/GO monoliths were created using a unidirectional freeze-drying process, and they had an amazing porosity for Pb^{2+} and Cu^{2+} adsorption. The hybrid monolith's compressive strength and porous structure were greatly enhanced by the addition of GO. Even after several recycling cycles, there was only a slight decrease in the adsorption capability of these materials, indicating their exceptional stability. Moreover, these substances are biodegradable and non-toxic. Because of their easy separation and large specific surface area, 3D graphene-CNT hybrid networks have been studied extensively. They may effectively remove heavy metal ions such as Cu^{2+}, Pb^{2+}, As^{3+}, and U^{6+}. As previously indicated, the hybrid aerogel of graphene and CNTs shown a potent ability to scavenge heavy metal ions through the function of binding agents. The main cause of this phenomena is because electrostatic interactions dominate how ions bind to porous hosts. It was shown that the increased number of oxygen-containing groups in the graphene/MWCNT hybrid aerogels after acid treatment helped to effectively bind the ions [155-158]. It's also important to remember that the addition of iron minerals to the graphene/CNT networks has the benefit of keeping the carbon nanostructures from aggregating. For comparison, the graphene aerogel's adsorption capacity alone is also provided. Moreover, the simple separation of trapped metals can be easily achieved through the application of their intrinsic magnetic characteristics [159]. In this context, Zhang et al. fabricated a graphene-CNT hybrid aerogel using a hydrothermal reduction method involving GO and CNTs, along with the incorporation of ferrous ions [160]. The quantity of ferrous sulphate and CNTs employed found to have a substantial impact on efficiency of metal elimination. Notably, the sample prepared with 30% CNTs exhibited higher efficiency compared to others with the same quantity of ferrous sulphate. Furthermore, a large-scale fabrication of FeO-graphene/CNT hybrid structures was successfully accomplished through a single step microwave synthesis method, demonstrating its exceptional versatility [161]. Because of its high aspect ratio and open porosity nanostructure, the HMs demonstrated effective exhibited effective arsenic from polluted water.

To effectively isolate and eliminate As^{3+} and Pb^{2+} contaminants from water, 3D graphene oxide (GO) membranes modified with glutathione-conjugated CNT were engineered. These membranes demonstrated the capability to capture not only As^{3+} and Pb^{2+} but also As^{5+} ions [162]. The remarkable elimination efficacy is credited to strong affinity of both As^{3+} and Pb^{2+} for glutathione, as they can readily bind to glutathione through thiol (-SH) interactions. Furthermore, the open-pore network of the CNTs-bridged GO membrane promoted the rapid diffusion of As^{5+}, Pb^{2+}, and As^{3+} ions within the 3D structure, thereby leading to a substantial increase in adsorption capacity. A 3D GO membrane bridged with a CNTs (without glutathione) was more effective in removing As^{3+}, As^{5+}, and Pb^{2+} (50% and 20%, respectively), but only 14% of As^{3+} and 10% of As^{5+} were effectively removed. Uranium has attracted a lot of attention because of its radioactive nature and the dangers it poses to the environment and human health. The removal of U^{6+} from graphene-CNTs aerogels was highly effective, and they had good monolayer sorption capabilities [163]. Diethylenetriamine functionalized CNTs dispersed in GO colloids were fabricated for the purpose of selectively extracting and analyzing trace levels of Pb^{2+}, Cr^{3+}, Mn^{2+}, and Fe^{3+} ions in wastewater using solid-phase extraction techniques [164].

Remarkably, suspensions of CNTs/GO sealed within dialysis bags demonstrated exceptional effectiveness in removing trace amounts of Gd^{3+} from water, all without the risk of re-pollution [165]. Hybrids with MCNTs: MGO ratios within the range of 1:8 to 1:2 exhibited a extraordinary synergistic effect. This outcome was accomplished by the inclusion of CNTs, which effectively prevented the restacking of GO nanosheets throughout the adsorption process. The hybrids with an MCNTs:MGO ratio of 1:6 displayed a remarkable theoretical maximum Gd^{3+} adsorption capacity of 534.76 mg per gram (at 60 minutes, pH 5.9, and a temperature of 303 K). This represented an impressive 86.42% increase compared to GO alone. Furthermore, in the fourth cycle, a notable sorption capacity of 347.83 mg per gram was maintained, underscoring its exceptional regeneration performance.

In this context, Huang et al. prepared 3D graphene/MWCNT hybrid nanocomposites by utilizing direct electrochemical reduction on GO-MWCNT nanocomposites. The glassy carbon electrode (GCE) modified with G-MWCNTs displayed exceptional sensitivity in the electrochemical detection of trace levels of Cd^{2+} and Pb^{2+} ions. It achieved a remarkable low detection concentration of 0.5 mg/L for Cd^{2+} and 0.5 mg/L for Pb^{2+}. The combined influence of MWCNTs and graphene not only boosted the preconcentration efficiency of metal ions but also expedited the electron transfer rate at the G-MWCNT/electrolyte interface, ultimately resulting in heightened sensitivity for detection [166].

5.1.4 Catalytic conversion of pollutants

Various techniques, including photocatalytic degradation, catalytic reduction, and adsorption have been employed to eliminate hazardous chemicals [167-169]. Effective pollutant conversion necessitates a high adsorption capacity. When a photocatalyst is exposed to visible and/or UV light, the energy from the light promotes electrons from the valence band to the conduction band, creating electron-hole pairs [170-172]. Subsequently, these generated electron-hole pairs migrate to the surface and initiate redox reactions with the adsorbate. Nonetheless, the catalytic efficacy of semiconductors faces significant limitations due to the rapid recombination of electron-hole pairs. In this context, CNT hybrids serve as a highly capable platform for the transformation of diverse contaminants using metal/metal oxide nanoparticles. As a result, CNT hybrids demonstrate substantial potential for the effective conversion of pollutants. CNT-based materials with excellent electrical conductivity and a spacious surface area were synthesized by introducing CNTs into both GO and rGO through the chemical vapor deposition process.

Wang *et al.* fabricated a TiO_2/MWCNT/ graphene using the solvothermal process. This hybrid exhibited enhanced photocatalytic efficacy compared to graphene/TiO_2 when degrading methylene blue and reducing Cr^{6+} under UV irradiation [173]. The photocatalytic activity was found to be contingent on the proportion of CNTs within the hybrid, with the optimal mass ratio determined as MWCNTs/TiO_2. CNTs acted as conduits for charge transport, leading to a notable reduction in the recombination rate of electron-hole pairs.

Qu *et al.* fabricated a heterostructure nanocomposite by combining oxygen-modified monolayer graphite-like C_3N_4 with GO and nitrogen-doped CNTs. This nanocomposite demonstrated exceptional activity in degrading Rhodamine 6G compared to both pure O-g-C_3N_4 and O-g-C_3N_4/GO under visible light irradiation [174]. Similarly, Yi *et al.* described the fabrication of graphene/CNTs-supported Cu nanocubes with Pt skin using a straightforward method involving electrodeposition and galvanic replacement [175].

Likewise, Tran *et al.* introduced an environmentally friendly and straightforward method for synthesizing a incompletely reduced GO/Fe/Ag /CNT hybrid within a 10 to 30 seconds, using microwave irradiation. When employed in the reduction of 4-nitrophenol with $NaBH_4$, the hybrid demonstrated remarkable catalytic efficacy. Leveraging its substantial surface area, GO effectively immobilized a greater number of silver nanoparticles, while the exceptional electrical conductivity of CNT/rGO, it can facilitate electron transfer when reducing 4-nitrophenol. With the presence of Fe nanoparticles and Fe_3C, easy recovery was facilitated through the assistance of an external magnet [176]. Moreover, Kotal *et al.* devised a graphene-assisted hybrid material composed of Co nanoparticles embedded in N-doped CNTs. The expansive surface area of graphene allowed for a more extensive decoration of ZIF-67, leading to increased growth of N-doped CNTs within the G/Co/NCNTs hybrid as compared to Co/NCNT variant. Increased nitrogen doping resulted in a band gap reduction and, consequently, improved photodegradation [177].

5.2 Biomedical applications

CNTs grasps the attention because their unique properties from many potential areas including research industry and academics. Because of their small size, electrical conductivity, and higher thermal and high strength, CNTs are valuable materials for biomedical research because of structural, electronic, and mechanical properties. Though, it quite challenging to employ CNTs in biomedical field due to purity concern, large-scale production and formulation and hydrophobic nature [178]. However, functionalization of CNTs with inorganic or organic moiety can be more efficient for the biomedical applications. Oskoueian *et al.* stated the functionalization of SWCNT with polyethylene glycol and tamoxifen, a potent anticancer drug, for the anticancer drug delivery application against human breast cancer cell lines [179]. The functionalization of the SWCNT is carried out by the oxidation with concentrated nitric and sulphuric acid and then it conjugated with a drug, tamoxifen (SWCNT-PEG-TAM). The comparative cytotoxic experimental data shows the improved cytotoxic activity against cancer cell up to 2.3 times with respect to single walled CNTs and functionalized CNTs with PEG. Ahemad et al. described the cytotoxic activity of functionalized single and multi-walled CNTs with TiO_2/Ag against the uterine cancer cell lines (SiHa) and normal cell lines (WRL68) [180]. This new conjugate of CNTs shows selective killing of cancerous cell viability up to ~80-70% whereas only ~10% normal cell was affected against the 200 to 400 µg/mL. Sol-gel method is used for the functionalization of CNTs, also, capable to show noteworthy bactericidal activity against gram positive and gram-negative bacteria. Yuan et al. [181] reviewed the different mechanistic pathways for persuading the cytotoxic activity against the cancer cell lines by employing functionalized CNTs which activates cascade of metabolic process and macrophages in cell *viz.*, ROS production, damage of lysosomes and protein, membrane rupture, *etc.* shown in Figure 07. Sheikhi et al. [182] theoretically studied the interaction of the new anticancer drug, Syndros with CNTs (6,6,6) nanotubes by the use of B3LYP/6-31G* level method for first time. The study reveals that, adsorption of drug molecule is sensitive on the CNTs (6,6,6) nanotubes due to their electronic properties. The bathochromic shift was calculated theoretically for the drug and nanotube complex when it compared with the absorption spectrum of drug. Also, the theoretical calculations shows the interaction between Syndros and CNTs (6,6,6) nanotubes. Therefore, CNTs can be used as drug delivery of the new anticancer drug Syndros in futuristic study.

The CNTs have been applied to a numerous type of biosensors, developing different aspects of their advantageous physicochemical properties. Chen et al. [183] depicted the synthesis of platinum microsphere decorated stretchy glucose sensor based on MWCNTs coated with carbonized silk fabric. This fabricated CNTs based glucose sensor is flexible, good conductor and stable in aqueous medium. The sensor shows the excellent sensitivity towards the H_2O_2 on decoration with the platinum. Moreover, after the modification of sensor with glucose oxidase shows high sensitivity (288.86 A/mM.cm^2) and stability towards the detection of glucose. Different phenolic compound also can be detected by using the electrodes modified with CNTs. Wee et al. [184] described the enhancement of sensing activity for the phenolic compound of tyrosinase immobilized enzyme electrode modified with CNTs. In order to prepare tyrosinase solutions and molecules for enzyme adsorption (EA), precipitation, and crosslinking (EAC and EAPC), tyrosinase solutions and molecules are prepared as required. For unmodified EA, EAC and EAPC electrodes shows sensing activity 34.01, 281.12 and 675.04 µA.mM^{-1} cm^{-2} respectively, whereas with modification with CNTs, sensing activity increases and found to be 146, 427, and 1160 µA.mM^{-1}cm^{-2} respectively. Singh et al. [185] reported a functionalization of caboxylated MWCNTs with indium tin oxide by electrophoretic deposition method for the detection of aflatoxin-B1 using electrochemical technique. The fabricated CNTs based immunosensor is highly sensitive i.e. 96.15 µA/ngmL^{-1}.cm^{-2} with the detection limit 0.08 ng/mL in the linear detection range of 0.25 to 1.375 ng/mL.

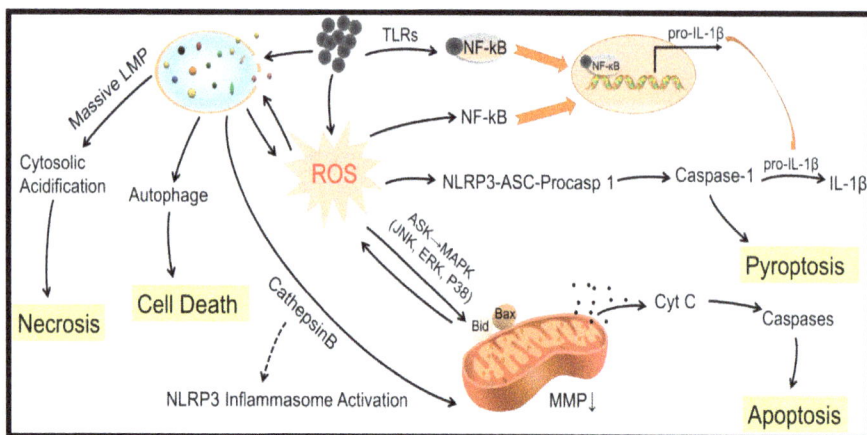

Figure 7: Different routes for the activation of the metabolic process and macrophages when it exposed to CNTs. This figure has been reproduced from ref [181] with the permission from Springer Nature, Copyright 2019. (http://creativecommons.org/licenses/by/4.0/)

Conclusions

In conclusion, the present work reported conventional and green synthesis of CNTs. The CNTs exhibits potential applications in biomedical and environmental because of their small size, high electrical conductivity, high thermal and strength, greater mechanical strength, high porosity, and greater surface area. Additionally, CNT's can produce highly interconnected structures with an effective mass transmission and internal high surface area. Similarly, the hybrid CNTs materials are exceptional materials because of their unappareled characteristics. The hybrid materials are smart materials because of its extraordinary optical, mechanical, sorption, magnetic, and conductivity properties. The CNT hybrids materials serve as a highly capable platform for the transformation of diverse contaminants using metal/metal oxide nanoparticles. The CNT-based nanocomposites have shown inspiring adsorption properties for the elimination of several contaminants and toxic metals ions. Consequently, CNTs has great potential for creation of exceptionally efficient materials in this field of environment and biomedical. The CNTs is futuristic materials, will be extensively used in future to tackle down many environment issues.

References

[1] Sharma, G. (2024). Biogenic carbon nanostructured materials for detection of cancer and medical applications: a mini review. *Hybrid Advances*, 100166.

[2] Gomathi, A., Vivekchand, S. R. C., Govindaraj, A., and Rao, C. N. R. (2005). Chemically bonded ceramic oxide coatings on carbon nanotubes and inorganic nanowires. *Adv. Mater.* 17, 2757–2761. http://dx.doi.org/10.1002/adma.200500539

[3] Zhao, X., He, X. D., Sun, Y., and Wang, L. D. (2011). Carbon nanotubes doped SiO_2/SiO_2-PbO double layer high emissivity coating. *Mater. Lett.* 65, 2592–2594. http://dx.doi.org/10.1016/j.matlet.2011.06.030

[4] Dillon, F. C., Moghal, J., Koós, A., Lozano, J. G., Miranda, L., Porwal, H., et al. (2015). Ceramic composites from mesoporous silica coated multi wall carbon nanotubes. *Microporous Mesoporous Mater.* 217, 159–166.http://dx.doi.org/10.1016/j.micromeso.2015.06.024

[5] Wu, Z., Gao, S., Chen, L., Jiang, D., Shao, Q., Zhang, B., et al. (2017). Electrically insulated epoxy nanocomposites reinforced with synergistic core shell SiO2 @MWCNTs and montmorillonite bifillers. *Macromol. Chem. Phys.* 218:1700357. http://dx.doi.org/10.1002/macp.201700357

[6] Whitsitt, E. A., and Barron, A. R. (2003). Silica coated single walled carbon nanotubes. *Nano Lett.* 3, 775–778. http://dx.doi.org/10.1021/nl034186m

[7] Chaudhary, R.G., Juneja, H.D., Gandhare, N.V., Gharpured, M.P. (2013). Synthesis, characterization and morphology behaviour of Mn (II), Co (II), Ni (II) and Cu (II) chelate polymer compounds based on chelating ligand, *Research Journal of Pharmaceutical, Biological and Chemical Sciences*, 4: 1625-1636.

[8] Ivnitski, D., Artyushkova, K., Rincón, R. A., Atanassov, P., Luckarift, H. R., and Johnson, G. R. (2008). Entrapment of enzymes and carbon nanotubes in biologically synthesized silica: glucose oxidase-catalyzed direct electron transfer. *Small* 4, 357–364. http://dx.doi.org/10.1002/smll.200700725

[9] Cui, W., Du, F., Zhao, J., Zhang, W., Yang, Y., Xie, X., et al. (2011). Improving thermal conductivity while retaining high electrical resistivity of epoxy composites by incorporating silica-coated multi-walled carbon nanotubes. *Carbon* N. Y. 49, 495–500. http://dx.doi.org/10.1016/j.carbon.2010.09.047

[10] Chen, L., Youji, L., Peng, X., Ming, L., and Mengxiong, Z. (2014). Carbon nanotube embedded mesoporous titania pore-hole inorganic hybrid materials with high thermal stability, improved crystallinity and visible-light driven photocatalytic performance. *Microporous Mesoporous Mater.* 195, 319–329. https://doi.org/10.1016/j.micromeso.2014.04.029

[11] Chaudhary, R.G., Juneja, H.D., Gharpure, M.P. (2013). Chelate polymer compounds with bis (bidentate) ligand: synthesis, spectral, morphological and thermal degradation studies, *Journal of the Chinese Advanced Materials Society*, 1: 121-133.https://doi.org/10.1080/22243682.2013.810396

[12] Deng, Y., Deng, C., Yang, D., Wang, C., Fu, S., and Zhang, X. (2005). Preparation, characterization and application of magnetic silica nanoparticle functionalized multi-walled carbon nanotubes. *Chem. Commun.* 44:5548–5550. https://doi.org/10.1039/B511683J

[13] Choi, W. S., Yang, H. M., Koo, H. Y., Lee, H. J., Lee, Y. B., Bae, T. S., & Jeon, I. C. (2010). Smart microcapsules encapsulating reconfigurable carbon nanotube cores. *Adv. Funct. Mater.*, 20(5): 820-825. https://doi.org/10.1002/adfm.200901739

[14] Mishra, R.K., Verma, K., Chaudhary, R.G., Lambat, T., Joseph, K., (2020). An efficient fabrication of polypropylene hybrid nanocomposites using carbon nanotubes and PET fibrils, *Materials Today: Proceedings*, 29 (3): 794-800. doi.org/10.1016/j.matpr.2020.04.753.

[15]   Mondal, A., Umekar M.S., Bhusari, G.S., Chouke, P.B., Lambat, T., Mondal, S.,, Ratiram G Chaudhary, R.G., Mahmood, S.H. (2021). Biogenic synthesis of metal/metal oxide nanostructured materials, *Curr. Pharm. Biotechnol*, 22(13):1782-1793. doi: 10.2174/1389201022666210111122911.

[16]   Sonkusare, V., Chaudhary, R.G., Bhusari, G., Rai, A.R., Juneja, H.D. (2018). Microwave-mediated synthesis, photocatalytic degradation and antibacterial activity of α-Bi_2O_3 Microflower/γ-Bi_2O_3 Microspindle. *Nano-Structures & Nano-Objects*, 13:121-131. doi.org/10.1016/j.nanoso.2018.01.002

[17]   Chaudhary, R.G., Sonkusare, V., Bhusari, G., Mondal, A., Shaik, D., Juneja, H.D. (2017). Microwave-mediated synthesis of spinel $CuAl_2O_4$ nanocomposites for enhanced electrochemical and catalytic performance. *Research on Chemical Intermediates*, 44: 239-2060. doi.org/10.1007/s11164-017-3213-z.

[18]   Wu, Z., Dong, F., Zhao, W., Wang, H., Liu, Y., and Guan, B. (2009). The fabrication and characterization of novel carbon doped TiO_2 nanotubes, nanowires and nanorods with high visible light photocatalytic activity. *Nanotechnology* 20:235701. DOI 10.1088/0957-4484/20/23/235701

[19]   Chaudhary, R.G., Sonkusare, V., Bhusari, G., Mondal, A., Potbhare, A.K., Sharma, R., Harjeet Juneja, H.D., Abdala, A.A. (2023). Preparation of Mesoporous ThO_2 nanoparticles: influence of calcination on morphology and visible-light-driven photocatalytic degradation of indigo carmine and methylene blue, *Enviornmental Research*, 222: 115363. doi.org/10.1016/j.envres.2023.115363

[20]   Aazam, E. S. (2014). Visible light photocatalytic degradation of thiophene using Ag–TiO_2/multi-walled carbon nanotubes nanocomposite. *Ceram. Int.* 40, 6705–6711. https://doi.org/10.1016/j.ceramint.2013.11.132

[21]   Potbhare, A. K., Umekar, M. S., Chouke, P. B., Bagade, M. B., Aziz, S. T., Abdala, A. A., & Chaudhary, R. G. (2020). Bioinspired graphene-based silver nanoparticles: Fabrication, characterization and antibacterial activity. *Materials Today: Proceedings*, 29, 720-725. https://doi.org/10.1016/j.matpr.2020.04.212

[22]   Chaudhry, A. U., Lonkar, S. P., Chaudhary, R. G., Mabrouk, A., & Abdala, A. A. (2020). Thermal, electrical, and mechanical properties of highly filled HDPE/graphite nanoplatelets composites. *Materials Today: Proceedings*, 29, 704-708. https://doi.org/10.1016/j.matpr.2020.04.168

[23]   Chaudhary, R. G., Potbhare, A. K., Aziz, S. T., Umekar, M. S., Bhuyar, S. S., & Mondal, A. (2021). Phytochemically fabricated reduced graphene Oxide-ZnO NCs by *Sesbania bispinosa* for photocatalytic performances. *Materials Today: Proceedings*, 36, 756-762. https://doi.org/10.1016/j.matpr.2020.05.821

[24]   Umekar, M. S., Chaudhary, R. G., Bhusari, G. S., Mondal, A., Potbhare, A. K., & Sami, M. (2020). Phytoreduced graphene oxide-titanium dioxide nanocomposites using *Moringa oleifera* stick extract. *Materials Today: Proceedings*, 29, 709-714. https://doi.org/10.1016/j.matpr.2020.04.169

[25]   Chaudhary, R. G. (2020). Graphene-based materials and their nanocomposites with metal oxides: Biosynthesis, electrochemical, photocatalytic and antimicrobial applications. *Materials Research Foundations*, 83, 79-116. https://doi.org/10.21741/9781644900970-4.

[26]   Umekar, M., Chaudhary, R., Bhusari, G., & Potbhare, A. (2021). Fabrication of zinc oxide-decorated phytoreduced graphene oxide nanohybrid via *Clerodendrum infortunatum*. Emerging Materials Research, 10(1), 75-84. https://doi.org/10.1680/jemmr.19.00175

[27]   Umekar, M. S., Bhusari, G. S., Bhoyar, T., Devthade, V., Kapgate, B. P., Potbhare, A. P., Chaudhary, R. G. & Abdala, A. A. (2023). Graphitic carbon nitride-based photocatalysts for environmental remediation of organic pollutants. *Current Nanoscience*, *19*(2), 148-169. https://doi.org/10.2174/1573413718666220127123935

[28]   Potbhare, A. K., Shrirame, T. S., Devthade, V., Yerpude, S. T., Umekar, M. S., Chaudhary, R. G., & Bhusari, G. S. (2022). Fabrications and applications of polymer–graphene nanocomposites for sustainability. In *Biogenic Sustainable Nanotechnology* (pp. 149-184). Elsevier https://doi.org/10.1016/B978-0-323-88535-5.00005-6

[29]   Bhilkar, P. R., Madankar, R. S., Shrirame, T. S., Utane, R. D., Potbhare, A. K., Yerpude, S., Chaudhary, R. G. (2022). Functionalized Carbon Nanomaterials: Fabrication, Properties, and Applications. *Mater. Res.*, 135, 55-82. https://doi.org/10.21741/9781644902172-4

[30]   Bright, W. (2000). *Note. Lang. Soc.* 29, 155–155.

[31]   Sahoo, N. G., Rana, S., Cho, J. W., Li, L., and Chan, S. H. (2010). Polymer nanocomposites based on functionalized carbon nanotubes. *Prog. Polym. Sci.* 35, 837–867. https://doi.org/10.1016/j.progpolymsci.2010.03.002

[32]   Gao, C., Guo, Z., Liu, J.-H., and Huang, X.-J. (2012). The new age of carbon nanotubes: an updated review of functionalized carbon nanotubes in electrochemical sensors. *Nanoscale* 4, 1948–1963. https://doi.org/10.1039/C2NR11757F

[33]   Rabti, A., Raouafi, N., and Merkoçi, A. (2016). Bio(Sensing) devices based on ferrocene–functionalized graphene and carbon nanotubes. *Carbon* N. Y. 108, 481–514. https://doi.org/10.1016/j.carbon.2016.07.043

[34]   Rao, G. P., Lu, C., and Su, F. (2007). Sorption of divalent metal ions from aqueous solution by carbon nanotubes: a review. *Separation Purif. Technol.* 58, 224–231. https://doi.org/10.1016/j.seppur.2006.12.006

[35]   Yang, S., Li, J., Shao, D., Hu, J., and Wang, X. (2009). Adsorption of Ni(II) on oxidized multi-walled carbon nanotubes: effect of contact time, PH, foreign ions and PAA. *J. Hazard. Mater.* 166, 109–116. https://doi.org/10.1016/j.jhazmat.2008.11.003

[36]   Zawisza, B., Skorek, R., Stankiewicz, G., and Sitko, R. (2012). Carbon nanotubes as a solid sorbent for the preconcentration of Cr, Mn, Fe, Co, Ni, Cu, Zn and Pbprior to wavelength-dispersive X-ray fluorescence spectrometry. *Talanta* 99, 918–923. https://doi.org/10.1016/j.talanta.2012.07.059

[37]   Ihsanullah, Abbas, A., Al-Amer, A. M., Laoui, T., Al-Marri, M. J., Nasser, M. S., et al. (2016). Heavy metal removal from aqueous solution by advanced carbon nanotubes: critical review of adsorption applications. *Separation Purif. Technol.* 157, 141–161. https://doi.org/10.1016/j.seppur.2015.11.039

[38]   Huang, J., Liu, Y., & You, T. (2010). Carbon nanofiber based electrochemical biosensors: A review. *Anal. Methods*, 2(3), 202-211. https://doi.org/10.1039/B9AY00312F

[39]   M. F. Yu, B. S. Files, S. Arepalli, and R. S. Ruoff, "Tensile loading of ropes of single wall carbon nanotubes and their mechanical properties," *Phys. Rev. Lett.*, vol. 84, no. 24, pp. 5552–5555, 2000. https://doi.org/10.1103/PhysRevLett.84.5552

[40]	S. Xie, W. Li, Z. Pan, B. Chang, and S. Lianfeng, "Mechanical and physical properties on carbon nanotube," *J. Phys. Chem. Solids*, vol. 61, no. 7, pp. 1153–1158, 2000. https://doi.org/10.1016/S0022-3697(99)00376-5

[41]	J. A. Elliott, J. K. W. Sandler, A. H. Windle, R. J. Young, and M. S. P. Shaffer, "Collapse of Single-Walled Carbon Nanotubes is Diameter Dependent," *Phys. Rev. Lett.*, vol. 92, no. 9, pp. 1–4, 2004. https://doi.org/10.1103/PhysRevLett.92.095501

[42]	Iijima, S., Helical microtubules of graphitic carbon. *Nature*, 1991. 354(6348): p. 56-58. http://dx.doi.org/10.1038/354056a0

[43]	Guldi, D.M. and N. Martín, Carbon nanotubes and related structures: synthesis, characterization, functionalization, and applications. 2010: John Wiley & Sons. http://dx.doi.org/10.1002/ange.201006930

[44]	Zheng, B., Y. Li, and J. Liu, CVD synthesis and purification of single-walled carbon nanotubes on aerogel-supported catalyst. *Appl. Phys. A.*, 2002. 74(3): p. 345-348. https://doi:10.1007/s003390201275

[45]	Ando, Y., et al., Growing carbon nanotubes. Materials today, 2004. 7(10): p. 22-29. https://doi.org/10.1016/S1369-7021(04)00446-8

[46]	Toubestani, D. H., Ghoranneviss, M., Mahmoodi, A., & Zareh, M. R. (2010, January). CVD growth of carbon nanotubes and nanofibers: big length and constant diameter. In *Macromolecular symposia* (Vol. 287, No. 1, pp. 143-147). Weinheim: WILEY-VCH Verlag. https://doi.org/10.1002/masy.201050120

[47]	Kang, Z., et al., One-step water-assisted synthesis of high-quality carbon nanotubes directly from graphite. *J. Am. Chem. Soc.*, 2003. 125(45): p. 13652-13653. https://doi.org/10.1021/ja037399m

[48]	Chen, Y., et al., The nucleation and growth of carbon nanotubes in a mechano-thermal process. *Carbon*, 2004. 42(8): p. 1543-1548. https://doi.org/10.1016/j.carbon.2004.02.003

[49]	Kusunoki, M., et al., Formation of self-aligned carbon nanotube films by surface decomposition of silicon carbide. *Philos. Mag. Lett.*, 1999. 79(4): p. 153-161. https://doi.org/10.1080/095008399177381

[50]	Viculis, L.M., J.J. Mack, and R.B. Kaner, A chemical route to carbon nanoscrolls. *Science*, 2003. 299(5611): p. 1361-1361. https://doi.org/10.1126/science.1078842

[51]	Luxembourg, D., G. Flamant, and D. Laplaze, Solar synthesis of single-walled carbon nanotubes at medium scale. *Carbon*, 2005. 43(11): p. 2302-2310. https://doi.org/10.1016/j.carbon.2005.04.010

[52]	Cho, W.S., et al., Synthesis of carbon nanotubes from bulk polymer. *Appl. Phys. Lett.*, 1996. 69(2): p. 278-279. https://doi.org/10.1063/1.117949

[53]	Mahanandia, P., et al., Synthesis of multi-wall carbon nanotubes by simple pyrolysis. *Solid State Commun.*, 2008. 145(3): p. 143-148. https://doi.org/10.1016/j.ssc.2007.10.020

[54]	Novoselova, I., et al., Electrolytic synthesis of carbon nanotubes from carbon dioxide in molten salts and their characterization. *Phys. E: Low-dimens. Syst. Nanostructures*, 2008. 40(7): p. 2231-2237. https://doi.org/10.1016/j.physe.2007.10.069

[55]	Zhang, Y., et al., Liquid phase synthesis of carbon nanotubes. *Physica B: Condensed Matter*, 2002. 323(1): p. 293-295. https://doi.org/10.1016/S0921-4526(02)01026-8

[56]   Szabó, A., et al., Synthesis methods of carbon nanotubes and related materials. *Materials*, 2010. 3(5): p. 3092-3140. https://doi.org/10.3390/ma3053092

[57]   Terrones, M., Science and technology of the twenty-first century: synthesis, properties, and applications of carbon nanotubes. *Annu. Rev. Mater. Res.*, 2003. 33(1): p. 419 501. https://doi.org/10.1146/annurev.matsci.33.012802.100255

[58]   Chen, G.Z., et al., Electrochemical investigation of the formation of carbon nanotubes in molten salts. *High Temp. Mater. Process.*, 1998. 2: p. 459-470. DOI 10.1007/s10008-012-1896-z

[59]   Ren, Z., Y. Lan, and Y. Wang, Aligned carbon nanotubes: physics, concepts, fabrication and devices. 2012: *Springer Science & Business Media*. ISBN: 9783642445392

[60]   Guo, T., et al., Self-assembly of tubular fullerenes. *The Journal of Physical Chemistry*, 1995. 99(27): p. 10694-10697. https://doi.org/10.1021/j100027a002

[61]   Rafique, M.M.A. and J. Iqbal, Production of carbon nanotubes by different routes-a review. *J. encapsulation adsorp. Sci.*, 2011. 1(02): p. 29. http://dx.doi.org/10.4236/jeas.2011.12004

[62]   Guo, T., et al., Catalytic growth of single-walled manotubes by laser vaporization. *Chem. Phys. Lett.*, 1995. 243(1): p. 49-54. https://doi.org/10.1016/0009-2614%2895%2900825-O

[63]   Söylev, D., Optimization of carbon nanotube properties by controlled amount oxidizers. 2011. http://hdl.handle.net/11147/3096

[64]   Kumar, M. and Y. Ando, Chemical vapor deposition of carbon nanotubes: a review on growth mechanism and mass production. *J. Nanosci. Nanotechnol.*, 2010. 10(6): p. 3739-3758. http://dx.doi.org/10.1166/jnn.2010.2939

[65]   Meyyappan, M., Carbon nanotubes: science and applications. 2004: CRC press. https://doi.org/10.1201/9780203494936

[66]   Öncel, Ç. and Y. Yürüm, Carbon nanotube synthesis via the catalytic CVD method: a review on the effect of reaction parameters. *Fuller. Nanotub. Carbon Nanostructures*, 2006. 14(1): p. 17-37. http://dx.doi.org/10.1080/15363830500538441

[67]   Boskovic, B.O., et al., Large-area synthesis of carbon nanofibres at room temperature. *Nat. mater.*, 2002. 1(3): p. 165-168. https://doi.org/10.1038/nmat755

[68]   Meyyappan, M., et al., Carbon nanotube growth by PECVD: a review. *Plasma Sources Sci. Technol.*, 2003. 12(2): p. 205. http://dx.doi.org/10.1088/0963-0252/12/2/312

[69]   Hu, C.-T., et al., Plasma-enhanced chemical vapor deposition carbon nanotubes for ethanol gas sensors. *Diam. Relat. Mater.*, 2009. 18(2): p. 472-477. http://dx.doi.org/10.1016/j.diamond.2008.10.057

[70]   Hofmann, S., et al., Low-temperature growth of carbon nanotubes by plasma-enhanced chemical vapor deposition. *Appl. Phys. Lett.*, 2003. 83(1): p. 135-137. http://dx.doi.org/10.1063/1.1589187

[71]   Queipo, P., et al., Aerosol Catalyst Particles for Substrate CVD Synthesis of Single-Walled Carbon Nanotubes. *Chem. Vap. Depos.*, 2006. 12(6): p. 364-369. https://doi.org/10.1002/cvde.200506446

[72]   Byeon, H., et al., Growth of ultra long multiwall carbon nanotube arrays by aerosol assisted chemical vapor deposition. *J. Nanosci. Nanotechnol.*, 2010. 10(9): p. 6116-6119. https://doi.org/10.1166/jnn.2010.2574

[73]   Abdullayeva, S., Musayeva, N., Frigeri, C., Huseynov, A., Jabbarov, R., Abdullayev, R., ... & Hasanov, R. (2015). Characterization of high quality carbon nanotubes synthesized via Aerosol-CVD. Journal: *J. Adv. Phys.*, 11. http://dx.doi.org/10.24297/jap.v11i3.6943

[74]   Ayala, P., et al., Cyclohexane triggers staged growth of pure and vertically aligned single wall carbon nanotubes. *Chem. Phys. Lett.*, 2008. 454(4): p. 332-336. http://dx.doi.org/10.1016/j.cplett.2008.02.041

[75]   Murakami, Y., et al., Characterization of single-walled carbon nanotubes catalytically synthesized from alcohol. *Chem. Phys. Lett.*, 2003. 374(1): p. 53-58. http://dx.doi.org/10.1016/S0009-2614(03)00687-0

[76]   Patole, S., et al., Optimization of water assisted chemical vapor deposition parameters for super growth of carbon nanotubes. *Carbon*, 2008. 46(14): p. 1987-1993. http://dx.doi.org/10.1016/j.carbon.2008.08.009

[77]   Sahoo, S.C., et al., Carbon nanoflake growth from carbon nanotubes by hot filament chemical vapor deposition. *Carbon*, 2014. 67: p. 704-711. http://dx.doi.org/10.1016/j.carbon.2013.10.062

[78]   Zhang, G., et al., Ultra-high-yield growth of vertical single-walled carbon nanotubes: Hidden roles of hydrogen and oxygen. *Proc. Natl. Acad. Sci. U. S. A.*, 2005. 102(45): p. 16141-16145. https://doi.org/10.1073/pnas.0507064102

[79]   Yabe, Y., et al., Synthesis of well-aligned carbon nanotubes by radio frequency plasma enhanced CVD method. *Diam. Relat. Mater.*, 2004. 13(4): p. 1292-1295. https://doi.org/10.1016/j.diamond.2003.11.067

[80]   Chen, S., et al., Growth of carbon nanotubes at low powers by impedance-matched microwave plasma enhanced chemical vapor deposition method. *J. Nanosci. Nanotechnol*, 2005. 5(11): p. 1887-1892. https://doi.org/10.1166/jnn.2005.437

[81]   Cantoro, M., et al., Catalytic chemical vapor deposition of single-wall carbon nanotubes at low temperatures. *Nano Lett.*, 2006. 6(6): p. 1107-1112. https://doi.org/10.1021/nl060068y

[82]   Queipo, P., et al., Aerosol Catalyst Particles for Substrate CVD Synthesis of Single-Walled Carbon Nanotubes. *Chem. Vap. Depos.*, 2006. 12(6): p. 364-369. https://doi.org/10.1002/cvde.200506446

[83]   Su, M., B. Zheng, and J. Liu, A scalable CVD method for the synthesis of single-walled carbon nanotubes with high catalyst productivity. *Chem. Phys. Lett.*, 2000. 322(5): p. 321-326. https://doi.org/10.1016/S0009-2614(00)00422-X

[84]   Maruyama, S., et al., Low-temperature synthesis of high-purity single-walled carbon nanotubes from alcohol. *Chem. Phys. Lett.*, 2002. 360(3): p. 229-234. https://doi.org/10.1016/S0009-2614(02)00838-2

[85]   Chen, Y., M. Conway, and J. Fitzgerald, Carbon nanotubes formed in graphite after mechanical grinding and thermal annealing. *Appl. Phys. A*, 2003. 76(4): p. 633-636.http://dx.doi.org/10.1007/s00339-002-1986-3

[86]   Bai, J., et al., Synthesis of SWNTs and MWNTs by a molten salt (NaCl) method. *Chem. Phys. Lett.*, 2002. 365(1): p. 184-188. http://dx.doi.org/10.1016/S0009-2614(02)01447-1

[87]   Vander Wal, R.L., L.J. Hall, and G.M. Berger, Optimization of flame synthesis for carbon nanotubes using supported catalyst. *J. Phys. Chem. B*, 2002. 106(51): p. 13122-13132. https://doi.org/10.11113/jt.v78.9595

[88] Vander Wal, R.L., G.M. Berger, and T.M. Ticich, Flame synthesis of carbon nanotubes using catalyst particles prepared by laser ablation. *Am. Chem. Soc., Div. Fuel Chem*, 2004. 49(2): p. 879. https://doi.org/10.1166/jnn.2003.201

[89] Braidy, N., M. El Khakani, and G. Botton, Effect of laser intensity on yield and physical characteristics of single wall carbon nanotubes produced by the Nd: YAG laser vaporization method. *Carbon*, 2002. 40(15): p. 2835-2842. http://dx.doi.org/10.1016/S0008-6223(02)00260-9

[90] Braidy, N., M. El Khakani, and G. Botton, Single-wall carbon nanotubes synthesis by means of UV laser vaporization. *Chem. Phys. Lett.*, 2002. 354(1): p. 88-92. http://dx.doi.org/10.1016/S0009-2614(02)00110-0

[91] Nishide, D., et al., High-yield production of single-wall carbon nanotubes in nitrogen gas. *Chem. Phys. Lett.*, 2003. 372(1): p. 45-50. http://dx.doi.org/10.1016/S0009-2614(03)00352-X

[92] Height, M.J., et al., Flame synthesis of single-walled carbon nanotubes. *Carbon*, 2004. 42(11): p. 2295-2307. http://dx.doi.org/10.1016/j.carbon.2004.05.010

[93] Height, M.J., J.B. Howard, and J.W. Tester. Flame Synthesis of Carbon Nanotubes. in *MRS Proceedings*. 2003. Cambridge Univ Press. doi:10.1557/PROC-772-M1.8

[94] Lee, S.J., et al., Large scale synthesis of carbon nanotubes by plasma rotating arc discharge technique. *Diam. Relat. Mater.*, 2002. 11(3): p. 914-917. https://doi.org/10.1016/S0925-9635(01)00639-2

[95] Wu, H., et al., One-pot synthesis of nanostructured carbon materials from carbon dioxide via electrolysis in molten carbonate salts. *Carbon*, 2016. 106: p. 208-217. https://doi.org/10.1016/j.carbon.2016.05.031

[96] Smeulders, D., et al., Rod milling and thermal annealing of graphite: Passing the equilibrium barrier. *J. Mater. Sci.*, 2005. 40(3): p. 655-662. https://doi.org/10.1007/s10853-005-6303-z

[97] Ando, Y., & Zhao, X. (2006). Synthesis of carbon nanotubes by arc-discharge method. *New Diam. Front. Carbon Technol.*, 16(3), 123-138.

[98] Wang, X. K., Lin, X. W., Dravid, V. P., Ketterson, J. B., & Chang, R. P. (1995). Carbon nanotubes synthesized in a hydrogen arc discharge. *Appl. Phys. Lett.*, 66(18), 2430-2432. https://doi.org/10.1063/1.113963

[99] Lee, S. J., Baik, H. K., Yoo, J. E., & Han, J. H. (2002). Large scale synthesis of carbon nanotubes by plasma rotating arc discharge technique. *Diam. Relat. Mater.*, 11(3-6), 914-917. https://doi.org/10.1016/S0925-9635(01)00639-2

[100] Long RQ, Lane RT. Carbon nanotubes as superior sorbent for dioxin removal. *J Am Chem Soc*. 2001;123,(9):2058–2059.1 https://doi.org/10.1021/ja003830l

[101] Tripathi, N., Pavelyev, V., & Islam, S. S. (2017). Synthesis of carbon nanotubes using green plant extract as catalyst: unconventional concept and its realization. Applied Nanoscience, 7, 557-566.

[102] Janas, D. (2020). From bio to nano: A review of sustainable methods of synthesis of carbon nanotubes. Sustainability, 12(10), 4115.

[103] Qu, J., Luo, C., Cong, Q., & Yuan, X. (2014). Recycling of the hyperaccumulator Brassica juncea L.: synthesis of carbon nanotube-Cu/ZnO nanocomposites. Journal of Material Cycles and Waste Management, 16, 162-166.

[104] Paul, S., & Samdarshi, S. K. (2011). A green precursor for carbon nanotube synthesis. New Carbon Materials, 26(2), 85-88.

[105] Hamid, Z. A., Azim, A. A., Mouez, F. A., & Rehim, S. A. (2017). Challenges on synthesis of carbon nanotubes from environmentally friendly green oil using pyrolysis technique. Journal of Analytical and Applied Pyrolysis, 126, 218-229.

[106] Patel, D. K., Kim, H. B., Dutta, S. D., Ganguly, K., & Lim, K. T. (2020). Carbon nanotubes-based nanomaterials and their agricultural and biotechnological applications. Materials, 13(7), 1679.

[107] Ghosh, P., Afre, R. A., Soga, T., & Jimbo, T. (2007). A simple method of producing single-walled carbon nanotubes from a natural precursor: Eucalyptus oil. Materials Letters, 61(17), 3768-3770.

[108] Suriani, A. B., Azira, A. A., Nik, S. F., Nor, R. M., & Rusop, M. (2009). Synthesis of vertically aligned carbon nanotubes using natural palm oil as carbon precursor. Materials Letters, 63(30), 2704-2706.

[109] Zhu, J., Jia, J., Kwong, F. L., Ng, D. H. L., & Tjong, S. C. (2012). Synthesis of multiwalled carbon nanotubes from bamboo charcoal and the roles of minerals on their growth. biomass and bioenergy, 36, 12-19.

[110] Lotfy, V. F., Fathy, N. A., & Basta, A. H. (2018). Novel approach for synthesizing different shapes of carbon nanotubes from rice straw residue. Journal of Environmental Chemical Engineering, 6(5), 6263-6274.

[111] Awasthi, K., Kumar, R., Tiwari, R. S., & Srivastava, O. N. (2010). Large scale synthesis of bundles of aligned carbon nanotubes using a natural precursor: turpentine oil. Journal of Experimental Nanoscience, 5(6), 498-508.

[112] Kumar, R., Singh, R. K., & Tiwari, R. S. (2016). Growth analysis and high-yield synthesis of aligned-stacked branched nitrogen-doped carbon nanotubes using sesame oil as a natural botanical hydrocarbon precursor. Materials & design, 94, 166-175.

[113] Kumar, R., Tiwari, R. S., & Srivastava, O. N. (2011). Scalable synthesis of aligned carbon nanotubes bundles using green natural precursor: neem oil. Nanoscale research letters, 6, 1-6.

[114] Beckett, P. (2003, April). Exploiting multiple functionality for nano-scale reconfigurable systems. In Proceedings of the 13th ACM Great Lakes symposium on VLSI (pp. 50-55). http://dx.doi.org/10.1145/764808.764822

[115] Hadjipaschalis I, Poullikkas A, Efthimiou V. Overview of current and future energy storage technologies for electric power applications. Renew. Sust. Energ. Rev. 2009;13(6–7):1513–1522. https://doi.org/10.1016/j.rser.2008.09.028

[116] Yan J, Fan Z, Wei T, et al. Carbon nanotube/MnO_2 composites synthesized by microwave-assisted method for supercapacitors with high power and energy densities. J. Power Sources. 2009;194(2):1202–1207. https://doi.org/10.1016/j.jpowsour.2009.06.006

[117] Jacobs CB, Peairs MJ, Venton BJ. Review: carbon nano tube based electrochemical sensors for biomolecules. *Anal. Chim. Acta*. 2010;662(2):105–127.1 https://doi.org/10.1016/j.aca.2010.01.009

[118] Yang N, Chen X, Ren T, et al. Carbon nanotube based biosensor. *Sens Actuator, B*. 2015;207:690–715.1 https://doi.org/10.1016/j.snb.2014.10.040

[119] Adeli M, Soleyman R, Beiranvanda Z, et al. Carbon nanotubes in cancer therapy: a more precise look at the role of carbon nanotube-polymer interactions. *Chem Soc Rev*. 2013;42(12):5231–5256.1 https://doi.org/10.1039/C3CS35431H

[120] Cinke M, Li J, Charles W, et al. CO_2 Adsorption in single-walled carbon nanotubes. *Chem Phys Lett*. 2003;376(5–6):761–766. https://doi.org/10.1016/S0009-2614(03)01124-2

[121] Qing. QL, Gao LL, Yu Z. Carbon nanotubes prepared from CO on pre-reduced La_2NiO_4 perovskite precursor. *Mater Res Bull*. 2001;36(3–4):471–477. https://doi.org/10.1016/S0025-5408(00)00480-3

[122] Park SJ, Lee DG. Performance improvement of micron-sized fibrous metal filters by direct growth of carbon nanotubes. *Carbon*. 2006;44(10):1930–1935. https://doi.org/10.1016/j.carbon.2006.02.005

[123] Park, J. H., Yoon, K. Y., Na, H., Kim, Y. S., Hwang, J., Kim, J., & Yoon, Y. H. (2011). Fabrication of a multi-walled carbon nanotube-deposited glass fiber air filter for the enhancement of nano and submicron aerosol particle filtration and additional antibacterial efficacy. *Sci. Total Environ.*, 409(19), 4132-4138. https://doi.org/10.1016/j.scitotenv.2011.04.060

[124] Shohreh F, Masoud Vesali N, Mona C, et al. Improving CO_2/CH_4 adsorptive selectivity of carbon nanotubes by functionalization with nitrogen containing groups. *Chem. Eng. Res. Des*. 2011;89(9):1669–1675. https://doi.org/10.1016/j.cherd.2010.10.002

[125] Somy A, Mehrnia MR, Delavari-Amrei H, et al. Adsorption of carbon dioxide using impregnated acti vated carbon promoted by zinc. *Int. J. Greenhouse Gas Control*. 2009;3(3):249–254. https://doi.org/10.1016/j.ijggc.2008.10.003

[126] Budaeva AD, Zoltoev EV. "Porous structure and sorp tion properties of nitrogen-containing activated carbon,". *Fuel*. 2010;89:2623–2627. https://doi.org/10.1016/j.fuel.2010.04.016

[127] Verlicchi P, Aukidy AM, Zambello E, Za. E Mbello. "Occurrence of pharmaceutical compounds in urban wastewater removal, mass load and environmental risk after a secondary treatment—a review,". *Sci Total Environ*. 2012;429:123–155.1 https://doi.org/10.1016/j.scitotenv.2012.04.028

[128] Augusto F, Carasek E, Silva RGC. New sorbents for extraction and microextraction techniques. *J Chromatogr A*. 2010;1217(16):2533–2542.1 https://doi.org/10.1016/j.chroma.2009.12.033

[129] Adenuga AA, Truong L, Tanguay RL, et al. Preparation of water soluble carbon nanotubes and assessment of their biological activity in embryonic zebrafish. *Int J Biomed Nanosci Nanotechnol*. 2013;3 (1/2):38–51. https://doi.org/10.1504%2FIJBNN.2013.054514

[130] Girardello, R., Tasselli, S., Baranzini, N., Valvassori, R., de Eguileor, M., & Grimaldi, A. (2015). Effects of carbon nanotube environmental dispersion on an aquatic invertebrate, Hirudo medicinalis. Plos one, 10(12), e0144361. https://doi.org/10.1371/journal.pone.0144361

[131] Wei C, Dai L, Roy A, et al. Multifunctional chemical vapor sensors of aligned carbon nanotube and polymer composites. *J. Am. Chem. Soc.* 2006;128(5):1412. https://doi.org/10.1021/ja0570335

[132] Attar S, Ranveer A. Carbon nanotubes and its environmental applications Journal of Environmental Science. *Comput Sci Eng Technol.* 2015;4:304–311.

[133] Kumar M, Ando Y. A simple method of producing aligned carbon nanotubes from an unconventional precursor – camphor. *Chem Phys Lett.* 2003;374(5– 6):521–5261. http://dx.doi.org/10.1016/S0009-2614(03)00742-5

[134] Kumar M, Ando Y. Single-wall and multi-wall carbon nanotubes from camphor - a botanical hydrocarbon. *Diam Relat Mater.* 2003;12(10–11):1845–1850. http://dx.doi.org/10.1016/S0925-9635(03)00217-6

[135] Khalid, A., Ahmad, P., Khan, A., Muhammad, S., Khandaker, M. U., Alam, M. M., ... & Emran, T. B. (2022). Effect of Cu doping on ZnO nanoparticles as a photocatalyst for the removal of organic wastewater. *Bioinorg. Chem. Appl., 2022.* https://doi.org/10.1155/2022/9459886

[136] S.Kabiri, D.N.H.Tran, T.Altalhi and D.Losic, Outstanding adsorption performance of graphene–carbon nanotube aero gels for continuous oil removal, *Carbon*, 2014, 80, 523–533. https://doi.org/10.1016/j.carbon.2014.08.092

[137] W. Wan, R. Zhang, W. Li, H. Liu, Y. Lin and L. Li, et al., Graphene–carbon nanotube aerogel as an ultra-light, com pressible and recyclable highly efficient absorbent for oil and dyes, *Environ. Sci.: Nano*, 2016, 3(1), 107–113. https://doi.org/10.1039/C5EN00125K

[138] W. Zhan, S. Yu, L. Gao, F. Wang, X. Fu and G. Sui, et al., Bioinspired Assembly of Carbon Nanotube into Graphene Aerogel with ''Cabbagelike'' Hierarchical Porous Structure for Highly Efficient Organic Pollutants Cleanup, *ACS Appl. Mater. Interfaces*, 2018, 10(1), 1093–1103. https://doi.org/10.1021/acsami.7b15322

[139] T.Tao, G.Li, Y. HeandP. Duan,Hybridcarbon nanotubes/ graphene/nickel fluffy spheres for fast magnetic separation and efficient removal of organic solvents from water, *Mater. Lett.*, 2019, 254, 440–443. https://doi.org/10.1016/j.matlet.2019.06.104

[140] J. Cai, J. Tian, H. Gu and Z. Guo, Amino carbon nanotube modified reduced graphene oxide aerogel for oil/water separation, *ES Mater. Manuf.*, 2019, 6(2), 68–74. http://dx.doi.org/10.30919/esmm5f611

[141] Shrirame, T. S., Khan, J. S., Umekar, M. S., Potbhare, A. K., Bhilkar, P. R., Bhusari, G. S., ... & Chaudhary, R. G. (2022). Graphene-Polymer Nanocomposites for Environmental Remediation of Organic Pollutants. *Metal Nanocomposites for Energy and Environmental Applications*, 321-349. http://dx.doi.org/10.1007/978-981-16-8599-6_14

[142] Z. Sui, Q. Meng, X. Zhang, R. Ma and B. Cao, Green synthesis of carbon nanotube–graphene hybrid aerogels and their use as versatile agents for water purification, *J. Mater. Chem.*, 2012, 22(18), 8767. https://doi.org/10.1039/C2JM00055E

[143] S. K. Das, J. Bhowal, A. R. Das and A. K. Guha, Adsorption behavior of rhodamine B on rhizopus o ryzae biomass, *Langmuir*, 2006, 22(17), 7265–7272. http://dx.doi.org/10.1021/la0526378

[144] L. Sun, C. Tian, L. Wang, J. Zou, G. Mu and H. Fu, Magnetically separable porous graphitic carbon with large surface area as excellent adsorbents for metal ions and dye, *J. Mater. Chem.*, 2011, 21(20), 7232–7239. https://doi.org/10.1039/C1JM10470E

[145]   B.S. Girgis, A. M. Soliman and N. A. Fathy, Development of micro-mesoporous carbons from several seed hulls under varying conditions of activation, *Microporous Mesoporous Mater.*, 2011, 142(2–3), 518–525. http://dx.doi.org/10.1016/j.micromeso.2010.12.044

[146]   L. Ai and J. Jiang, Removal of methylene blue from aqueous solution with self-assembled cylindrical graphene carbon nanotube hybrid, *Chem. Eng. J.*, 2012, 192, 156–163. http://dx.doi.org/10.1016/j.cej.2012.03.056

[147]   Aziz, S. T., Ummekar, M., Karajagi, I., Riyajuddin, S. K., Siddhartha, K. V. R., Saini, A., ... & Dutta, A. (2022). A Janus cerium-doped bismuth oxide electrocatalyst for complete water splitting. *Cell Reports Physical Science*, 3(11). http://dx.doi.org/10.1016/j.xcrp.2022.101106

[148]   M. Kotal and A. K. Bhowmick, Multifunctional hybrid materials based on carbon nanotube chemically bonded to reduced graphene oxide, *J. Phys. Chem.* C, 2013, 117(48), 25865–25875. http://dx.doi.org/10.1021/jp4097265

[149]   K. Goh, W. Jiang, H. E. Karahan, S. Zhai, L. Wei and D. Yu, et al., All-carbon nanoarchitectures as high-performance separation membranes with superior stability, *Adv. Funct. Mater.*, 2015, 25(47), 7348–7359. http://dx.doi.org/10.1002/adfm.201502955

[150]   S. Wang, H. Sun, H.-M. Ang and M. Tade´, Adsorptive remediation of environmental pollutants using novel graphene-based nanomaterials, *Chem. Eng. J.*, 2013, 226, 336–347. http://dx.doi.org/10.1016/j.cej.2013.04.070

[151]   D. Zhao, Y. Wang, S. Zhao, M. Wakeel, Z. Wang and R. S. Shaikh, et al., A simple method for preparing ultra light graphene aerogel for rapid removal of U(VI) from aqueous solution, *Environ. Pollut.*, 2019, 251, 547–554. https://doi.org/10.1016/j.envpol.2019.05.011

[152]   H. Gao, Y. Sun, J. Zhou, R. Xu and H. Duan, Mussel inspired synthesis of polydopamine-functionalized gra phene hydrogel as reusable adsorbents for water purification, *ACS Appl. Mater. Interfaces*, 2013, 5(2), 425–432. https://doi.org/10.1021/am302500v

[153]   Y. Lei, F. Chen, Y. Luo and L. Zhang, Synthesis of three dimensional graphene oxide foam for the removal of heavy metal ions, *Chem. Phys. Lett.*, 2014, 593, 122–127. http://dx.doi.org/10.1016/j.cplett.2013.12.066

[154]   X. Mi, G. Huang, W. Xie, W. Wang, Y. Liu and J. Gao, Preparation of graphene oxide aerogel and its adsorption for Cu^{2+} ions, *Carbon*, 2012, 50(13), 4856–4864. http://dx.doi.org/10.1016/j.carbon.2012.06.013

[155]   M. Liu, C. Chen, J. Hu, X. Wu and X. Wang, Synthesis of magnetite/graphene oxide composite and application for cobalt(II) removal, *J. Phys. Chem. C*, 2011, 115(51), 25234–25240. http://dx.doi.org/10.1021/jp208575m

[156]   G. Zhao, X. Ren, X. Gao, X. Tan, J. Li and C. Chen, et al., Removal of Pb(II) ions from aqueous solutions on few layered graphene oxide nanosheets, *Dalton Trans.*, 2011, 40(41), 10945–10952. https://doi.org/10.1039/C1DT11005E

[157]   N.Zhang, H. Qiu, Y. Si, W. Wang andJ. Gao, Fabrication of highly porous biodegradable monoliths strengthened by graphene oxide and their adsorption of metal ions, *Carbon*, 2011, 49(3), 827–837. http://dx.doi.org/10.1016/j.carbon.2010.10.024

[158]   B. Xiao and K. Thomas, Competitive adsorption of aqu eous metal ions on an oxidized nanoporous activated carbon, *Langmuir*, 2004, 20(11), 4566–4578. https://doi.org/10.1021/la049712j

[159] M. Benitez, D. Das, R. Ferreira, U. Pischel and H. Garcı´a, Urea-containing mesoporous silica for the adsorption of Fe(III) cations, *Chem. Mater.*, 2006, 18(23), 5597–5603. http://dx.doi.org/10.1021/cm061287n

[160] A. Liu, K. Hidajat, S. Kawi and D. Zhao, A new class of hybrid mesoporous materials with functionalized organic monolayers for selective adsorption of heavy metal ions, *Chem. Commun.*, 2000, (13), 1145–1146. https://doi.org/10.1039/B002661L

[161] L. Li, G. Zhou, Z. Weng, X.-Y. Shan, F. Li and H.-M. Cheng, Monolithic Fe₂O₃/graphene hybrid for highly efficient lithium storage and arsenic removal, *Carbon*, 2014, 67, 500–507. http://dx.doi.org/10.1016/j.carbon.2013.10.022

[162] M. Zhang, B. Gao, X. Cao and L. Yang, Synthesis of a multifunctional graphene–carbon nanotube aerogel and its strong adsorption of lead from aqueous solution, *RSC Adv.*, 2013, 3(43), 21099–21105. https://doi.org/10.1039/C3RA44340J

[163] S. Vadahanambi, S.-H. Lee, W.-J. Kim and I.-K. Oh, Arsenic removal from contaminated water using three dimensional graphene-carbon nanotube-iron oxide nanostructures, *Environ. Sci. Technol.*, 2013, 47(18), 10510–10517. http://dx.doi.org/10.1021/es401389g

[164] B. P. Viraka Nellore, R. Kanchanapally, F. Pedraza, S. S. Sinha, A. Pramanik and A. T. Hamme, et al., Bio conjugated CNT-bridged 3D porous graphene oxide membrane for highly efficient disinfection of pathogenic bacteria and removal of toxic metals from water, *ACS Appl. Mater. Interfaces*, 2015, 7(34), 19210–19218. https://doi.org/10.1021%2Facsami.5b05012

[165] Z. Gu, Y. Wang, J. Tang, J. Yang, J. Liao and Y. Yang, et al., The removal of uranium(VI) from aqueous solution by graphene oxide–carbon nanotubes hybrid aerogels, *J. Radioanal. Nucl. Chem.*, 2015, 303(3), 1835–1842. http://dx.doi.org/10.1007/s10967-014-3795-5

[166] H. Huang, T. Chen, X. Liu and H. Ma, Ultrasensitive and simultaneous detection of heavy metal ions based on three-dimensional graphene-carbon nanotubes hybrid electrode materials, *Anal. Chim. Acta*, 2014, 852, 45–54. https://doi.org/10.1016/j.aca.2014.09.010

[167] Tomar, R., Abdala, A. A., Chaudhary, R. G., & Singh, N. B. (2020). Photocatalytic degradation of dyes by nanomaterials. *Mater. Today: Proc.*, 29, 967-973. http://dx.doi.org/10.1016/j.matpr.2020.04.144

[168] Chouke, P. B., Shrirame, T., Potbhare, A. K., Mondal, A., Chaudhary, A. R., Mondal, S., Chaudhary, R. G. (2022). Bioinspired metal/metal oxide nanoparticles: A road map to potential applications. *Mater. Today Adv.*, 16, 100314. https://doi.org/10.1016/j.mtadv.2022.100314

[169] Umekar, M. S., Bhusari, G. S., Potbhare, A. K., Mondal, A., Kapgate, B. P., Desimone, M. F., & Chaudhary, R. G. (2021). Bioinspired reduced graphene oxide based nanohybrids for photocatalysis and antibacterial applications. *Curr. Pharm. Biotechnol.*, 22(13),1759-1781. http://dx.doi.org/10.2174/1389201022666201231115826

[170] Chaudhary, R.G., Mahmood, S., Jotania, R.B. (2023). Green nanomaterials for clean and sustainable environment. Curr. Nanosci. 19(6): 746-747. doi.org/10.2174/1573413719062305511103527

[171] Chouke, P. B., Dadure, K. M., Potbhare, A. K., Bhusari, G. S., Mondal, A., Chaudhary, K., ... & Masram, D. T. (2022). Biosynthesized δ-Bi₂O₃ nanoparticles from *Crinum*

viviparum flower extract for photocatalytic dye degradation and molecular docking. *ACS Omega*, 7(24): 20983-20993. https://doi.org/10.1021%2Facsomega.2c01745

[172]   Sonkusare, V. N., Chaudhary, R. G., Bhusari, G. S., Mondal, A., Potbhare, A. K., Mishra, R. K., Abdala, A. A. (2020). Mesoporous octahedron-shaped tricobalt tetroxide nanoparticles for photocatalytic degradation of toxic dyes. *ACS Omega*, 5(14), 7823-7835. https://doi.org/10.1021%2Facsomega.9b03998

[173]   C. Wang, M. Cao, P. Wang, Y. Ao, J. Hou and J. Qian, Preparation of graphene–carbon nanotube-TiO$_2$ compo sites with enhanced photocatalytic activity for the removal of dye and Cr(VI), *Appl. Catal., A*, 2014, 473, 83–89. http://dx.doi.org/10.1016/j.apcata.2013.12.028

[174]   L. Qu, G. Zhu, J. Ji, T. Yadav, Y. Chen and G. Yang, et al., Recyclable visible light-driven Og-C$_3$N$_4$/graphene oxide/ N-carbon nanotube membrane for efficient removal of organic pollutants, ACS *Appl. Mater. Interfaces*, 2018, 10(49), 42427–42435. https://doi.org/10.1021/acsami.8b15905

[175]   G. Yi, Z. Chang, G. Liu and L. Yang, In situ fabrication of copper nanocubes with platinum skin on 3D graphene carbon nanotubes hybrid for efficient methanol electrooxidation, *Int. J. Electrochem. Sci.*, 2019, 14, 7232–7240. https://doi.org/10.20964/2019.08.25

[176]   X. T. Tran, M. Hussain and H. T. Kim, Facile and fast synthesis of a reduced graphene oxide/carbon nanotube/ iron/silver hybrid and its enhanced performance in cata lytic reduction of 4-nitrophenol, *Solid State Sci.*, 2020, 100, 106107. https://doi.org/10.1016/j.solidstatesciences.2019.106107

[177]   M. Kotal, A. Sharma, S. Jakhar, V. Mishra, S. Roy and S. C. Sahoo, et al., Graphene-Templated Cobalt Nanopar ticle Embedded Nitrogen-Doped Carbon Nanotubes for Efficient Visible-Light Photocatalysis, *Cryst. Growth Des.*, 2020, 20(7), 4627–4639. https://doi.org/10.1021/acs.cgd.0c00430

[178]   Simon, J., Flahaut, E. and Golzio, M., 2019. Overview of carbon nanotubes for biomedical applications. *Materials*, 12(4), p.624. https://doi.org/10.3390/ma12040624

[179]   Oskoueian, Arshin, Khamirul Amin Matori, Saadi Bayat, Ehsan Oskoueian, Farhad Ostovan, and Meysam Toozandehjani. "Fabrication, characterization, and functionalization of single-walled carbon nanotube conjugated with tamoxifen and its anticancer potential against human breast cancer cells." *J. Nanomater.* 2018 (2018): 1-13. https://doi.org/10.1155/2018/8417016

[180]   Ahmed, Duha S., Mustafa KA Mohammed, and Mohammad R. Mohammad. "Sol–gel synthesis of Ag-doped titania-coated carbon nanotubes and study their biomedical applications." *Chem.l Pap.*74, no. 1 (2020): 197-208. https://doi.org/10.1007/s11696-019-00869-9

[181]   Yuan, X., Zhang, X., Sun, L., Wei, Y., & Wei, X. (2019). Cellular toxicity and immunological effects of carbon-based nanomaterials. *Part. Fibre Toxicol.*, 16(1), 1-27. https://doi.org/10.1186/s12989-019-0299-z

[182]   Sheikhi, Masoome, Siyamak Shahab, Mehrnoosh Khaleghian, and Rakesh Kumar. "Interaction between new anti-cancer drug syndros and CNT (6, 6-6) nanotube for medical applications: geometry optimization, molecular structure, spectroscopic (NMR, UV/Vis, excited state), FMO, MEP and HOMO-LUMO investigation." *Appl. Surf. Sci.* 434 (2018): 504-513. https://doi.org/10.1016/j.apsusc.2017.10.154

[183]   Chen, Chao, Rui Ran, Zhiyu Yang, Ruitao Lv, Wanci Shen, Feiyu Kang, and Zheng-Hong Huang. "An efficient flexible electrochemical glucose sensor based on carbon nanotubes/carbonized silk fabrics decorated with Pt microspheres." *Sens. Actuators B: Chem.* 256 (2018): 63-70. https://doi.org/10.1016/j.snb.2017.10.067

[184]   Wee, Youngho, Seunghwan Park, Young Hyeon Kwon, Youngjun Ju, Kyung-Min Yeon, and Jungbae Kim. "Tyrosinase-immobilized CNT based biosensor for highly-tt 6`sensitive detection of phenolic compounds." *Biosens. Bioelectron.* 132 (2019): 279-285. https://doi.org/10.1016/j.bios.2019.03.008

[185]   Singh, Chandan, Saurabh Srivastava, Md Azahar Ali, Tejendra K. Gupta, Gajjala Sumana, Anchal Srivastava, R. B. Mathur, and Bansi D. Malhotra. "Carboxylated multiwalled carbon nanotubes-based biosensor for aflatoxin detection." *Sens. Actuators B: Chem.* 185 (2013): 258-264. https://doi.org/10.1016/j.snb.2013.04.040

Green Synthesis and Emerging Applications of Frontier Nanomaterials   Materials Research Forum LLC
Materials Research Foundations 169 (2024) 231-248   https://doi.org/10.21741/9781644903278-9

Chapter 9

Zirconium and rhodium-based nanomaterials: Green synthesis and emerging applications

Md. Ahad Ali[1,2], Md. Abu Bin Hasan Susan[1,3*]

[1]Department of Chemistry, University of Dhaka, Dhaka 1000, Bangladesh

[2]Department of Chemistry, Jashore University of Science and Technology, Jashore 7408, Bangladesh

[3]Dhaka University Nanotechnology Center, University of Dhaka, Dhaka 1000, Bangladesh

* susan@du.ac.bd

Abstract

Zirconium (Zr) and rhodium (Rh)-based nanomaterials (NMs) have gained a surge of interest as potential candidates for new applications. There have been numerous attempts to synthesize NMs of this kind with optimum sizes, shapes, and morphologies. Most often, the successful routes for synthesis of these NMs involve hazardous chemicals and byproducts to render them unfriendly to human health and environment. Novel, more efficient, and sustainable methods for synthesizing NMs with higher stability and biocompatibility are thus critically sought for. Green chemistry has been exploited to reduce negative impacts and improve environmental benignity, and sustainability of the techniques. The application of biomolecules like fucoidans, enzymes, proteins, carbohydrates, polyphenols, alkaloids, terpenoids, flavonoids, saponins, amino acids, etc. has been proven useful as a green source in synthesis of such NMs. This chapter focuses on green syntheses of Zr and Rh-based NMs. The developments so far have been summarized and their uses in emerging application have been highlighted through an extensive review of literature. Finally, future directions and prospects of green methods have been pointed out for diverse applications.

Keywords

Green Synthesis, Zirconium, Rhodium, Nanomaterial, Emerging Application

Contents

1.   Introduction

Nanotechnology deals with the design, synthesis, and applications of functional materials which have at least one physical dimension in the 1–100 nm range. The reason of the evolution of this field is the diverse properties possessed by such functional materials. They have a great chemical, optical, mechanical, thermal, electrical, and magnetic properties which come from the fact that NMs have extremely high surface area and functionality. But the main problem is the preparation of these nanosized materials and prevention of their aggregations. In the last decade, various methods are developed for designing NMs of this kind. The most useful type is the bottom-up wet chemical synthesis of NMs. This bottom-up techniques not only successfully design NMs by atom-to-atom fabrication up to controllable size distribution but also minimize the defects on the surface as compared to the other top-down techniques. These synthetic methods are generally costly, non-environmentally friendly, and chemicals used during the process and byproducts have toxic effect on human and animals.

In the era of fourth industrial revolution, one of the greatest challenges is to save our environment for sustaining our lives. The term 'Green Chemistry' comes into play from the 20th century when world leaders agreed to save the environment. As the field of nanotechnology is growing rapidly, 'Green Chemistry' has also been the demand of time. Numerous green methods have so far been developed to synthesize NMs. When comparing the conventional filigree of artificial routes with NMs, the introduction of modern synthetic techniques has made it possible to produce superior NMs through a simple green synthesis, which would help eliminate harsh processing conditions by enabling synthesis at physiological temperature, pressure, and pH at a very little expense. The most interesting part of these green synthesis is that NMs synthesized using different methods have variable task-specific applications.

Zr is a silvery gray metal that is exceptionally durable, malleable, ductile, and shiny. It is comparable to titanium in terms of physical and chemical features. Zr has remarkable durability under adverse conditions. Zr has the same degree of toughness as copper but is much more lightweight than steel. One of the major sources of Zr is the zircon mineral widely found in sands. Zr can also be found as a byproduct in the processing of the titanium and tin from their corresponding minerals. The nanoparticles (NPs) of Zr are mainly synthesized from the reduction of the inorganic and organic chlorine compounds of Zr.

Rh, a hardly found element in the earth crust and the most precious metal found in nature, has a great importance as a potential catalyst. Although in the ancient time Rh was used only as a shiny metal for preparation of jewelries, it has become a potential candidate for application in different fields. The main feature of Rh is its stability; even in a very harsh condition like at high temperature and pressure they are not susceptible to undergo oxidation and can withstand in acidic or alkaline conditions. Besides their other properties like small electrical and thermal resistivity as well as high melting and boiling point brought the attraction of scientific community towards them.

There are many ways to synthesize Rh NPs. These include: reduction of Rh salts, chemical vapor deposition, thermal decomposition of Rh-based complexes, electrodeposition, etc. In each of these methods most of the chemicals used, for example the reducing agents and the NPs stabilizer, and even the decomposition products, are not environmentally friendly. Therefore, a green method of synthesis of NPs has been given extra importance in scientific research.

2. Zr-based NPs

There are different types of Zr-based NPs. Of them notables are pure metallic Zr, zirconia (ZrO_2), Zr phosphate, Zr-based metal organic framework, Zr-doped metals, Zr-based nanocomposites.

2.1 Green synthesis and applications of metallic Zr NPs

There are various types of green media which are used as media for synthesis of metallic Zr NPs. The plant-based media are the most promising media for synthesis of NPs firstly because of the presence of different compounds with multiple functional groups like reducing sugars, flavonoids, alkaloids, terpenoids, polyphenols, saponins, amino acids, etc. that can act as bioreducing, biochelating, biostabilizing, and biocapping agents. Secondly, plants provide the opportunity to prevalence, local availability, low cost, etc. Thirdly, plant extracts are easily controllable and extraction of the multiple compounds are also facile using green solvents like water, ethanol, etc. Different parts of plants like stems, barks, leaves, flowers, fruits, fruit peels, roots, gums, etc. provide the opportunity to use a single plant with multiple properties.

A general plant-based biosynthesis of Zr NPs have multiple steps to follow. First one is the extraction of active compounds from different parts of plants. In the green procedure polar compounds like polyphenols, sugars, gallic acids, etc. are extracted in water while for non-polar compounds are extracted using ethanol. Secondly, the precursor generally Zr nitrate or Zr chloride is added to the medium for the reaction to be carried out. Diverse conditions like heating or irradiation are applied at this step. When the reaction is completed, the contents are centrifuged and dried. A general scheme of metallic NPs synthesis is shown in Figure 1. Table 1 summarizes biosynthesis of Zr NPs using different plant extracts.

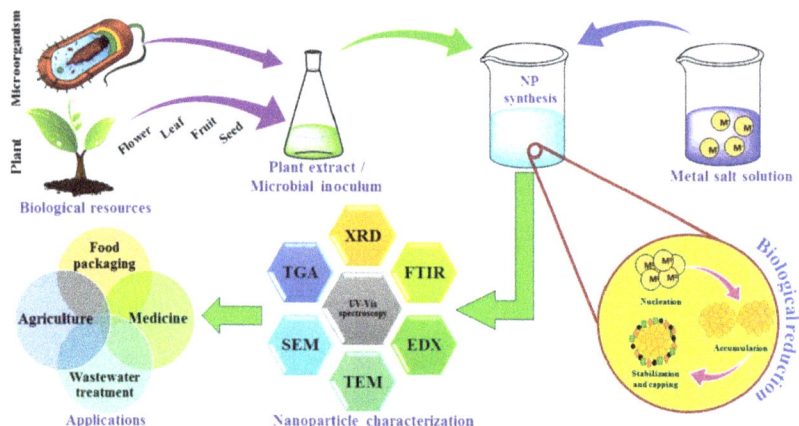

Figure 1. Synthesis of metallic Zr NPs using green routes and their characterizations and applications [1]. (Reprinted from [1]).

Another prospective green synthetic route for Zr NPs is fungus mediated biosynthesis. Extracellular biosynthesis of Zr NPs from $ZrCl_4$ in fungus of *penicillium* family results in spherical NPs having particle size below 100 nm [7].

2.2   Green synthesis and applications of ZrO_2 NPs

Like metallic Zr NPs, synthesis of ZrO_2 NPs is possible in plants, bacteria, fungi, microalgae, etc. Plant-based compounds and microorganism based secreted bioactive metabolites are potent to synthesize and stabilize the ZrO_2 NPs. Synthetic route of ZrO_2 NPs is similar to that of metallic Zr. The only difference is that in ZrO_2 after drying crude product must be calcined to get optimum crystallites.

Bacteria can release bioorganic compounds during their metabolism process which can act as bioreducing, biocapping and biostabilizing agents for synthesis of ZrO_2 NPs. The mechanism of synthesis follows two major steps, biosorption of the precursor materials and bioreduction to the product. The former step initiates in the cell wall where the bacteria secrete biomolecules like proteins, polysaccharides, etc. which provides negatively charged surface functionalities to adsorb bivalent zirconium ion a mainly via supramolecular interactions like electrostatic attraction, ion exchange, hydrogen bonding, coordination bonding, etc. The latter step chemical reaction occurs where zirconia ZrO_2 is formed which is stabilized by in the cellular environment by biocapping agents like proteins, amines, etc. Extremophilic *Acinetobacter sp.* KCSI1, *Enterobacter sp.* strain RNT10 have been successfully used for synthesis of bio ZrO_2 NPs [8,9]. Monoclinic and tetragonal ZrO_2 NPs having particle size of 5-25 nm have also been synthesized by using bacteria.

Table 1. Synthesis of Metallic Zr NPs Using Plant Extracts

Plant Species	Plant Tissues	Precursor	Product	Particle Size (nm)	Potential Properties	Ref.
Punica granatum	Fruit peel		Zr NPs	20-60	Antibacterial, antifungal, antioxidant	[2]
Allium sativum	Bulb	ZrO. (NO$_3$)$_2$	Zr NPs	554	Antibacterial, antimicrobial	[3]
Zingiber officinale	Stem	ZrO. (NO$_3$)$_2$	Zr NPs	406	Antibacterial, antimicrobial	[3]
Sphagneticola trilobata	Leaf	ZrO. (NO$_3$)$_2$	Zr NPs	20-100	Antifungal, antituberculosis, antimalarial	[4]
Camellia sinensis	Leaf	ZrO$_2$	Zr NPs		Antibacterial	[5]
Citrus limon	Fruit peel	ZrCl$_2$	Zr NPs		Antioxidant, anticancer	[6]
Citrus limon	Fruit juice	ZrCl$_2$	Zr NPs		Antioxidant, anticancer	[6]

Another pathway of synthesis of ZrO$_2$ is through fungi. Most of the fungi absorb micronutrient from soil through nutrient cycling where microorganisms present in fungi decompose organic matters in soil by secretion of digestion enzymes [10]. Some of the fungi have been used to minimize the heavy metal contamination of soil due to their extraordinary capability to absorb heavy metal ions [11]. Based on these advantageous features fungi is considered as a potential medium to prepare NPs. Extracellular secretion of *Penicilium* fungal species was used for biosynthesis of ZrO$_2$ [7]. *Penicillium purpurogenome* PTCC 5212, *Penicillium aculeatum* PTCC 5167, and *Penicillium notatum* PTCC 5074 can be successfully used to prepare spherical NPs with particle size less than 100 nm. The secreted biomolecules present in the supernatant of fungi provides the role of reducing and stabilizing agents. Most of the fungi that have potential to fabricate ZrO$_2$ have cationic proteins minimizes the adsorption of positive Zr^{2+} ions on it. Negative complex ions of Zr may be advantageously used to minimize this problem. ZrF$_6^{2-}$ was used as a precursor to synthesize ZrO$_2$ in *Fusarium oxysporum* fungi [12]. Positive pole of proteins binds negative complex ion of Zr via electrostatic force of attraction without imposing toxicity in the host fungi.

Algae having fucoidans can act as a medium for synthesis of NPs. Fucoidans are long chain monosaccharides having sulfur esteric group that can be potential bioreducing group for synthesis of ZrO$_2$. A seaweed *sargassum wightii* was used to synthesize ZrO$_2$ which could successfully fabricate ZrO$_2$ having particle size of 4.8 nm with tetragonal morphology [13].

Plant-based materials are the most advantageous materials among the other materials for green synthesis of ZrO$_2$ due to the versatility and low toxicity of plant materials. Besides plant mediated syntheses are easily controllable and manipulatable and require much less time for completion. As mentioned earlier different parts of plants have different functionalities. *Table 2* summarizes plant extract mediated synthesis of ZrO$_2$ NPs.

Table 2. Synthesis of ZrO₂ NPs Using Plant Extracts

Plant Species	Plant Tissue	Precursor	Morphology	Particle Size (nm)	Potential Properties	Ref.
Eucalyptus globulus	Leaf	$ZrOCl_2$	spherical and rod like tetragonal	9.6	anticancer	[14]
Sapindus mukorossi	Fruit	$ZrOCl_2$	monoclinic	13-26		[15]
Zingiber officinale	Root	$ZrOCl_2.8H_2O$	monoclinic	295-583	antibacterial	[16]
Laurus nobilis	Leaf	$ZrOCl_2.8H_2O$	spherical monoclinic and tetragonal	20-100	antimicrobial	[17]
Guettarda speciosa	Leaf	$ZrOCl_2.8H_2O$	spherical tetragonal	6-9	antibacterial	[18]
Azadirachta indica	Gum	$Zr(OCH(CH_3)_2)_4$			antibacterial antifungal, antimicrobial	[19]
Murraya koenigii	Leaf	$(Zr(NO_3)_4$	spherical monoclinic	27	antibacterial	[20]
Enicostemma littorale	Plant extract	$ZrOCl_2.8H_2O$	tetragonal nanoflakes	8-15	anticancer	[21]
Annona reticulata	Leaf	$ZrOCl_2.8H_2O$	spherical	13-20	antibacterial	[22]
Mentha piperita L.	Leaf	$ZrOCl_2.8H_2O$	cubic ZrO_2	7-8	adsorption	[23]
Ficus benghalensis	Leaf	$ZrOCl_2.8H_2O$	spherical monoclinic and tetragonal	10-18	photocatalytic	[24]

Microwave-assisted synthesis of ZrO_2 is another prospectus green method where plant or animal originated materials are used as capping and reducing agents. A Zr precursor is mixed with a green reducing and capping agent with the help of water and exposed to microwave to prepare gel which is then calcined to get required ZrO_2 powder. For example, using anhydrous citric acid as capping and reducing agents, tetragonal ZrO_2 NPs having particle size of 5-10 nm were obtained [25]. For honey as a capping agent similar tetragonal ZrO_2 NPs with particle size of 26 nm were obtained [26]. Cubic ZrO_2 NPs with 70-77 nm particle size were obtained when L-serine was used as capping and reducing agents [27].

2.3 Zr-based metal organic frameworks

A group of extremely crystalline materials known as organic frameworks (MOFs) are made up of metal nodes and organic linkers. MOFs have emerged as potential prospects for drug delivery, catalysis, gas storage, and solar cells, due to their thermal and chemical durability, permanent porosity, high surface area, and adaptable structure and functionality. Due to their outstanding chemical and mechanical stability, MOFs based on Zr have gained a lot of attention. One of the most studied Zr MOFs is UiO-66, which was first reported by Cavaka $et\ al$ [28]. It is composed of nodes of $Zr_6(O)_4(OH)_4$ and linkers of benzene dicarboxylate. Tetravalent Zr(IV) ions have the largest coordination number and a strong affinity for oxygen donor ligands, which prevents displacement and increase the stability of Zr-based MOFs. The functionalization of this extremely stable MOF creates a variety of new application possibilities. Utilizing functionalized ligands during synthesis is one such method. Numerous attempts to synthesize Zr-based MOFs using more environmentally friendly techniques failed due to low crystallinity and reproducibility. Recent years have experienced the introduction of the modulator method, in which synthesis of MOFs involved ligands with one coordination site. These ligands compete with linkers for complexation with metal cations and change the kinetics of formation of crystals. Acetic acid turned a lot of attention as a modulating agent due to their good solubility in water. Another problem with water soluble modulators is the capability of crystallization, in most of the cases water solubility costs the crystallinity of the products. Synthesis of UiO-66, however, depends on the presence of water since the structure of $Zr_6(O)_4(OH)_4$ is made up of OH and -O bridges. The ligand, 1,2,4,5-benzenetetracarboxylic acid was for the first time used for synthesizing UiO-66(COOH)$_2$ in aqueous solution without the use of any modulators by Yang $et\ al.$ [29], But this method was found to be non-reproducible. The first green approach was the synthesis using acetic acid modulator. Both zirconium nitrate and zirconium chloride were found to be successful to prepare nano-sized crystalline MOFs with UiO-66 morphology [30,31]. Another problem with Zr-based MOFs is the external hydrophobicity which limits their application specially in biomedical fields. A hybrid of Zr-based MOFs with hydrogels provides a suitable solution of this problem. Zr-MOF-alginate composite materials can be prepared using green methods [32]. The researches on green synthesis of Zr-MOFs are rare in the literatures. But plant and animal-based proton donors can be used as a water-soluble modulation in synthesis of Zr-based MOFs, which will not only make the process greener and sustainable but also will open the door of imposing multiple functionalities in the MOFs.

2.4   Green synthesis and applications of other Zr-based NMs

Zirconium phosphate is one of a prospectus Zr compound to be used in potential applications. For example, $Aegle\ marmelos$ fruit extract gives NPs with average crystallite size of 38 nm with for cubic crystalline lattice and can serve as an antimicrobial agent and photocatalyst [33]. The layered structure of Zr opens the door of its usage in drug delivery. Nanoencapsulation of insulin inside the layers of zirconium phosphate for oral delivery application is a proven fact. In vitro cytotoxicity of zirconium phosphate demonstrated no toxicity, even pH dependent release and high structural stability make it a potential candidate for oral insulin administration instead of injection [34].

2.5   Applications and prospects

Plant and microorganism-based synthesis products possess antioxidant, antibacterial, antifungal, antimicrobial, and anticancer activities. These characteristics made them a potential candidate in pharmaceuticals. In dentistry, they can be used for multifunctional applications like anti-biofilm preparation and biointegration. The most important characteristics of these NPs is their biocompatibility. It makes them a potential candidate for cancer therapy and drug delivery.

ZrO_2 is mostly familiar for its excellent biomedical applications specially in dentistry as antibacterial, antimicrobial, and antifungal agent. Besides ZrO_2 NPs have also been applied in orthopedic implants, hip arthroplasty, radio mollifying agent, one tissue engineering, bone resorption, etc. The most emerging applications of ZrO_2 are in the field of adsorption, catalysis, sensing, fuel cell, environmental remediation. Further details of these applications are discussed below.

2.5.1 Environmental remediation

2.5.1.1 Adsorption

One of the most pressing concerns facing the world today is environmental contamination, which calls for several urgent steps. Particularly, the presence of new pollutants like antibiotics and pharmaceutical drugs can degrade water quality and present a number of potential risks to both human beings and aquatic lives. Numerous studies documented the emergence of antibiotic resistant genes in hospital effluents, rivers, sediments; and facilities for treatment of wastewater. As a result, it is essential to treat antibiotics extracted from aqueous media using efficient procedures. Adsorption may be the greatest remediation technique given the chemical stability and non-degradability of antibiotics, as well as its many benefits, such as high efficacy and a quick and customizable kinetic process. However, choosing and creating new, cutting-edge materials for water purification remains difficult.

Due to the ease of synthesis of ZrO_2 NPs with small size, favorable adsorption active sites, high surface area, and porous structure, they offer the prospect of using them as adsorbents for dyes, toxins, heavy metal ions, pharmaceutical byproducts, and other organic and inorganic pollutants. So, ZrO_2 NPs can be effective materials for polluted water and soil remediation. 1 g of ZrO_2 NPs prepared by *Pseudomonas aeruginosa* can successfully remove 526.32 mg of tetracycline in only 15 min at pH 6 and the adsorption follows Langmuir isotherm model [35]. Tetragonal ZrO_2 NPs prepared from *Euclea natalensis* root extract can successfully remove tetracycline up to 30.45 mg/g [36]. As(V) and As(III) ions are adsorbed on ZrO_2 NPs with adsorption capacity of ~3.6 mg/g for As(V) and ~0.6 mg/g for As(III) at the low arsenic equilibrium concentration of 0.01 mg/L, and 29 mg/g and >47 at the high arsenic equilibrium concentration. As(V) and As(III) are adsorbed on the surface of ZrO_2 nanoparticle through inner sphere complexing mechanism. The immobilization of the NPs on glass cloth surface may be one of the feasible methods of application in wastewater treatment with better separation of NPs from the treated water [37].

Numerous textile dyes are developing into chronic contaminants in addition to antibiotic pollution of water. According to estimates, a textile dyes are released annually into water bodies in a significant amount to cause various detrimental effects on human health. The metabolism processes of the aquatic lives and photosynthetic efficiency are known to be negatively impacted by their dismissal of solar light penetration. Similar to antibiotics, chemical stability of majority of the dyes makes biological techniques like phytoremediation to break down challenging. ZrO_2 and Zr-based doping NMs are potential semiconducting materials with band gaps in UV-visible region of electromagnetic spectrum. They also have good thermal and chemical stability and adsorption capability to be effective for this particular application. The optical band gap and adsorption capability are highly dependent on particle size, morphology, and surface functionality of the prepared Zr-based NMs. The application of these semiconducting NPs could be a great solution for this problem. Semiconducting NPs have both adsorption efficiency and photocatalytic capabilities which made them potential candidate for dye adsorption and degradation. Pericarp extract of *Sapindus mukorossi* plant can synthesize highly crystalline tetragonal ZrO_2 NPs (average size 5 nm), that can adsorb methylene blue at 94 % removal efficiency exhibiting adsorption capacity of 23.26 mg/g [35]. *Lagerstroemia speciosa* leaf extract mediated synthesized ZrO_2 shows photocatalytic effect to degrade methyl orange with an efficiency of 94.58 % [38].

2.5.1.2 Catalysis

NPs have exceptionally high surface to volume ratio to render them highly effective in catalysis. In the last few decades, the application of NMs as catalyst increased exponentially. ZrO_2-based NMs are potential candidates in the field of catalysis. ZrO_2-based NMs are mainly applied in esterification and transesterification reactions. Zr NPs have already been used in transesterification of cooking oils, soyabean oils, tannery waste fat materials, etc. to prepare biodiesels. But almost all the Zr NPs used in these applications are prepared by non-environmentally friendly methods. It can be a great opportunity for the scientific community to apply Zr NPs prepared by green methods as catalysts in esterification and transesterification reactions.

Beside heterogeneous catalysis, Zr-based NMs are also effective in photocatalytic applications due to their semiconducting behavior. Synthesis of Zr-based NMs, as stated earlier, have the potential to tune each of these due to the presence of biological compounds associated in the synthetic procedures. The green Zr-based NMs have been used for catalytic photodegradation of organic pollutant, pharmaceuticals, and dyes from wastewater. Biosynthesis from *Ficus benghalensis* leaf extract yielded ZrO_2 NPs with optical band gap of 5.9 to 4.3 eV with good photodegradation efficiency [24]. In the presence of ultraviolet radiation, Mg-doped ZrO_2 microspheres made from *Aloe vera* leaf extract exhibited the photocatalytic activity against Rhodamine B (93%). Mg-doping creates oxygen vacancy in the active site of surface of the catalyst which is mainly responsible for this high catalytic activity [39]. Besides doping by main group metals, transition metal doping is also an effective means to increase the catalytic activity of ZrO_2. Samarium-doped ZrO_2 prepared in *Leucus aspera* extract shows great catalytic activity towards degradation of acid green dye in presence of sunlight [40]. Cu-doped ZrO_2 NPs prepared form *Commelina diffusa* extract is a highly efficient and ultrafast (1-150 s) photocatalyst that can degrade organic dyes like 2,4-dinitrophenylhydrazine, Congo red, nigrosine, methyl orange [41]. Ag-doped ZrO_2 synthesized in *Ageratum conyzoides* can also degrade these dyes along with *para*-nitrophenol with a great catalytic efficiency [42]. The major problem associated with this kind of nanocatalysts is the separation and reuse of catalyst for water after its use. To solve this problem magnetic nanocatalysts have been fabricated. $Ag/Fe_3O_4/ZrO_2$ nanocomposite was successfully prepared which can degrade methyl orange and *para*-nitrophenol with a good catalytic efficiency [43].

2.5.2 Sensing

Advantageous physicochemical and electrochemical characteristics, such as high surface area, low band gap, good thermal and chemical stability, and fluorescent nature have made ZrO_2-based NPs attractive for sensing applications. Zr-based NMs have already been used in as an electrochemical sensor for different pharmaceutical byproducts, pesticides, toxins, heavy metals, and even diverse organic compounds with different functionalities. Nanocomposites based on Zr have also been successfully applied in biosensing for sensing glucose and other biological molecules. Again Zr-based NMs used in sensing are mainly synthesized without taking green approaches. The Zr-based NMs prepared by green methods possess various surface functionality, biocompatibility, and morphology tuning opportunity which can be even more effective for use in sensors and biosensors.

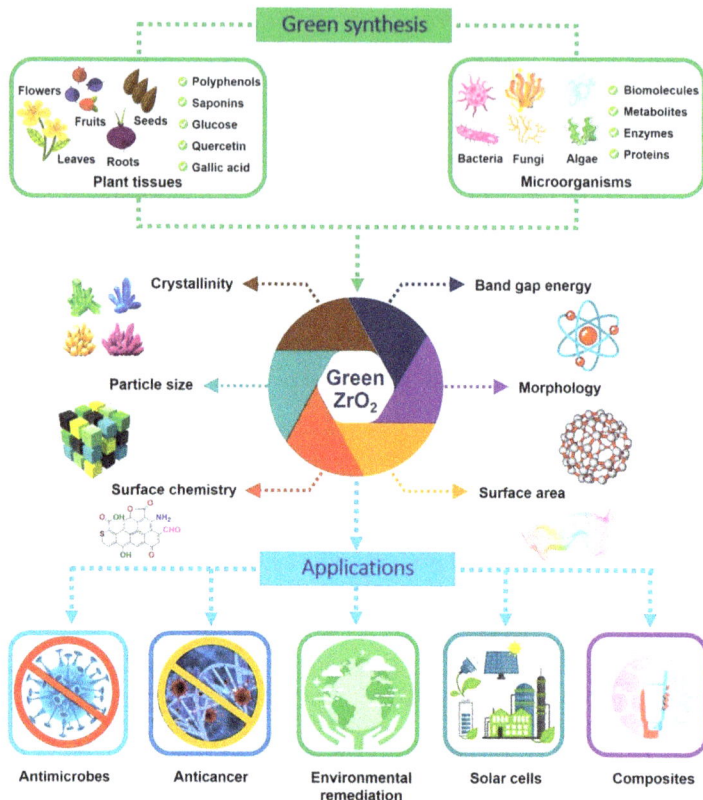

Figure 2. Green syntheses and applications of ZrO₂. Reproduced with permission from [45]. Copyright 2022, Springer.

2.5.3 Solar cell

A number of commercially available semiconductors, including silicon, have demonstrated outstanding performance as solar cells, but the manufacturing processes are often costly. Moreover, these procedures are based on chemically created protocols, which hinders sustainable development. The biosynthesis of inexpensive materials for dye-sensitized solar cells is therefore appealing. Green ZrO_2 NPs are a potential electronic nanomaterial to harvest solar energy. ZrO_2 NPs have been synthesized through a chemical process and have been found to be highly efficient at absorbing and reflecting sunlight. In addition to being corrosion resistant and thermally stable,

ZrO_2 NPs are ideal for use in solar cells. A further advantage of their small size is that they can be easily incorporated into the design of solar cells. ZrO_2 NPs synthesized using *Glorisa superba* plant extract was used to enhance energy conversion efficiency by combining it with opuntia dye [44]. Using lemon juice, cubic-phase ZrO_2 NPs with 21 nm size were synthesized. The prepared NPs with electrical conductivity and activation energy of 0.0034 S/cm and 31.015 kJ/mol showed relatively good electronic properties as potential electrolytes.

2.5.4 Future prospects

It has become increasingly popular to synthesize biodegradable and low-cost ZrO_2 NPs from biological resources because they offer many advantages over chemical methods. Aside from their large surface area, small size, good dispersion, diverse surface chemistry, and sufficient band gap energy, green ZrO_2 NPs have many unique properties to make them useful in biomedical and environmental fields. The main advantage of using biological organic compounds in synthesis of ZrO_2 is the availability and diversity of biological materials and their tremendous capability to control crystallinity and morphology of the NPs. It is also possible to create nanostructure-based binary or ternary composites from such compounds. In addition, the use of biological organic compounds for synthesis results in lower cost and higher efficiency compared to other methods. A number of high-tech manufacturing industries may benefit from these superior properties of ZrO_2 NPs as they may be used to develop high efficiency solar cells for energy harvesting, advanced nanosensors for smart agriculture, and electronic nanodevices. Such NMs may also find applications in biomedical fields, such as cancer therapy, biocompatible smart implants, and drug delivery systems. Moreover, their superior mechanical and chemical properties make them suitable for use versatile applications. In several environmental domains, such as the treatment of developing persistent organic pollutants, identification of poisonous gas, anti-inflammatory drugs, and pharmaceutical wastes, and personal care products, the possibility of using green ZrO_2 NPs is largely untapped. It is anticipated that green ZrO_2 NPs will considerably help to the development of new environmental cleanup strategies due to their exceptional features.

3.   Green synthesis and applications of Rh-based NMs

3.1   Green synthesis

There are different types of Rh-based NMs with a great interest. These include: pure metallic Rh, Rh-based MOFs, Rh-based solid solution, Rh-based nanocomposites, etc. Synthesis of Rh NPs can be accomplished using a variety of green methods, including reduction of hydrogen in water as a solvent, reduction of ethanol in an ethanol-water mixture, and others. It has already been demonstrated that Rh NPs can be synthesized and stabilized successfully, and that their deposition onto TiO_2 under air at ambient condition, without any gas treatment or calcination, is an effective green catalyst for a variety of organic reactions [46]. Hydrothermal method can be used effectively to synthesize Rh NPs. A hydrothermal method simply uses supercritical water as a green reaction medium and has been successful to synthesize Rh having well dispersed crystalline NPs with an average particle size of 3.7 nm [47]. An "eco-friendly" reaction pathway was developed consisting of ionic liquids and supercritical CO_2 by synthesizing a simple catalyst and separating the products efficiently. In order to prepare Rh NPs, metal precursors were reduced in different ionic liquids and extracted by a supercritical carbon dioxide to remove the metal ligands. In comparison to other catalysts tested, Palladium-1-butyl-3-methylimidazolium hexafluorophosphate displayed the best performance [48]. Plant-based green synthesis of metallic Rh NPs has been carried out via *Aspalathus Linearis* natural extract from RhCl₃ as a raw material. This study revealed that quasi-monodisperse spherical and amorphous Rh NPs were successfully synthesized in the 0.8-1.6 nm size range [49]. The major problem of plant-based synthesis is the amorphous nature of the product which abridged its potentiality. But this method provides a novel route for synthesis of Rh NPs. Further investigations are required to get optimum results. Besides study on the green synthesis of Rh-based other NMs like nanocomposites, solid solutions, MOFs are yet to be unfolded. Research on Rh continues to present a number of challenges. The following points may be considered as future challenges for more efficient use of Rh. The first and most important one is the development of facile green fabrication methods for Rh NMs with precise control over their size, shape, crystallinity, and morphology. Besides An in-depth investigation of the characterization and mechanisms is necessary. An enhanced theoretical simulation and in situ characterization of the compounds would enable revealing more information about the mechanisms involved in their synthesis as well as their interactions with one another. The cost of Rh precursors is always a significant factor. Therefore, the minimization of Rh content in Rh-based NMs can be a cost-effective and promising strategy.

3.2   Emerging applications and prospects of Rh based NMs

3.2.1   Energy

The most emerging use of Rh-based NMs is in the hydrogen and oxygen evolution reactions. Hydrogen evolution reaction (HER) activity is higher for Rh NPs in 1.0 M KOH solution compared to platnum NPs. Meanwhile, Rh NPs has lower charge-transfer resistance and smaller Tafel slope. Rh nanosheets show even better HER activity. Rh and its oxides dispersed in carbon nitride nanosheets are effective catalysts in both HER and oxygen evolution reaction under visible light.

3.2.2   Biomedical applications

Theranostic performance is improved in hypoxic tumor microenvironments using biocompatible Rh-based nanotheranostic systems. The nanotheranostics exhibit improved tumor-targeting efficiency and allow for the simultaneous delivery of therapeutic and imaging agents. This enables precise monitoring of tumor responses to treatment and improved overall therapeutic efficacy [50]. Morphology of Rh-based NMs greatly affects the efficacy of cancer therapy. Of the three types of morphology, nanoshells, nanoframes, and nanoplates, the porous nanoplates show the better result in cancer therapeutics. Besides cancer therapeutics, Rh-based NMs are also useful in sonodynamic therapy of rheumatoid arthritis. It has been shown that Rh NPs are able to kill activated synovial inflammatory cells because of their non-invasive nature and their capacity to penetrate deep into tissues [51].

3.2.3   Biosensing

Rh nanosheets demonstrate ultrahigh peroxidase activity, exhibiting high sensitivity and selectivity for colorimetric biosensing of H_2O_2 and xanthine. Rh-based paper sensors have been developed that are simple, rapid, and sensitive to ascorbic acid measurements. The sensor consists of a gold NP-labeled redox hydrogel, which is used to detect the ascorbic acid in a sample. The results obtained were accurate and reproducible, making this sensor an effective and efficient method for ascorbic acid detection [52].

3.2.4 Prospects and challenges

Although Rh-based NMs have been applied in different emerging fields, almost all the synthetic methods for the used Rh-based NMs are non-environmentally friendly. Some green approaches are already taken, but the resulting NMs have been found to be much less effective for potential applications. This is because of the weak crystallinity and uncontrolled morphology of the prepared green NMs. But systematic optimization of the synthesis procedure of Rh-based NMs may result in desired NMs for potential applications. Biological synthesis is a highly potential route to synthesize NMs with different sizes, shapes, and functionalities.

Conclusions

This chapter summarizes green synthesis, characteristics, emerging applications, and prospects of Zr and Rh-based NMs. Green synthesis of Zr and Rh-based NMs mainly use biological compounds derived from different plant tissues and microorganisms like bacteria, fungi, algae, etc. Bioactive compounds derived from these sources act as bioreducing, biochelating, biostabilizing, and biocapping agents. These materials help the conversion of the precursor to NMs controlling the course of the reaction, regulate the product morphology, and add functionality to the product. Different plant materials consisting of different sets of bioactive molecules result in variability in the functionality and morphology of the product which leads to various potential applications. Emerging applications of Zr-based NMs are in the industries related to biomedicine, environmental remediation, energy, etc. While Rh-based NMs are prospective in catalysis, energy, and biomedicine. Zr-based NMs synthesized through green approach have been proved to be useful in such industries but green synthesis of Rh-based NMs is rare. But both set of NMs show potential prospects of using green methods for synthesis focusing on their emerging applications.

Acknowledgment

The authors acknowledge financial support from the DU-UGC, Bangladesh research grant for the themed research. MAH also thanks Bose Centre for Advanced Research in Sciences, University of Dhaka, Bangladesh for a research fellowship.

References

[1] P.K. Dikshit, J. Kumar, A.K. S. Sadhu, S. Sharma, S. Singh, P.K. Gupta, B. S. Kim, Green synthesis of metallic nanoparticles: applications and limitations. Catalysts, 11(2021) 902. https://doi.org/10.3390/catal11080902

[2] T. P. Chau, G. R. Veeraragavan, M. Narayanan, A. Chinnathambi, S. A. Alharbi, B. Subramani, S. Pikulkaew, Green synthesis of Zirconium nanoparticles using Punica granatum (pomegranate) peel extract and their antimicrobial and antioxidant potency. Environ. Res. 209(2022) 112771. https://doi.org/10.1016/j.envres.2022.112771

[3] M. A. Chowdhury, N. Hossain, M. G. Mostofa, M. R. Mia, M. Tushar, M. M. Rana, M. H. Hossain, Green synthesis and characterization of zirconium nanoparticle for dental implant applications, Heliyon, 9(2023) 1. https://doi.org/10.1016/j.heliyon.2022.e12711

[4] S. Kazi, S. Nirwan, S. Kunde, Green synthesis, characterization and bio-evaluation of zirconium nanoparticles using the dried biomass of Sphagneticola trilobata plant leaf, Bio. Nano. Sci. 12(2022) 731-740. https://doi.org/10.1007/s12668-022-01006-9

[5] S. Agarwal, S. Maiti, S. Rajeshkumar, V. Agarwal, M. Deshmukh, D. Ganapathy, Green synthesis and characterization of strontium and zirconium nanoparticles from green tea leaf extracts and studying their antimicrobial activity and anti-inflammatory activity against oral pathogens. J. Pharm. Negat. (2022) 539-543.

[6] R. H. Salih, S. H. Ahmed, R. S. Hameed, I. H. T. Al-Karkhi, Evaluation of a new green zirconium nanoparticle from lemon and peel extract antioxidant and anticancer activity. Med. Legal Update, 21(2) (2021) 977-981. https://doi.org/10.37506/mlu.v21i2.2810

[7] A. R. G. Ghomi, M. Mohammadi-Khanaposhti, H. Vahidi, F. Kobarfard, M. A. S. Reza, H. Barabadi, Fungus-mediated extracellular biosynthesis and characterization of zirconium nanoparticles using standard penicillium species and their preliminary bactericidal potential: a novel biological approach to nanoparticle synthesis, IJPR 18(4) (2019) 2101.

[8] S. P. Suriyaraj, G. Ramadoss, K. Chandraraj, R. Selvakumar, One pot facile green synthesis of crystalline bio-ZrO2 nanoparticles using Acinetobacter sp. KCSI1 under room temperature. Mater. Sci. Eng. C 105(2019) 110021. https://doi.org/10.1016/j.msec.2019.110021

[9] T. Ahmed, H. Ren, M. Noman, M. Shahid, M. Liu, M. A. Ali, B. Li, Green synthesis and characterization of zirconium oxide nanoparticles by using a native Enterobacter sp. and its antifungal activity against bayberry twig blight disease pathogen Pestalotiopsis versicolor. NanoImpact 21(2021) 100281. https://doi.org/10.1016/j.impact.2020.100281

[10] M. Riaz, M. Kamran, Y. Fang, Q. Wang, H. Cao, G. Yang, X. Wang, Arbuscular mycorrhizal fungi-induced mitigation of heavy metal phytotoxicity in metal contaminated soils: A critical review. J. Hazard. Mater. 402(2021) 123919. https://doi.org/10.1016/j.jhazmat.2020.123919

[11] J.A. Ferreira, S. Varjani, M.J. Taherzadeh, A Critical Review on the Ubiquitous Role of Filamentous Fungi in Pollution Mitigation. Curr. Pollut. Rep. 6(2020) 295-309. https://doi.org/10.1007/s40726-020-00156-2

[12] V. Bansal, D. Rautaray, A. Ahmad, M. Sastry, Biosynthesis of zirconia nanoparticles using the fungus Fusarium oxysporum. J. Mater. Chem. 14(22) (2004) 3303-3305. https://doi.org/10.1039/b407904c

[13] M. Kumaresan, K. V. Anand, K. Govindaraju, S. Tamilselvan, V. G. Kumar, Seaweed Sargassum wightii mediated preparation of zirconia (ZrO2) nanoparticles and their antibacterial activity against gram positive and gram-negative bacteria. Microb. Pathog. 124(2018) 311-315. https://doi.org/10.1016/j.micpath.2018.08.060

[14] S. Balaji, B. K. Mandal, S. Ranjan, N. Dasgupta, R. Chidambaram, Nano-zirconia-evaluation of its antioxidant and anticancer activity. J. Photochem. Photobiol. B, Biol. 170(2017) 125-133. https://doi.org/10.1016/j.jphotobiol.2017.04.004

[15] A. K. Arora, L. Chauhan, P. Kumar, Synthesis and Characterization of Zirconium Oxide Nanoparticles using Sapindus mukorossi (Soapnut) as natural surfactant, A green synthetic approach. AJRC 16(1) (2023) 79-82. https://doi.org/10.52711/0974-4150.2023.00013

[16] R. Thyagarajan, G. Narendrakumar, V. Rameshkumar, M. Varshiney, Green synthesis of zirconia nanoparticles based on ginger root extract: optimization of reaction conditions, application in dentistry. Res. J. Pharm. Technol. 15(11) (2022) 5314-5320. https://doi.org/10.52711/0974-360X.2022.00895

[17] T.P. S. Kandasamy, A. Chinnathambi, Synthesis of zirconia nanoparticles using Laurus nobilis for use as an antimicrobial agent. Appl. Nanosci. 13(2023) 1337-1344. https://doi.org/10.1007/s13204-021-02041-w

[18] N. Muthulakshmi, A. Kathirvel, R. Subramanian, M. Senthil, Biofabrication of zirconia nanoparticles: synthesis spectral characterization and biological activity evaluation against pathogenic bacteria. Biointerface Res. Appl. Chem. 13(2023) 190. https://doi.org/10.33263/BRIAC132.190

[19] S. A. Korde, P. B. Thombre, S. S. Dipake, J. N. Sangshetti, A. S. Rajbhoj, S. T. Gaikwad, Neem gum (Azadirachta indicia) facilitated green synthesis of TiO2 and ZrO2 nanoparticles as antimicrobial agents. Inorg. Chem. Commun. 153(2023) 110777. https://doi.org/10.1016/j.inoche.2023.110777

[20] P. Chelliah, S. M. Wabaidur, H. P. Sharma, H. S. Majdi, D. A. Smait, M. A. Najm, W. C. Lai, Photocatalytic Organic Contaminant Degradation of Green Synthesized ZrO2 NPs and Their Antibacterial Activities. Separations 10(3) (2023) 156. https://doi.org/10.3390/separations10030156

[21] P. Sumathi, N. Renuka, R. Subramanian, G. Periyasami, M. Rahaman, P. Karthikeyan, Prospective in vitro A431 cell line anticancer efficacy of zirconia nanoflakes derived from Enicostemma littorale aqueous extract. Cell Biochem. Funct. (2023). https://doi.org/10.1002/cbf.3822

[22] K. Selvam, C. Sudhakar, T. Selvankumar, Photocatalytic degradation of malachite green and antibacterial potential of biomimetic-synthesized zirconium oxide nanoparticles using Annona reticulata leaf extract. Appl. Nanosci. 13(2023) 2837-2843. https://doi.org/10.1007/s13204-021-02148-0

[23] I.M.A. Hasan, H. Salah El-Din, A.A. AbdElRaady, Peppermint-Mediated Green Synthesis of Nano ZrO2 and Its Adsorptive Removal of Cobalt from Water. Inorganics 10(2022) 257. https://doi.org/10.3390/inorganics10120257

[24] H.M. Shinde, T.T. Bhosale, N.L. Gavade, Biosynthesis of ZrO2 nanoparticles from Ficus benghalensis leaf extract for photocatalytic activity. J. Mater. Sci: Mater Electron 29(2018) 14055-14064. https://doi.org/10.1007/s10854-018-9537-7

[25] R. Dwivedi, A. Maurya, A. Verma, R. Prasad, K. S. Bartwal, Microwave assisted sol-gel synthesis of tetragonal zirconia nanoparticles. J. Alloys Compd. 509(24) (2011) 6848-6851. https://doi.org/10.1016/j.jallcom.2011.03.138

[26] B.S. Bukhari, M. Imran, M. Bashir, Honey mediated microwave assisted sol-gel synthesis of stabilized zirconia nanofibers. J. Sol-Gel Sci. Technol. 87(2018) 554-567. https://doi.org/10.1007/s10971-018-4749-0

[27] S. Manjunatha, M. S. Dharmaprakash, Microwave assisted synthesis of cubic Zirconia nanoparticles and study of optical and photoluminescence properties. J. Lumin. 180(2016) 20-24. https://doi.org/10.1016/j.jlumin.2016.07.055

[28] J. H. Cavka, S. Jakobsen, U. Olsbye, N. Guillou, C. Lamberti, S. Bordiga, K. P. Lillerud, A new zirconium inorganic building brick forming metal organic frameworks with exceptional stability. J. Am. Chem. Soc. 130(42) (2008) 13850-13851. https://doi.org/10.1021/ja8057953

[29] Y. Yang, Y. Xia, Polycarboxyl metal-organic framework UiO-66-(COOH)2 as efficient desorption/ionization matrix of laser desorption/ionization mass spectrometry for selective enrichment and detection of phosphopeptides. J. Nanopart. Res. 21(2019) 1-12. https://doi.org/10.1007/s11051-018-4445-6

[30] B. Bueken, N. Van Velthoven, T. Willhammar, T. Stassin, I. Stassen, D. A. Keen, T. D. Bennett, Gel-based morphological design of zirconium metal-organic frameworks. Chem. Sci. 8(5) (2017) 3939-3948. https://doi.org/10.1039/C6SC05602D

[31] M. N. Nimbalkar, B. R. Bhat, Facile green synthesis of zirconium-based metal-organic framework having carboxylic anchors. Mater. Today: Proc. 9(2019) 522-527. https://doi.org/10.1016/j.matpr.2018.10.371

[32] S. E. Klein, J. D. Sosa, A. C. Castonguay, W. I. Flores, L. D. Zarzar, Y. Liu, Green synthesis of Zr-based metal-organic framework hydrogel composites and their enhanced adsorptive properties. Inorg. Chem. Front. 7(24) (2020) 4813-4821. https://doi.org/10.1039/D0QI00840K

[33] H. N. Deepakumari, V. L. Ranganatha, G. Nagaraju, R. Prakruthi, C. Mallikarjunaswamy, Facile green synthesis of zirconium phosphate nanoparticles using Aegle marmelos: antimicrobial and photodegradation studies. Mater. Today: Proc. 62(2022) 5169-5173. https://doi.org/10.1016/j.matpr.2022.02.579

[34] A. Díaz, A. David, R. Pérez, M. L. González, A. Báez, S. E. Wark, J. L. Colón, Nanoencapsulation of insulin into zirconium phosphate for oral delivery applications. Biomacromolecules, 11(9) (2010) 2465-2470. https://doi.org/10.1021/bm100659p

[35] B. Debnath, M. Majumdar, M. Bhowmik, K. L. Bhowmik, A. Debnath, D. N. Roy, The effective adsorption of tetracycline onto zirconia nanoparticles synthesized by novel microbial green technology. J. Environ. Manage., 261(2020) 110235. https://doi.org/10.1016/j.jenvman.2020.110235

[36] A. F. V. da Silva, A. P. Fagundes, D. L. P. Macuvele, E. F. U. de Carvalho, M. Durazzo, N. Padoin, H. G. Riella, Green synthesis of zirconia nanoparticles based on Euclea natalensis plant extract: Optimization of reaction conditions and evaluation of adsorptive properties. Colloids Surf. A Physicochem. Eng. Asp. 583(2019) 123915. https://doi.org/10.1016/j.colsurfa.2019.123915

[37] C. Hang, Q. Li, S. Gao, J. K. Shang, As(III) and As(V) adsorption by hydrous zirconium oxide nanoparticles synthesized by a hydrothermal process followed with heat treatment. Ind. Eng. Chem. Res. 51(1) (2012) 353-361. https://doi.org/10.1021/ie202260g

[38] V. S. Saraswathi, K. Santhakumar, Photocatalytic activity against azo dye and cytotoxicity on MCF-7 cell lines of zirconium oxide nanoparticle mediated using leaves of Lagerstroemia

speciosa. J. Photochem. Photobiol. B, Biol. 169(2017) 47-55.
https://doi.org/10.1016/j.jphotobiol.2017.02.023

[39] L. Renuka, K. S. Anantharaju, S. C. Sharma, H. P. Nagaswarupa, S. C. Prashantha, H. Nagabhushana, Y. S. Vidya, Hollow microspheres Mg-doped ZrO2 nanoparticles: green assisted synthesis and applications in photocatalysis and photoluminescence. J. Alloys Compd. 672(2016) 609-622. https://doi.org/10.1016/j.jallcom.2016.02.124

[40] K. Gurushantha, K. S. Anantharaju, S. C. Sharma, H. P. Nagaswarupa, S. C. Prashantha, K. V. Mahesh, H. Nagabhushana, Bio-mediated Sm doped nano cubic zirconia: Photoluminescent, Judd-Ofelt analysis, electrochemical impedance spectroscopy and photocatalytic performance. J. Alloys Compd. 685(2016) 761-773. https://doi.org/10.1016/j.jallcom.2016.06.105

[41] S. M. Hamad, S. A. Mahmud, S. M. Sajadi, Z. A. Omar, Biosynthesis of Cu/ZrO2 nanocomposite using 7-hydroxy-4'-methoxy-isoflavon extracted from Commelina diffusa and evaluation of its catalytic activity. Surf. Interfaces 15(2019) 125-134. https://doi.org/10.1016/j.surfin.2019.02.008

[42] M. Maham, M. Nasrollahzadeh, S. M. Sajadi, Facile synthesis of Ag/ZrO2 nanocomposite as a recyclable catalyst for the treatment of environmental pollutants. Compos. B. Eng. 185(2020) 107783. https://doi.org/10.1016/j.compositesb.2020.107783

[43] A. Rostami-Vartooni, A. Moradi-Saadatmand, M. Bagherzadeh, M. Mahdavi, Green synthesis of Ag/Fe3O4/ZrO2 nanocomposite using aqueous Centaurea cyanus flower extract and its catalytic application for reduction of organic pollutants. Iran. J. Catal. 9(1) (2019) 27-35.

[44] R. Vennila, P. Kamaraj, M. Arthanareeswari, M. Sridharan, G. Sudha, S. Devikala, K. Rajeshwari, Biosynthesis of ZrO nanoparticles and its natural dye sensitized solar cell studies. Mater. Today: Proc. 5(2) (2018) 8691-8698. https://doi.org/10.1016/j.matpr.2017.12.295

[45] T. V. Tran, D. T. C. Nguyen, P. S. Kumar, A. T. M. Din, A. A. Jalil, D. V. N. Vo, Green synthesis of ZrO2 nanoparticles and nanocomposites for biomedical and environmental applications: A review. Environ. Chem. Lett. (2022) 1-23. https://doi.org/10.1007/s10311-021-01367-9

[46] C. Hubert, A. Denicourt-Nowicki, P. Beaunier, A. Roucoux, TiO2-supported Rh nanoparticles: From green catalyst preparation to application in arene hydrogenation in neat water. Green Chem. 12(7) (2010) 1167-1170. https://doi.org/10.1039/c004079g

[47] Y. Lee, S. Jang, C. W. Cho, J. S. Bae, S. Park, K. H. Park, Recyclable rhodium nanoparticles: green hydrothermal synthesis, characterization, and highly catalytic performance in reduction of nitroarenes. J. Nanosci. Nanotechnol. 13(11) (2013) 7477-7481. https://doi.org/10.1166/jnn.2013.7903

[48] F. Jutz, J. M. Andanson, A. Baiker, A green pathway for hydrogenations on ionic liquid-stabilized nanoparticles. J. Catal. 268(2) (2009) 356-366. https://doi.org/10.1016/j.jcat.2009.10.006

[49] E. Ismail, M. Kenfouch, M. Dhlamini, Green Biosynthesis of Rhodium Nanoparticles Via Aspalathus Linearis Natural Extract. J. Nanomater. Mol. Nanotechnol. 6(2017) 2. https://doi.org/10.4172/2324-8777.1000212

[50] M. Ding, Z. Miao, F. Zhang, J. Liu, X. Shuai, Z. Zha, Z. Cao, Catalytic rhodium (Rh)-based (mesoporous polydopamine) MPDA nanoparticles with enhanced phototherapeutic efficiency for overcoming tumor hypoxia. Biomater. Sci. 8(15) (2020) 4157-4165. https://doi.org/10.1039/D0BM00625D

[51] S. Kang, W. Shin, M. H. Choi, M. Ahn, Y. K. Kim, S. Kim, H. Jang, Morphology-controlled synthesis of rhodium nanoparticles for cancer phototherapy. ACS nano 12(7) (2018) 6997-7008. https://doi.org/10.1021/acsnano.8b02698

[52] S. Cai, W. Xiao, H. Duan, X. Liang, C. Wang, R. Yang, Y. Li, Single-layer Rh nanosheets with ultrahigh peroxidase-like activity for colorimetric biosensing. Nano Res. 11(2018) 6304-6315. https://doi.org/10.1007/s12274-018-2154-1

Green Synthesis and Emerging Applications of Frontier Nanomaterials   Materials Research Forum LLC
Materials Research Foundations 169 (2024) 249-274   https://doi.org/10.21741/9781644903278-10

Chapter 10

Quantum dots: Green synthesis, characterizations and applications

Yogita Sahu[1], Sunita Sanwaria[1], R.M. Patel[2]*, Md. Abu Bin Hasan Susan[3], Ajaya K. Singh[1,4]*

[1] Department of Chemistry, Govt. V. Y. T. PG. Autonomous, College, Durg, Chhattisgarh, 491001, India

[2]Hemchand Yadav University, Durg (C.G.) 490009, India

[3]Department of Chemistry and Dhaka University Nanotechnology Center, University of Dhaka, Dhaka 1000, Bangladesh

[4] School of Chemistry & Physics, University of KwaZulu-Natal, Durban, South Africa

*ajayaksingh_au@yahoo.co.in

Abstract

Quantum dots (QDs) are a fascinating domain of modern research. The nanoscale dimension of QD semiconductors determines their optical and electrical characteristics to make them appealing for diverse applications, which inter alia include: electronics, biotechnology, energy, and photonics. Syntheses of QDs commonly use chemical, green synthesis, and microwave-assisted methods. Characterizations using electron microscopy, x-ray diffraction; spectroscopy, etc. provide insights into their unique properties. QDs find applications in displays, biological imaging, sensing, photovoltaics, solid-state lighting, and hold promise for quantum computing. Research is underway to enhance their properties, deal with toxicity issues, and explore new applications. QDs have so far been a remarkable category of nanomaterials with enormous potential for scientific and technological breakthroughs.

Keyword

Quantum dots; Green Method, Bioinspired Synthesis, Chemical Method, Spectral Characterization

Contents

1.   Introduction

In the discipline of nanotechnology, remarkable strides have been made in the manipulation of matter at the nanoscale, resulting in the development of an extensive array of nanoparticles (NPs) with distinct properties and functionalities. Nanoparticles are tiny structures of carbon derivatives, metals, metal oxides, semiconductors, etc. (Figure 1). They typically have sizes ranging from 1-100 nanometers (nm) in at least one dimension [1-2]. In response to the worldwide trend towards sustainability, a new and promising research field called "green nanotechnology" (GN) has emerged, presenting a renewed perspective. GN aims at addressing critical global issues, such as sustainable energy, clean water access, improved healthcare, and environmental preservation [3].

This concept has gained momentum due to the recognition of the detrimental effects of unsustainable practices inherited from the industrial revolution that continue to impact humanity today. As a result, the concept of green processes, encompassing synthesis, production, and disposal, has emerged as an alternative to non-green technologies, emphasizing the possibility of achieving cleaner, safer, and more efficient methods that have a neutral or positive environmental impact [4]. Within the scientific community, a significant challenge lies in unravelling the mysteries of the nanoscale world. A notable breakthrough in this endeavour involves the utilization of nanodots (NDs) as fundamental building materials for the development of cutting-edge technologies [5-6]. In nanotechnology studies nanodots have the size 1-10 nm [7]. They are basically two types: carbon dots (CDs) and quantum dots (QDs) (Figure 2). In the early 1980s, Alexey Ekimov observed unusual optical properties in semiconductor nanocrystals embedded in a glass matrix. These nanocrystals, termed "colloidal QDs" exhibited quantum confinement effects and a change in the size brough about significant change in optical behavior [8]. Ekimov's work involved synthesizing semiconductor nanocrystals of varying sizes and compositions exploiting colloidal chemistry. He studied their optical properties using spectroscopic methods and explained their results in terms of quantum mechanical phenomena such as quantum confinement.

Due to their amazing electrical and optical capabilities at the nanoscale, QDs often referred to as nanocrystals, are semiconductor particles that have captivated the attention of researchers and engineers [10]. To understand QDs, it is important to delve into their composition and structure. Typically, QDs are produced from semiconducting materials like cadmium selenide, lead sulfide, or indium arsenide [11]. These materials are chosen for their ability to confine electrons and holes within the QD structure. In simple terms, charge carriers, such as electrons and holes, are essential to the behaviour of semiconductors [12-13]. Xu et $al.$, synthesized CDs, spherical carbon NPs with diameters of less than 10 nm, where the carbon atoms are sp^2 hybridized [14].

Since charge carriers are quantum mechanically confined inside the limited small dimensions, QDs display size-dependent properties. This leads to discrete energy levels similar to those observed in atoms [15-20]. The QDs are capable of emitting radiation of specific frequencies as dictated by their size when excited by energy, such as through electrical or optical means. This characteristic, sometimes referred to as size-dependent fluorescence or size-tuneable emission, enables exact control over the colour of the emitted light [21-22]. The Stark effect enables the tuning of the emission wavelength of QDs. By manipulating the energy levels of the charge carriers confined inside the QD, this effect provides control over the wavelength of the emitted light [23]. QDs possess several advantageous features, including excellent photochemical stability, high quantum yield (QY), and broad absorption spectrum [24]. These make them valuable in applications such as biomedical imaging, where their brightness, small size, and tuneable emission properties are utilized for precise labelling and detection of biological structures. QDs have diverse applications in fields like optoelectronics, where they enhance colour accuracy and efficiency in displays such as televisions and monitors. They also hold promise in areas like solar cells, where their broad absorption spectra enable efficient light harvesting [25-27].

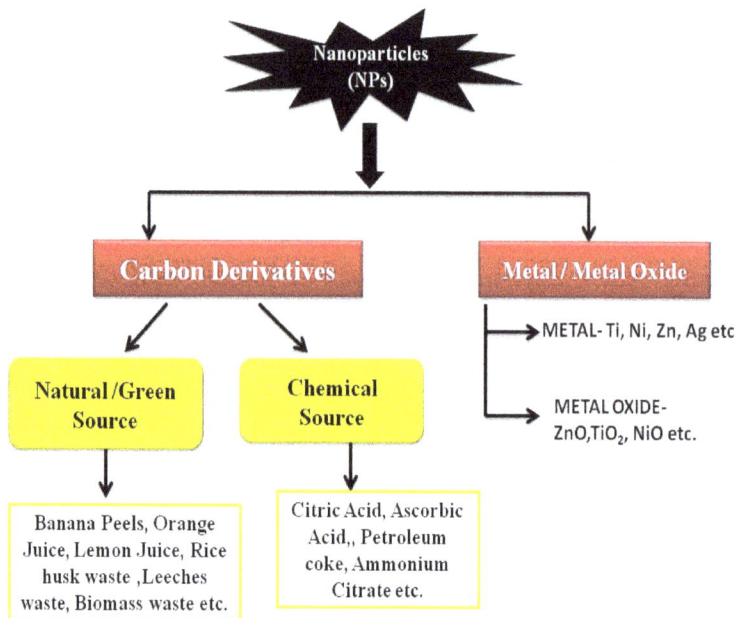

Figure 1: Materials for synthesis of NPs

2.   Preparation Methods of QDs:

Preparation methods of QDscan be broadly categorized into chemical methods and green methods (Fig. 1). Here are detailed explanations of each category:

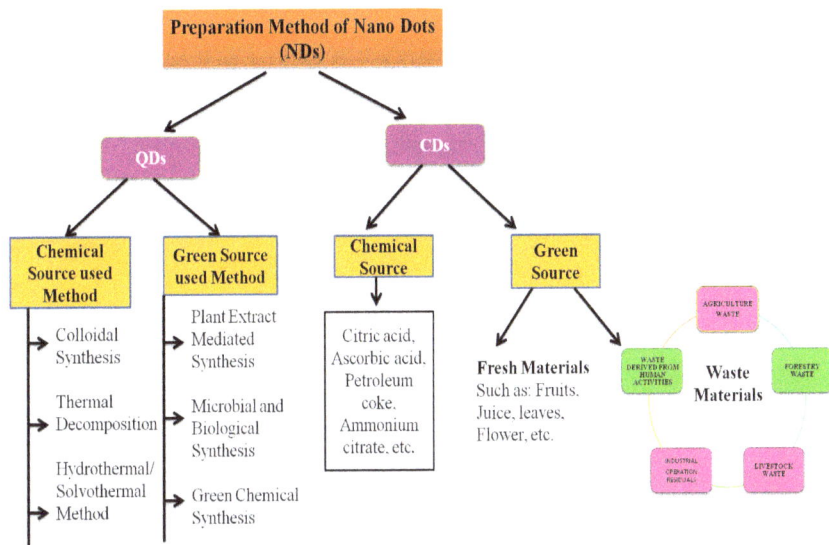

Figure 2: Preparation method of QDs

2.1 Chemical methods

Chemical methods are widely used and involve various chemical reactions and precursors. Some commonly used chemical methods for QD synthesis include:

2.1.1 Colloidal synthesis

In this method, QDs are nucleated and grow in a colloidal solution. Metal salts or metal-organic compounds are mixed as precursors with a stabilizing agent in a solvent and heated to a specific temperature. The growth of QDs is initiated by controlled nucleation. By adjusting parameters including temperature, precursor concentration, and time, the size as well as properties of QDs may be tuned [28–30]. Zhang et al. [31] reported colloidal synthesis of stable $CH_3NH_3PbI_3$ QDs. While previous studies faced challenges with degradation, they achieved air-stable QDs using ligand-assisted precipitation synthesis with acetonitrile as a solvent. The resultant QDs exhibited good stability, high photoluminescence QY, and tuneable emission. Lin et al. [32] used a simple method for synthesizing photoluminescent MoS_2 QDs. Using oleyl amine as a reducing agent, they prepared MoS_2 QDs from $(NH_4)_2MoS_4$ precursor, which exhibited fluorescence at 575 nm with a QY of 4.4%. The QDs had a spherical morphology, uniform thickness of approximately 3 nm, and showed size-dependent tuneable photoluminescence. They could be transferred to aqueous solution, where they well-dispersed and showed low toxicity and maintained their photoluminescent properties. The QDs were effectively employed as probes for optical cellular imaging in real-time in 293T cells (Fig. 3).

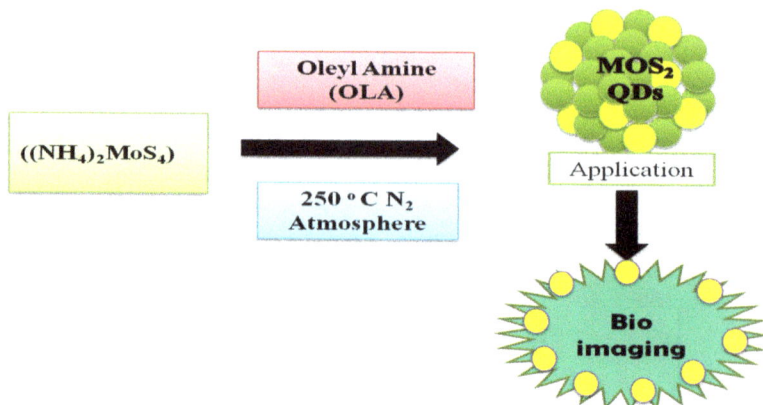

Figure 3: Preparation of MoS₂ QDs from (NH₄)₂MoS₄

2.1.2 Thermal decomposition

In this method, organometallic precursors are heated at high temperatures to induce thermal decomposition. The precursors decompose to give rise to QDs. The reaction conditions are carefully controlled to obtain QDs with desired sizes and compositions [32]. Chen et al. [33] focused on the biomedical applications of QDs based on cadmium-free Ag-In-S (AIS) chalcopyrite as well as core/shell structures derived from them (AIS/ZnS). Optical properties of the QDs were remarkable and they showed low toxicity. Thermal decomposition gave high-quality AIS and AIS/ZnS QDs. Silver acetate, indium acetate, and oleic acid were mixed in dodecanethiol and heated at 170 °C to yield AIS QDs with a QY of 13%. By subsequently growing a ZnS shell on the AIS QDs, the researchers prepared AIS/ZnS QDs with an enhanced QY of 41%. To facilitate their application in cellular imaging, the AIS/ZnS QDs were incorporated into the core of polymeric micelles. Cellular imaging experiments demonstrate specific internalization of chlorotoxin-conjugated QD-micelles into U-87 brain tumour cells. The simple surface/interface chemistry and scalable synthesis of these QDs, along with their use in phase transfer and bioconjugation, present new opportunities. Furthermore, the versatility of QD-micelles holds promise for broader applications in diagnosis and therapy based on cells or tissues. A novel method to develop metal sulphide semiconductor (such as CdS, Bi₂S₃, and Sb₂S₃) sensitized TiO₂ films was introduced by Lutz et al. [34]. The method provides a simple route for the controlled thermal decomposition of a metal xanthate precursor with a single source on the surface of a mesoporous metal oxide. The metal sulphide deposit has grown successfully. The development of cutting-edge materials with improved performance for solar energy conversion applications has been made possible by these findings. The effect of thermal induction on the composition and structure of cubic ZnS/Zn(OH)₂ QDs was investigated by Nistor et al. [35]. The QDs were tightly aggregated and of core/shell structure. By employing multifrequency electron paramagnetic resonance, the researchers analyzed the changes occurring in the shell of these QDs. The ε-Zn(OH)₂ shell played very dominant role in the structural modifications and decomposition behaviour of the core-shell QDs under thermal conditions. The study emphasizes the significance of nanoscale dimensions

Green Synthesis and Emerging Applications of Frontier Nanomaterials   Materials Research Forum LLC
Materials Research Foundations 169 (2024) 249-274   https://doi.org/10.21741/9781644903278-10

and disorder in influencing the transformation process, thereby contributing to the comprehension and design of core-shell QDs systems.

2.1.3 Hydrothermal/solvothermal synthesis

QDs are synthesized in a solvent at high temperatures/pressures using hydrothermal/solvothermal method. The reaction is performed in a sealed vessel with the precursors dissolved in the solvent. The production of QDs with controlled size and crystallinity is facilitated by the high temperature and pressure [36]. Ren et al. [37] successfully prepared MoS_2 QDs with uniform size distribution using a novel route based on hydrothermal-solvothermal cutting integrated method. The resultant QDs were approximately 2.6 nm in size on average and showed a few-layer thickness. The QDs displayed excellent photoluminescence properties. The method is cost-effective and efficient, offering potential advancements in the preparation of QDs derived from other layered materials. Wu et al. [38] successfully fabricated composite coaxial nanoarrays consisting of CuO NPs sensitized ZnO nanorods using a combined hydrothermal−solvothermal method. By varying the reaction time, they were able to regulate the number of CuO-NPs on the ZnO nanorod surface. They also investigated the mechanism of synthesis of the ZnO/CuO nanoarray. Zn vacancies on the surface of ZnO nanorods accumulated Cu^{2+} ions from the solution. The hydroxy groups in the solution interacted with these captured ions to generate $Cu(OH)_2$ to serve as the first nucleation sites for surface development of CuO-NPs. Analyses of energy band and optical absorption characteristics showed that CuO functioned as an electron acceptor and a supplementary absorption material in polymer–inorganic hybrid solar cells. CuO also demonstrated promise in dye-sensitized solar cells as a material for harvesting light. These findings open up possibilities for utilizing the ZnO/CuO nanoarray for solar cells with enhanced performance. Sajinovic et al. [39] developed a methodology for incorporating CdS QDs into a polystyrene matrix. This was achieved through a phase transfer process, transferring the CdS QDs from an aqueous to an organic phase. The polystyrene matrix and the surface of the CdS QDs developed chemical interactions. Radiationless recombination was effectively suppressed by blocking the surface states and enhanced the band-to-band luminescence of the QDs.

2.2   Green method

Green methods aim to reduce the environmental impact of synthesis of QDs by utilizing environmentally friendly materials and techniques. These methods often involve natural extracts or biomolecules as stabilizing/capping agents [40]. Some commonly used green methods for QDs synthesis include. Microbial and biological synthesis methods offer sustainable and eco-friendly approaches. Microorganisms, such as bacteria or fungi, or biological systems, such as enzymes or proteins are used to synthesize QDs. These methods offer advantages such as mild reaction conditions, minimal toxicity, and the potential for production in mass quantities [41].

2.2.1   Microbial synthesis

Microbial synthesis involves the use of microorganisms to produce QDs. These include bacteria, fungi, or yeast. The microorganisms can be engineered to express specific proteins or enzymes that facilitate the synthesis of QDs [42-44]. The general steps involved in microbial synthesis are as follows:

2.2.1(i). Microorganism selection

Suitable microorganisms with the ability to produce QDs or express relevant proteins/enzymes are selected.

2.2.1(ii). Cultivation

The selected microorganisms are cultured in a suitable growth medium under controlled conditions, providing the necessary nutrients and environmental factors for their growth.

2.2.1(iii). Introduction of precursors

The culture medium is supplemented with precursor compounds containing the desired metals. The microorganisms take up these precursors, and intracellular enzymatic reactions occur to form QDs.

2.2.1(iv). Nucleation and growth

The reduced metal ions within the microorganisms serve as the venters for nucleation, initiating formation and development of QDs. The usual conditions of the reaction are optimized tune QDs with desirable size and properties.

2.2.1(v). Harvesting and purification

The QDs-containing microorganisms are harvested and processed to separate the QDs from the cellular biomass. Techniques like centrifugation, filtration, or cell lysis can be employed for purification. Bao et al. [45] demonstrated a method to efficiently synthesize CdTe QDs using a bacterium, *Escherichia coli*. The method allowed tuneable fluorescence emission, resulting in CdTe QDs with superior optical properties tuneable with size. The QDs exhibited good crystallinity. A surface protein capping layer could be identified, which not only maintained the viability of cells (92.9%) in high QD concentrations of up to 2 μM but also contributed to fluorescence emission. Upon functionalization with folic acid, QDs allowed successful in vitro imaging of cultured cervical cancer cells. Using foreign genes generating a CdS binding peptide. Mi et al. [46] prepared CdS QDs in genetically modified *Escherichia coli*. After employing cell lysis and freezing-thawing, QDs and bacteria were separated effectively and purified using an anion-exchange resin. This approach offers the potential for producing biocompatible QDs that are suitable for bio-labelling and imaging applications described in Figure 4.

Green Synthesis and Emerging Applications of Frontier Nanomaterials   Materials Research Forum LLC
Materials Research Foundations 169 (2024) 249-274   https://doi.org/10.21741/9781644903278-10

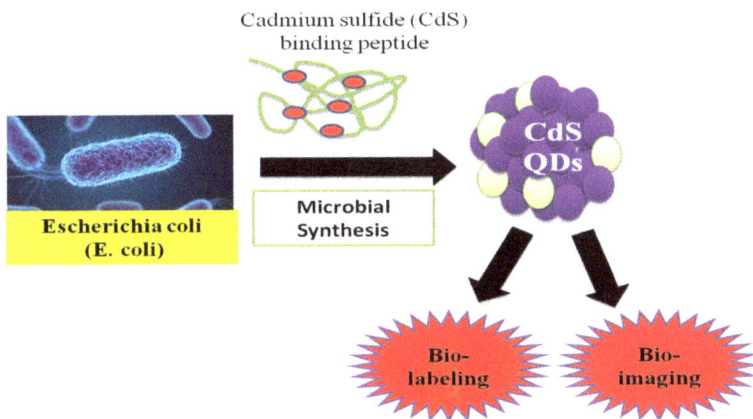

Figure 4: Preparation of CdS QDs following microbial synthesis

2.2.2   Biological synthesis

Biological synthesis methods involve the use of biological systems, such as enzymes, proteins, or biomolecules, for the synthesis of QDs. These biological systems can be extracted from living organisms or produced through recombinant DNA technology [47]. The steps involved in biological synthesis are as follows:

2.2.2(i). Biomolecule extraction or production

Biomolecules, such as enzymes or proteins, with the ability to catalyze the synthesis of QDs are extracted from natural sources or produced through genetic engineering techniques.

2.2.2(ii). Reaction setup

The extracted or produced biomolecules are mixed with precursor solutions containing metal ions. The reaction conditions are optimized for efficient synthesis of QDs.

2.2.2(iii). Nucleation and growth

The biomolecules act as catalysts and promote nucleation and growth of QDs. They adjust size, shape, and composition of the QDs.

2.2.2(iv). Characterization and purification

Fluorescence (FL) and UV-Vis spectroscopy, XRD, and electron microscopy (EM) are conveniently used to characterize the synthesized QDs for determining their properties. Purification methods, including centrifugation, dialysis, or chromatography, may be employed to remove impurities or unreacted precursors. The microbial and biological synthesis are advantageous since they involve eco-friendly production processes, potential scalability, and mild reaction conditions. The choice of microorganisms, enzymes, or proteins, as well as the optimization of reaction conditions, play crucial roles in obtaining QDs with desired properties. Mal et al. [48] reported metal chalcogenide QDs, including metal sulphides, selenides, and tellurides. QDs have been widely utilized for imaging cells and tracking macromolecules. Various methods encompassing physics, chemistry, and biology have been devised to produce many different kinds of QDs. Preparation of biological QDs follows green chemistry principles by avoiding or using as little as possible hazardous chemicals, high pressures, heat, and byproducts.

2.2.3 Plant extract-mediated synthesis

This approach utilizes plant extracts containing bioactive compounds, such as flavonoids, polyphenols, or proteins, as reducing and stabilizing agents. The plant extracts can be obtained by grinding plant tissues and then used to reduce metal ions into QDs. Furthermore, the synthesised QDs are stabilised by the bioactive compounds included in the plant extract, which makes them appropriate for a range of uses [49].

2.2.3(i). Plant material selection

The first step is to select a suitable plant material. Plants rich in bioactive compounds, such as green tea, Aloe vera, or grapefruit, are commonly used for their reducing and stabilizing properties. The plant material is collected and thoroughly washed to remove any contaminants.

2.2.3(ii). Preparation of plant extract

The plant material is ground or blended to obtain a fine powder or pulp. It is then mixed with a suitable solvent, such as distilled water or ethanol, and subjected to extraction methods such as maceration or sonication. This process helps extract the bioactive compounds from the plant material. The resulting mixture is then filtered or centrifuged to remove any solid particles.

2.2.3(iii). Synthesis of QDs

The plant extract is mixed with precursor solutions containing metal salts, such as cadmium chloride ($CdCl_2$) or zinc acetate ($Zn(CH_3COO)_2$). The mixture is then subjected to reduction of the meta ions by the bioactive compounds in the plant extract typically at an elevated temperature.

2.2.3(iv). Nucleation and growth

The reduced metal ions act as nucleation centres, initiating the development of QDs. The conditions are carefully controlled. The bioactive compounds in the plant extract not only serve as reducing agents but also help in stabilizing the formed QDs, preventing their aggregation.

2.2.3(v). Characterization and purification

The synthesized QDs are characterized using standard techniques to determine their size, composition, and optical properties. The QDs may undergo purification steps, such as centrifugation or dialysis, to eliminate any impurities or residual precursors [50].

Pugazhenthiran et al. [51] prepared monodispersed Ag QDs with sizes below 5 nm. Sweet lime (Citrus limetta) peel (SLP) extract is used in the method as a green reducing agent (Fig. 5). The resultant Ag QDs have a face-centered cubic structure. The surface plasmon resonance phenomenon is responsible for the absorption peak that characterises the optical properties of Ag

QDs, which is located at around 415 nm. Excellent photoluminescence quenching is shown by Ag QDs, suggesting a low rate of recombination and extended lifetime of photoexcited electrons.

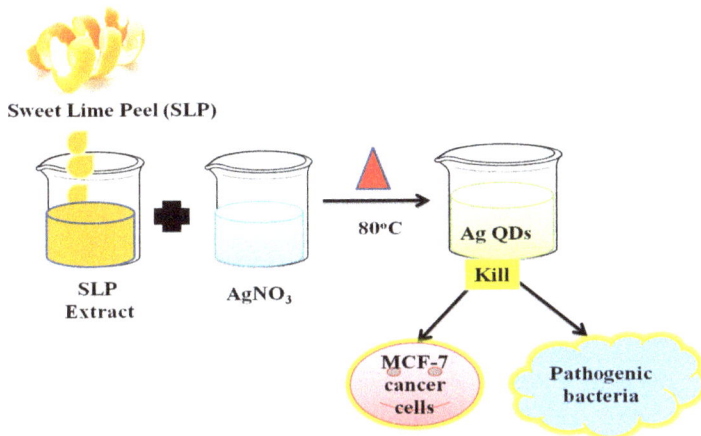

Figure 5:   Preparation of Ag QDs following Plant Extract-Mediated   Synthesis

Using Barbated Skullcup herb as a reducing agent, Wang et al. [52] reported extracellular synthesis of Au NPs (Fig. 6). Rapid reduction of the gold ions formed well-dispersed NPs (5 to 30 nm). At 540 nm, the NPs displayed a distinctive absorption peak. When modified on a glassy carbon electrode, they improved the electronic transmission rate, suggesting potential applications in electrochemical sensing and catalysis.

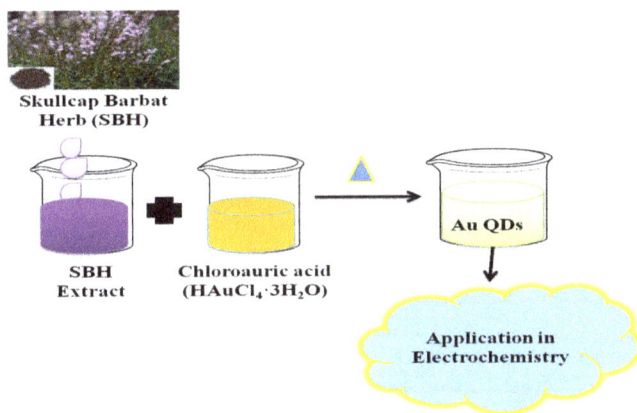

Figure 6: Preparation of Au QDs via Skullcap barbat herbs

2.2.4 Microwave-assisted synthesis

This method is rapid and efficient. It reduces reaction times and increases yield by using microwave irradiation to accelerate the synthesis [53–54].

2.2.4(i). Selection of precursors

The first step is to select suitable precursor compounds containing the desired metals for synthesis of QDs. These precursors can be metal salts or metal-organic compounds.

2.2.4(ii). Preparation of reaction mixture

The precursor compounds are dissolved in a coordinating solvent or a mixture of solvents. The concentration of the precursors and the solvent composition may vary depending on the desired QD properties.

2.2.4(iii). Microwave irradiation

The reaction mixture is placed in a microwave reactor that is capable of generating microwave energy. The reactor is sealed to prevent any loss of volatile components during the synthesis process.

2.2.4(iv). Reaction optimization

The reaction parameters, including the power level, irradiation time, and temperature, are carefully optimized to ensure efficient QD synthesis. The reaction mixture is quickly heated by the microwave energy.

2.2.4(v). Cooling and quenching

After the desired reaction time, the microwave irradiation is stopped, and the mixture is cooled down to stabilize the formed QDs and prevent further growth.

2.2.4(vi). Characterization and purification

The synthesized QDs are characterized using standard techniques to determine their size, composition, and optical properties. Purification methods, such as centrifugation or filtration, may be employed to remove any impurities or unreacted precursors. Microwave irradiation in provides several advantages. The rapid and uniform heating achieved by microwaves allows for shorter reaction times and improved control over QDs formation. Additionally, microwave-assisted synthesis often leads to higher yields and improved reproducibility compared to conventional heating methods. Liu et al. [55] has developed a novel detection method using titanium carbide QDs (Ti_3C_2 QDs) prepared through a microwave-assisted method. Curcumin and hypochlorite (ClO^-) can be detected simultaneously by employing colorimetric and ratio fluorescence methods because of this unique approach. The Ti_3C_2 QDs exhibit fluorescence emission within the 350–600 nm range, with a maximum (430 nm) matching with the absorption maximum of curcumin. This favourable spectral overlap facilitates fluorescence resonance energy transfer (FRET). Upon addition of ClO^-, the oxidation of phenolic and methoxy groups of curcumin leads to the restoration of Ti_3C_2 fluorescence of QDs. The designed probe enables the naked-eye detection of curcumin and ClO^-. The detection limits and linear detection ranges for curcumin and ClO^- are provided, making this study the first of its kind to employ Ti_3C_2 QDs in a dual-channel method.

Mendez et al. [56] synthesised C-doped TiO_2 and ZnO hybrid materials by means of microwave-assisted solvothermal synthesis. The optoelectronic and textural characteristics of the hybrid were affected by temperature and reaction time. Controlled calcination yielded C-doped QDs of TiO_2 and ZnO nanostructured spheres. It turned out that the synthesis conditions of TiO_2-C and ZnO-C films affected their photoelectrochemical activity, and the bandgap energy experienced a red shift. The values of the ZnO-C and TiO_2-C were 3.13 eV and 3.02 eV, respectively, lower than those of bare semiconductors due to C-doping. The films showed the potential as photoelectrodes for QD-sensitized solar cells.

Zhu et al. [57] synthesized SnO_2 QDs using a Zhu et al. [57] a microwave-assisted method. SnO_2 QDs wrapped in organic components and measuring approximately 2-4 nm in size were successfully synthesized at 160 °C in under 0.5 min. After that, QDs wrapped in organic components were annealed for 4 h at 400 °C in the air. This produced highly crystalline SnO_2 particles that were between 5 and 10 nm in size. The annealed SnO_2 QDs exhibited exceptional gas sensitivity in response to ethanol vapour. In particular, when subjected to 300 ppm of ethanol vapour, the sensors showed a relative resistance of 215, which is defined as the ratio of the resistance of the sample in air to that in the tested gases.

2. Green chemical method

Green chemical synthesis methods provide environmentally friendly approaches for the preparation of QDs by utilizing non-toxic and sustainable materials. These methods are aimed at minimizing the environmental impact of synthesis of QDs while lowering the amount of hazardous chemicals used [58-60]. Here is a detailed description of the green chemical synthesis method for QDs:

Selection of Non-Toxic Precursors: In green chemical synthesis, non-toxic and environmentally friendly precursor compounds are chosen. These precursors may include metal salts, metal-organic frameworks (MOFs), or other non-toxic metal sources.

Solvent Selection: Green solvents, such as water or non-hazardous organic solvents, are preferred for the synthesis. Water is particularly advantageous due to its non-toxicity, low cost, and wide availability.

Ligand Exchange or Surface Modification: To regulate the surface chemistry of the QDs and increase stability, ligand exchange or surface modification techniques are employed. Non-toxic ligands, such as thiol-based compounds or bio-derived molecules, can be used to replace toxic ligands and improve the biocompatibility of the QDs.

Microwave-Assisted or Low-Temperature Synthesis: Green chemical synthesis methods often utilize microwave-assisted or low-temperature synthesis conditions to minimize energy consumption and reduce the formation of harmful byproducts. Microwave irradiation can provide rapid and efficient heating, while low-temperature synthesis permits precise control over the QD growing process.

Catalysts and Reducing Agents: Green catalysts and reducing agents are employed to facilitate the formation of QDs without the use of toxic or hazardous compounds. For example, natural extracts or biomolecules can act as reducing agents or catalysts to facilitate metal ions to be reduced to QDs.

Characterization and Purification: The QDs are characterized using standard spectroscopic, electron microscopic, and diffraction techniques to determine their size, composition, and optical properties. Purification methods, such as centrifugation or filtration, may be employed to remove any impurities or unreacted precursors. Green chemical synthesis methods for QDs offer several advantages, including the use of non-toxic precursors, environmentally friendly solvents, and reduced energy consumption. These methods contribute to sustainable and eco-friendly nanomaterial synthesis and are particularly valuable for biomedical and environmental applications. Bozetine et al. [61] described a simple and single step synthetic method for producing ZnO/CQD nanocomposites (NCs). Under irradiation of visible light at ambient temperature, the photocatalytic activity of the ZnO/CQDs NCs has been evaluated for the degradation of a model organic pollutant, rhodamine B. When the NCs were compared to ZnO particles prepared under the same conditions of experimentation, they remarkably demonstrated a highly efficient photodegradation capability. The ZnO/CQDs NCs showed great potential for a catalytic system with high efficiency. Baruah et al. [62] reported a an environmentally benign method for successful synthesis of ZnS QDs. The process was simple, time-efficient, cost-effective, and reproducible. The ensuing QD thin films (QDTF) were suited for solar cell applications because of their unique surface morphologies, which allowed for the efficient absorption and conversion of harmful UV light into useful visible light. In a variety of antibacterial investigations, the produced ZnS QDs (Size= 2 to 6 nm) showed both biocompatibility and strong antibacterial activity. Crucially, there were no negative impacts of the QDs on mammalian cells, highlighting their potential for diverse biomedical applications.

3.　Characterization of QDs

The QDs are characterized by various methods. These methods offer important details regarding the size, shape, composition, optical properties, and structural features of QDs [63-66]. Some commonly used characterization techniques for QDs are described in detail (Fig. 7):

UV-Vis Spectroscopy: The technique is frequently employed to analyse the optical characteristics of QDs. It evaluates how QDs in the visible (Vis) and ultraviolet (UV) regions of the electromagnetic spectrum absorb and emit light. Details on bandgap, absorption peak wavelength, and QY of QDs can be obtained by UV-Vis spectroscopy.

Fluorescence Spectroscopy: FL investigates the fluorescence properties of QDs. It measures the emission spectrum when the QDs are subject to excitation with a specific wavelength of radiation. This technique helps in evaluating the emission peak wavelength, intensity, and stability of fluorescence of QDs.

Transmission Electron Microscopy: QDs at the nanoscale can be observed thanks to TEM, a high-resolution imaging method. It offers comprehensive details regarding the distribution, size, and shape of QDs. TEM also allows for the examination of crystal lattice structures and the determination of interplanar distances within the QDs.

X-ray Diffraction: QD composition and crystal structure are examined using XRD. It aids in determining the crystallographic phases and offers details on how the atoms are arranged within the QDs. Crystallographic features such lattice spacing, crystal size, and orientation can be determined in XRD patterns.

Fourier Transform Infrared Spectroscopy: The surface chemistry and functional groups of QDs are investigated using FTIR spectroscopy. It offers details on the molecular vibrations and chemical bonding within the QDs. FTIR spectra can be used to identify surface ligands, capping agents, or any modifications made to the QDs.

Dynamic Light Scattering: The size distribution and solvodynamic diameter of QDs in a colloidal suspension can be determined using the DLS approach. It monitors fluctuations in light scattering brought on by movement of QDs resulting from Brownian motion. DLS is particularly useful for assessing the stability and aggregation behaviour of QDs in solution.

Energy-Dispersive X-ray Spectroscopy: EDS is used for determining the elemental composition of QDs. It detects and analyzes the characteristic X-rays emitted by the elements present within the QDs. EDS can be combined with TEM or scanning electron microscopy (SEM) to determine the elemental composition and stoichiometry of QDs.

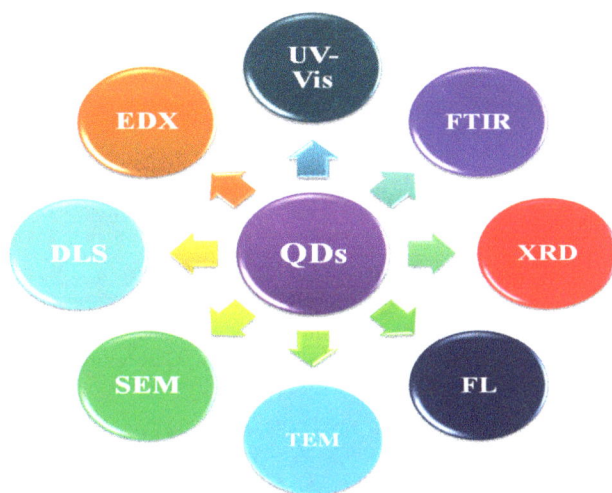

Figure 7: Characterization Techniques of QDs

4.   Applications and future perspectives

Because of their unique characteristics, QDs have attracted a lot of interest and are considered promising materials for an assortment of applications [67-74]. Here are some key applications of QDs and future perspectives (Fig. 8).

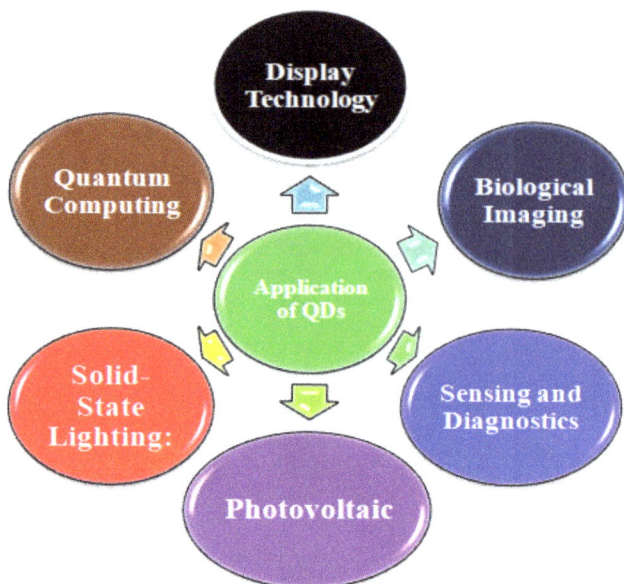

Figure 8: Application of QDs in difference Fields

Display Technology: QDs are being used in display technology, specifically in developing displays of high-quality, vibrant displays. To improve the colour gamut and efficiency of LCD and LED displays, as well as their colour accuracy and brightness, QDs can be employed as colour converters.

Biological Imaging: The tuneable emission wavelengths, high brightness, and photostability make QDs ideal for applications such as cell labelling, in vivo imaging, and tracking of biomolecules. QDs can provide enhanced resolution and imaging capabilities compared to traditional organic dyes.

Sensing and Diagnostics: QDs have shown potential in sensing and diagnostics applications. They can be used as sensitive probes for detecting biological analytes, such as proteins, DNA, and pathogens. The unique optical properties of QDs allow for highly sensitive and multiplexed detection, enabling advancements in medical diagnostics and environmental monitoring.

Photovoltaics: In the area of solar cell or photovoltaic technology, QDs have shown possibilities. Incorporation of QDs into solar cells causes enhancement of light absorption, enables multi-junction cells, and improves energy conversion efficiency. QDs offer the potential for less expensive, flexible, and high-performance solar cell devices.

Solid-State Lighting: QDs have the capacity to entirely transform the lighting industry. They can be applied in light-emitting diodes (LEDs) as phosphor materials to provide excellent white light.

QD-based lighting offers advantages such as energy efficiency, colour tunability, and improved colour rendering.

Quantum Computing: QDs hold promise in quantum computing, where the quantum properties of particles are harnessed for advanced computational tasks. QDs can be utilized as qubits, the fundamental units of information in quantum computing. Their long coherence times and potential for scalability make them attractive for quantum information processing. Some basic comparison described in Table 1.

Table 1:   Comparison between QDs and CDs

QDs vs. Carbon Dots	Source	Method	Cytotoxicity/ Detection Limit Range	Reference
CdTe QDs	Cadmium chloride (CdCl$_2$) sodium borohydride, tellurium powder, phosphate buffered saline (PBS) tablets	Electrocehmic al method	157 ± 31 µg/mL by MTT cytotoxicity assay	75
InPZnS QDs	InPZnS alloy core and a thin ZnS shell	Heating Up Method	Exposure to 70 nM of QDs over a period of 72 hours resulted in cytotoxic effects on environmental health	76
Indium-based QDs(CFQDs)	The In and Zn content can be increased by improving the composition and implementing a stronger outer shell.	Heating Up Method	The administration of QDs at doses of 12.5 mg/kg and 50 mg/kg led to the accumulation of QDs primarily in the liver and spleen.	77
InZnP and InZnPS QD	Indium myristate, ZnSt$_2$	Precipitation Method	6.25-200 nM at 24 h exposure	78
CQDs	Zirconium chloride trinitrophenol (TNP), 2,4-dinitrophenol (DNP) 4-nitrophenol (4-NP), and phenol (PHE), Polyvinylpyrrolid	Hydrothermal Method	0.01–20.0 µM at a low detection limit of 3.5 nM	79

	one (PVP) acetic acid (HAc), citric acid (CA)			
CQDs	Gelatin	Hydrothermal Mthod	$1-75 \ \mu mol \ L^{-1}$	**80**
CQDs	Carbohydrate	Hydrothermal Method	$0 \ to \ 1 \times 10^3$	**81**

The thorough literature review reveals that synthesis and characterization studies of CDs and QDs and their applications have received an upsurge of interest. A systematic representation of CDs and their applications (published articles from 1980 to 2024) is summarized in Figure 9.

The future of QDs involves ongoing research and development to enhance their properties and expand their applications. Efforts are being made to improve synthesis methods of QDs, enhance their stability, and reduce their toxicity. In addition, researchers are exploring hybrid systems, such as combining QDs with other nanomaterials or integrating QDs into flexible and wearable devices. As the field progresses, QDs possess the capacity to benefit a variety of industries such as computing, electronics, healthcare, and energy. However, it is crucial to address concerns regarding their toxicity, environmental impact, and large-scale synthesis to ensure their safe and sustainable implementation.

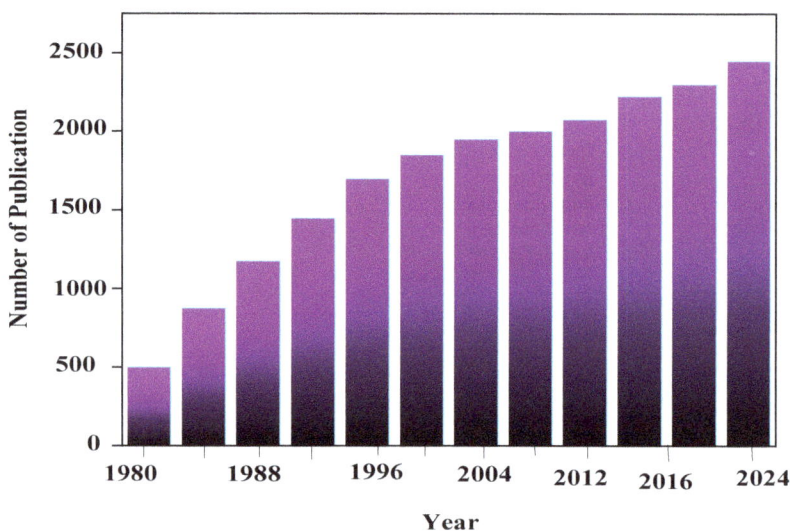

Figure: 9   Publication Graph of QDs (Source http://www.sciencedirect.com, http://www.wiley.com, http://www.acs.com, http://www.rsc.com)

Conclusion

Quantum dots have emerged as versatile nanomaterials with remarkable properties. They turned out to be very appealing for a variety of domains due to their outstanding quantum yields, tuneable emission wavelengths, great photostability, and distinct optical and electrical features dependent on the size in the nanoscale. The extensive research and development efforts in synthesis, characterization techniques, and surface modifications of QDs have contributed to significant advancements in the field. Researchers have made substantial progress in improving stability of QDs, enhancing their biocompatibility, and reducing their toxicity, addressing key concerns associated with their use. The applications of QDs span across diverse areas. They have been successfully integrated to display technology to improve brightness and color quality, in biological imaging for precise cell labeling and tracking, in sensing and diagnostics for sensitive detection of biomolecules, in photovoltaics for efficient energy conversion, and in solid-state lighting for high-quality illumination. Looking ahead, the future of QDs holds immense potential. Ongoing research aims to further enhance their properties, explore novel applications, and address challenges related to large-scale synthesis, environmental impact, and safety considerations. Hybrid systems, combining QDs with other nanomaterials, and integration into flexible and wearable devices are promising avenues for future exploration. As the field progresses, QDs have the potential to revolutionize various industries and contribute to advancements in electronics, healthcare, energy, and computing. However, it is essential to continue research efforts in order to understand and mitigate any potential risks associated with their use. Overall, QDs have established themselves as a remarkable class of nanomaterials with exciting prospects for technological innovation and scientific discoveries. With further advancements and interdisciplinary collaborations, QDs now possess the ability to profoundly impact how many different fields develop in the future and drive advancements in nanotechnology and beyond.

References

[1] M. C. Daniel, D. Astruc, Gold Nanoparticles: Assembly, Supramolecular Chemistry, Quantum-Size-Related Properties, and Applications toward Biology, Catalysis, and Nanotechnology, Chemical Reviews 104 (2004) 293-346. https://doi.org/10.1021/cr030698+

[2] T.P. Yadav, R.M. Yadav, D.P. Singh, Mechanical milling: a top down approach for the synthesis of nanomaterials and nanocomposites, Nanoscience and Nanotechnology 2 (2012) 22-48. https://doi.org/10.5923/j.nn.20120203.01

[3] V. A. Basiuk, E.V. Basiuk, Green Processes for Nanotechnology. 1st ed. Mexico DF, MX: Springer, (2015) 434. https://doi.org/10.1007/978-3-319-15461-9

[4] D. Hines, P.V. Kamat, Recent Advances in Quantum Dot Surface Chemistry. ACS Applied Materials & Interfaces, 6 (5) (2014) 3041-3057. https://doi.org/10.1021/am405196u

[5] A. S. Karakoti, R. Shukla, R. Shanker, S. Singh, Surface Functionalization of Quantum Dots for Biological Applications. Advance Colloid Interface Science 215(2015)28-45. https://doi.org/10.1016/j.cis.2014.11.004

[6] A.K. Potbhare, S.K. Tarik Aziz, M.M. Ayyub, A. Kahate, R. Madankar, S. Wankar, A. Dutta, A.A. Abdala, H.M. Sami, R. Adhikari, R.G. Chaudhury, Bioinspired Graphene-based metal oxide nanocomposites for photocatalytic and electrochemical performances: an updated review, Nanoscale Advances, 6 (2024) 2539-2568. https://doi.org/10.1039/D3NA01071F

[7] P. Pierobon, G. Cappello, Quantum dots to tail single bio-molecules inside living cells, Advanced Drug Delivery Reviews 64 (2012) 167-178. https://doi.org/10.1016/j.addr.2011.06.004

[8] A. I. Ekimov, A.L. Efros, A.A. Onushchenko, Quantum Size Effect in Semiconductor Microcrystals. Solid State Communication, 56 (1985)921-924. https://doi.org/10.1016/S0038-1098(85)80025-9

[9] W. Liu, S. Zhang, L. Wang, C. Qu, C. Zhang, L. Hong, L. Yuan, Z. Huang, Z. Wang, S. Liu, G. Jiang, CdSe Quantum Dot (QD)-Induced Morphological and Functional Impairments to Liver in Mice. PLoS ONE 6 (2011) 24406. https://doi.org/10.1371/journal.pone.0024406

[10] R.G. Chaudhary, A.K. Potbhare, P.B. Chouke, A.R. Rai, R.K. Mishra, M.F. Desimone, A.A. Abdala, Graphene-based materials and their nanocomposites with metal oxides: biosynthesis, electrochemical, photocatalytic and antimicrobial applications, Magnetic Oxides and Composites II, Materials Research Foundation, 83 (2020), pp. 79-116. https://doi.org/10.21741/9781644900970-4

[11] K. Shivaji, S. Mani, P. Ponmurugan, C. De Castro, M. Lloyd Davies, M. Balasubramanian, S. Pitchaimuthu, Green Synthesis Derived CdS Quantum Dots Using Tea Leaf Extract: Antimicrobial, Bioimaging and Therapeutic Applications in Lung Cancer Cell. ACS Applied Nano Materials 4 (2018) 1683-1693. https://doi.org/10.1021/acsanm.8b00147

[12] N. Chen, Y. He, Y. Su, X. Li, Q. Huang, H. Wang, X. Zhang, R. Tai, C. Fan, The cytotoxicity of cadmium-based quantum dots. Biomaterials 33 (2012) 1238-44. https://doi.org/10.1016/j.biomaterials.2011.10.070

[13] K.G. Li, J.T. Chen, S.S. Bai, X. Wen, S.Y. Song, Q. Yu, J. Li, Y. Q. Wang, Intracellular oxidative stress and cadmium ions release induce cytotoxicity of unmodified cadmium sulfide quantum dots. Toxicology in vitro: an international journal published in association with BIBRA 23 (2009)1007-13. https://doi.org/10.1016/j.tiv.2009.06.020

[14] Y. Sahu, A.Hashmi, R. Patel, A.K. Singh, M.A.B. H. Susan, S.A.C. Carabineiro, Potential Development of N-doped carbon dots and metal- Oxide Carbon dot Composites for Chemical and Biosensing, Nanomaterials 12(2022) 3434. https://doi.org/10.3390/nano12193434

[15] L. Pavesi, L. Dal Negro, C. Mazzoleni, G. Franzo, F. Priolo, Optical gain in silicon nanocrystals. Nature 408 (2000) 440-444. https://doi.org/10.1038/35044012

[16] Y. He, S. Su, T. Xu, Y. Zhong, J.A. Zapien, J. Li, C. Fan, S.T. Lee, "Silicon nanowires-based highly-efficient SERS-active platform for ultrasensitive DNA detection," Nano Today 6 (2011) 122-130. https://doi.org/10.1016/j.nantod.2011.02.004

[17] Y. Zhong, F. Peng, X. Wei, Y. Zhou, J. Wang, X. Jiang, Y. Su, S. Su, S. T. Lee, Y. He, Microwave Assisted Synthesis of Biofunctional and Fluorescent Silicon Nanoparticles Using Proteins as Hydrophilic Ligands. Angewandte Chemie International edition, 51 (2012) 8485-8489. https://doi.org/10.1002/anie.201202085

[18] A.K. Potbhare, M.S. Umekar, P.B. Chouke, M.B. Bagade, S.T. Aziz, A. A. Abdala, R.G. Chaudhary, Bioinspired graphene-based silver nanoparticles: Fabrication, characterization and antibacterial activity. Materials Today: Proceedings, 29 (2020) 720-725. https://doi.org/10.1016/j.matpr.2020.04.212

[19] M.S. Umekar, G.S. Bhusari, A.K. Potbhare, A. Mondal, B.P. Kapgate, M.F. Desimone, R.G. Chaudhary, Bioinspired reduced graphene oxide based nanohybrids for photocatalysis and antibacterial applications. Curr. Pharm. Biotechnol., 22(13) (2021) 1759-1781. https://doi.org/10.2174/1389201022666201231115826

[20] Y. Su, X. Wei, F. Peng, Y. Zhong, Y. Lu, S. Su, T. Xu, S.T. Lee, Y. He, Gold nanoparticles-decorated silicon nanowires as highly efficient near-infrared hyperthermia agents for cancer cells destruction. Nano Letters 12 (2012) 1845-1850. https://doi.org/10.1021/nl204203t

[21] Y. Zhong, F. Peng, F. Bao, S. Wang, X. Ji, L. Yang, Y. Su, S.T. Lee, Y. He, Large-Scale Aqueous Synthesis of Fluorescent and Biocompatible Silicon Nanoparticles and Their Use as Highly Photostable Biological Probes journal of the American Chemical Society 135 (2013) 8350-8356. https://doi.org/10.1021/ja4026227

[22] Z. M. Alvanda , H. R. Rajabia, A. Mirzaeib, A. Masoumiaslc , H. Sadatfarajid, Rapid and green synthesis of cadmium telluride quantum dots with low toxicity based on a plant-mediated approach after microwave and ultrasonic assisted extraction: Synthesis, characterization, biological potentials and comparison study, Materials Science & Engineering C 98 (2019) 535-544. https://doi.org/10.1016/j.msec.2019.01.010

[23] S. Angizi, A. Hatamie, H. Ghanbari, A. Simchi, Mechanochemical Green Synthesis of Exfoliated Edge-Functionalized Boron Nitride Quantum Dots: Application to Vitamin C Sensing through Hybridization with Gold Electrodes, ACS Applied Materials & Interfaces 34 (2018) 28819-28827 https://doi.org/10.1021/acsami.8b07332

[24] P.R. Bhilkar, R.S. Madankar, T.S. Shrirame, R.D. Utane, A.K. Potbhare, S. Yerpude, R.G. Chaudhary, Functionalized carbon nanomaterials: fabrication, properties, and applications. Materials. Research Foundation, 135 (2022) pp. 55-82.

[25] G. B. Smith, C.G. Granqvist, Green Nanotechnology: Introduction and Invitation. In: Smith G.B., Granqvist C.G., editors. Green Nanotechnology: Solutions for Sustainabil- ity and Energy in the Built Environment. 1st ed. Boca Raton, FL: Taylor & Francis Group 2010) 1-4.

[26] W. J. Parak, A.E. Nel, P.S. Weiss, Grand Challenges for Nanoscience and Nanotechnol-ogy. ACS Nano 9 (2015) 6637-6640. https://doi.org/10.1021/acsnano.5b04386

[27] M. S. Umekar, G.S. Bhusari, T. Bhoyar, V. Devthade, B.P. Kapgate, A. K. Potbhare, R.G. Chaudhary, A.A. Abdala, Graphitic carbon nitride-based photocatalysts for environmental remediation of organic pollutants. Current Nanoscience, 19(2) (2023)148-169. https://doi.org/10.2174/1573413718666220127123935

[28] Y. Pu, F. Cai, D. Wang, J.X. Wang, J. F, Chen, Colloidal Synthesis of Semiconductor Quantum Dots toward Large-Scale Production: A Review. Industrial & Engineering Chemistry Research, 57(6) (2018) 1790-1802. https://doi.org/10.1021/acs.iecr.7b04836

[29] S. D. Hiremath, M. Banerjee, A. Chatterjee, Review of 2D MnO2 Nanosheets as FRET-Based Nanodot Fluorescence Quenchers in Chemosensing Applications. ACS Applied Nano Materials, 5 (2022) 17373-17412. https://doi.org/10.1021/acsanm.2c03470

[30] J. Zhang, W. Xu, L. Wang, Z. Zheng, F. Liu, P. Yu, G. Yang, Colossal Vacancy Effect of 2D CuInP2S6 Quantum Dots for Enhanced Broadband Photodetection. Crystal Growth & Design, 23 (2023)1259-1268. https://doi.org/10.1021/acs.cgd.2c01404

[31] F. Zhang, S. Huang, P. Wang, X. Chen, S. Zhao, Y. Dong, H. Zhong, Colloidal Synthesis of Air-Stable CH3NH3PbI3 Quantum Dots by Gaining Chemical Insight into the Solvent Effects. Chemistry of Materials, 29(8) (2017) 3793-3799. https://doi.org/10.1021/acs.chemmater.7b01100

[32] H. Lin, C. Wang, J. Wu, Z. Xu, Y. Huang, C. Zhang, Colloidal synthesis of MoS2 quantum dots: size-dependent tunable photoluminescence and bioimaging. New Journal of Chemistry, 39 (2015) 8492-8497. https://doi.org/10.1039/C5NJ01698C

[32] D. S. Sajinović, Z.V. Saponjić, N. Cvjetićanin, M. Marinović-Cincović, J. M. Nedeljkovic, Synthesis and characterization of CdS quantum dots-polystyrene composite. Chemical Physics Letters, 329(2000)168-172. https://doi.org/10.1016/S0009-2614(00)00990-8

[33] S. Chen, M. Ahmadiantehrani, N.G. Publicover, K.W. Jr. Hunter, X. Zhu, Thermal decomposition based synthesis of Ag-In-S/ZnS quantum dots and their chlorotoxin-modified micelles for brain tumor cell targeting. RSC Advances, 5 (2015) 60612-60620. https://doi.org/10.1039/C5RA11250H

[34] T. Lutz, A. MacLachlan, A. Sudlow, J. Nelson, M.S. Hill, K.C. Molloy, S.A. Haque, Thermal decomposition of solution processable metal xanthates on mesoporous titanium dioxide films: a new route to quantum-dot sensitised heterojunctions. Physical Chemistry Chemical Physics, 14(2012)16192-16196. https://doi.org/10.1039/c2cp43534a

[35] S.V. Nistor, D. Ghica, M. Stefan, L.C. Nistor, "Sequential Thermal Decomposition of the Shell of Cubic ZnS/Zn(OH)2 Core-Shell Quantum Dots Observed With Mn2+ Probing Ions." Journal of Physical Chemistry C, 117 (2013) 22017-22028. https://doi.org/10.1021/jp4063093

[36] W. Fan, H. Zhang, W. Lu, X. Li. "Synthesis of ZnO/CuO composite coaxial nanoarrays by combined hydrothermal-solvothermal method and potential for solar cells." Journal of Composite Materials, 49 (2015) 1201-1209. https://doi.org/10.1177/0021998314541311

[37] Ren, X., Zhang, Y., Li, H., Wang, H., & Chen, X. (Year). Uniform MoS2 Quantum Dots Prepared via a Hydrothermal-Solvothermal Cutting Integrated Route. Journal of Nanoscience and Nanotechnology, 18 (2018) 7071-7075.

[38] F. Wu, H. Zhang, X. Li, Synthesis of ZnO/CuO composite coaxial nanoarrays by combined hydrothermal-solvothermal method and potential for solar cells, Journal of Composite Materials 49 (2014) 1-6. https://doi.org/10.1177/0021998314541311

[39] D. Sajinovic, Z. Saponjic, N. Cvjeticanin, M. Marinovic-Cincovic, Synthesis and characterization of CdS quantum dots-polystyrene composite. Chemical Physics Letters 329 (2000) 168-172. https://doi.org/10.1016/S0009-2614(00)00990-8

[40] R.G. Chaudhary, A.K. Potbhare, S.T. Aziz, M.S. Umekar, S.S. Bhuyar, A. Mondal, Phytochemically fabricated reduced graphene Oxide-ZnO NCs by Sesbania bispinosa for photocatalytic performances. Materials Today: Proceedings, 36 (2021) 756-762. https://doi.org/10.1016/j.matpr.2020.05.821

[41] R. Mahle, P. Kumbhakar, D. Nayar, T. N. Narayanan, K. K. Sadasivuni, C. S. Tiwary, R. Banerjee. Current advances in bio-fabricated quantum dots emphasizing the study of mechanisms to diversify their catalytic and biomedical applications. Dalton Transactions, 50(2021) 14062-14080. https://doi.org/10.1039/D1DT01529J

[42] Mi.Congcong, Y. Wang, J. Zhang, H. Huang, L. Xu, S. Wang. Biosynthesis and characterization of CdS quantum dots in genetically engineered Escherichia coli. Journal of Biotechnology, 153 (2011) 125-132. https://doi.org/10.1016/j.jbiotec.2011.03.014

[43] R. P. Singh, A.R. Rai, A.A. Abdala, R.G. Chaudhary, Biogenic Sustainable Nanotechnology: Trends and Progress, Elsevier, Amsterdam, Netherland, 2022, pp.1-391.

[44] Y. Cheng ,S.D. Ling, Y. Geng, Y. Wang, J. Xu. Microfluidic synthesis of quantum dots and their applications in bio-sensing and bio-imaging." Nanoscale Advances, 3 (2021) 2180-2195. https://doi.org/10.1039/D0NA00933D

[45] H. Bao, Z. Lu, X. Cui, Y. Qiao, J. Guo, J.M. Anderson, C.M. Li. Extracellular microbial synthesis of biocompatible CdTe quantum dots. Acta Biomaterialia, 6, 9 (2010) 3534-3541. https://doi.org/10.1016/j.actbio.2010.03.030

[46] C. Mi, Y. Wang, J. Zhang, H. Huang, L. Xu, S. Wang, X. Fang, J. Fang, C. Mao, S. Xu/ Biosynthesis and characterization of CdS quantum dots in genetically engineered Escherichia coli. Journal of Biotechnology 153, 3-4 (2011)125-132. https://doi.org/10.1016/j.jbiotec.2011.03.014

[47] T. Mohammad, H. Alijani, P. Faris, E. Salarkia, M. Naderifar, M. R. Akbarizadeh, N. Hashemi, S. Iravani, A. T. Jalil, M. M. Saleh, A. Fathi, M. Khatami , Plant-mediated synthesis of sphalerite (ZnS) quantum dots, Th1-Th2 genes expression and their biomedical applications, South African Journal of Botany, 155 (2023) 127-139 https://doi.org/10.1016/j.sajb.2023.01.041

[48] J. Mal, Y.V. Nancharaiah, E. D. van Hullebusch, P.N. L. Lens, Metal chalcogenide quantum dots: biotechnological synthesis and applications." RSC Advances, 6 (2016) 41477-41495. https://doi.org/10.1039/C6RA08447H

[49] S. Ahmed, M. Ahmad, B. L. Swami, S. Ikram, A review on plants extract mediated synthesis of silver nanoparticles for antimicrobial applications: A green expertise, Journal of Advanced Research 7(2016) 17-28. https://doi.org/10.1016/j.jare.2015.02.007

[50] Z. M. Alvand, H. R. Rajabi, A. Mirzaei, A. Masoumiasl, Ultrasonic and microwave assisted extraction as rapid and efficient techniques for plant mediated synthesis of quantum dots: green synthesis, characterization of zinc telluride and comparison study of some biological activities, New Journal of Chemistry 43 (2019) 15126-15138. https://doi.org/10.1039/C9NJ03144H

[51] N. Pugazhenthiran , S. Murugesan , T. Muneeswaran c, S. Suresh d, M. Kandasamy, H. Valdés, M. Selvaraj, A. Dennyson Savariraj, R.V. Mangalaraja, Biocidal activity of citrus limetta peel extract mediated green synthesized silver quantum dots against MCF-7 cancer cells and pathogenic bacteria, Journal of Environmental Chemical Engineering, 9 (2021)105089 https://doi.org/10.1016/j.jece.2021.105089

[52] Y. Wang, X. He, K. Wang, X. Zhang, W. Tan, Barbated Skullcup herb extract-mediated biosynthesis of gold nanoparticles and its primary application in in electrochemistry, Colloids Surf B Biointerfaces, 73(2009) 75-79. https://doi.org/10.1016/j.colsurfb.2009.04.027

[53] R.K. Singh a, R. Kumar b, D.P. Singh c, R. Savu d, S.A. Moshkalev d, Progress in microwave-assisted synthesis of quantum dots (graphene/carbon/semiconducting) for

bioapplications: a review, Materials Today Chemistry 12 (2019) 282-314. https://doi.org/10.1016/j.mtchem.2019.03.001

[54] A. Kumar, Y. Kuang, Z. Liang, X. Sun, Microwave chemistry, recent advancements, and eco-friendly microwave-assisted synthesis of nanoarchitectures and their applications: a review, Materials Today Nano 11 (2020) 100076. https://doi.org/10.1016/j.mtnano.2020.100076

[55] M. Liu, Y. Bai, Y. He, J. Zhou, Y. Ge, J. Zhou, G. Song, Facile microwave-assisted synthesis of Ti3C2 MXene quantum dots for ratiometric fluorescence detection of hypochlorite, Microchimica Acta volume 188 (2021) 15. https://doi.org/10.1007/s00604-020-04668-y

[56] J. R. R. Mendez, J. Matos, F. Luis, C. Ruiz, Ana C. González-Castillo, G. B. Yáñez, Microwave-assisted synthesis of C-doped TiO2 and ZnO hybrid nanostructured materials as quantum-dots sensitized solar cells, Applied Surface Science, 434, 15 (2018) 744-755 https://doi.org/10.1016/j.apsusc.2017.10.236

[57] L. Zhu, M. Wang, T. K. Lam, C. Zhang, H. Du, B. Li, Y. Yao, Fast microwave-assisted synthesis of gas-sensing SnO2 quantum dots with high sensitivity Sensors and Actuators B: Chemical, 236 (2016) 646-653. https://doi.org/10.1016/j.snb.2016.04.173

[58] A. Barras, S. Cordier, M. R. Das R. Boukherroub, Fast photocatalytic degradation of rhodamine B over [Mo6Br8(N3)6]2− cluster units under sun light irradiation, Applied Catalysis B: Environmental, 123-124 (2012) 1-8 https://doi.org/10.1016/j.apcatb.2012.04.006

[59] R. Devarapalli, M. V. Shelke, S. Cordier, S. Szunerits, Rabah Boukherroub, One-pot synthesis of gold nanoparticle/molybdenum cluster/graphene oxide nanocomposite and its photocatalytic activity, Applied Catalysis B: Environmental, 130-131 (2013) 270-276 https://doi.org/10.1016/j.apcatb.2012.11.017

[60] M. S. Umekar, R.G. Chaudhary, G.S. Bhusari, A. Mondal, A.K. Potbhare, M. Sami, Phytoreduced graphene oxide-titanium dioxide nanocomposites using Moringa oleifera stick extract. Materials Today: Proceedings, 29 (2020) 709-714. https://doi.org/10.1016/j.matpr.2020.04.169

[61] H.Bozetine, Q. Wang, A. Barras, M. Li, T. Hadjersi, S. Szunerits, R. Boukherroub, Green chemistry approach for the synthesis of ZnO-carbon dots nanocomposites with good photocatalytic properties under visible light, Journal of Colloid and Interface Science, 1 (2016) 286-294 https://doi.org/10.1016/j.jcis.2015.12.001

[62] J. M. Baruah, S. Kalita, J. Narayan, Green chemistry synthesis of biocompatible ZnS quantum dots (QDs): their application as potential thin films and antibacterial agent, International Nano Letters, (2019)149-159. https://doi.org/10.1007/s40089-019-0270-x

[63] J. Drbohlavova, V. Adam, R. Kizek, J. Hubalek."Quantum Dots - Characterization, Preparation and Usage in Biological Systems. International Journal of Molecular Sciences 10(2009) 656-673. https://doi.org/10.3390/ijms10020656

[64] A. Foerster, A. Nicholas , A. Besley. "Quantum Chemical Characterization and Design of Quantum Dots for Sensing Applications", Journal of Physical Chemistry A, 126(2022) 2899-2908. https://doi.org/10.1021/acs.jpca.2c00947

[65] B. K. John, T. Abraham, B. Mathew. A Review on Characterization Techniques for Carbon Quantum Dots and Their Applications in Agrochemical Residue Detection, Journal of Fluorescence 32 (2022) 449-471. https://doi.org/10.1007/s10895-021-02852-8

[66] M. Umekar, R.G. Chaudhary, G.S. Bhusari, A.K. Potbhare, Fabrication of zinc oxide-decorated phytoreduced graphene oxide nanohybrid via Clerodendrum infortunatum. Emerging Materials Research, 10(1) (2021) 75-84. https://doi.org/10.1680/jemmr.19.00175

[67] A.A. H. Abdellatif, M.A. Younis, M. Alsharidah,O. Al Rugaie, H.M. Tawfeek, Biomedical Applications of Quantum Dots: Overview, Challenges, and Clinical Potential, 17(2022) 1951-1970. https://doi.org/10.2147/IJN.S357980

[68] W.A. A. Mohamed, H. A. El-Gawad, S. Mekkey , H. Galal , H. Handal , H. Mousa, Ammar Labib, Quantum dots synthetization and future prospect applications, Journal Nanotechnology Reviews 10 (2021)1926-1940. https://doi.org/10.1515/ntrev-2021-0118

[69] X. Gong, Z. Yang ,G. Walters, R. Comin, Z. Ning, E. Beauregard, V. Adinolfi, O. Voznyy, E.H. Sargent, Highly efficient quantum dot near-infrared light-emitting diodes. Nature Photon. 10(2016) 253-257. https://doi.org/10.1038/nphoton.2016.11

[70] G. J. Supran, K.W. Song, G.W. Hwang, R. E Correa, J. Scherer, E.A.Dauler, Y. Shirasaki,M.G. Bawendi, V. Bulovi? 2015. High-Performance Shortwave-Infrared Light-Emitting Devices Using Core-Shell (PbS-CdS) Colloidal Quantum Dots. Advance Materials 27(2015) 1437-1442. https://doi.org/10.1002/adma.201404636

[71] S.A.Veldhuis, P.P. Boix, N. Yantara, M. Li, T.C. Sum, N. Mathews, S.G. Mhaisalkar, 2016. Perovskite Materials for Light-Emitting Diodes and Lasers. Advance Material 28(2016)6804-6834. https://doi.org/10.1002/adma.201600669

[72] J. Song, J. Li, X. Li, L. Xu, Y. Dong, H. Zeng, 2015. Quantum Dot Light-Emitting Diodes Based on Inorganic Perovskite Cesium Lead Halides (CsPbX3). Advance material 27(2015)7162-7167. https://doi.org/10.1002/adma.201502567

[73] X. Dai, Z. Zhang, Y. Jin, Y.Niu, H.Cao, X. Liang, L.Chen, J. Wang, X. Peng, Solution-processed, high-performance light-emitting diodes based on quantum dots. Nature 515(2014) 96-99. https://doi.org/10.1038/nature13829

[74] O.Chen, J. Zhao,V.P. Chauhan, J. Cui, C. Wong, D.K. Harris,H. Wei H, Han, D. Fukumura, R.K. Jain, 2013. Compact high-quality CdSe/ CdS core/shell nanocrystals with narrow emission linewidths and suppressed blinking. Nature Materials 12(2013) 445-451. https://doi.org/10.1038/nmat3539

[75] T. O'Hara, B. Seddon, A. O'Connor, S. McClean, B. Singh, E. Iwuoha, X. Fuku, E. Dempsey, Quantum dot nanotoxicity investigations using human lung cells and TOXOR electrochemical enzyme assay methodology. ACS Sensors 2 (2017) 165-171. https://doi.org/10.1021/acssensors.6b00673

[76] M. Allocca, L. Mattera, A. Bauduin, B. Miedziak, M. Moros, L. De Trizio, A. Tino, P. Reiss, A. Ambrosone, C. Tortiglione, An integrated multilevel analysis profiling biosafety and toxicity onduced by indium- and cadmium-based quantum dots in vivo. Environmental Science and Technology, 53(2019) 3938-3947. https://doi.org/10.1021/acs.est.9b00373

[77] E. Yaghini, H. Turner, A. Pilling, I. Naasani, A.J. MacRobert, In vivo biodistribution and toxicology studies of cadmium-free indium-based quantum dot nanoparticles in a rat model. Nanomedicine Nanotechnology Biology Medicine 14(2018) 2644-2655. https://doi.org/10.1016/j.nano.2018.07.009

[78] A. Tarantini, K.D.Wegner,F. Dussert, G. Sarret, D. Beal, L. Mattera, C. Lincheneau, O. Proux, D. Truffier-Boutry, C. Moriscot, Physicochemical alterations and toxicity of InP alloyed quantum dots aged in environmental conditions: A safer by design evaluation. NanoImpact. 14 (2019) 100168. https://doi.org/10.1016/j.impact.2019.100168

[79] J.M. Yang, X.W. Hu,Y.M. Liu, W. Zhang, Fabrication of a carbon quantum dots-immobilized zirconium-based metal-organic framework composite fluorescence sensor for highly sensitive detection of 4-nitrophenol. Microporous Mesoporous Materials 274(2019)149-154. https://doi.org/10.1016/j.micromeso.2018.07.042

[80] X. Ren, J. Wei, J. Ren, L. Qiang, F.Tang, X. Meng, A sensitive biosensor for the fluorescence detection of the acetylcholinesterase reaction system based on carbon dots. Colloids Surfaces B Biointerfaces. 125 (2014) 90-95. https://doi.org/10.1016/j.colsurfb.2014.11.007

[81] G. Jiang, T. Jiang, H. Zhou, J. Yao, X. Kong, Preparation of N-doped carbon quantum dots for highly sensitive detection of dopamine by an electrochemical method. RSC Advance 5 (2015) 9064-9068. https://doi.org/10.1039/C4RA16773B

Green Synthesis and Emerging Applications of Frontier Nanomaterials   Materials Research Forum LLC
Materials Research Foundations 169 (2024) 275-314   https://doi.org/10.21741/9781644903278-11

Chapter 11

Ti-based nanomaterials and their potent applications

Alpa Shrivastava[1], Meena Chakraborty[2], Sunita Sanwaria[3], Ajaya Kumar Singh[3,4]*

[1]Indira Gandhi Govt PG College Vaishali Nagar Bhilai, Distt. Durg (C.G.), India

[2]Govt College Bori, Distt. Durg (C.G.), India

[3]Govt. V.Y.T. PG Autonomous College Durg (C.G.), India

[4]School of Chemistry & Physics, University of KwaZulu-Natal, Durban, South Africa

Abstract

Nanoparticles (NPs) have become center point of interest for researchers all over the world because of its enthralling properties and wide range of application in various fields such as chemical energy, healthcare industry, cosmetic industry etc. Titanium is a transition metal which is non-toxic, economical, biologically inactive, chemically inert, having large surface area, good catalytic activity and great absorption capacity in the near-IR region. Therefore, areas of applications and new possibilities are increasing. At nanoscale we can exploit physical properties different than macroscale Titanium. Green synthesized Ti-based NPs research is gaining momentum as better pollution free alternative. This review will focus on methods of preparation, characteristics, application, benefits and challenges in use of Ti-based NMs in different fields.

Keywords

Titanium Oxide NPs, Titanium Nitride, Titanium Carbide NPs, Nanoparticles in Anticancer Therapy, Nanoparticles Antimicrobial Property

Contents

1.   Introduction

Nanotechnology is combination of techniques and process for manipulating atoms and molecules at nanoscale to obtain nano sized materials with novel functionalities and enhanced features. Nanoscience has created revolutionary change in the world of research because of its low-cost production, environment friendly nature, great strength, long-lasting performance, low weight and specific functioning. Because of these unique properties, NMs have wide range of applications in the field of medicine, electronics, agriculture, food and textile industry, energy production and environment conservation. In nature, open fire, volcanic activities, precipitation and bioreduction are the activities which produces NMs. Bacteria such as *Lactobacilus sp., Schwanella sp., Stapylococus aureus*, yeast such as *Hansenula anomala, Sacharomyces boulardii, Sacharomyces,* fungi such as *Verticillium sp., Aspergillus terreus, Fusarium oxysporum* and algae such as *Botryococcus sp., Coelastrum sp., Coelastrum sp.* are the examples of natural reducers [1], Whereas there is a huge list of NMs of carbon, gold, silicon, titanium and many other elements, which are made by human through top-down approaches like electron beam, photo ion beam or X-ray lithograph cutting, etching, grinding, ball milling or bottom-up approaches like precipitation, hydrothermal synthesis, template assisted sol-gel, electrodeposition [2] (**Figure 1**).

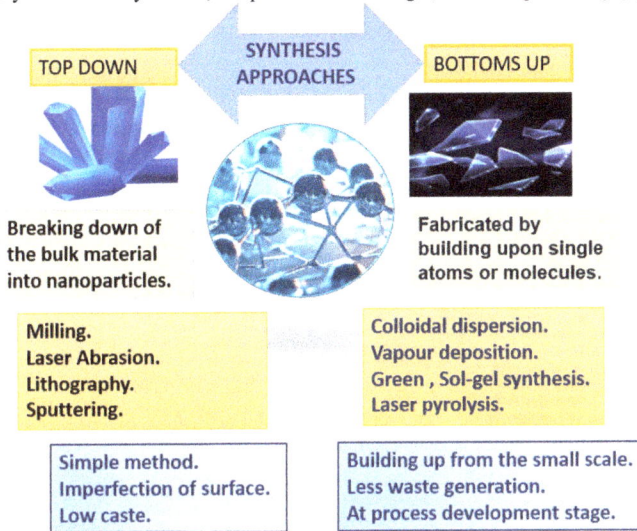

Figure1: Nanoparticle synthesis approaches

In the world of nanoscience, Titanium-based NPs (Ti NPs) has drawn attention of researchers because it is non-toxic, low in cost, chemically inert, enhanced surface, appreciable catalytic activity, absorption capacity in infrared region, biologically inactive material [3]. These properties made Ti-based NPs suitable for bio-applications compared to other heavy metals such as gold, silver, tungsten, bismuth etc. [4]. Moreover, Ti-based nanomaterials (NMs) are more appropriate for a variety of applications due to their customizable composition, structure, and size modifications [5,6].

Ti-based NPs broadly include oxides, carbides, nitrides, sulphides, hydrides of titanium, along with organic coordination compounds of titanium, and their NCs [7]. Titanium carbide based NMs possess great wear resistance, good conductivity and narrow range particle size distribution. They are useful as coating material for drills, tools, saw blades, etc., alloy additives, surface materials, abrasion resistance and heat control in spacecraft [8].

Titanium nitride can be prepared by the reaction of carbamide with chloroform and titanium chloride or by pulsed laser ablation of aqueous and organic liquids [9,10]. At par hardness and chemical stability at high temperature, absorption in IR region and UV shielding makes it useful for various applications [11]. It is possible to create TiH_2 NPs by heating TiH_2 powder in octylamine and then wet grinding. TiH_2 NPs can be used as conductive ink because of their air stable property [12].

Because of their high quantum efficiency, surface area, chemical inertness, and stability, $TiO2$ NPs find applications in environmental photocatalysis, photoelectricity, dye-sensitized solar cells, biomedical implants, cancer therapy, medications, microbial assay, pigments in paints and polymers, dental care, cosmetics, and agriculture [13-16]. Because of the photogenerated holes' great oxidizing power and their chemically inactive and benign nature, TiO_2 nanoparticles can be employed as a photocatalyst for the breakdown of dangerous substances [17]. The three polymorphic forms of titanium dioxide are anatase, rutile, and brookite. Brookite is only stable at low temperatures and is very unstable at room temperature. On the other hand, rutile is the stable form of TiO_2 and is frequently used as a white pigment in paint. Anatase, on the other hand, has a crystalline structure and is used as a photocatalyst [13, 17-18]. Nanostructured titanium dioxide can be prepared by methods like sol-gel, hydrothermal, solvothermal, liquid-phase low-temperature, microwave, template, chemical vapor deposition, sonochemical, electrodeposition, etc. This chapter will summarize the recent advances in preparation properties and applications of Ti-based NMs.

2.   Potential of TiO_2-based NMs

Titanium oxide TiO_2 is available naturally and also produced synthetically, due to wide range of application (Table1). Its features, which include low cost, non-toxicity, resistance to corrosion, higher refractive index, photocatalytic activity, and ability to absorb UV radiation, have made it indispensable in a variety of applications, including food preparation, cosmetics, sunscreens, plastics, paints, inks, and paper; food coloring and food additives; air purification; cars, electric appliances, and medication administration [19-20] Studies revealed that about 25% of matter in nano-database had reference of $TiO2$ NPs[21]. Along with all these benefits, NPs also have negative impact on environment [22]. Excessive use and large production can cost a lot to air, water and soil component of environment [23-24] hence, proper decontamination is essential in applied field.

Table 1. Potential application of TiO_2 based NMs

Field	Applications	Reference
Agriculture	Removal of toxic pollutants & heavy metals from soil & water	[25]
Medical	In the treatment of cancer	[26]
Cosmetic	Sunscreen	[27]
Food	Additives & coloring agents	[28]
Transportation	Light & side mirror of the vehicle, removal of oxides of nitrogen	[29]
Infrastructure	Paints, Tiles, Glass coatings	[30]
Electrical	Solar batteries	[31]
Air purification	Degradation of pollutants	[32]

Green Synthesis and Emerging Applications of Frontier Nanomaterials   Materials Research Forum LLC
Materials Research Foundations 169 (2024) 275-314   https://doi.org/10.21741/9781644903278-11

3.   Preparation of TiO₂-based NMs

TiO2-based NMs can be produced chemically or biologically, and each process has pros and cons of its own.

Table 2. Reported methods for synthesis of TiO₂-based NMs

Synthesis procedure	Synthesis Technique	Advantages	Disadvantages	References
Chemical	Sol-gel method Solvo-thermal method Co-precipitation method Hydrothermal method Micelle and Inverse Micelle Method	Easy synthesis and on size and shape control of NPs	Expensive, required high temperature, pressure and energy, eco-toxicity and environmental sustainability issues, time taking, low production	[33-36]
Biological	From the extract of green plants, from cellulose fibers, albumen, starch, and other biological sources; from bacterial and fungal extracts, etc.	Eco-friendly, cost effective, dispersible structure.	-	[37-41]

Various methods used for preparation of TiO₂ NPs are as follows:

3.1   Chemical methods

3.1.1   Sol-gel method

This wet method is effective, low cost and widely used for preparation of metal NPs. In this process TiO₂ NPs are prepared by hydrolysis and polycondensation reactions of titanium alkoxide, which leads to form oxopolymers and lead to formation of oxide network. The process is dependent on numerous reaction parameters, including Temperature, pH of the medium, titanium alkoxide reactivity, solvent type, and rate of poly-condensation and hydrolysis. [42-45].

3.1.2   Solvo-thermal method

Solvothermal synthesis process refers to process involving reaction in non-aqueous solvents [42]. High boiling point organic solvents enhance synthesis possibility at elevated temperatures with higher control on size, shape and crystal formation [42,46]. This method is suitable for preparing nanostructures with narrow size distribution and disparity [42,46-47]. This process uses several solvents at different pressure ranges to produce nanoscale metals, ceramics, semiconductors, and polymers.

3.1.3 Hydrothermal method

The size of NPs, the specific surface area, the area-to-volume ratio, and the temperature all depend on the wet preparation procedure that entails dissolving and recrystallizing under predetermined pressure and temperature [42-43]. Titanium oxide-based nanowires, nanotubes and nanorods can be prepared by this method. Aligned Titania fibrilliar structure with desired length has been reported using this preparation system [48].

3.1.4 Micelle and inverse micelle method

When surfactant molecules group together, long chain micelles with a hydrophilic head and a hydrophobic tail are formed. Reverse micelles are globular aggregates created when surfactants self-assemble in a polar solvent, while normal micelles are those created in water [49-50]. In the range of 1–50 nm and 200 nm, respectively, TiO_2 forms micelles with Triton X–100 and AOT (bis(2-ethylhexyl) sulfosuccinate sodium). It is discovered that the surfaces of TiO_2 and Triton X-100 micelle are smooth, but the surfaces of TiO_2 and AOT micelle are somewhat rough [51].

3.1.5 Microwave method

High frequency electromagnetic energy can be used to process the dielectric substance. Microwave heating occurs at frequencies in the range of 900 to 2450 MHz. In order to transfer energy from the microwave to the substance, ionic component mobility induces a current to flow inside the material at lower microwave frequencies. At higher microwave frequencies, energy absorption primarily occurs from dipole molecules, which are reoriented by a microwave electric field. [13]. Microwave synthesis has been employed for synthesis of various TiO_2 NPS [52-53]. When titanium slags were activated by microwave radiation and Na_2CO_3 was used as additive, they produced 43.5 nm sized rutile TiO_2 powder with highest value of crystallinity (99.21 %) [53].

3.1.6 Template method

Template method is easy and cost-effective process involving preparing NMs with controlled morphology and structure. The template method can be used for synthesis of nanorods, nanotubes, or porous materials of titanium dioxide [54]. Selvaraj et al., prepared TiO_2 NPs by using soft templet method from titanium isopropoxide. Cationic surfactants molecules, such as CTAB (cetyl trimethylammonium bromide), SDS (sodium dodecyl sulfate), and DTAB (dodecyl trimethylammonium bromide), were used as templates. A 4:1 volume ratio of deionized water and ethanol was used to dissolve the required amount of CTAB. Dropwise addition of Titanium isopropoxide with continuous stirring resulted in gel formation and separated by centrifugation and calcinated to obtain soft template. Same procedure can be adopted for the preparation of TiO_2 NPs by using SDS and DTAB as templates. The size of the synthesized NPs of TiO_2 was 10 - 14 nm [54-55].

3.1.7 Vapour deposition method

Any procedure that condenses materials in the vapour state into a solid state is referred to as the vapour deposition method. Usually, these procedures are employed to create coatings to alter the wear resistance, corrosion resistance, mechanical, electrical, thermal, and optical characteristics of different substrates [13]. These methods have drawn a lot of interest recently as a way to create various TiO_2-based NPs. The majority of vapour deposition processes take place in vacuum chambers. The process known as physical vapour deposition (PVD) is what happens when there is no chemical reaction. If not, the phrase chemical vapour deposition, or CVD, is employed [56–58]. On glass substrates, Zhang & Li synthesized pure anatase using the CVD technique. Using this technique, they created 300 nm-thick TiO_2 (anatase phase) thin films on a quartz glass substrate [56]. Carbon-based TiOx-DLC coatings were prepared on silicon substrates by Jedrzejczak et al. utilizing CVD and a capacitive linked radiofrequency of plasma discharge in a vacuum system [57]. High performance TiC NP was created by Cao et al. by dissolving Ti powder in melted salt and reacting it with diamond NPs to create TiC NPs. NPs smaller than 10 mm can be prepared easily and affordably with this approach [59].

3.1.8 Electrochemical method

Anandgaonker et al., prepared tetragonal structured TiO_2 NPs by electrochemical method using tetra propyl ammonium bromide salt as stabilizing agent in an organic medium of tetra hydro furan and acetonitrile by optimizing current density [60]. The size of NPs can be controlled by current density, solvent polarity, distance between electrodes and concentration of stabilizers. In a scalable green electrochemical approach, electrolyte solution containing a surfactant in water reduced use of toxic chemicals. Three surfactants—tetrapropylammonium bromide (TPAB), tetrabutylammonium bromide (TBAB), and cetyltrimethylammonium bromide (CTAB)—were used in the synthesis of e-smart TiO_2 nanomaterial (e-TiO_2/CTAB nanorod). This unique nanostructure can transform into polar and nonpolar solvents [61].

3.1.9 Oxidation methods

These techniques entail anodization or the use of oxidants to oxidise titanium metal. The process of anodizing titanium sheet in 0.5% hydrogen fluoride at a voltage of 10–20 V results in aligned TiO_2 nanotubes, the diameter of which, can be adjusted [62]. In a different investigation, anodized titanium plates were heat treated at 500 °C for six hours in an oxygen atmosphere, which produced crystallised TiO_2 nanotubes [63]. TiO_2 nanorods have also been shown to develop when hydrogen peroxide is directly applied to titanium metal.

3.1.10 Sonochemical method

The high-surface area ranges of NMs covered by the ultrasonic approach includes transition metals, carbides, oxides, alloys, and colloids. The synthesis of titania-based NMs can be prepared using the highly selective and advanced sonochemical process. The power and frequency of the sonicator bath at varying temperatures determines the preparation. By adjusting the sonicator bath's power according to temperature, we can produce the required NMs product [13]. Titania nanoparticles were created by vigorously swirling pellets dissolved in NaOH at room temperature and then irradiating them in an ultrasonic bath. The required product was obtained by centrifuging the resulting precipitates, repeatedly washing and decanting them with deionized water, and finally drying them [42, 64-65].

3.2 Biological and green methods

Chemical methods tried for preparation of nanomaterials of TiO_2 can harm the environment, requires high energy and are cost, whereas biological methods are much more economic and environment friendly. Principles involving green methods include use of less hazardous materials, design of safe chemicals, use of safe solvents, reduction of derivatives and pollution prevention. Simple method from plant preparation is represented in **figure 2**.

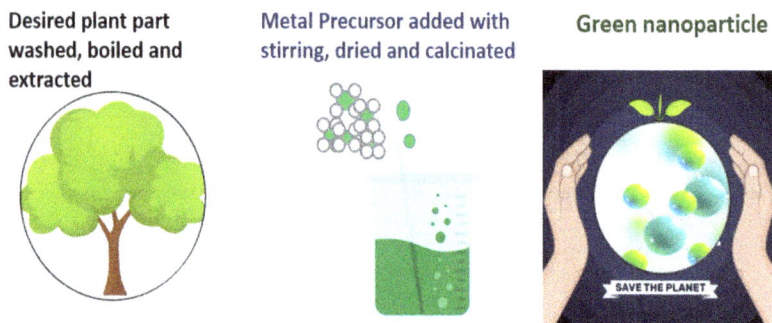

Desired plant part washed, boiled and extracted

Metal Precursor added with stirring, dried and calcinated

Green nanoparticle

SAVE THE PLANET

Figure: 2. Green synthesis of NPs.

3.2.1 Preparation from extract of green plants

Proteins, carbohydrates, enzymes, phenolic acids, and alkaloids found in green plants are available for use in the reduction and stabilization processes involved in the manufacture of TiO_2 NPs. researchers described making TiO_2 NPs using a cinnamon powder extract [66], from leaves of *Syzygium cumini* [67] from *Moring oleifera* extract [68], from orange peel extract [69] seed extract from *Cucurbita pepo* [70] and *Trigonella foenum* leaves [38]. TiO_2 NPs prepared was stable, cost effective, non-toxic with decent light absorption and refractive index [71].

3.2.2 Preparation from extract of bacterial and fungal

A large group of bacteria can be utilized to prepare NMs and alteration of genetic code and culture possibilities make feasible tool for the process. For the biogenesis of TiO_2 NPs, bacterial and fungal extracts have also been used. The metabolites included in microbial biomass can aid in the stabilization and bio-reduction of metal oxide, particularly TiO_2 NPs [37]. Furthermore, fungal extracts have drawn more attention than bacterial extracts because of their advantages over bacterial extracts, including simple extraction, cost-effective large-scale manufacture, and enhanced surface area of synthetic TiO_2. Since fungi contain the enzymes and metabolites necessary for the reduction of bulk salts into constituents of ions, it is possible to generate different sized TiO_2 nanoparticles (NPs) from fungal extract. When making TiO_2 nanoparticles using bacterial extract, baker's yeast [72], *Streptomyces sp.* [73], *Acinetobacter baumannii*, *Aeromonas hydrophila*, *Lactobacillus sp.* and *Bacillus subtilis bacteria* were demonstrated for synthesis of TiO_2.

3.2.3   Preparation from other biological sources

In addition to plants and microorganisms, other biological sources are also used in the synthesis of TiO_2 NPs. For instance, it has been claimed that starch can be utilized to easily produce TiO_2 with strong photocatalytic properties in an environmentally acceptable way [39]. Similarly, cellulose fibers were used to create uniform TiO_2 nanowire. It also showed the advantages of regaining cellulose fibers without changing their shape [40]. In a different study, Bao et al. synthesized TiO_2 with less environmental impact by using the regulated form and crystal phases of albumen [42]. Another benefit of green synthesis is that eggshells were used as a biological material in a simple and biomimetic way to form a hierarchical tube network of TiO_2. In a different environmentally friendly experiment, rutile TiO_2 was produced using chitosan and β-cyclodextrin, but anatase TiO_2 was created using soluble starch and shown more photocatalytic activity, indicating a higher level of usefulness and environmental friendliness [74].

4.   Properties of TiO_2 NMs

Many attractive features of TiO_2 NPs such as high light-conversion efficiencies [75], chemical stability, thin film transparency, and economic viability have increased its application in various fields of environmental remedies and health care. Different properties of TiO_2 NMs are as follows:

4.1   Crystal properties

There are three different crystalline structures in TiO2 particles, anatase, brookite and rutile [76]. While brookite has an orthorhombic structure, anatase and rutile phases have tetragonal crystal structures. [77-79]. At 0^0 K, the rutile phase is less stable than the anatase phase, whereas the anatase octahedral structure is somewhat larger than that of rutile [80]. The anatase phase is more advantageous for solar cell applications due to its low density, high electron mobility, and low dielectric constant [77]. When the temperature rises, the anatase phase changes into the rutile phase, which is more stable than the anatase phase for particles larger than 14 nm [81].

4.2   Optical properties

TiO_2 NPs are preferred choice to be used in optical devices pertaining to mechanical, crystalline and aqueous durability, porous and particle morphology great transparency in the visible region [82-83]. When the particle size is reduced several nanometers scale, due to the large surface-to-volume ratio, optical properties can be altered as required [84]. The smaller size of TiO_2 NPs determines the transparency of UV visible light. Of the three crystalline phases of TiO_2 NPs, the anatase phase has the lowest ability for oxygen adsorption, increasing the degree of hydroxylation and, to a lesser extent, raising the Fermi level, which increases the reactivity of the material towards light [42]. The energy of UV radiation is sufficient to cross the band gap between the valance band and conduction band when it is incident on the anatase phase. As a result, an excited electron moves from the valence band into the conduction band, creating a positive hole in the valence band. The strong photocatalytic characteristics of the anatase phase are caused by these charge carrier electrons and holes. [85]. The rutile crystalline phase of TiO_2 NPs is considered as poor photo catalyst because in their structure, large amount of negatively charged electrons and positively charged holes gets recombined with each other [86], but it showed pretty good refractive index and optical absorptivity, which made it useful for optical communication devices [87].

4.3　Electrochemical properties

The electrochemical properties of TiO_2 NPs are significantly influenced by their individual particle size and crystal structure. The indirect energy band gaps of the rutile, anatase, and brookite phases are 3.02, 3.2, and 2.96 eV, respectively [88-90]. According to studies, the Fermi level is around 0.1 eV lower than the anatase. Between anatase and rutile, the anatase phase has a reduced electron effective mass, which increases charge carrier mobility and makes it appropriate for the manufacturing of optoelectronic devices. [84].

4.4　Thermal properties of TiO_2 NPs

Thermal conductivity of a substance is its ability to conduct heat through it. Research shows that TiO_2 NPs fluids have heat transfer abilities. Study conducted on rutile and anatase phase of TiO_2 NPs found a relationship between structure and thermodynamic properties. The development of both rutile and anatase NPs is mostly ascribed to low index surfaces of crystal structures, which are recognized for their stability. Thermodynamic features can be attributed to total energy, surface energy, radial distribution function, and sphericity of the NPs. In comparison to rutile, anatase is more thermodynamically stable at the nanocrystalline phase, as shown by the difference in surface energies of these two phases. Because titanium's great orderliness and coordination decrease with temperature, the surface energy of both phases drops as the temperature rises in a vacuum [91].

In a different investigation, Alias and Ani discovered that TiO2 NPs in an ethylene glycol (EG) base fluid demonstrate good heat conductivity when gum arabic is used as a surfactant. They discovered that the presence of surfactant improved the stability of nanofluid samples by altering the surface area of TiO_2 NPs, and that Brownian motion's effect increased thermal conductivity by raising NP concentration and temperature [92]. Properties of TiO_2 NPs are summerized in Table 3.

Table 3: Properties of TiO_2 NPs

Property	Anatase	Rutile	Brookite	References
Crystal structure	Tetragonal	Tetragonal	Orthorhombic	[77-79]
Molecule (cell)	2	2	4	[85,93]
Density (g cm^{-3})	3.79	4.13	3.99	[85,93]
Energy band gap (eV)	3.2	3.02	2.96	[88-90]
Optical properties	Good photo catalyst	Poor photo catalyst	-	[85-86]
Conductivity type	n-type	n-type	n-type	[85,93]
Thermal properties	Heat conducting fluid	Heat conducting fluid	-	[91]

5.   Applications of Titanium NPs

Nanotechnology field is fast emerging with newer applications are being reported. Easily availability, non-toxicity and stability has promoted scientists to discover new forms of titanium NPs. Ti NPs are suitable for industries dealing with toothpaste, pharmaceuticals, paint coatings, inks, food processing, cosmetics and textile. Major research is centered around titanium dioxide although other molecules like titanium nitride, titanium carbide etc. also find importance in academician's and industry interest. Some applications of TiO_2 NPs is graphically represented in figure 3.

Global TiO_2 NPs market is valued to be US\$ 242.2 million in 2023 and it is anticipated to grow at CGAR of 6.2% to reach 442 million US\$ by 2033. TIN NPs is also gaining importance with application in plastic packaging materials, fuel cells, therapeutic thermal applications [94].

High purity, strong wear resistance, a narrow particle size distribution, and a large specific surface area are attributes of nano TiC. Furthermore, there is good electrical conductivity in the product. The major application for Nano Ti Care is as a coating material for tools, saw blades, drills, etc. Applications of TiC NPs are represented in **Figure 4**. Titanium nitride (TiN) attracts attention due to its resonance located in the visible and near-infrared range works as an alternative plasmonic material to gold and has been widely used in microelectronics, biomedical devices and food-contact applications. TiC and TiN applications are listed in table 5.

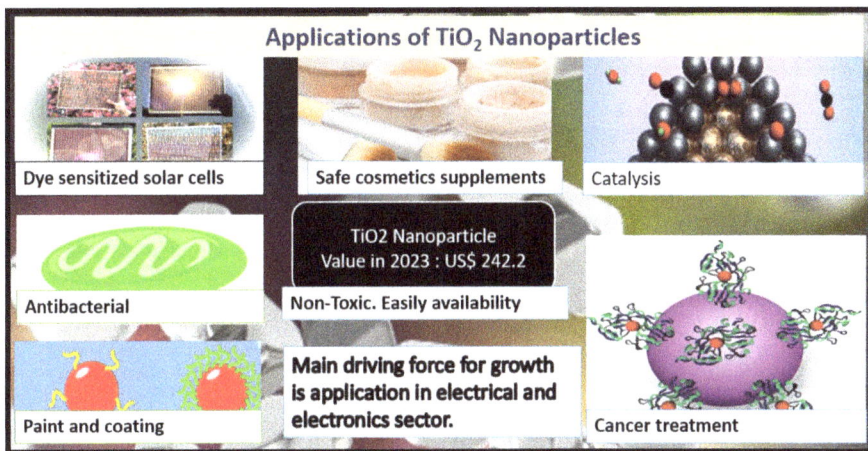

Applications of TiO₂ Nanoparticles

Dye sensitized solar cells	Safe cosmetics supplements	Catalysis
Antibacterial	TiO2 Nanoparticle Value in 2023 : US\$ 242.2 Non-Toxic. Easily availability	
Paint and coating	**Main driving force for growth is application in electrical and electronics sector.**	Cancer treatment

Figure 3: Applications of TiO₂ NPs

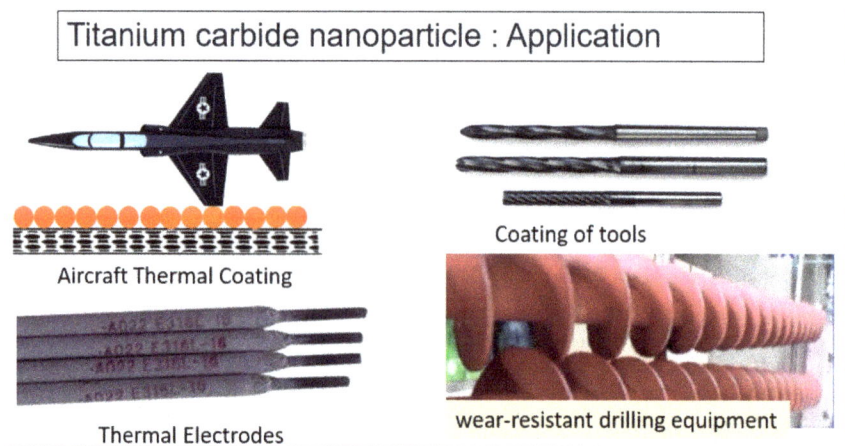

Figure 4: Application of TiC NPs

5.1   Application as nanomedicine in anticancer therapy

Nanotechnology supports administration of targeted therapeutic, imaging system and provide the required patterning of surfaces morphology for drug delivery and release. Thousands of biomolecules are in research stage alone or in combination for desired therapeutic and biological targeted therapy. Apart from other applications cancer therapy is important aspect of Ti NPs, these NPs are safe, cost effective, control post-operative infections, easily available and proven multifunctional in oncology. NPs identify and bind exclusively to cancer cells when they form a bond with the drug carrier to target organs. When concentrated visible light is sent towards the impacted area, the localized Ti NPs respond by generating free oxygen radicals (FOR), which disrupt mitochondrial processes and send out a signal to begin the death of cancer cells. Mechanism of action of NPs of titanium is presented in figure 4.

Near infrared-mediated phototherapy involving Ti NPs is a promising alternative to radiotherapy with less side effects to the body [95]. The unique tumor imaging and treatment properties can be attributed to enhanced imaging due to large surface accumulation of Ti NPs in affected organ [96], good photoacoustic imaging properties for detection, conversion of oxygen into promoting ROS-mediated cancer therapy. Ti NPs can be applied alone or in combination for variety of treatment methods like photo-thermal therapy, photodynamic and sonodynamic therapy, drug release, cell imaging, biosensors for biological assay. Kim et al., 2019b reported sonodynamic chemotherapy which exhibits ultrasound mediated drug delivery via ROS generation from sonosensitization. DOX -coordinated titanium dioxide NPs capsuled with polymeric phenyboronic acid facilitated DOX release by ROS under US radiation. Combination of DOX and phenyboronic acid worked for tumor cells growth inhibition. Success of coordinated NPs system pertains to optimum blood compatibility, ultrasound stimulation for drug release and targeted therapy [97].

Chemotherapy for cancer treatment requires balancing effective anticancer activity and toxicity to human cells for prescribed medicine dose. It's interesting to note that in the presence of an acidic milieu, Ti NPs connect to cancer cells quickly [98-99]. pH of human body is 7.4 and extracellular tumor pH is 6 to 5, suitable for release of daunorubicin through drug delivery system. This concept of hiding drugs with NPs delivery systems can decrease the toxicity of drugs [100]. pH-responsive drug delivery protocol by Zhang et al., suggested controlled release of chemotherapeutic Daunorubicin drug, based on TiO_2 enhance chemotherapeutic efficiency [99].

TiO_2 NPs are proving to be alternate system for photosinsitizing or imaging cancer therapy. Reed et al., utilized TiO_2 NPs coated glycoprotein transferrin –as photosensitizing system in Cerenkov radiation induced therapy, and reported more effective for suppressing multiple myeloma cell line [101]. Xian et al. reported TiS_2 NPs provided good contrast in optical imaging process with enhanced sensitivity and resolution [102], Similar results were shown by TiN [103], Ti_3C_2 [104], black TiO_2[105]. Nb-doped TiO_2 also facilitated efficient imaging by absorbance in near IR [106].

Combined or coating form of NPs are being utilized to increase efficacy of treatments. Au-TiO_2 NPs coordinated with DOX showed good diffusion into the tumor's vicinity with cancer cell loss up to 70 %, Confirmed by aptosis studies [107].

In an additional trial, the doxorubicin loading content (~25.0–35.0%) of the hyaluronic acid modified NMs and DOX loaded nano-MOFs was attributed to the creation of hollow structures and π-π stacking bonds. The complex demonstrated non-toxicity and biocompatibility both in vivo and in vitro, showing the ability to evade lysosomal degradation, enhanced intracellular drug accumulation for anticancer efficaciousness, and reduced DOX side effects [108]. It has also been reported that titanium nanosheets can be used as a very stable and biosafe DOX medication carrier for coordinated photochemotherapy of cancer [109]. According to Abdel Ghany et al., the treatment of human malignant amniotic cells (WISH) and breast cancer cells (MDA-MB-231 and MCF-7) with erlotinib and vorinostat loaded TiO2 NPs suppressed cell proliferation [110].

In a dose-specific drug, Ag-doped TiO2 NPs decreased the viability of human liver cancer (HepG2) cells [111]. Better photothermal efficiency and lower cost are found in the optimized SiO2-coated TiN compared to commercial gold, which is commonly used in cancer treatments [112]. Copper acetate and copper acetylacetonate, two cancer therapeutic compounds, are efficient against various malignant cells. In a study, Lopez et al. found that the copper complexes and nano-TiO2 loaded individually and in combination are more cytotoxic than commonly used cis-platin [113]. For use as drug carrier, functionalized TiO_2 NPs loaded with, copper complexes and cell viability tests showed survival of cell lines up to 90% resulting in good alternative for treatment. Novel single near infrared laser-induced black titania nanotherapy used in photodynamic and photothermal treatment of mice bearing tumors, showed suppression within 2 days with no recurrence up to 20 days' investigation [114]. Surface-coordinated PEG and TIN NPs demonstrated adequate blood system stability, and when exposed to an 808 nm laser, the mouse tumor region's temperature rose quickly enough to effectively kill cancer cells [115]. Some suggested research work in cancer therapy is summarized in table no 5.

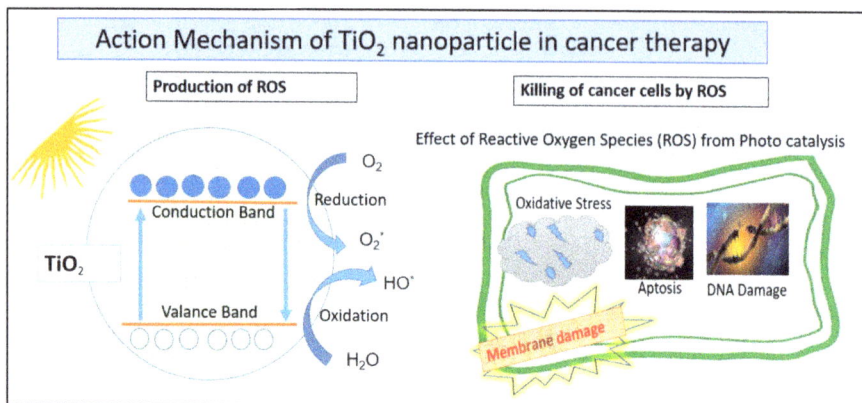

Fig 5: Action mechanism of TiO₂ NPs in cancer therapy

Table 4: Titanium NPs in cancer therapy

SN	Type of Therapy	Used Ti NPs form	Research results	Reference
1	Photo-thermal therapy of tumors	PEGylated TiO_2 NPs in vivo trials in animals	Average tumor size in the mice sharply decreased.	[116]
2	Triple combination therapy triggered by NIR (photodynamic, photothermal, and chemical.	Composite Doxorubicin hydrochloride @ TiO_2-x system	Tumor growth is significantly inhibited and the solid tumor is ablating both in vitro and in vivo.	[117]
3	Light-based cancer therapy called Cerenkov radiation induced therapy (CRIT)	25 nm diameter TiO_2 NPs with the glycoprotein transferrin.	In vitro improved cell killing for multiple myeloma cells from 23% to 57%.	[101]
4	Cancer therapeutic applications	TiO_2 NPs at 50 and 100 µg/ml concentrations	displayed HT29 cells' highest level of cytotoxicity. HT29 cell apoptosis was verified.	[118]
5	Breast and Skin Cancer treatment	TiO_2 NPs	enhanced UV-A radiation's ability to kill skin and breast cancer cells.	[119]
6	Threaupitic induce G2/M n breast cancer cell cycle arrest	Vorinostat-loaded titanium oxide NPs (anatase).	Capable to induce cells to commit apoptosis.	[110]
7	Cancer therapy	TiO_2 NPs	Significant increase in apoptosis varying between 25-40% by 24 h.	[120]
8	Cytotoxicity and apoptosis on the highly malignant cancer cells.	Silver-modified Nanostructured TiO_2	Inhibition of cell proliferation on cultured breast cancer and aptosis.	[121]

9	Chemotherapeutic effects against cancer	TiO_2 NPs	increased ROS levels and autophagy blockage in AGS stomach cancer cells, which promoted the apoptotic effects of chemotherapy.	[122]
10	Therapeutic effect against cancer and antibacterial activity	TiO_2 NPs prepared using cynodon dactylon leaf extract	growth of E. coli is inhibited, and A549 lung carcinoma is more resistant to cancer-causing agents.	[71]
11	Breast Cancer Inhibition	TiO_2 NPs synthetized using aqueous leaf extract of the tropical medicinal shrub Zanthoxylum armatum as a reducing agent	Breast anticancer activity demonstrated both ex vivo in mammary carcinoma cells and in vivo (mice). Cost-effective, efficient, and safer than DOX.	[123]
12	Targeted Brain Cancer Therapy	1,2-epoxy-3-isopropoxypropane-capped TiO2 NPs covalently conjugated with antibody.	directing attention away from healthy brain cells and toward malignancy. started the process of generating ROS.	[124]
13	Human ovarian carcinoma cells A2780 in vitro therapy	TiO_2 NPs	Increasing cellular viability is beneficial for specific cancer treatment.	[125]
14	Examining the biological effects on endothelial cell lines from the umbilical vein and colorectal cancer.	TiO_2 NPs	HCT116 cells showed improved overall cell survival along with decreased levels of Bcl-2 and Caspase 3. cancer cells' programmed death.	[126]
15	Localized cancer treatment	Titania (TiO_2) nanotube arrays in 3D titanium wire-based implants created by nano-engineering.	Apoptosis and a regression in tumor burden within the first three days of implant.	[127]
16	Sonodynamic-immunotherapy for tumor therapy	$CaCO_3$@Pt-TiO_2 NCs	To overcome tumor hypoxia, increased ROS quantum yield of TiO_2 promoted the breakdown of H_2O_2.	[109]
17	Medication Delivery Systems for Cancer Treatment	Diamond-Shaped Mesoporous Titania Nanobricks.	Superior tumor inhibition, improved drug loading, pH responsiveness, and excellent biocompatibility.	[128]

5.2   Titanium NPs in antimicrobial activity

Development of resistance of microorganism bacteria and viruses has emphasized continual research and discovery of new potential and active molecules. NMs are actively included in counter strategies [129]. The NPs as antimicrobial agents, can penetrate bacterial membranes, disorientate biofilm and suitable drug carrier for antibiotics [130]. TiO_2 NPs showed inhibition of bacteria due to microorganism comparable nanometer scale size and oxidizing power for disrupting bacterial cell functions [131]. Understanding mechanism and protocol of microbial death by NPs is yet to clearly established [132]. A few proposed mechanisms of action are stress, oxidative assault, cell wall disruption, and damage to DNA. Naturally occurring antioxidants such as ascorbic acid, carotene, and tocopherol, as well as enzymatic antioxidant defense systems like catalases and superoxide dismutase, inhibit lipid peroxidation and the effects of Ti NPs generated ROS radicals like $OH^{2\cdot-}$ and $OH^{\cdot}$ result in redox reactions that ultimately lead to the death of microorganisms by changing and destroying potential structures like cell walls, cell membranes, DNA, enzymes, and metabolic pathways [133].

Increase in ROS production is supporting mechanism for these therapies. Antimicrobial photocatalysis of TiO_2 NPs supported with sodium iodide resulted in threefold increase in antimicrobial properties [134]. Reactive oxygen species produced by Ti NPs include superoxide, oxygen radicals, and hydroxide radicals, which causes bacterial cell walls to become disoriented and inhibit their ability to breathe [135]. Gram-positive and gram-negative bacteria of Staphylococcus aureus and Escherichia coli were shown to be effectively suppressed by the NPs [60]. Jatoi et al., reported cellulose acetate nanofibers with Ag NPs and TiO_2 NPs frame proved good bactericidal for E. coli and S. aureus [136]. Similarly, several researchers reported leakage of the intracellular contents of bacterial cell by endocytosis and toxicity of NPs [137-138]. Drawback with metal NPs is toxicity in target organs after release of toxic NPs with adverse health effect [139].

In a greener approach, TiO_2 NPs modified with herbal plants Withania somnifera (Ashwagandha), Eclipta prostrata (Karisalankanni) and Glycyrrhiza glabra (Athimathuram) treated against Staphylococcus aureus and Streptococcus mutants, damaged cell walls of screened bacteria [140]. Comparable studies on the antibacterial action of green produced Ti NPs have used Trigonella foenum-graecum leaf extract [38].

New tests have been conducted to see if NMs may function as implants on their own or in conjunction with traditional materials. Additionally, the implants need to be antimicrobial and biocompatible to avoid post-operative infections that could result in implant failure. [141]. Bone/dental implants of TiO_2 nanotubes embedded on Ti6Al4V and structured on Silver NPs enhanced antimicrobial property. Biocidal activity of S. aureus decreased upto 80% and growth was hampered by this composite implant [142]. Stainless steel antimicrobial coating of TiO_2 NPs / Trimethoxy(propyl)silane showed better hydrophobicity [143]. ZnO-decorated titanium carbide NPs faciliated degradation of 99.9% Methylene blue dye in 40-minute time frame with excellent antibacterial activity [144]. Silver doped Ti NPs are reported to be good additives for dental polymer application [145], poly methyl methacrylate acrylic resins embedded with SiO_2 and Ti NPs on cariogenic bacteria [146] new salicyl hydrazido chitosan derivatives coated with titanium dioxide nanoparticles: anti-biofilm activity [147], antimicrobial spray containing TiO_2 NPS into rats [148]. Inhibition of drug resistant bacteria on biofilm of metal NPs on implants is important parameter to counter infection and drug resistance in hospitals [149]. In bone therapy, Coating of TiO_2 NPs of metal implants in human body controlled bacterial growth on implants [150]. TiO_2 NPs restricted bacterial biofilm formation in the glass surface [151]. Ti NPs are excellent antimicrobial additives for ambulatory and nonambulatory medical devices. Silver coated NPs is effective against various strains found to act on DNA damage generating bacterial cytotoxicity [152]. After 30 minutes of treatment, the photokilling of two strains of Escherichia coli by radiation in anatase TiO_2 NPs suspension was described [153]. When potassium iodide, a harmless inorganic salt, is added to TiO_2 NPs stimulated by UVA, it can kill gram positive and negative bacteria six times faster and produce a long-lasting oxidized iodine moiety that continues to kill bacteria even after the light source is turned off [154].

5.3   Titanium NPs in agricultural applications

Nanotechnology in agriculture has gained importance in the last decade. Nanotechnology research has led to the development of efficient agro agents and nutrient delivery, less pesticide usages [155]. TiO_2 NPs increased chlorophyll synthesis, photosynthesis and plant dry weight in spinach [156]. Controlled hydroponic experiment for mineral uptake and chlorophyll content, reported increased uptake of micro and macro nutrient on maze plant [157]. Toxicity control in plant mass is another important application of Ti NPs. In an important experiment foliar spray of TiO_2 NPs decreased toxicity by reducing Cd content upto 2.5% and increased dismutase and transferase activity and regulated metabolic pathways [158]. When rice was treated with 500 mg of TiO_2 NPs kg−1, it grew longer and produced more biomass; however, when the dose was raised, it also produced more H_2O_2, lipid peroxidation, and electrolyte leakage [159]. Wang et al., suggested that the photocatalysis of the TiO_2 NPs driven by pulsed discharge plasma was capable of 88% of the p-nitrophenol removal efficiency. Enhancing the pulsed discharge voltage increased the removal efficiency of the pollutant [160]. In similar experiment TiO_2 NPs have been applied to degrade high molecular weight polycyclic aromatic hydrocarbons from soil contamination [161]. Žabar et al., reported photodegradation of pesticides imidacloprid and thiamethoxam in a photoreactor, with TiO_2 immobilized glass slides [162]. Recent reports suggested remediation of contaminated soil with nano-TiO_2, by light therapy or restriction of uptake of soil pollutants. Some examples include photodegradation of diphenylarsinic acid [163], Photocatalytic degradation of mecoprop and (4-chloro-2-methylphenoxy) acetic acid herbicides using nitrogen-doped TiO_2 suspensions [164], Photodegradation of 4-chlorophenoxyacetic acid using visible light-activated N-doped TiO_2 [165]. TiO_2 NPs used a robust sorption process to minimize the arsenic exposure of rice seedlings in a field experiment by 40–90%. Furthermore, TiO_2 NPs reduced the amount of iron plaque by 50–63% [166]. The risk of harm to health was decreased by reducing arsenic retention. It was discovered that silicon (Si) and titanium dioxide (TiO_2 NPs) effectively prevented arsenic poisoning in rice and enhanced plant development [167]. Qi et al., sprayed tomato plants with TiO_2 and under harsh light condition with improvement seen in photosynthesis [168]. Barley plants treated with Ti NPs resulted in longer vegetative phase growth and increased photosynthetic activity [169]. Similar investigations confirm a higher biomass output and grain yield in treated plants, based on these evidences. Less $nTiO_2$ was available at higher concentrations ($nTiO_2$ 500 mg kg−1) due to the tendency to produce agglomerates [170].

5.4   Titanium NPs in food preservation

Materials based on natural gas show promise in preserving and extending the shelf life of food due to their antioxidant and antibacterial properties. NPs in food preservation have been the subject of numerous research studies [171]. White button mushroom (Agaricus bisporus) is healthy nutrient rich but with low shelf life. Oxidation suppression of this mushroom is essential to preserve freshness and nutrition values. TiO_2 NP and chitosan films with active coating materials, thymol and tween as food preservatives increased shelf life up to 12 days [172]. TiO_2 NPs and litchi peel extract incorporated in chitosan structure proved to be novel packaging and coating for apple, inhibited respiration rate and weight loss [173]. It was found that employing environmentally friendly Chitosan, TiO_2 NPs, and sodium tripolyphosphate, cucumber shelf life could be extended by up to 21 days [174]. The development of active and biosafe packaging utilizing nanoparticles has been spurred by environmental concerns pertaining to plastic packaging. Because of their superior photocatalyzing, antibacterial, and biocompatibility qualities, TiO_2 NPs are a stable and reasonably priced metal oxide for biodegradable food packaging [175].

5.5 Titanium NPs in fuel cell application

An opening has emerged for the application of Ti NPs in energy storage devices due to recent research. Wide band gap semiconductors (transition-metal oxide) such as titanium dioxide (TiO_2) are non-toxic, extremely stable, photoactive, and absorb light in the ultraviolet (UV) spectrum. [176]. TiO_2 NPs enhance surface area for interface in photo voltaic cell [177]. TiO_2 NPs embedded into polystyrene (PS) with in various ratio was found to be optimum for solar cell application [178]. In one investigation, the thickness of the TiO_2 coating and the electrode materials' nanostructure enabled brief ionic diffusion paths that increased specific capacitance. [179]. Dhungel and Park reported photoelectrode of TiO_2 film casted on F doped glass surface showed high energy conversion up to 7.2% in dye sensitized solar cells [180]. Bixa orellana seed extract combined with green produced TiO_2 nanoparticles displayed a high photovoltaic conversion of 2.97% [181]. Increases in solar-to-electrical power conversion efficiency of up to 6% were noted in dye-sensitive solar cells produced with TiO_2 nanoparticles [87].

Fuel cells with proton exchange membranes have great promise for both fixed and mobile power applications. At the cathode, where O_2 molecules are reduced by electrons, an oxygen reduction reaction takes place. Durability characteristics confirmed that 30 nm Ti NPs catalyzed oxygen reduction activity after 5000 potential cycles from 0.6 to 1.0 V vs. RHE [182]. Pure TiN NPs generated using a unique thermal plasma arc discharge process demonstrated good energy storage capacity in a fresh experiment, with specific capacitances of 192.8 and 435.1 F/g showing promise for supercapacitor applications [183].

It has been demonstrated that Ti NPs enhance heat transfer in heat exchangers, heat sinks, and radiators. Compared to basic liquids, the application of nanofluid shown an increase in heat transfer rate. With the use of TiN nanofluid, solar vapor can be produced without the need for black tubes or panels or other sunlight absorbers, and the system's excess thermal capacitance can be decreased [184]. Doping of TiN thin films with N increased capacitance by up to three times [185]. Titanium nitrides are a suitable active material for the fuel cell electrode due to their high physical features, which include a melting point of up to 3000 ^{0}C, good hardness and resistance, electrical conductivity, and stability [186].

Titanium nitride nanocomposites (NCs) have superior electrochemical performance because of possible base material synergy [187]. The focus on titanium nitride as super capacitor electrodes is related to its mechanical strength, cost-effectiveness, and high electrical conductivity, which is comparable to that of metals (4×103-5×10 4 S cm-1) [188]. The reversible faradaic charge-storage process and electrostatic charge storage are responsible for TiN's charge-storing properties [9]. Three-dimensional nanostructured TiN NPs were deposited on a sintered electrode using an electrophoretic technique, which minimized structural and chemical damage and decreased retention capability losses [189].

In the wavelength range of 0.3-0.9 μm, the TiN-nanopatterns/dielectric/TiN stack metamaterial showed about 93% light absorption. Furthermore, TiN layer patterning can change the emission wavelength [190]. For various combinations, the photoanode made of biosynthesized TiO2 NPs demonstrated current conversion efficiencies of 6.64%, 2.66%, and 18%, respectively [191]. In addition to demonstrating three reversible redox peaks with the formal potential of +0.23, +0.45, and +0.78V, the electrochemical co-polymerization of aniline and nano-TiC produced Polyaniline/titanium carbide NPs also revealed electrocatalytic activity for the reduction of nitrite [192].

Titanium carbide (Tin+ 1Cn) nanosheets are appropriate for usage in high energy and temperature applications because they have brittle ceramic characteristics and metal-like heat and electricity conduction [193–194]. For high charge storage in capacitance applications, the layered electrochemically active surface of Ti_2C and Ti_3C_2 is a feature. Because of its potential for sustainability, microbial fuel cells (MFCs), which use microorganisms to break down organic matter and generate electricity, are the subject of intense research. According to a recent study, biogenic TiO_2 NPs can greatly reduce biofouling and enhance the performance of MFCs in the oxygen reduction reaction (ORR) at a cheap cost [195].

5.6 Titanium NPs in paints and coatings application

Ti NPs commercial growth is increasing with demand in Paints and coating industry in automobiles and buildings. Low friction, anticorrosion ability and high temperature resistance is essential requirement for coating properties. Nanomaterial provides improved characteristics to paints like water/dirt repellent, UV-protection, antimicrobial resistance, scratch resistance increased lifespan. However, studies have shown that the use of photocatalytic Ti NPs in organic paints leads to the degradation of the binder by UV irradiation [196]. For this reason, the rutile form is preferred in coatings for UV-protection. Properties of rutile like high UV opacity and lower photoactivity is reason behind preference over anatase form. Size parameter of 90 nm is more effective than 70 nm for paint industry [197]. In a different experiment, the authors found that the surface of nano-TiO2 coated with mixed oxides of silicon, aluminum, zirconium, and phosphorous to an isocyanate acrylic clear coating increased the coating's lifetime and that UV blocking was more successful for particles smaller than 25 nm [198].

Ti NPs modified the silane coupling agent (CH2=C(CH3) COOC3H6Si(OCH3)3) to improve its mechanical properties, including binding force, toughness, and hardness, while reducing its dry toughness by 73% [199]. When compared to PEO coating, the hardness of nanoparticles with coatings is two times higher, and wear resistance rises by 2.2 times [200]. Zhang and Lin et al. [201] found an improvement in antibacterial and anti-corrosion properties for titanium dioxide-polytetrafluoroethylene NCs coating on stainless steel surface.

TiO2-modified paint usage above 3.5 weight percent in a concentration-controlled trial led to chalking and yellowing flaws in dried form [202]. By adding water-dispersible copper (Cu)/titanium dioxide (TiO2) colloidal dispersion to commercial styrene-acrylic latex paint and stirring at a high speed, an environmentally friendly and long-lasting antibacterial coating was created that demonstrated the capacity to maintain antibacterial efficacy due to ROS and Cu-ion release behavior over prolonged periods of time [203]. To stop the development of infectious diseases like COVID-19, Mohite et al. proposed coating public areas with smart paint photocatalytic pathogenic disinfectant [204]. TiN NP was used to make a thermoplasmonic paint with notable photo-thermal capabilities. The paint heated up in two hours, indicating that it is suitable for low-grade thermal energy applications on conductive surfaces, such as those found in radioisotope thermoelectric generators.

5.7 Titanium NPs in wastewater treatment

Nanotechnology has an important role in environmental issues related to water. Application of nanotechnology in desalination is a hopeful concept in addressing drinking water scarcity issue. Nanotechnology is increasingly researched for wastewater treatment techniques like remediation sensing, water treatment and different biological systems [205-208]. Similar to this, doped Ti NPs have demonstrated enhanced bactericidal effects because of a decreased rate at which photogenerated electron hole pairs recombine. TiO_2 NPs' photocatalytic activity can be efficiently increased by doping them with the right amounts of Sn and F. TiO_2 that was synthesized with more dopants provided the largest zone of inhibition against S. aureus and E. coli, measuring 38 and 35 mm, respectively [209]. Irshad et al. [23] proposed that TiO_2 NPs could be an effective way to remove Cd from wastewater in a selective manner. Studies on the use of titanium-based NCs materials to remove arsenic from water have been reported. These materials include hybrid titania nanostructures such as Mn_2O_3/TiO_2 [210], composites of titania nanotubes and CNTs [211], adsorbents based on titania, such as kaolinite, montmorillonite, and red mud [212], TiO_2 NCs based on V_2O_5 [213], and Fe_2O_3/TiO_2-SiO_2 ternary NCs [214].

Table 5: Applications of NPs: TiC and TiN.

Application broader aspect	Nanoparticle type	Result reported	Ref.
Sensing Applications	Gold- NPs-Decorated Titanium Nitride Electrodes	decreased the gap between nanocolumns from about 23 to approximately 15 nm, increasing the electrodes' electroactivity.	[215]
Electrochemical investigation	2D carbon nanosheets covered a TiN composite anode	shown shortened ionic transport channels and had outstanding electrochemical characteristics with a high reversible capacity of 170 mA h g−1 and 149 mA h g−1 after 5000.	[173]
Photovoltaic application	Titanium nitride based plasmonic NPs	Enhanced route length for bowtie-shaped nanoplate on 30 nm Si_3N_4 with a maximum scattering cross-section of 4.58 Wm−2 17. About 30% of the light was absorbed at its greatest effectiveness.	[216]
Cancer therapy	Synthesized and PEG-coated NPs.	PEGylation extended NP circulation, decreased cell damage and hemotoxicity, increased tumor delivery, and showed no signs of toxicity or organ harm.	[217]

Photothermal and solar energy application	TiN NPs	780 nm laser light was used for illumination, and it took five minutes to heat and cool. Promising material for local heating applications was found to have strong heat dispersion.	[218]
Supercapacitive performance evaluation	Sandwich-like structure of ZnO NPs-decorated two-dimensional Ti_3C_2	85% of the capacitance was preserved at 5 A g−1 after 10,000 cycles, demonstrating outstanding cycling stability and an improved specific capacitance of 120 F g−1 at 2 mV s−1.	[163]
Carbon dioxide adsorption process	TiC nanopowder	Significant adsorption capacities of TiC nanopowder with moderate heat of adsorption of around 30e35 kJ mol⁻1.	[8]
Electromagnetic wave absorption	Hybrids of TiC nanowires/paraffin	Demonstrated a unique capacity to absorb electromagnetic waves in the X-band (8.2–12.4 GHz).	[219]
Photocatalytic bacteriostatic applications	TiO_2 grafted 2D-TiC nanosheets	Outstanding antibacterial properties against Staphylococcus aureus and Pseudomonas aeruginosa.	[220]
Pressure sensors.	polyvinyl alcohol - polyvinyl pyrrolidinone - titanium carbide NCs	Electrical resistance of NCs decreased with an increase of the pressure.	[221]
Nanoscale electromagnetic energy transfer	TiN NPs	Applied for surface plasmon polariton propagation in visible and telecommunication wavelength ranges even at high temperature.	[222]
Bone scaffold	Photocrosslinked salecan composite hydrogel embedding TiC NPs	Non-toxic, cell proliferation and good cytocompatibility.	]223]
Cancer theranostics	$GdW_{10}@Ti_3C_2$ composite nanosheets	Effectively eradicated tumor without further reoccurrence during the observation period	[224]
Methanol fuel cell application	Titanium carbide-derived carbon as a novel support for platinum catalysts	Oxygen reduction increased by 18% in comparison to carbon black, even without optimization of the catalyst and support.	[225]
humidity sensors	Polyvinyl alcohol PVA Polyacrylic acid blend doped TiN NPs	By absorbing water molecules, the polymer chain relaxes and aligns, making room for charge carriers and enhancing the composite's sensing response.	[226]

6.   Other titanium NPs and applications

$BaTiO_3$ NPs is proven to be suitable material for biomedical applications and in wearable bioelectronics [227]. Numerous crystalline phases of nanodots, NPs, nanobowls, nanowires, nanocubes, and nanorods of $BaTiO_3$ have been created. Research on antimicrobial property of $BaTiO_3$ shows significant reduction in gram-positive and gram-negative bacterial growth and have pronounced effect on the biofilm formation against clinical isolates of P. aeruginosa and S. aureus.

at a very low concentration [228]. Another nanostructured material of Titanium TiH_2 is proven to be promising for conductive nano ink applications. Conductive inks in printed electronic applications are used for printing, metal oxides, conductive polymers, or carbon NMs with some limitations of air stability limits. In a study TiH_2 ink was cured using pulsed light processing after printing to yield conductive traces which can be further used in printed circuit boards, catalysts, coating etc. [12].

7.   Challenges and future prospect

In recent years, the primary usage of titanium nanostructures has been in cleaning the environment and creating hydrogen and photcatalysis. Although field applications of these nanostructured particles have its limitations. One of the biggest problems faced with these nanostructures is their efficiency, sensitivity and compatibility for wide scale usage. Structured Ti NPs could lead to new outcomes in improvement of nanomedicine, antimicrobial applications, water treatment, paints and coatings etc. The future lies in modifying and investigating for better understanding of new era of applications. UV light requirement for photocatalysis application is hinder point in full scale application in energy sector. Combining with other material may be the solution for this drawback. Areas for future research may be pollutant degradation, full scale medical implant applications, safe coating application, packaging etc.

Conclusion

The nanomaterial family of Titanium is the focal point of nanoresearch due to availability, cost effectiveness, easy preparation, stability and potential application possibilities. Biosafety and inertness are the basis of new potential application areas in medical and packaging industry. Usages in food addition and food coloring has been banned by some countries due to health issue reports. Applications as drug carrier, fuel and solar cells are in continual growing stage. However full toxicity and long-term profile is yet to fully explored. Plant based synthesis and green application may be future research area of academicians and industry.

References

[1] S.Griffin, , M.I, Masood. M.J Nasim, M.Sarfraz, A.P, Ebokaiwe, K.H.Schäfer, C.M.Keck, C. Jacob, Natural Nanoparticles: A Particular Matter Inspired by Nature. Antioxidants (Basel). (2017). 29.7(1):3. https://doi.org/ 10.3390/antiox7010003

[2] J. P. Singh, M. Kumar, A. Sharma, A. Pandey, G., Chae, K. H., & S. Lee, Bottom-up and top-down approaches for MgO. Sonochemical Reactions. (2020)10.5772/intechopen.91182

[3]. K.Lan, Y. Liu, W. Zhang, Y. Liu,A. Elzatahry, R. Wang, R., ... & D. Zhao, Uniform ordered two-dimensional mesoporous TiO_2 nanosheets from hydrothermal-induced solvent-confined monomicelle assembly. Journal of the American Chemical Society, (2018). 140(11), 4135-4143

[4] K. Kim, K. T., M.Y. Eo, T. T. H Nguyen, & S.M. Kim, General review of titanium toxicity. International Journal of Implant Dentistry. (2019a). 5(1). 1-12

[5] B., M. Malekshahi, K.A. Nemati, L. Fatholahi, Z. Malekshahi, A review on synthesis of nano-TiO_2 via different methods. Journal of nanostructures, (2013). 3(1). 1-9

[6] W. Zhou, W. Li, W., Wang, J. Q., Qu, Y., Yang, Y., Y. Xie & Zhao, D. Ordered mesoporous black TiO_2 as highly efficient hydrogen evolution photocatalyst. Journal of the American Chemical Society. (2014). 136(26). 9280-9283

[7] X. Wang, X. Zhong, & L. Cheng, Titanium-based nanomaterials for cancer theranostics. Coordination Chemistry Reviews. (2021). 430. 213662

[8] S. Ghosh, P. Ranjan, A. Kumaar, R. Sarathi, & S. Ramaprabhu, Synthesis of titanium carbide nanoparticles by wire explosion process and its application in carbon dioxide adsorption. Journal of Alloys and Compounds. (2019)794. 645–653. https://doi.org/10.1016/j.jallcom.2019.04.299

[9] S. Tang, Q. Cheng, J. Zhao, J. Liang, C. Liu, Q. Lan, Q. Cao, Y.-C., Liu, J. Preparation of titanium nitride nanomaterials for electrode and application in energy storage. Results Phys. 7. (2017)1198–1201

[10] A.A., Popov, G.V. Tikhonowski, P.V. Shakhov, E.A. Popova-Kuznetsova , G.I. Tselikov, R.I.Romanov, A.M. Markeev, S.M.Klimentov, A.V. Kabashin, Synthesis of Titanium Nitride Nanoparticles by Pulsed Laser Ablation in Different Aqueous and Organic Solutions. Nanomaterials. (2022)12(10):1672. doi.org/10.3390/nano12101672

[11] U. Guler, S. Suslov, A.V. Kildishev, A. Boltasseva & V.M. Shalaev, Colloidal plasmonic titanium nitride nanoparticles: properties and applications. Nanophotonics, (2015).4(3). 269-276

[12] E.B. Secor, N.S.Bell, M.P.Romero, R.R. Tafoya, T.H. Nguyen, & T.J.Boyle, Titanium hydride nanoparticles and nanoinks for aerosol jet printed electronics. Nanoscale. (2022).14(35). 12651-12657

[13] A. Kumar & G. Pandey,Different methods used for the synthesis of TiO_2 based nanomaterials: A review. Am. J. Nano Res. Appl, (2018). 6(1), 1-10

[14] B. Niu, X. Wang, K. Wu, X. He, & R. Zhang, Mesoporous titanium dioxide: Synthesis and applications in photocatalysis, energy and biology. Materials. (2018). 11(10). 1910

[15] S.T. Yerpude, A.K. Potbhare, P. Bhilkar, A.R. Rai, R.P. Singh, A.A. Abdala, R. Adhikari, R. Sharma, R.G. Chaudhary, Biomedical and clinical applications of platinum-based nanohybrids: An update review, Environmental Research, 231 (2023) 116148

[16] M.S. Umekar, G.S. Bhusari, T. Bhoyar, V. Devthade, B.P. Kapgate, A.P. Potbhare, R.G. Chaudhary & A.A. Abdala, Graphitic carbon nitride-based photocatalysts for environmental remediation of organic pollutants, Current Nanoscience, 19 (2) 2023, 148-169

[17] H.Zhang, & J.F. Banfield, Structural characteristics and mechanical and thermodynamic properties of nanocrystalline TiO_2. Chemical reviews. 114(2014). 19. 9613-9644

[18] M.T.Noman, M.A. Ashraf, & A. Ali, Synthesis and applications of nano-TiO_2: A review. Environmental Science and Pollution Research, 26, (2019)3262-3291

[19] G. Nagpal, R. Chaudhary, R.G. Chaudhary & N.B. Singh, Emerging trends of nanotechnology in cosmetics, Applications of emerging nanomaterials and nanotechnology, Materials Research Forum LLC, 2023, 148, pp.127-169

[20] J. Shi, D. Yang,Z. Jiang, Y. Jiang, Y. Liang, Y. Zhu, Simultaneous size control and surface functionalization of titania nanoparticles through bioadhesion-assisted bio-inspired mineralization. J. Nanopart. Res. 14 (9). (2012) 1120

[21] M.E. Vance, T. Kuiken, E.P. Vejerano, S.P. McGinnis, Jr M.F. Hochella, D. Rejeski, M.S.Hull, Nanotechnology in the real world: Redeveloping the nanomaterial consumer products inventory. Beilstein J. Nanotech 6. (2015) 1769–1780

[22] M. N. Khan, M. Mobin, Z.K. Abbas, K.A. AlMutairi, Z.H. Siddiqui, Role of nanomaterials in plants under challenging environments. Plant Physio. Biochem. 110. (2017). 194–209

[23] M.A. Irshad, M.B. Shakoor, S. Ali, R. Nawaz, M. Rizwan, Synthesis and application of titanium dioxide nanoparticles for removal of cadmium from wastewater: kinetic and equilibrium study. Water Air Soil Pollut. 230 (2019)12, 278

[24] R.G. Chaudhary, G.S. Bhusari, A.D. Tiple, A.R. Rai, S.R. Somkuvar, A.K. Potbhare, T.L. Lambat, P.P. Ingle & A.A. Abdala, Metal/metal oxide nanoparticles: toxicity, applications, and future prospects, Current Pharmaceutical Design, 2019, 25, 4013-4029.

[25] M.A. Irshad, M. A., Nawaz, R., U.R. Rehman, M. Z., Adrees, M., Rizwan, M., Ali, S., ... & Tasleem, S.Synthesis, characterization and advanced sustainable applications of titanium dioxide nanoparticles: A review. Ecotoxicology and environmental safety. 212. (2021) 111978

[26] G. Ou, Z Li, D. Li, L. Cheng, Z. Liu, H. Wu, Photothermal therapy by using titanium oxide nanoparticles. Nano Res. (2016), 9(5):1236-1243. https://doi.org/10.1007/s12274-016-1019-8

[27] R. Ghamarpoor, A. Fallah, M. Jamshidi, Investigating the use of titanium dioxide (TiO$_2$) nanoparticles on the amount of protection against UV irradiation. *Sci Rep* 13, (2023) 9793. https://doi.org/10.1038/s41598-023-37057-5

[28] M.H..Ropers & H. Terrisse, & Mercier-Bonin, Muriel & H. Bernard. (2017). Titanium Dioxide as Food Additive. 10.5772/intechopen.68883

[29] S. Chandren & N. Zulfemi, Titania nanoparticles coated on polycarbonate car headlights for self-cleaning purpose. Journal of Physics: Conference Series. 1321. (2019) 022032. 10.1088/1742-6596/1321/2/022032

[30] M.L.Coutinho, J.P. Veiga, M.F. Macedo, A.Z. Miller, Testing the Feasibility of Titanium Dioxide Sol-Gel Coatings on Portuguese Glazed Tiles to Prevent Biological Colonization. *Coatings, 10*, (2020)1169. https://doi.org/10.3390/coatings10121169

[31] S. Paul, M.A. Rahman, S.B. Sharif, J.H. Kim, S.E.T. Siddiqui, M.A.M Hossain. TiO$_2$ as an Anode of High-Performance Lithium-Ion Batteries: A Comprehensive Review towards Practical Application. *Nanomaterials, 12*, (2022) 2034. https://doi.org/10.3390/nano12122034

[32] El Sharkawy, H.M., Shawky, A.M., Elshypany, R. *et al.* Efficient photocatalytic degradation of organic pollutants over TiO$_2$ nanoparticles modified with nitrogen and MoS$_2$ under visible light irradiation. *Sci Rep* **13**, (2023) 8845. https://doi.org/ https://doi.org/10.1038/s41598-023-35265-7

[33] R. Sharma, A. Sarkar, R. Jha,A. K. Sharma, D. Sharma. Sol-gel–mediated synthesis of TiO_2 nanocrystals: structural, optical, and electrochemical properties. Int. J. Appl. Ceram. Technol. 17 (2020) 3. 1400–1409

[34] V.M. Ramakrishnan, M. Natarajan, A. Santhanam, V. Asokan, D. Velauthapillai. Size controlled synthesis of TiO_2 nanoparticles by modified solvothermal method towards effective photo catalytic and photovoltaic applications. Mater. Res. Bull. 97. (2018) 351–360

[35] N. Horti, M. Kamatagi, N.Patil, S. Nataraj, M. Sannaikar, S. Inamdar. Synthesis and photoluminescence properties of titanium oxide (TiO_2) nanoparticles: effect of calcination temperature. Optik 194, (2019) 163070

[36] Z. Wang,A.A. Haidry, L.Xie, A. Zavabeti, Z. Li, W. Yin, R.L. Fomekong, B. Saruhan, Acetone sensing applications of Ag modified TiO_2 porous nanoparticles synthesized via facile hydrothermal method. Appl. Sur. Sci. (2020)533. 147383

[37] A. Mondal, M.S. Umekar, G.S. Bhusari, P.B. Chouke, T. Lambat, S. Mondal, R.G. Chaudhary, S.H. Mahmood, Biogenic synthesis of metal/metal oxide nanostructured materials, Curr. Pharm. Biotechnol. 22 (13) 2021, 1782-1793.

[38] M.S. Umekar, R.G. Chaudhary, G.S. Bhusari, A. Mondal, A.K. Potbhare, M. Sami, Phytoreduced graphene oxide-titanium dioxide nanocomposites using *Moringa oleifera* stick extract, Materials Today: Proceedings, 2020, 29, 709-714

[39] S.S. Muniandy, N.H.M. Kaus, Z.T. Jiang, M. Altarawneh, H.L. Lee. Green synthesis of mesoporous anatase TiO_2 nanoparticles and their photocatalytic activities. RSC Adv. 7 (2017) 76. 48083–48094

[40] N.S. Venkataramanan, K. Matsui, H. Kawanami,Y. Ikushima. Green synthesis of titania nanowire composites on natural cellulose fibers. Green Chem. 9 (2007) 1, 18–19

[41] S.J. Bao,C. Lei, M.W. Xu, C.J.Cai, C. D.Z. Jia. Environment-friendly biomimetic synthesis of TiO_2 nanomaterials for photocatalytic application. Nanotechnology. 23 (20). (2012)205601

[42] A.A. Jawad, R.M. Lua'i, R.M.Lua'I , N.H. Safir, S.A. Jawad & A.K. Abbas,Synthesis Methods and Applications of TiO_2 based Nanomaterials. Al-Nahrain Journal of Science, 25(4), (2022).1-10

[43] Kretzschmar, A. L., & Manefield, M. The role of lipids in activated sludge floc formation. AIMS Environmental science, 2(2). (2015).122-133

[44] G. Scholz & E. Kemnitz. Sol-gel synthesis of metal fluorides: reactivity and mechanisms. In Modern Synthesis Processes and Reactivity of Fluorinated Compounds. (2017).609-649. Elsevier

[45] C. Hintze, K. Morita,R. Riedel, E. Ionescu & G. Mera. Facile sol–gel synthesis of reduced graphene oxide/silica nanocomposites. Journal of the European Ceramic Society, 36(12), (2016). 2923-2930

[46] A. Zdravkov, J. Kudryashova, A. Kanaev, A. Povolotskiy, A. Volkova, E. Golikova & N.A. Khimich. A new solvothermal route to efficient titania photocatalyst. Materials Chemistry and Physics. (2015) 160, 73-79

[47] S. Kurajica, I. Minga, I. Grčić, V. Mandić & M. Plodinec. The utilization of modified alkoxide as a precursor for solvothermal synthesis of nanocrystalline titania. Materials Chemistry and Physics. 196, (2017).194-204

[48] H. Peng, G. Li & Z. Zhang. ynthesis of bundle-like structure of titania nanotubes. Materials Letters. 59 (10). (2005).1142-1145

[49] P. Kluson, H. Luskova, O. Solcova, L. Matejova & T. Cajthaml. Lamellar micelles-mediated synthesis of nanoscale thick sheets of titania. Materials Letters. 61(14-15). (2007).2931-2934

[50] S. Elbasuney. Sustainable steric stabilization of colloidal titania nanoparticles. Applied Surface Science. (2017).409, 438-447

[51] E. Stathatos, P. Lianos, F. Del Monte, D. Levy & D. Tsiourvas. Formation of TiO$_2$ nanoparticles in reverse micelles and their deposition as thin films on glass substrates. Langmuir. 13(16). (1997).4295-4300

[52] G. Cabello, R.A. Davoglio & E.C. Pereira. Microwave-assisted synthesis of anatase-TiO$_2$ nanoparticles with catalytic activity in oxygen reduction. Journal of Electroanalytical Chemistry. (2017).794. 36-42

[53] J. Kang, L. Gao, M. Zhang, J. Pu, L. He, R. Ruan, R., ... & Chen, G. Synthesis of rutile TiO$_2$ powder by microwave-enhanced roasting followed by hydrochloric acid leaching. Advanced Powder Technology. 31(3). (2020).1140-1147

[54] P. Selvaraj, A. Roy, H. Ullah, P. Sujatha Devi, A.A. Tahir, T.K. Mallick, & S. Sundaram, Soft-template synthesis of high surface area mesoporous titanium dioxide for dye-sensitized solar cells. International Journal of Energy Research. 43(1). (2019).523-534

[55] I.F. Mironyuk, L.M. Soltys, T.R. Tatarchuk, & K.O.Savka,Methods of titanium dioxide synthesis. Physics and Chemistry of Solid State, 21(3). (2020).462-477

[56] Q. Zhang, & C. Li, High temperature stable anatase phase titanium dioxide films synthesized by mist chemical vapor deposition. Nanomaterials. 10(5). (2020b).911

[57] A. Jedrzejczak, D. Batory, M. Prowizor, M. Dominik, M. Smietana,M. Cichomski , ... & M. Dudek. Titanium (IV) isopropoxide as a source of titanium and oxygen atoms in carbon based coatings deposited by Radio Frequency Plasma Enhanced Chemical Vapour Deposition method. Thin Solid Films, (2020). 693, 137697

[58] M.N.Subramaniam, P.S. Goh, W.J. Lau, A.F. Ismail, M. Gürsoy & M. Karaman. Synthesis of Titania nanotubes/polyaniline via rotating bed-plasma enhanced chemical vapor deposition for enhanced visible light photodegradation. Applied Surface Science. (2019)484. 740-750

[59] C. Cao, W. Liu, A. Javadi H. Ling, X. Li. Scalable Manufacturing of 10 nm TiC Nanoparticles through Molten Salt Reaction. Procedia Manufacturing. 10, (2017),634-640. https://doi.org/10.1016/j.promfg.2017.07.066

[60] P. Anandgaonker, G. Kulkarni, S. Gaikwad, S., A. Rajbhoj. Synthesis of TiO$_2$ nanoparticles by electrochemical method and their antibacterial application. Arabian Journal of Chemistry .12. (8). (2019).1815-1822. https://doi.org/10.1016/j.arabjc.2014.12.015

[61] Q.D. Mai, H.A. Nguyen, N.N. Huyen *et al.* Large-Scale Green Electrochemical Synthesis of Smart Titanium Dioxide Nanomaterials: Controlled Morphology and Rotatable Surface Ligands via Tuning Electrolyte Structures. *J. Electron. Mater.* 52, (2023).5884–5900. https://doi.org/10.1007/s11664-023-10550-3

[62] C. Xiaobo. Titanium dioxide nanomaterials and their energy applications. Chinese Journal of Catalysis. 30(8). (2009).839-851

[63] O.K. Varghese, D. Gong, M. Paulose, C.A. Grimes E.C. Dickey. Crystallization and high-temperature structural stability of titanium oxide nanotube arrays. Journal of Materials Research, 18. 1. (2003).156-165

[64] H. Arami, M. Mazloumi,R. Khalifehzadeh & S.K. Sadrnezhaad. Sonochemical preparation of TiO_2 nanoparticles. Materials Letters. (2007)61. 23-24. 4559-4561

[65] J.M. Wu, S. Hayakawa, K. Tsuru & A. Osaka, Nanocrystalline titania made from interactions of Ti with hydrogen peroxide solutions containing tantalum chloride. Crystal growth & design. 2(2). (2002)147-149

[66] G. Nabi, W. Raza, M. Tahir, M.Green synthesis of TiO_2 nanoparticle using cinnamon powder extract and the study of optical properties. J. Inorg. Organomet. Polym. Mater. (2019) 1–5

[67] S.N. Kumar, A. Zeenat, M. Pradeep Kumar, K. Pradeep. Green synthesis of TiO_2 nanoparticles from Syzygium cumini extract for photo-catalytic removal of lead (Pb) in explosive industrial wastewater. Green Process. Synth. 9 (1), (2020) 171–181

[68] V. Patidar, P. Jain. Green synthesis of TiO_2 nanoparticle using moringa oleifera leaf extract. Int. Res. J. Eng. Technol. 4. (2017)470–473

[69] A.M. Amanulla, R. Sundaram, Green synthesis of TiO_2 nanoparticles using orange peel extract for antibacterial, cytotoxicity and humidity sensor applications. Mater. Today Proc. 8, (2019). 323–331

[70] J.M. Abisharani, S. Devikala, R.D. Kumar, M. Arthanareeswari, P. Kamaraj. Green synthesis of TiO_2 nanoparticles using Cucurbita pepo seeds extract. Mater. Today Proc. 14. (2019)302–307

[71] D. Hariharan, K. Srinivasan, L.C. Nehru. Synthesis and characterization of TiO_2 nanoparticles using cynodon dactylon leaf extract for antibacterial and anticancer (a549 cell lines) activity. J Nanomed Res. (2017) 5(6):1–6. https://doi.org/ 10.15406/jnmr.2017.05.00138

[72] M. Peiris, T. Gunasekara, P. Jayaweera, S. Fernando. TiO_2 nanoparticles from Baker's yeast: a potent antimicrobial. J. Microbiol. Biotechnol. 28 (10). (2018).1664–1670

[73] G.K. Ağçeli, H. Hammachi, S.P. Kodal et al. A Novel Approach to Synthesize TiO_2 Nanoparticles: Biosynthesis by Using Streptomyces sp. HC1. J Inorg Organomet Polym 30, (2020).3221–3229. https://doi.org/10.1007/s10904-020-01486-w

[74] S. J. Bao, S. Lei,M.W. Xu.C.J. Cai , C.J. Cheng, C.M. Li, Environmentally-friendly biomimicking synthesis of TiO_2 nanomaterials using saccharides to tailor morphology, crystal phase and photocatalytic activity. **CrystEngComm**, (2013),**15**, 4694-4699

[75] S.D. Mo and W.Y. Ching. Electronic and optical properties of three phases of titanium dioxide: rutile, anatase, and brookite. Physical Review B (1995). 51: 13023–13032.

[76] Y. Hoang, H. Zung and N.H.B. Trong. Structural properties of amorphous TiO_2 nanoparticles. The European Physical Journal D (2007) 44: 515–524.

[77] S, M, Gupta and M. Tripathi. A review of TiO_2 nanoparticles. Chinese Science Bulletin (2011).56: 1639.

[78] A.I. Kingon, J.P. Maria and S.K. Streiffer. Alternative dielectrics to silicon dioxide for memory and logic devices. Nature. 406. (2000). 1032–1038.

[79] W. Li, C. Ni, H. Lin et al. Size dependence of thermal stability of TiO_2 nanoparticles. Journal of Applied Physics. 96. (2004). 6663–6668.

[80] D. Dambournet, I. Belharouak and K. Amine. Tailored preparation methods of TiO2 anatase, rutile, brookite: mechanism of formation and electrochemical properties. Chemistry of Materials 22. (2010). 1173–1179.

[81] Q. Zhang, L. Gao and J. Guo. Effects of calcination on the photocatalytic properties of nanosized TiO_2 powders prepared by $TiCl_4$ hydrolysis. Applied Catalysis B 26. (2000). 207–215.

[82] G. Govindasamy, P. Murugasen and S. Sagadevan. Investigations on the synthesis, optical and electrical properties of TiO2 thin films by chemical bath deposition (CBD) method. Materials Research. 19(2016). 413–419.

[83] P.D. Christy, N.S.N. Jothi, N. Melikechi and P. Sagayaraj. Synthesis, structural and optical properties of well dispersed anatase TiO_2 nanoparticles by non-hydrothermal method. Crystal Research and Technology 44 (2009).484–488.

[84] Y. Zhao, C. Li, X. Liu, F. Gu, H. Jiang, W. Shao, L. Zhang, Y. He. Synthesis and optical properties of TiO2 nanoparticles, Materials Letters,61,1,2007,79-83. https://doi.org/10.1016/j.matlet.2006.04.010

[85] O.U. Akakuru, Z.M. Iqbal & A. Wu. TiO_2 Nanoparticles: Applications in Nanobiotechnology and Nanomedicine. In TiO2 nanoparticles: properties and applications. Eds Wu, A., Ren, W., Wiley Wch. (2020).1-66

[86] K.Nakata and A. Fujishima. TiO_2 photocatalysis: design and applications. Journal of Photochemistry and Photobiology C. 13. (2012)169–189.

[87]. S. Shaikh, R. Mane, B. Min et al. D-sorbitol-induced phase control of TiO_2 nanoparticles and its application for dye-sensitized solar cells. Sci Rep. 6. (2016). 20103. https://doi.org/10.1038/srep20103

[88] A.K. Tripathi, M.K.Singh, M.C. Mathpal, et al. Study of structural transformation in TiO_2 nanoparticles and its optical properties. Journal of Alloys and Compounds. 549 (2013). 114–120.

[89] A. Gogos, K. Knauer, and T.D. Bucheli, Nanomaterials in plant protection and fertilization: current state, foreseen applications, and research priorities. Journal of Agricultural and Food Chemistry. 60 (2012). 9781–9792.

[90] S.G.Kumar, and L.G. Devi, Review on modified TiO2 photocatalysis under UV/visible light: selected results and related mechanisms on interfacial charge carrier transfer dynamics. The Journal of Physical Chemistry A 115(2011). 13211–13241

[91] G.O.U. Okeke, Physico-thermal properties of TiO_2 nanoparticles using molecular dynamics simulations with relevance to thermal conductance of nanofluids, Thesis, University of Leeds. (2013)

[92] H. Alias, M. C. Ani. Thermal characteristic of nanofluids containing titanium dioxide nanoparticles in ethylene glycol. Chemical Engineering Transactions. 56. (2017)1459-1464 doi:10.3303/CET1756244

[93] I. Ali, M. Suhail, Z.A. Alothman & A. Alwarthan, Recent advances in syntheses, properties and applications of TiO_2 nanostructures. RSC advances. 8(53). (2018)30125-30147

[94] U. Guler, A. Kildishev, A. Boltasseva and V. Shalaev. Titanium nitride nanoparticles for therapeutic applications. Conference on Lasers and Electro-Optics (CLEO) - Laser Science to Photonic Applications. USA. (2014).1-2, doi: 10.1364/CLEO_QELS.2014.FM1K.4

[95] L. Cheng, C. Wang, L.Z. Feng, K. Yang, Z. Liu. Functional Nanomaterials for Phototherapies of Cancer. Chem. Rev. 114, (2014).10869−10939

[96] Yu. Cao, W. Tingting, D. Wenhao, D. Haifeng, Z. Xueji. TiO2 Nanosheets with Au Nanocrystals Decorated Edge for Mitochondria-Targeting Enhanced Sonodynamic Therapy. Chemistry of Materials, Acs. chemmater. (2019). https://doi:10.1021/acs.chemmater.9b03430

[97] S. Kim, S. Im, E. Park, J. Lee, C. Kim, T. Kim & W.J. Kim. Drug-loaded titanium dioxide nanoparticle coated with tumor targeting polymer as a sonodynamic chemotherapeutic agent for anti-cancer therapy. Nanomedicine: Nanotechnology. Biology and Medicine, (2019b).102110. https://doi:10.1016/j.nano.2019.102110

[98] R. Kim. Recent advances in understanding the cell death pathways activated by anticancer therapy. Cancer. Apr 15;103(8). (2005).1551-60. https://doi.org/ 10.1002/cncr.20947

[99] T. Shrirame, P. Bhilkar, A. Chaudhary, A. Rai, R. Singh, P. Dhongle, S. Thakare, A. Abdala, R. Chaudhary, Magnetic Nanoparticles: Fabrications and applications in cancer therapy and diagnosis, magnetic nanoparticles for biomedical applications, Materials Research Forum, 143 (2023) pp.199-232.

[100] M. Ferrari, A. Barker. Downing G, Cancer nanotechnology: opportunities and challenges. Nanobiotechnology. 1. (2005).129. https://doi.org/10.1038/nrc1566

[101] N. Reed, R. Raliya, R. Tang, B. Xu, N. Mixdorf, S. Achilefu & P. Biswas. Electrospray Functionalization of Titanium Dioxide Nanoparticles with Transferrin for Cerenkov Radiation Induced Cancer Therapy. ACS Applied Bio Materials. (2019). https://doi.org/10.1021/acsabm.8b00755

[102] Q.S. Xian, T. Shen, L. Liu, A. Cheng, Z. Liu. Two-dimensional TiS2 nanosheets for in vivo photoacoustic imaging and photothermal cancer therapy. Nanoscale. 7. (2015).6380–6387. https://doi.org/10.1039/C5NR00893J

[103] C. Wang, C. Dai, Z. Hu, H. Li, L. Yu, H. Lin, HJ. Bai, Y. Chen. Photonic cancer nanomedicine using the near infrared-II biowindow enabled by biocompatible titanium nitride nanoplatforms, Nanoscale Horiz. 4 (2019a).415–425

[104] X. Han, J. Huang, X. Jing, D. Yang, H. Lin, Z. Wang & Y. Chen. Oxygen-deficient black titania for synergistic/enhanced sonodynamic and photoinduced cancer therapy at near infrared-II biowindow. Acs Nano, 12(5), (2018).4545-4555

[105] M. Wang, Y. Zhao, Chang, B. Ding, X. Deng, S. Cui, Z. Hou, J. Lin. Azo Initiator Loaded Black Mesoporous Titania with Multiple Optical Energy Conversion for Synergetic Photo-Thermal-Dynamic Therapy A.C.S. Appl. Mater. Interfaces 11. (2019b).47730–47738

[106] K. Gao, W. Tu, X. Yu, F. Ahmad, X. Zhang, W. Wu, X. An. X. Chen, X, W. Li, W-doped TiO$_2$ nanoparticles with strong absorption in the NIR-II window for photoacoustic/CT dual-modal imaging and synergistic thermoradiotherapy of tumors. Theranostics. 9(18) (2019).5214-5226. https://doi.org/10.7150/thno.33574

[107] M.W. Akram, F. Raziq, M. Fakhar-e-Alam, M.H. Aziz, K.S. Alimgeer, M. Atif, M. Amir, A. Hanif, W.A. Farooq. Tailoring of Au-TiO$_2$ nanoparticles conjugated with doxorubicin for their synergistic response and photodynamic therapy applications, Journal of Photochemistry and Photobiology A: Chemistry. 84. (2019).112040. https://doi.org/10.1016/j.jphotochem.2019.112040

[108] J.L. Song, Z.Q. Huang, J. Mao, W.J. Chen, B. Wang, F.W. Yang, S.H. Liu, H.J. Zhang, L.P. Qiu, J.H. Chen. A facile synthesis of uniform hollow MIL-125 titanium-based nanoplatform for endosomal esacpe and intracellular drug delivery, Chemical Engineering Journal, 396. (2020). 125246. doi.org/10.1016/j.cej.2020.125246

[109] X. Yuan, Y. Zhu, S. Li, Y. Wu, Z. Wang, R. Gao, S. Luo, J. Shen, J. Wu, and L. Ge. Titanium nanosheet as robust and biosafe drug carrier for combined photochemo cancer therapy. Journal of Nanobiotechnology 20 (2022). 154. doi.org/10.1186/s12951-022-01374-0

[110] S. Abdel-Ghany, S. Raslan, H. Tombuloglu et al. Vorinostat-loaded titanium oxide nanoparticles (anatase) induce G2/M cell cycle arrest in breast cancer cells via PALB2, upregulation. Biotech. 10. 9 (2020). 407. https://doi.org/ 10.1007/s13205-020-02391-2

[111] M. Ahamed, M.A.M. Khan, M.J. Akhtar, H.A. Alhadlaq, A. Alshamshan, Ag-doping regulates the cytotoxicity of TiO$_2$ nanoparticles via oxidative stress in human cancer cells. Sci. Rep. 7. (2017). 17662

[112] P.M. Gschwend, S. Conti, A. Käch, C. Maake, & S.E. Pratsinis, Silica-coated TiN particles for killing cancer cells. ACS Applied Materials & Interfaces. (2019). doi:10.1021/acsami.9b07239

[113] T. López, E. Ortiz, P. Guevara, E. Gómez, E., & O. Novaro, Physicochemical characterization of functionalized-nanostructured-titania as a carrier of copper complexes for cancer treatment. Materials Chemistry and Physics.146(1-2). (2014). 37–49. https://doi:10.1016/j.matchemphys. 2014.02

[114] J. Mou, T. Lin, F. Huang, et al. Black titania-based theranostic nanoplatform for single NIR laser induced dual-modal imaging-guided PTT/PDT. Biomaterials 84: (2016). 13–24. https://doi: 10.1016/j.biomaterials.2016.01.009

[115] W. He, K. Ai, K., C. Jiang, Y. Li, X. Song, & L. Lu. Plasmonic titanium nitride nanoparticles for in vivo photoacoustic tomography imaging and photothermal cancer therapy. Biomaterials. 132. (2017). 37–47. https://doi:10.1016/j.biomaterials.2017.04.007

[116] M. A. Behnam, F. Emami, Z. Sobhani, A.R. Dehghanian. The application of titanium dioxide (TiO_2) nanoparticles in the photo-thermal therapy of melanoma cancer model. Iran J Basic Med Sci. 21. (2018). 1133-1139. https://doi.org/ 10.22038/IJBMS.2018.30284.7304

[117] W. Guo, F. Wang, D. Ding, C. Song, C. Guo & S. Liu. TiO_2–x Based Nanoplatform for Bimodal Cancer Imaging and NIR-Triggered Chem/Photodynamic/Photothermal Combination Therapy. Chemistry of Materials. 29(21). (2017). 9262–9274. https://doi.org/ 10.1021/acs.chemmater.7b03241

[118] F. Shokrolahi, E. Aliasgari, A. Mirzaie. Cytotoxic Effects of Titanium Dioxide Nanoparticles on Colon Cancer Cell Line (HT29) and Analysis of Caspase-3 and 9 Gene Expression Using Real Time PCR and Flow Cytometry. Iran South Med J. 21.6. (2019). 426-438

[119] M.Y. Bilkan, Z. Çiçek, A.G.C. Kurşun, M. Özler, M.A. Eşmekaya. Investigations on Effects of Titanium Dioxide (TiO_2) Nanoparticle in Combination with UV Radiation on Breast and Skin Cancer Cells. (2022). Research Square. https://doi.org/10.21203/rs.3.rs-2084950/v1

[120] H. Naghoosi, M.A. Saremi . Titanium Dioxide Nanoparticle Can Induce Apoptosis in Cancer Cells. Autumn. 5. 19 (2020) 6-18

[121] N. Lagopati, A. kotsinas, D. veroutis, K. evangelou and Others. Biological Effect of Silver-modified Nanostructured Titanium Dioxide in Cancer. Cancer genomics & proteomics (2021). 18: 425-439 doi:10.21873/cgp.20269

[122] S. Azimeea,M. Rahmatic, H. Fahimi, M.A. Moosavi. TiO_2 nanoparticles enhance the chemotherapeutic effects of 5-fluorouracil in human AGS gastric cancer cells via autophagy blockade. Life Sciences 248 (2020). 117466. https://doi.org/10.1016/j.lfs.2020.117466

[123] H. Iqbal, A. Razzaq, B. Uzair, N. Ul. Ain,S. Sajjad, SN.A. Althobaiti, A.E. Albalawi, B. Menaa, M. Haroon, M. Khan. et al. Breast Cancer Inhibition by Biosynthesized Titanium Dioxide Nanoparticles Is Comparable to Free Doxorubicin but Appeared Safer in BALB/c Mice. Materials. 14. (2021). 3155

[124] E.A. Rozhkova, I. Ulasov, B. Lai, Dimitrijevic, N. M., Lesniak, M. S., & Rajh, T. A High-Performance Nanobio Photocatalyst for Targeted Brain Cancer Therapy. Nano Letters, (2009). 9(9). 3337–3342. https://doi.org/10.1021/nl901610f

[125] V. Bernard, V. Mornstein. The viability of ovarian carcinoma cells a2780 affected by titanium dioxide nanoparticles and low ultrasound intensity. Lékař a technika, Vol. 46. 1. (2016) 21-24

[126] K.N. Rahmani, Y. Rasmi, A. Abbasi, N. Koshoridze, A. Shirpoor, G. Burjanadze, E. Saboory. Bio-Effects of TiO_2 Nanoparticles on Human Colorectal Cancer and Umbilical Vein

Endothelial Cell Lines. Asian Pac J Cancer Prev. 26. (2018). 19(10):2821-2829. https://doi.org/ 10.22034/APJCP.2018.19.10.2821

[127] G. Kaur, T. Willsmore, K. Gulati, I. Zinonos, Y. Wang, M. Kurian, S. Hay, D. Losic, A. Evdokiou. Titanium wire implants with nanotube arrays: A study model for localized cancer treatment, Biomaterials. 101. (2016). 176-188. https://doi.org/10.1016/j.biomaterials.2016.05.048

[128] Y. Wang, Q. Wang, C. Zhang, Synthesis of Diamond-Shaped Mesoporous Titania Nanobricks as pH-Responsive Drug Delivery Vehicles for Cancer. Chemistry select Therapy. 4, (2019c). 28.8225-8228. doi.org/10.1002/slct.201900992

[129] M.J. Hajipour, K.M. Fromm, Akbar, A. Ashkarran, et al. Antibacterial properties of nanoparticles. Trends Biotechnol.30.10. (2012). 499–511. https://doi.org/10.1016/j.tibtech.2012.06.004

[130] Z. Pang, R. Raudonis, B.R. Glick, B, T.J. Lin, Z. Cheng, Antibiotic resistance in Pseudomonas aeruginosa: mechanisms and alternative therapeutic strategies. Biotechnol Adv. (2019). https://doi.org/10.1016/ j.biotechadv.2018.11.013

[131] M.G. Vincent, N.P. John, P.M. Narayanan, C. Vani, S. Murugan. In vitro study on the efficacy of zinc oxide and titanium dioxide nanoparticles against metallo beta-lactamase and biofilm producing Pseudomonas aeruginosa. J. Appl. Pharm Sci.4(7) (2014). 41–46. https://doi.org/10.7324/ JAPS.2014.40707

[132] A. Kuback, M.S. Diez, D. Rojo et al. Understanding the antimicrobial mechanism of TiO_2-based nanocomposite films in a pathogenic bacterium. Scientific Reports 4. (2014. 4134

[133] J. Kiwi, S. Rtimi. Mechanisms of the antibacterial effects of TiO_2 -FeO x under solar or visible light: Schottky barriers versus surface plasmon resonance. Coatings. (2018). 8:391. https://doi.org/ 10.3390/coatings8110391

[134] X. Wu, Y.Y. Huang, Y. Kushida. et al. Broad-spectrum antimicrobial photocatalysis mediated by titanium dioxide and UVA is potentiated by addition of bromide ion via formation of hypobromite. Free Radical Biology and Medicine 95: (2016). 74–81

[135] A. Mukhopadhyay, S. Basak, J.K. Das, S.K. Medda, K. Chattopadhyay & G. De. Ag–TiO_2 nanoparticle codoped SiO_2 films on ZrO_2 barrier-coated glass substrates with antibacterial activity in ambient condition. ACS Applied Materials & Interfaces. 2(9). (2010). 2540–2546

[136] A. W. Jatoi, I.S. Kimc, Q.Q. Nid. Cellulose acetate nanofibers embedded with AgNPs anchored TiO_2 nanoparticles for long term excellent antibacterial applications. Carbohydrate Polymers. 207. (2019). 640-649. https://doi.org/10.1016/j.carbpol.2018.12.029

[137] P.B. Chouke, A.K. Potbhare, N. P. Meshram, M.M. Rai, K.M. Dadure, K. Chaudhary, A.R. Rai, M.F. Desimone, R.G. Chaudhary & D.T. Masram, Bioinspired NiO nanospheres: Exploring in-vitro toxicity using Bm-17 and L. rohita liver cells, DNA degradation, docking and proposed vacuolization mechanism, ACS Omega, 7 (8) 2022, 6869−6884

[138] A.K. Potbhare, R.G. Chaudhary, P.B. Chouke, S. Yerpude, A. Mondal, V.N. Sonkusare, A.R. Rai, H.D. Juneja. Phytosynthesis of nearly monodisperse CuO nanospheres using

Phyllanthus reticulatus/Conyza bonariensis and its antioxidant/antibacterial assays. Materials Science and Engineering: C 99 (2019): 783-793

[139] P.V. Asha Rani, Low Kah Mun, G., M.P. Hande and S. Valiyaveettil, S. Cytotoxicity and genotoxicity of silver nanoparticles in human cells. ACS Nano (2009). 3: 279–290

[140] P. Maheswari,S. , S.Harish, M. Navaneethan, C. Muthamizhchelvan ,S Ponnusamy, Y. Hayakawa, Bio-modified TiO_2 nanoparticles with Withania somnifera, Eclipta prostrata and Glycyrrhiza glabra for anticancer and antibacterial applications, Materials Science and Engineering: C. 108 .(2020).110457. https://doi.org/10.1016/j.msec.2019.110457

[141] A. Connaughton, A. Childs, S. Dylewski, & V.J. Sabesan. Biofilm disrupting technology for orthopedic implants: what's on the horizon? Frontiers in 497 Medicine (Lausanne). (2014). 1, 22

[142] F.U. Gunputh, H. Le, K. Lawton, A. Besinis, C. Tredwin & R.D. Handy. Antibacterial properties of silver nanoparticles grown in situ and anchored to titanium dioxide nanotubes on titanium implant against Staphylococcus aureus, Nanotoxicology, 14:1. (2020). 97-110. https://doi.org/ 10.1080/17435390.2019.1665727

[143] S.M. Emarati & M. Mozammel. Efficient one-step fabrication of superhydrophobic nano-TiO2/TMPSi ceramic composite coating with enhanced corrosion resistance on 316L. Ceramics International. 46(2). (2020). 1652–1661. https://doi.org/10.1016/j.ceramint.2019.09.13

[144] T. Naz, A. Rasheed, S. Ajmal, N. Sarwar. T. Rasheed, M.M.Baig, M.S. Zafar, D.J. Kang, G. Dastgeer. A facile approach to synthesize ZnO-decorated titanium carbide nanoarchitectures to boost up the photodegradation performance,Ceramics International, Volume 47. 23. (2021). 33454-33462. doi.org/10.1016/j.ceramint.2021.08.252

[145] C. Chambers, S.B. Stewart, B. Su, H.F. Jenkinson, J.R. Sandy, A.J. Ireland. Silver doped titanium dioxide nanoparticles as antimicrobial additives to dental polymers, Dental Materials. 33.3. (2017). e115-e123.https://doi.org/10.1016/j.dental.2016.11.008

[146] A. Sodagar, S. Khalil, M.Z. Kassaee, A.S. Shahroudi, B. Pourakbari, A. Bahador. Antimicrobial properties of poly (methyl methacrylate) acrylic resins incorporated with silicon dioxide and titanium dioxide nanoparticles on cariogenic bacteria. J Orthod (2016). Sci. 5(1):7-13. https://doi.org/ 10.4103/2278-0203.176652.

[147] N.Y. Elmehbad, N.A. Mohamed, N.A. Abd El-Ghany. Evaluation of the antimicrobial and anti-biofilm activity of novel salicylhydrazido chitosan derivatives impregnated with titanium dioxide nanoparticles, International Journal of Biological Macromolecules. 205. (2022). 719-730. https://doi.org/10.1016/j.ijbiomac.2022.03.076

[148] W. McKinney,M. Jackson, T.M. Sager, T. J.S. Reynolds, B. T. Chen, A. Afshari, K. Krajnak, S. Waugh. etc., Pulmonary and cardiovascular responses of rats to inhalation of a commercial antimicrobial spray containing titanium dioxide nanoparticles. International Forum for Respiratory Research, (2012). 447-457. 24.7. doi.org/10.3109/08958378.2012.685111

[149] J. Del-Pozo, M. Crumlish, H.M. Ferguson, J.F. Turnbull. A retrospective cross-sectional study on candidatus arthromitus associated rainbow trout gastroeuterities (RTGE) in the UK. Aquacultrue. (2009). 290:22-7

[150] Y.H. Tsuang, J.S. Sun, Y.C. Huang. et al. Studies of photokilling of bacteria using titanium dioxide nanoparticles. Artificial Organs (2008). 32: 167–174

[151] N.G. Chorianopoulos, D.S. Tsoukleris, E.Z. Panagou, P. Falaras, G. Nychas. Use of titanium dioxide (TiO₂) photocatalysts as alternative means for Listeria monocytogenes biofilm disinfection in food processing. Food Microbiol. . (2010). 28:164-70

[152] F. Martinez-Gutierrez, P.L. Olive, A. Banuelos, E. Orrantia, N. Nino, E.M. Sanchez, … Av-Gay. Synthesis, characterization, and evaluation of antimicrobial and cytotoxic effect of silver and titanium nanoparticles. Nanomedicine: Nanotechnology, Biology and Medicine. 6.5. (2010). 681–688. https://doi.org/10.1016/j.nano.2010.02.001

[153] M. Bonnet, C. Massard, P.P. Veisseire, O. Camares, K.O. Awitor, Environmental Toxicity and Antimicrobial Efficiency of Titanium Dioxide Nanoparticles in Suspension. Journal of Biomaterials and Nanobiotechnology. (2015). 06 (03).213 - 224. ff10.4236/jbnb.2015.63020ff. ffhal-01829427f

[154] Y.Y. Huang, H. Choi. Y. Kushida, B. Bhayana, Y. Wang, MR. Hamblin, Broad-Spectrum Antimicrobial Effects of Photocatalysis Using Titanium Dioxide Nanoparticles Are Strongly Potentiated by Addition of Potassium Iodide. Antimicrob Agents Chemother. 22; (2016). 60(9):5445-53. https://doi.org/ 10.1128/AAC.00980-16

[155] S. Lyu, X. Wei, J. Chen, C. Wang, X. Wang, D. Pan, Titanium as a beneficial element for crop production. – Frontiers in Plant Science. (20178: 597. https://doi.org/ 10.3389/fpls.2017.00597

[156] L Zheng, F. Hong, S. Lu, C. Liu. Effect of nano- TiO₂ on strength of naturally aged seeds and growth of spinach. – Biological Trace Element Research. (2005). 104: 83-92. https://doi.org/ 10.1385/BTER:104:1:083

[157] H. Dağhan. Effects of TiO₂ nanoparticles on maize (Zea mays L.) growth, chlorophyll content and nutrient uptake. Applied Ecology and Environmental Research 16.5. (2018).:6873-6883. https://doi.org/10.15666/aeer/1605_68736883

[158] J. Lian, L. Zhao, J. Wu, H. Xiong, Y. Bao, A. Zeb, J. Tang, W. Liu. Foliar spray of TiO₂ nanoparticles prevails over root application in reducing Cd accumulation and mitigating Cd-induced phytotoxicity in maize (Zea mays L.). Chemosphere. (2020). 239. .124794. https://doi.org/10.1016/j.chemosphere.2019.124794

[159] S.P.T. Waani, S. Irum, I. Gul, K. Yaqoob,M.S. Khalid, M.A. Manzoor, U., Noor, T., Ali S., Rizwan, M., Arshad, M., TiO₂ nanoparticles dose, application method and phosphorous, levels influence genotoxicity in Rice (Oryza sativa L.), soil enzymatic activities and plant growth. Ecotoxicology and Environmental Safety. (2021). 213.111977. doi.org/10.1016/j.ecoenv.2021.111977

[160] T.C. Wang, N. Lu, J. Li and Y. Wu. Plasma-TiO₂ catalytic method for high-efficiency remediation of p-nitrophenol contaminated soil in pulsed discharge. Environmental Science and Technology. (2011). 45: 9301–9307

[161] B. Karnchanasest, Santisukkasaem. A preliminary study for removing Phenanthrene and Benzo(a)Pyrene from soil by nanoparticles. J Applied Sciences 7. (2007) 3317–3321

[162] R. Žabar, T. Komel, J. Fabjan et al. Photocatalytic degradation with immobilised TiO_2 of three selected neonicotinoid insecticides: imidacloprid, thiamethoxam and clothianidin. Chemosphere. (2012). 89: 293–301

[163] A.N. Wang, Y. Teng, X.F. Hu. et al. Diphenylarsinic acid contaminated soil remediation by titanium dioxide (P25) photocatalysis: degrada.tion pathway, optimization of operating parameters and effects of soil properties. The Science of the Total Environment (2016a). 541: 348–355

[164] B.F. Abramović, D.V. Šojić, V.B. Anderluh, N.D. Abazović, M.I. Čomor. Nitrogen-doped (2021). TiO_2 suspensions in photocatalytic degradation of mecoprop and (4-chloro-2-methylphenoxy) acetic acid herbicides using various light sources,Desalination. 244. (2009). 1–3.293-302. https://doi.org/10.1016/j.desal.2008.06.008

[165] A. Abdelhaleem, W. Chu. Photodegradation of 4-chlorophenoxyacetic acid under visible LED activated N-doped TiO_2 and the mechanism of stepwise rate increment of the reused catalyst,Journal of Hazardous Materials.338, (2017). 491-501. https://doi.org/10.1016/j.jhazmat.2017.05.056

[166] X. Wu, J. Hu, F. Wu, X. Zhang, B. Wang, Y. Yang, G. Shen, J. Liu, S. Tao, X. Wang, Application of TiO2 nanoparticles to reduce bioaccumulation of arsenic in rice seedlings (Oryza sativa L.): A mechanistic study. Journal of Hazardous Materials. 405. (2021). 124047. https://doi.org/10.1016/j.jhazmat.2020.124047

[167] T. Kiany, L. Pishkar, N. Sartipnia, A. Iranbakhsh, G. Barzin. Effects of silicon and titanium dioxide nanoparticles on arsenic accumulation, phytochelatin metabolism, and antioxidant system by rice under arsenic toxicity. Environ Sci Pollut Res Int. 2022 May;29(23):34725-34737. https://doi.org/ 10.1007/s11356-021-17927-z.

[168] M. Qi, Y. Liu,T. Li. Nano-TiO_2 improve the photosynthesis of tomato leaves under mild heat stress. Biological Trace Element Research. (2013). 156(1-3):323-328. https://doi.org/10.1007/s12011-013-9833-2

[169] S.M. Dofing. Phenological development–yield relationships in spring barley in a subarctic environment. Canadian Journal of Plant Science. (1995). 75(1):93-97. https://doi.org/ doi:10.4141/ cjps95-015

[170] A. Mattiello, and L. Marchiol. Application of Nanotechnology in Agriculture: Assessment of TiO2 Nanoparticle Effects on Barley. Chapter 2 Book. Application of Titanium Dioxide. Eds: Magdalena Janus. (2016). https://doi.org/ 10.5772/intechopen.68710

[171] N.A. Sagar, N. Kumar, R. Choudhary, V.K. Bajpai, H. Cao, S. Shukla, S. Pareek, Prospecting the role of nanotechnology in extending the shelf-life of fresh produce and in developing advanced packaging, Food Packaging and Shelf Life. 34.(2022). 100955. doi.org/10.1016/j.fpsl.2022.100955

[172] E. Khojah, R. Sami, M. Helal, A. Elhakem, N. Benajiba, M. Alharbi, M.S. Alkaltham, Effect of Coatings Using Titanium Dioxide Nanoparticles and Chitosan Films on Oxidation

during Storage on White Button Mushroom. Crystals. 11. (2021). 603.
https://doi.org/10.3390/cryst11060603

[173] Z. Liu, M. Du, H. Liu, K. Zhang, X. Xu, K. Liu, Q. Liu. Chitosan films incorporating litchi peel extract and titanium dioxide nanoparticles and their application as coatings on watercored apples. Progress in Organic Coatings. 151. (2021). 106103. https://doi.org/10.1016/j.porgcoat.2020.10610

[174] M. Helal, R. Sami, E. Khojah et al. Evaluating the coating process of titanium dioxide nanoparticles and sodium tripolyphosphate on cucumbers under chilling condition to extend the shelf-life. Sci Rep 11, (2021). 20312 https://doi.org/10.1038/s41598-021-99023-3

[175] M.A. Sani, M. Maleki. Eghbaljoo-Gharehgheshlaghi, H., Khezerlou, A., Mohammadian, E., Liu, Q., Jafari S.M. Titanium dioxide nanoparticles as multifunctional surface-active materials for smart/active nanocomposite packaging films,Advances in Colloid and Interface Science,Volume 300 . (2022). 102593. doi.org/10.1016/j.cis.2021.102593

[176] R. Abazari A.R. Mahjou S.A. Sanati. Facile and efficient preparation of anatase titania nanoparticles in micelle nanoreactors: morphology, structure, and their high photocatalytic activity under UV light illumination. RSC Adv. 4. (2014). 56406. https://doi.org/10.1039/C4RA10018B

[177] J. Carbajo, A. Tolosana-Moranchel, J.A. Casas, et al. Analysis of photoefficiency in TiO_2 aqueous suspensions: effect of titania hydrodynamic particle size and catalyst loading on their optical properties. Applied Catalysis B (2018). 221: 1–8

[178] Hamzah, Maytham & Jabbar, Abdullah, S. Mezan, A Tuama, M. Agam. Fabrications of PS/TiO2 nanocomposite for solar cells applications. AIP Conference Proceedings. (2019). 2151. 020011. 10.1063/1.5124641

[179] D.Chen, Q. Wang, R. Wang, G. Shen. Ternary oxide nanostructured materials for supercapacitors: A review. J. Mater. Chem. A. 3. (2015). 10158–10173

[180] S.K. Dhungel, J.G. Park, Optimization of paste formulation for TiO_2 nanoparticles with wide range of size distribution for its application in dye sensitized solar cells. Renewable Energy 35 (2010). 2776-2780

[181] I.C. Maurya, S. Singh, S. Senapati, P. Srivastava & L. Bahadur. Green synthesis of TiO2 nanoparticles using Bixa orellana seed extract and its application for solar cells. Solar Energy. 194. (2019). 952–95. https://doi.org/10.1016/j.solener.2019.10.090

[182] G.R. Mirshekari & A.P. Shirvanian. Electrochemical behavior of titanium oxide nanoparticles for oxygen reduction reaction environment in PEM fuel cells. Materials Today Energy, 9, (2018). 235–239. https://doi.org/ 10.1016/j.mtener.2018.05.015

[183] L. Kumaresan, H. Amir, G. Shanmugavelayutham, C. Viswanathan, Plasma assists titanium nitride and surface modified titanium nitride nanoparticles from titanium scraps for magnetic properties and supercapacitor applications,Ceramics International.Volume 48. (2022). Issue 20. .30393-30406. https://doi.org/10.1016/j.ceramint.2022.06.317

[184] S. Ishii, R.P. Sugavaneshwar & T. Nagao. Titanium Nitride Nanoparticles as Plasmonic Solar Heat Transducers. The Journal of Physical Chemistry C. (2016). 120(4). 2343–2348. https://doi.org/10.1021/acs.jpcc.5b09604

[185] A. Achour, M. Chaker, H. Achour, A. Arman, M. Islam, M. Mardani, … T. Brousse. Role of nitrogen doping at the surface of titanium nitride thin films towards capacitive charge storage enhancement. Journal of Power Sources, 359. (2017). 349–354. https://doi.org/10.1016/j.jpowsour.2017.05.07

[186] J. Zhang, H. Hu, X. Liu, D.S. Li. Development of the applications of titanium nitride in fuel cells. Materials Today Chemistry. 11. (2019a). 42–59. https://doi.org/10.1016/j.mtchem.2018.10.005

[187] N. Parveen, M.O. Ansari, S.A. Ansari, P. Kumar, Nanostructured Titanium Nitride and Its Composites as High-Performance Supercapacitor Electrode Material. Nanomaterials. 13. (2023). 105. https://doi.org/10.3390/nano13010105

[188] A. Achour, R.L. Porto, M.A. Soussou, M. Islam, M. Boujtita,K.A. Aissa, K.A., Le Brizoual, L., Djouadi, A., T. Brousse, Titanium nitride films for micro-supercapacitors: Effect of surface chemistry and film morphology on the capacitance. J. Power Sources. 300. (2015). 525–532

[189] Z. Gonzalez, J. Yus, R. Moratalla B. Ferrari, Electrophoretic deposition of binder-free TiN nanoparticles to design 3D microstructures. The role of sintering in the microstructural robustness of supercapacitor electrodes.Electrochimica Acta,Volume 369,(2021),https://doi.org/10.1016/j.electacta.2020.137654

[190] H. Wang, Q. Chen, L.Wen, S. Song, X. Hu and G. Xu, Titanium-nitride-based integrated plasmonic absorber/emitter for solar thermophotovoltaic application. Photon. Res. 3, (2015). 329-334.

[191] E.T. Bekele, E.A. Zereffa, N.S. Gultom. D.H. Kuo, B.A. Gonfa, F.K. Sabir, "Biotemplated Synthesis of Titanium Oxide Nanoparticles in the Presence of Root Extract of Kniphofia schemperi and Its Application for Dye Sensitized Solar Cells", International Journal of Photoenergy. 6648325. (2021). 12. https://doi.org/10.1155/2021/6648325

[192] Z. Su, D. Pan, H. Han, M. Lin, X. Hu, X. Wu, Synthesis, Properties and Application of Polyaniline/Titanium Carbide Nanoparticles Modified Electrode. Int. J. Electrochem. Sci., (2015). Vol. 10

[193] R. Syamsai, P. Kollu, S.K. Jeong, et al. Synthesis and properties of 2D-titanium carbide MXene sheets towards electrochemical energy storage applications. Ceram Int. (2017). 43: 13119–13126

[194] M. Mariano, O. Mashtalir, F.Q. Antonio, et al. Solution-processed titanium carbide MXene films examined as highly transparent conductors. Nanoscale; (2016). 8: 16371–16378

[195] A. Kumar, T. Siddiqui, S. Pandit, A. Roy, A. Gacem, A., A.A. Souwaileh, A.S. Mathuriya, T. Fatma, P. Sharma, S. Rustagi, et al. Application of Biogenic TiO_2 Nanoparticles as ORR Catalysts on Cathode for Enhanced Performance of Microbial Fuel Cell. *Catalysts, 13,* (2023). 937. https://doi.org/10.3390/catal13060937

[196] T. Marolt, A.S. Škapin J. Bernard, P. Živec, M. Gaberšček, Photocatalytic activity of anatase-containing facade coatings. Surface and Coatings Technology.206. (2011). 6.2011.1355-1361.https://doi.org/10.1016/j.surfcoat.2011.08.053

[197] N.S. Allen, M. Edge, A. Ortega,C.M. Liauw, J. Stratton & R.B. McIntyre, Behaviour of nanoparticle (ultrafine) titanium dioxide pigments and stabilisers on the photooxidative stability of water based acrylic and isocyanate based acrylic coatings. Polymer Degradation and Stability. (2002). 78(3). 467–478. https://doi.org/10.1016/s0141-3910(02)00189-1.

[198] N.S. Allen, R. McIntyre, J.M. Kerrod, Hill, C., & Edge, M. Photo-Stabilisation and UV Blocking Efficacy of Coated Macro and Nano-Rutile Titanium Dioxide Particles in Paints and Coatings. Journal of Polymers and the Environment (2018). doi:10.1007/s10924-018-1298-0.

[199] L. Ying, Y. Wu, C. Nie, C. Wu, G. Wang, Improvement of the Tribological Properties and Corrosion Resistance of Epoxy–PTFE Composite Coating by Nanoparticle Modification. Coatings. 11. (2021). 10. doi.org/10.3390/ coatings11010010

[200] D.V. Mashtalyar, S.L. Sinebryukhov, I.M. Imshinetskiy, A.S. Gnedenkov, K.V. Nadaraia, A. Ustinov & S.V. Gnedenkov, Hard wearproof PEO-coatings formed on Mg alloy using TiN nanoparticles. Applied Surface Science. (2019). 144062. https://doi.org/10.1016/j.apsusc.2019.144062

[201] S. Zhang, X. Liang, G.M. Gadd, Q. Zhao, Advanced titanium dioxide-polytetrafluorethylene (TiO$_2$-PTFE) nanocomposite coatings on stainless steel surfaces with antibacterial and anti-corrosion properties. Applied Surface Science, 490. (2019b). 231-241. doi.org/10.1016/j.apsusc.2019.06.070

[202] A. Khataee, L. Moradkhannejhad, V. Heydari, B. Vahid and S.W. Joo, "Self-cleaning acrylic water-based white paint modified with different types of TiO$_2$ nanoparticles", Pigment & Resin Technology. (2016), 45 (1). 24-29. https://doi.org/10.1108/PRT-09-2014-0070

[203] S. Chen, Y. Guo, H. Zhong, S. Chen, J. Li, Z. Ge, J. Tang, J. Synergistic antibacterial mechanism and coating application of copper/titanium dioxide nanoparticles, Chemical Engineering Journal. 256.(2014)238-246. https://doi.org/10.1016/j.cej.2014.07.006

[204] V.S. Mohite, M.M. Darade, R.K. Sharma, S.H. Pawar, Nanoparticle Engineered Photocatalytic Paints: A Roadmap to Self-Sterilizing against the Spread of Communicable Diseases. Catalysts. 12. (2022). 326. doi.org/ 10.3390/catal12030326

[204] P.B. Chouke, T. Shrirame, A.K. Potbhare, A. Mondal, A.R. Chaudhary, S. Mondal, S.R. Thakare, E. Nepovimova, M. Valis, K. Kuca, Bioinspired metal/metal oxide nanoparticles: A road map to potential applications, Materials Today Advances, 16 (2022) 100314

[206] V.N. Sonkusare, R.G. Chaudhary, G.S. Bhusari, A.R. Rai, H.D. Juneja, Microwave-mediated synthesis, photocatalytic degradation and antibacterial activity of α-Bi2O3 microflowers/novel γ -Bi$_2$O$_3$ microspindles, Nano-Structures & Nano-Objects, 13 (2018), 121-131

[207] V.N. Sonkusare, R.G. Chaudhary, G.S. Bhusari, A. Mondal, A.K. Potbhare, R.K. Mishra, H.D. Juneja A.A. Abdala Mesoporous Octahedron-Shaped Tricobalt Tetroxide Nanoparticles for Photocatalytic Degradation of Toxic Dyes, ACS Omega, 5 (2020), 7823-7835

[208] M.S. Nagmote, A.R. Rai, R. Sharma, M.F. Desimone, R.G. Chaudhary, N.B. Singh, Bioremediation of heavy metals using microorganisms, CRC Press, 2024, pp. 168-190

[209] K. Ancy, C. Vijilvani, M.R. Bindhu, S.J.S.c Bai, K.S. Almaary, T.M. Dawoud, ., ... M.S. Alfadul, Visible light assisted photocatalytic degradation of commercial dyes and waste water

by Sn–F co-doped titanium dioxide nanoparticles with potential antimicrobial application. Chemosphere, 277, (2021). 130247. https://doi.org/10.1016/j.chemosphere.2021.13

[210] T.A. Saleh, V.K. Gupta, Functionalization of tungsten oxide into MWCNT and its application for sunlight-induced degradation of rhodamine B, J. Colloid Interface Sci. 362. (2011).337-334

[211] N.R. Nicomel, K. Folens, P.V.D. Voort, G.D. Laing, Technologies for arsenic removal fromWater: current status and future perspectives, Int. J. Environ. Res. Public Health. (2015) 13 (1) 62

[212] X. DingG. Li, S. Zhang, J. Chen, J. Yuan, Preparation and characterization of hydrophobic TiO2 pillared clay: the effect of acid hydrolysis catalyst and doped Pt amount on photocatalytic activity, J. Colloid Interface Sci. 320 (2008) 501–507

[213] L. Xiea, Z. Zheng, S. Weng, J. Huanga et al., Morphology engineering of V_2O_5/ TiO_2 nanocomposites withenhanced visible light-driven photofunctions for arsenic removal, Environ. Appl. Catal. B 184 (2016) 347–354

[214] M. Sadegh, M. Irandoust, F. Khorshidi, M. Feyzi, Removal of Arsenic (III) from natural contaminated water using magnetic nanocomposite: kinetics and isotherm studies, J. Iran. Chem. Soc. (2016)7 (13) 1175–1188

[215] R.K. Khan, A.A. Farghaly, T.A., Silva, D. Ye & M.M. Collinson, Gold Nanoparticle-Decorated Titanium Nitride Electrodes Prepared by Glancing Angle Deposition for Sensing Applications. ACS Applied Nano Materials. (2019). https://doi.org/10.1021/acsanm.8b02354

[216] N. Akhtary and A. Zubair, Titanium nitride based plasmonic nanoparticles for photovoltaic application. (2023). 2. (7) Optics Continuum 1702. https://doi.org/10.1364/OPTCON.493184

[217] I. Zelepukin, A. Popov, V. Shipunova, G. Tikhonowski, A.B. Mirkasymov, et al. Laser-synthesized TiN nanoparticles for biomedical applications: Evaluation of safety, biodistribution and pharmacokinetics. Materials Science and Engineering: C.120. (2021). 111717.

[218] U. Guler, J.C. Ndukaife, G.V Naik, A.G.A.N., Nnanna, A.V. Kildishev, V.M. Shalaev, A. Boltasseva, Local heating with titanium nitride nanoparticles. (2013). CLEO: Technical Digest

[219] X. Yuan, L. Cheng, L. Kong, X. Yin, & L. Zhang, Preparation of titanium carbide nanowires for application in electromagnetic wave absorption. Journal of Alloys and Compounds. 596. (2014). 132–139. https://doi.org/10.1016/j.jallcom.2014.01.022

[220] H. Feng, W. Wang, M. Zhang, S. Zhu, Q. Wang,J. Liu & S. Chen, 2D titanium carbide-based nanocomposites for photocatalytic bacteriostatic applications. Applied Catalysis B: Environmental. 266. (2020). 118609. https://doi.org/10.1016/j.apcatb.2020.118609

[221] A. Hashim, M.A. Habeeb, A. Khalaf, A. Hadi, Synthesis of Novel (Polymer Blend-Titanium Carbide) Nanocomposites and Studying their Characterizations for Piezoelectric Applications. Journal of University of Babylon. Pure and Applied Sciences. (2018). Vol. (26-6)

[222] V.I. Zakomirnyi, I.L. Rasskazov, V.S. Gerasimov, A.E. Ershov, S.P. Polyutov, S.V. Karpov & H. Ågren, Titanium nitride nanoparticles as an alternative platform for plasmonic waveguides in the visible and telecommunication wavelength ranges. Photonics and Nanostructures - Fundamentals and Applications. 30. (2018). 50–56. https://doi.org/10.1016/j.photonics.2018.04.0

[223] X. Hu, Y. Wang, M. Xu, L. Zhang, J. Zhang, J., W. Dong, Development of photocrosslinked salecan composite hydrogel embedding titanium carbide nanoparticles as cell scaffold. International Journal of Biological Macromolecules 123. (2019). 549–557. https://doi.org/10.1016/j.ijbiomac.2018.11.125.

[224] Zong, L., H. Wu, H. Lin, et al. A polyoxometalate-functionalized two-dimensional titanium carbide composite MXene for effective cancer theranostics. Nano Res. 11, (2018). 4149–4168 doi.org/10.1007/s12274-018-2002-3

[225] A. Schlange, A.R. dos Santos, B. Hasse, B. J.M. Etzold, U. Kunz & T. Turek, Titanium carbide-derived carbon as a novel support for platinum catalysts in direct methanol fuel cell application. Journal of Power Sources. 199. (2012). 22–28. https://doi.org/10.1016/j.jpowsour.2011.09.107

[226] A. Hashim, Z. Hamad, A. Hashim & Z. Hamad, Humidity Sensing Performance of Polymer Blend-Titanium Nitride Nanocomposites: Structural, Electrical, and Optical Properties. 19. (2021). 893–903

[227] Sood A., Desseigne, M., Dev, A., Maurizi, L., Kumar, A., et al., A Comprehensive Review on Barium Titanate Nanoparticles as a Persuasive Piezoelectric Material for Biomedical Applications: Prospects and Challenges. Small, (2023). 19 (12). 2206401.

[228] A.A. Shah, A. Khan, S. Dwivedi, J. Musarrat, A. Azam, Antibacterial and Antibiofilm Activity of Barium Titanate Nanoparticles. Materials Letters.229 (2018) .130-133.https://doi.org/10.1016/j.matlet.2018.06.107

Green Synthesis and Emerging Applications of Frontier Nanomaterials   Materials Research Forum LLC
Materials Research Foundations 169 (2024) 315-330   https://doi.org/10.21741/9781644903278-12

Chapter 12

Recent advancements in supercapacitors of bismuth oxide nanomaterials

A.P. Angre[1], P.S. Gaikar[2], P.A. Patil[3], R.G. Chaudhary[4], S. Mondal[4], S.H. Mahmood[5], and T.L. Lambat[6,*]

[1]Department of Physics, Ramnarain Ruia Autonomous College, Matunga, Mumbai 400019, Maharashtra, India

[2]Department of Physics, Rayat Shikshan Sanstha's Karmaveer Bhaurao Patil College Vashi, Navi Mumbai- 400703, Maharashtra, India

[3]Department of Chemistry, Changu Kana Thakur Arts, Commerce & Science, College, New Panvel (Autonomous), Raigad, 410206, Maharashtra, India

[4]Post Graduate Department of Chemistry, Seth Kesarimal Porwal College, Kamptee 441001, Maharashtra, India

[5]Department of Physics, The University of Jordan, Amman 11942, Jordan.

[6]Department of Chemistry, Manoharbhai Patel College of Arts, Commerce & Science, Deori-441901,Gondia, Maharashtra, India

* lambatges@gmail.com

Abstract

Bismuth oxide nanomaterials (NMs) are effective materials for supercapacitor and battery technologies due to their redox behavior, superior charge storage capability, and being eco-friendly. Accordingly, we have witnessed a growing interest in these NMs applications and commercialization of effective energy storage devices. Recent studies have fused on the investigation of their supercapacitive properties *via* electrochemical techniques in an effort to understand their charge storage mechanism and improve their electrochemical performances. In this chapter, the article provides information on the chemical method preparation of bismuth oxide (Bi_2O_3) for supercapacitor applications. We present a summary of energy storage devices, structure and bismuth oxide electrode materials available in the market. Furthermore, the key challenges and future perspectives of bismuth oxide for energy storage applications are discussed.

Keywords

Bismuth Oxide NMs, Supercapacitor, Electrochemical Technique, Cyclic Voltammetry, Charge-Discharge Measurement

Contents

1.   Introduction

Society is mostly facing problems in the arena of energy storage, so there is a vast need for clean, effective, and sustainable energy for the latest energy storage and conversion devices and technologies [1–6]. Nowadays, supercapacitors are fascinating and capable energy storage devices because of their features like ecological friendliness, high power density, extended stability, more prolonged and outstanding reversibility, and the potential to provide moderate energy density compared to conventional batteries and capacitors [7-10, 4]. Basically, supercapacitors have three types, electrical double-layer capacitors (EDLC), pseudocapacitors and hybrid capacitors , shown in Fig 1. [10–14].

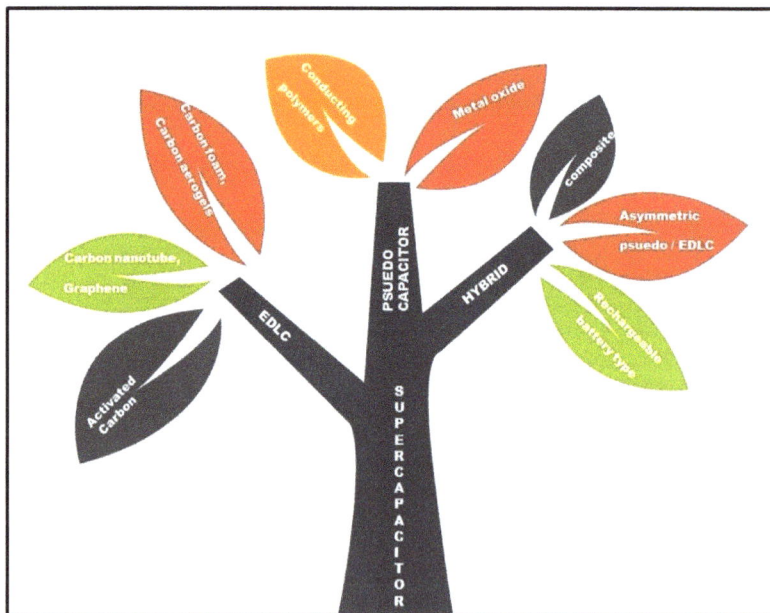

Figure 1. Schematic diagram of types of supercapacitors

Charge separation occurs at the electrode/electrolyte interface in electrical double-layer capacitors. Carbon-based materials such as carbon nanotubes, single-walled/multi-walled carbon nanotubes, graphene, fullerene, etc. are examples of electrical double-layer capacitors [15-16, 2]. However, materials like transition metal oxides/hydroxides for pseudocapacitors should have several oxidation states for efficient oxidation-reduction charge transfer [17, 18]. The Ru-based NMs were reported to have good specific capacitance, outstanding reversibility and significant life cycles [19, 20]. However, the high relative cost and limited availability of Ru-based NMs have driven the search for new cost-effective alternatives that can be readily available and exhibit similar or improved electrochemical performance [1, 21, 22]. Accordingly, several NMs were investigated for that purpose, including, and not limited to, NiO [23], Mn_3O_4 [24], Co_3O_4 [25] and Bi_2O_3 [26]. Bismuth-based NMs like sulfides, selenides, oxides, ferrites, etc., have been broadly used as biosensors, gas sensors, catalysts, optical materials, supercapacitors, [27–32]. The major applications of bismuth oxide is shown in Fig. 2. Specifically, Bi_2O_3 was found to exibit favorable characteristics for supercapacitor electrode material due to its high electrochemical stability, high redox reversibility, wide band gap, outstanding ionic conductivity, and relatively high power and capacity [33]. Consequently, the characteristics of bismuth hydroxides and oxides as potential candidates for supercapacitor or ultracapacitor applications have been reported by many investigators.

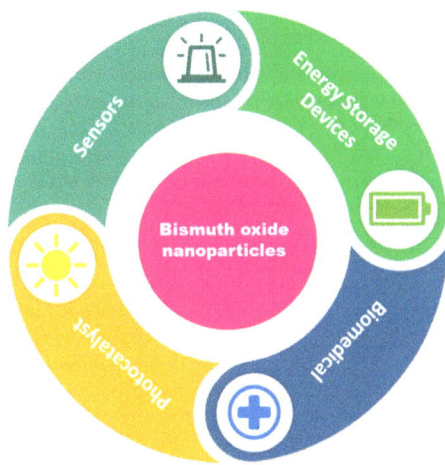

Figure 2. Major applications of Bismuth oxide nanoparticles

2. Bismuth Oxide Structure

Bismuth is the heaviest group-V elements revealing semimetal behaviour with a narrow band overlap. It has several virtues and benefits for practical applications such as being non-hazardous, abundant, and cost-effective. Bismuth oxide (Bi_2O_3) is one of the most important compounds of bismuth for industrial applications; it is an essential p-type metal-oxide-semiconductor [34]. The important developments in nanotechnology have also attracted attention to the nano-nature and applicability of bismuth, which features shift from semimetal to semiconductor as a consequence of the reduction in its particle size [35]. There exist five crystallographic polymorphs of bismuth oxide (Bi_2O_3), namely, α-Bi_2O_3 (monoclinic), β-Bi_2O_3 (tetragonal), γ-Bi_2O_3 (body-centered cubic), δ-Bi_2O_3 (face-centered cubic), and ε-Bi_2O_3. (triclinic). The α-Bi_2O_3 is stable at room temperature (RT), and trnsforms ino the other stable (δ-Bi_2O_3) phase when heated to above 730 °C. When cooled, the high-temperature δ-Bi_2O_3 phase transforms back to the α-phase, and may pass through other intermediate metastable phases (sohwn in Table 1) depending on the cooling rate. Transformation of the δ-phase to the metastable β-phase occurs at about 850 °C, which subsequently transforms to the α-phase at about 303 °C upon further cooling. The metastable γ-phase develops at 639 °C upon cooling the δ-phase at a lower rate, and this metastable phase transforms to the stable α-phase at 500 °C upon further cooling. The γ-phase can persist down to room temperature by slow cooling, and the different metastable phases can be stabilized by doping the bismuth oxide with impurities [36]. **Table 1.** Bismuth oxide phases structure, space group, crystalline system and stability range [37].

3. Properties and application of bismuth oxide-based materials

Bi_2O_3 has been explored broadly due to its unique electrical and optical properties like large band gap, dielectric permittivity, refractive index, good carrier mobility and outstanding photoconductivity and photoluminescence [34].

Because of these unique characteristics, Bi_2O_3 has been considered for usage in numerous domains such as optoelectronics, functional ceramics, photoelectric materials, gas sensors, supercapacitors, fuel cells, photocatalysts, and high-temperature superconductors [38-39]. In addition, Bi_2O_3 is one of the main constituents in the engineering of transparent ceramic glass. In the industrial field of manufacturing technologies based on bismuth oxide, optical fibre has progressed significantly due to the variability of superior optical fibre production [27].

Phase	Structure	Space Group	Crystalline System	Stability Range
α- Bi_2O_3		$P2_1/c$	Monoclinic	R.T.–768 K
β- Bi_2O_3		$P\bar{4}2_1/c$	Tetragonal	748-918 K
γ- Bi_2O_3		I23	Body Centered Cubic	909-1023 K
δ- Bi_2O_3		$Fm\bar{3}m$	Cubic (fcc)	908-1163 K

4. Synthesis of bismuth oxide by chemical method

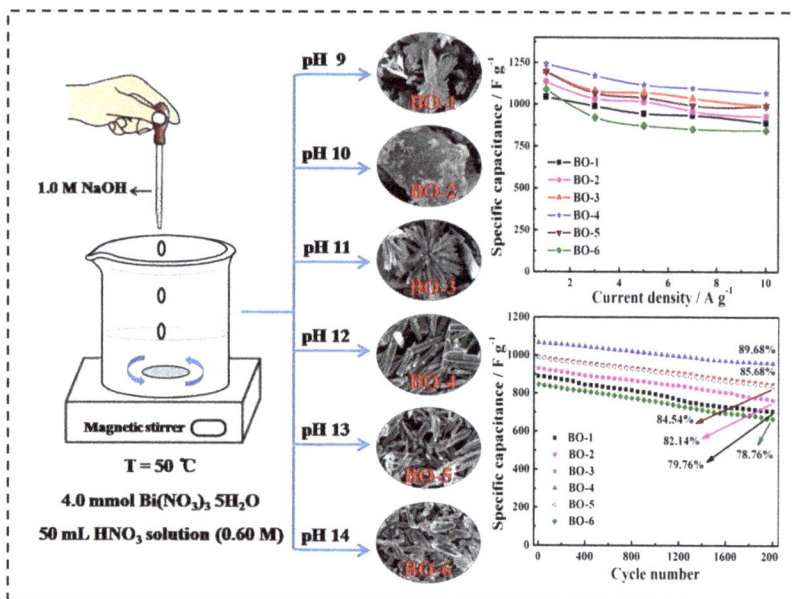

Figure 3. Synthesised bismuth oxide compounds with different morphologies by a simple chemical precipitation process under varying pH. Adopted with permission from ref 40. Copyright 2019, Elsevier.

Bismuth oxide is synthesized by various chemical and physical methods like Co-precipitation, sol-gel, hydrothermal, ball milling, electrodposition. Their lot The results of experimental work revealed that the synthesis methods and conditions can strongly impact the structure and phase composition of Bi_2O_3. Fig. 3 depicted Bi_2O_3 electrodes synthesised by co-precipitation process with varying pH valuesThe pH value is critical for controlling the chemical and physical processes involved in synthesis. As a result, the morphology of the Bi_2O_3 obtained for pH 9 to pH 14 resulted in distinct outcomes for supercapacitive properties [40]. Generally, Bi_2O_3 is synthesized by the oxidation of bismuth metal at 1073 K. It can be produced by thermal decomposition of hydroxides or carbonates in bismuth salt solutions. Calcination of these powders gives fine particles of Bi_2O_3. The spray pyrolysis method is also used to produce nanoscale Bi_2O_3 particles. The properties of Bi_2O_3 are strongly influenced by characteristics like particle size, morphological structure, and phase purity. Switzer and co-workers synthesized crystalline Bi_2O_3 thin films by electrodeposition methods on a gold substrate. Fig. 4 demonstated that SEM image of Bi_2O_3 thin film synthesized by SILAR Method.

Green Synthesis and Emerging Applications of Frontier Nanomaterials   Materials Research Forum LLC
Materials Research Foundations 169 (2024) 315-330   https://doi.org/10.21741/9781644903278-12

Figure 4. SEM image of Bi_2O_3 thin film by SILAR method

Killedar and co-workers have prepared Bi_2O_3 NMs using the spray pyrolysis method. Similarly, Lokhande et al. have reported the deposition of Bi_2O_3 thin films using the spray pyrolysis method [21]. Leontie and co-workers obtained Bi_2O_3 thin films prepared by thermal oxidation of bismuth thin films by oxidation heating and cooling temperatures [19, 34]. Metikos-Hukovic has synthesized Bi_2O_3 thin films by anodic oxidation of bismuth with n-type and p-type electrical conductivity. Based on synthesis technology, the electrical conductivity of Bi_2O_3 may change by five orders of magnitude, while its band gap may change from 2 to 3.96 eV [21, 41]. Representative scanning electron microscopy (SEM), Energy Dispersive X-ray (EDX) and transmission electron microscopy (TEM) images of bismuth oxide coral spherical nanoparticles synthesized by hydrothermal method are shown in Fig. 5 a,b and c respectively [42].

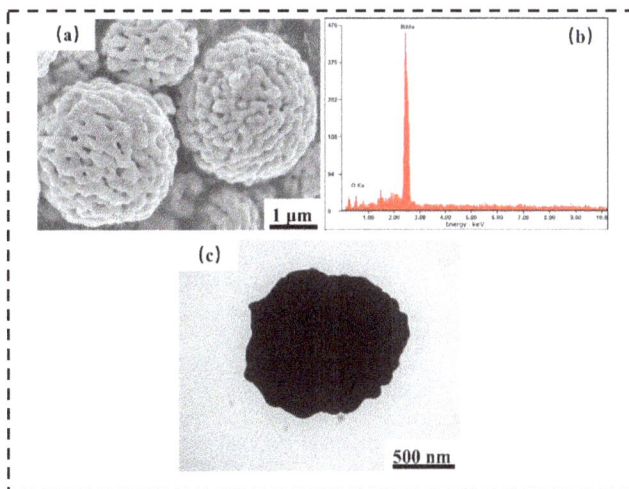

Figure 5. Bismuth Oxide nanoplates a) SEM b)EDX c) TEM Images Adopted with permission from ref 42. Copyright 2015, Elsevier.

5.   Bismuth oxide for supercapacitor applications

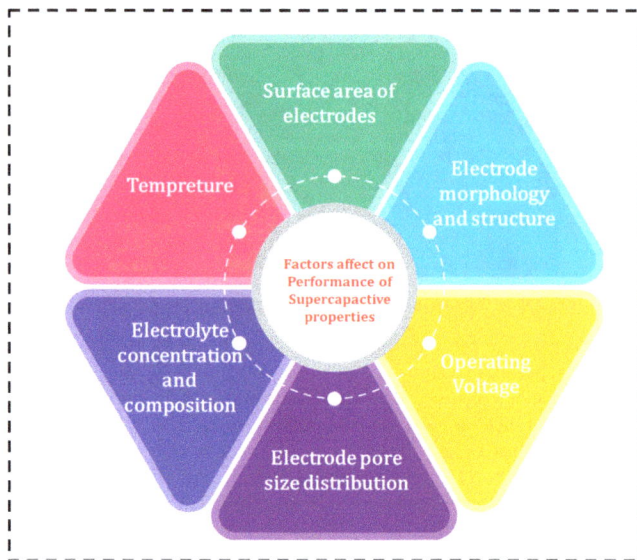

Figure 6. Factor affects the performance of supercapacitive properties of electrodes

Fig. 6 displays the supercapacitive characteristics of the electrodes with respect to a number of parameters, including morphology, structure, operating voltage, temperature, concentration and composition of electrolytes, and surface area, morphology, and structure of the electrodes. All of the aforementioned factors affect bismuth oxide and bismuth oxide based nanocomposite. Table 2. tabulated Supercapacitive performance of the Bismuth oxode based electrodes. The supercapacitive characteristics.In the field of supercapacitor applications, Bi-compounds exhibited relatively high not only capacitance but also performance rate. Bismuth Oxides are synthesized by various methods for supercapacitor applications. Gujar *et al.* synthesized ~39 nm sized nanoparticle of Bi_2O_3 on Cu substrate at room temperature by an electrodeposition method. The synthesized Bi_2O_3 thin film was successfully used as a supercapacitor electrode and exhibited a SC of 98 $F.g^{-1}$ at a scan rate of 20 $mV.s^{-1}$ in 1 M NaOH electrolyte [26]. Tong and co-workers prepared Bi_2O_3 on Ti plates by using the electrochemical deposition method. Electrochemical deposition of Bi_2O_3 nanobelts was carried out in a solution of Bismuth Nitrate, sucrose and Na_2-EDTA with a current density 40 $mA.cm^{-2}$ for 1 hour. The supercapacitive performance of the nanobelts was investigated with cyclic voltammetry; it exhibited a SC of 250 $F.g^{-1}$ at a scan rate of 100 $mV.s^{-1}$ in 1.0 M Na_2SO_4 aqueous electrolyte. The electrochemical reactions corresponding to the formation of Bi_2O_3 were described as follows.

$$NO_3^- + H_2O + 2e^- \longrightarrow NO_2^- + 2\ OH^-$$

$$Bi^{3+} + 2OH^- \longrightarrow Bi(OH)_2^+$$

$$2\ Bi\ (OH)_2^+ + 2\ OH^- \longrightarrow Bi_2O_3 + 3H_2O$$

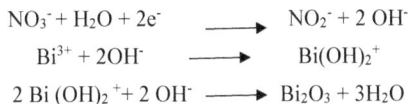

Shinde et al. synthesized flower-type Bi_2O_3 thin films using the chemical bath deposition method. Well-arranged nano-platelets were grown onto Ni-Foam and exhibited a SC found 577 $F.g^{-1}$ at a current density 1 $mA.cm^{-2}$ [20]. Ambare et.al. synthesized yellowish cubic Bi_2O_3 inter-connected upright standing nanoplates onto an elastic Ni foam using a spray pyrolysis aqueous route at 523 K with a constant spray rate ~10 ml. min^{-1}. The steps of the growth mechanism of synthesis of Bi_2O_3 are nucleation, coalescence and collision, particle formation, and growth [41]. The possible chemical reaction that takes place is as follows.

$$2Bi\ (NO_3)_2\ .6H_2O + 2H_2O + HNO_3 \xrightarrow{\ 523K\ } Bi_2O_3 + 3N_2 \uparrow + 8\ H_2 \uparrow + 11\ O_2 \uparrow$$

The cyclic voltammetry (CV) pattern of Bi_2O_3 electrode in 1 M Na_2SO_4 electrolyte exhibited a SC of 322.5 $F.g^{-1}$ at 5 $mV.s^{-1}$ scan rate in the potential range –0.8 to 1.9 V vs Ag/AgCl. As the scan rate increased, the current density and potential window increased, and the curves shifted to a positive potential end. The redox peaks revealed improvement for high scan rates because Bi_2O_3 supports a higher scan rate for higher redox activity. Due to a decrease in internal resistance, material species' charge and mass transfer resistance decreased. Hence, sustaining oxidation-reduction transition at a high scan rate can be challenging. Cubic Bi_2O_3 electrode SC decreased as the cycles increased; after 2500 cycles, it exhibited a SC of 223 $F.g^{-1}$ which was retained up to 5000 cycles. The stable SC is due to the excellent chemical stability and crystallinity of the Bi_2O_3 electrode, where ionic transportation is quicker with minor diffusion lengths.

Qiu $et\ al.$ used the oxidative metal vapor transport deposition technique to prepare ultrathin metastable nanowires like β-Bi_2O_3. The charge-discharge measurements revealed a SC of 691.3 $F.g^{-1}$ at a current density 2.0 $A.g^{-1}$ and an outstanding retention rate of about 75.5% over 3000 cycles at a current density of 10.0 $A.g^{-1}$[33]. CV and GCD images displayed for Bi_2O_3 electrodes in Fig. 7 (A) and (B) respectively [33]. Sun and co-workers synthesized hierarchical 3D Bi_2O_3 nanoplates with nanowires curled and connected. Bi_2O_3 electrodes exhibited outstanding electrochemical performance with 832 $F.g^{-1}$ SC at 1 $A.g^{-1}$ current density with a potential window 0 to -1.2 V in 6 M KOH aqueous electrolyte. The electrode retained 90% of its initial capacitance over 3000 cycles at 20 $A.g^{-1}$ current density. The asymmetric Bi_2O_3 and carbon nanotube supercapacitor was charged by 2 V DC power supply for 5 s, and a light-emitting diode driven by the supercapacitor kept lighting for about 20 s. [43]. Yuan and co-workers reported that a binder-free $Bi_2O_{2.33}$ was obtained on Ni-foam using a single-step chemical precipitation method assisted with TBAB. Flower-type $Bi_2O_{2.33}$ exhibited a maximum SC value of 557 $F.g^{-1}$ at 1 $mA.cm^{-2}$ with capacitance retention of 85% observed after 2000 charge-discharge cycles. [27, 44]. Zeng and co-workers prepared hierarchical nanobelts Bi_2O_3 using the electrodeposition method, and the average SC value was 250 $F.g^{-1}$, which revealed that the performance of the supercapacitor was also influenced by surface morphology [45].

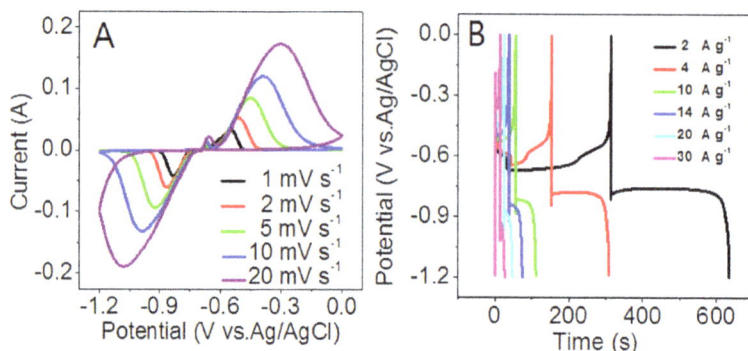

Figure 7. (A) CV collected at 1-20 mV.s⁻¹ scan rates (B) Galvanostatic charge-discharge curves at 1-30 A.g⁻¹ current densities for Bi_2O_3 electrodes [33]. Adopted with permission from ref 33. Copyright 2018, Elsevier.

Table 2. Supercapacitive performance of the bismuth oxode based electrodes

Sr.No.	Electrode	Specific Capacitance	Cyclic stability	Reference
1.	Bi_2O_3	1227.5 F g⁻¹ @ 2 mV s⁻¹	93.29% SC after 3500 cycles	[46]
2.	$Sr_3Bi_2O_6$	1228.7 F g⁻¹ @ 1A g⁻¹	75.1% SC @ 3000 cycles	[47]
3.	$BiMnO_3$	1500F g⁻¹ @ 20mA cm⁻²	90% SC@ 5000 cycles	[48]
4.	Bi_2O_3	366 F g⁻¹ @ 1A g⁻¹	87% SC over 2000 cycles	[49]
5.	$BiFeO_3$	461.9 F g⁻¹ @ 1A g⁻¹	94.4% SC over 2000 cycles	[50]
6.	Carbon quantum dot-Bi_2O_3	343 C g⁻¹ @ 0.5 A g⁻¹	100% SC over 2500 cycle	[51]
7.	Bi_2O_3- MnO_2	431.18 F g⁻¹ @ 5 mV s⁻¹	92% SC at 5000 cycle	[52]
8.	Bi_2O_3	421.76 F g⁻¹ @ 10 mA cm⁻²	60% SC @ 1000 cycles	[53]
9.	Bismuth ferrite (BFO)	678 F g⁻¹ @ 5 mV s⁻¹	90.5% SC after 3000 cycles.	[54]
10.	Bi_2O_3 /nitrogen-doped carbon dots	1046 F g⁻¹ @ 1A g⁻¹	83.5 % SC at 1500 cycle	[55]
11.	Bi-MOF	896.1 C g⁻¹ @ 0.5 A g⁻¹	71.2 SC after 2000 cycles	[56]
12.	Bi_2WO_6	131 F g⁻¹ @ 1 A g⁻¹	89% SC at 5000 cycles	[57]

13.	BiOCl	379 $F \cdot g^{-1}$ @ 1.25 A g^{-1}	80% SC over 5000 cycles	[58]
14.	Bi@BiVO₄	1105 F g^{-1} @ 1 A g^{-1}	-	[59]

Conclusion and future directions

In conclusion, by synthesizing chemically stable and mechanically robust thin films of Bi_2O_3 electrodes using simple, cheap and eco-friendly chemical processes, supercapacitors with superior specific capacitance, maximum energy density, and long cycle life were realized. The potential of Bi_2O_3 electrodes for applications in energy storage devices was motivated by their electrochemical properties and non-toxic nature. It has been found that Bi_2O_3 follows a redox behaviour that resembles a battery-type behaviour. Additionally, Bi_2O_3 materials function in the negative potential window from (-1–0 V) in aqueous alkaline electrolytes, which is different for various electrolytes. The difference in surface morphologies and specific surface area of the nanocomposites influence the charge–storage capability and specific capacitance of the Bi_2O_3 electrodes. Different electrolytes must be examined to extend the range of working potential windows in the upcoming years. The Bi_2O_3 electrochemical system needs further experimentation on mass transfer at the electrode/electrolyte interface and the kinetic analysis of the movement of ions for a better understanding of the reaction kinetics. This may push the supercapacitor technology forward, thus paving the way for next- generation energy storage devices.

Conflicts of Interest

The authors declare that they have no conflicts of interest.

References

[1] Xia X, Zhang Y, Chao D, Guan C, Zhang Y, Li L, Ge X, Inguez Bacho I, Tu J, Fan HJ (2014) Solution synthesis of metal oxides for electrochemical energy storage applications. Nanoscale 6:5008. https://doi.org/10.1039/c4nr00024b

[2] Forouzandeh P, Kumaravel V, Pillai S (2020) Electrode Materials for Supercapacitors: A Review of Recent Advances. Catalysts 10:969. https://doi.org/10.3390/CATAL10090969

[3] Gaikar PS, Gaikwad SL, Mahadule RK, Wakde GC, Angre AP, Bandekar AS, Arjunwadkar PR (2016) β-Cobalt Hydroxide as an Efficient Electrode for Electrochemical Supercapacitor Application. J Nanoeng Nanomanufacturing 6:157–160. https://doi.org/10.1166/jnan.2016.1278

[4] Li N, Li X, Yang C, Wang F, Li J, Wang H, Chen C, Liu S, Pan Y, Li D (2016) Fabrication of a flexible free-standing film electrode composed of polypyrrole coated cellulose nanofibers/multi-walled carbon nanotubes composite for supercapacitors. RSC Adv 6:86744–86751. https://doi.org/10.1039/C6RA19529F

[5] A. Mondal, H. Tanaya Das, S. Mondal, V. N. Sonkusare, R. G. Chaudhary (2023) Emerging nanomaterials in energy storage. Emerging Applications of Nanomaterials, 141, 294-326. https://doi.org/10.1016/j.electacta.2012.10.088

[6] Chen S, Xing W, Duan J, Hu X, Qiao S (2013) Nanostructured morphology control for efficient supercapacitor electrodes. J Mater Chem A, 1:2941–2954. https://doi.org/10.1039/c2ta00627h.

[7] S.K. Tarik Aziz, M. Awasthi, S. Guria, M. Umekar, I. Karajagi, S. K. Riyajuddin, K. V. R. Siddhartha, A. Saini, A.K. Potbhare, R.G. Chaudhary, V. Vishal, P. C. Ghosh, A. Dutta, Electrochemical water splitting by a bidirectional electrocatalyst, *STAR Protocols*, **2023**, 4, 102448.

[8] Shinde NM, Shinde P V., Mane RS, Ho Kim K (2021) Solution-method processed Bi-type nanoelectrode materials for supercapacitor applications: A review. Renew Sustain Energy Rev 135:110084. https://doi.org/10.1016/J.RSER.2020.110084

[9] Wang Y, Zhang L, Hou H, Xu W, Duan G, He S, Liu K, Jiang S (2020) Recent progress in carbon-based materials for supercapacitor electrodes: a review. J Mater Sci 2020 561 56:173–200. https://doi.org/10.1007/S10853-020-05157-6

[10] Gaikar P, Pawar SP, Mane RS, Nuashad M, Shinde D V (2016) Synthesis of nickel sufide as a promising electrode material for pseudocapacitor application. RSC Adv 6:112589–112593. https://doi.org/10.1039/C6RA22606J

[11] Iro ZS, Subramani C, Dash SS (2016) A Brief Review on Electrode Materials for Supercapacitor. Int J Electrochem Sci 11:10628–10643. https://doi.org/10.20964/2016.12.50

[12] Zhang F, Zhang T, Yang X, Zhang L, Leng K, Huang Y, Chen Y (2013) A high-performance supercapacitor-battery hybrid energy storage device based on graphene-enhanced electrode materials with ultrahigh energy density. Energy Environ Sci 6:1623–1632. https://doi.org/10.1039/c3ee40509e

[13] Sawangphruk M, Kaewsongpol T (2012) Direct electrodeposition and superior pseudocapacitive property of ultrahigh porous silver-incorporated polyaniline films. Mater Lett 87:142–145. https://doi.org/10.1016/j.matlet.2012.07.103

[14] A.K. Potbhare, SKT Aziz, Mohd. M. Ayyub, A. Kahate, R. Madankar, S. Wankar, A. Dutta, A. Abdala, S.H. Mohmood, R. Adhikari, R.G. Chaudhury, Bioinspired Graphene-based metal oxide nanocomposites for photocatalytic and electrochemical performances: An Updated Review, Nanoscale Advances, 2024. doi.org/10.1039/D3NA01071F.

[15] Bhilkar, P. R., Madankar, R. S., Shrirame, T. S., Utane, R. D., Potbhare, A. K., Yerpude, S., Chaudhary, R. G. (2022). Functionalized carbon nanomaterials: fabrication, properties, and applications. Mater. Res., 135, 55-82. https://doi.org/10.21741/9781644902172-4

[16] Gaikar PS, Angre AP, Wadhawa G, Ledade P V., Mahmood SH, Lambat TL (2022) Green synthesis of cobalt oxide thin films as an electrode material for electrochemical capacitor application. Curr Res Green Sustain Chem 5:100265. https://doi.org/10.1016/j.crgsc.2022.100265

[17] Shrirame, T.S., Khan, J.S., Umekar, M.S., Potbhare, A.K., Bhilkar, P.R., Bhusari, G.S., Chaudhary, R.G. (2022). Graphene-polymer nanocomposites for environmental remediation of organic pollutants. Metal nanocomposites for energy and environmental applications, 321-349. http://dx.doi.org/10.1007/978-981-16-8599-6_14.

[18]   De B, Banerjee S, Verma KD, Pal T, Manna P, Kar K (2020) Transition Metal Oxides as Electrode Materials for Supercapacitors. Springer Ser Mater Sci 302:89–111. https://doi.org/10.1007/978-3-030-52359-6_4

[19]   Yong-gang W, Xiao-gang Z (2004) Preparation and electrochemical capacitance of RuO 2 / TiO 2 nanotubes composites. 49:1957–1962. https://doi.org/10.1016/j.electacta.2003.12.023

[20]   Shinde NM, Xia QX, Yun JM, Singh S, Mane RS, Kim KH (2017) A binder-free wet chemical synthesis approach to decorate nanoflowers of bismuth oxide on Ni-foam for fabricating laboratory scale potential pencil-type asymmetric supercapacitor device. Dalt Trans 46:6601–6611. https://doi.org/10.1039/C7DT00953D

[21]   A.K. Potbhare, P. Bhilkar, S. Yerpude, R. Madankar, R. Shingda, R. Adhikari, R.G. Chaudhary, Nanomaterials as photocatalyst, Application of Emerging Nanomaterials and Nanotechnology, 148 (2023) 304-333.

[22]   Hu C, Huang Y, Chang K (2002) Annealing effects on the physicochemical characteristics of hydrous ruthenium and ruthenium ± iridium oxides for electrochemical supercapacitors. 108:117–127

[23]   Chouke, Prashant B., Ajay K. Potbhare, Nitin P. Meshram, Manoj M. Rai, Kanhaiya M. Dadure, Karan Chaudhary, Alok R. Rai, Martin F. Desimone, Ratiram G. Chaudhary, and Dhanraj T. Masram. Bioinspired NiO nanospheres: Exploring in vitro toxicity using Bm-17 and L. rohita liver cells, DNA degradation, docking, and proposed vacuolization mechanism. ACS Omega, 7 (2022): 6869-6884.

[24]   Ali G, Yusoff M, Ng Y, Lim H, Feng C (2015) Potentiostatic and Galvanostatic Electrodeposition of Manganese Oxide for Supercapacitor Application: A Comparison Study. Curr Appl Phys 15:1143–1147. https://doi.org/10.1016/j.cap.2015.06.022

[25]   Gaikar PS, Navale ST, Jadhav V V, Shinde P V, Dubal DP, Arjunwadkar PR, Stadler FJ, Naushad M, Ghfar AA, Mane RS (2017) A simple wet-chemical synthesis , reaction mechanism , and charge storage application of cobalt oxide electrodes of different morphologies. Electrochim Acta 253:151–162. https://doi.org/10.1016/j.electacta.2017.09.039

[26]   V.N. Sonkusare, R.G. Chaudhary, G.S. Bhusari, A.R. Rai, H.D. Juneja, Microwave-mediated synthesis, photocatalytic degradation and antibacterial activity of α-Bi2O3 microflowers/novel γ -Bi2O3 microspindles, Nano-Structures & Nano-Objects, 13 (2018), 121-131.

[27]   Devi N, Ray SS (2020) Performance of bismuth-based materials for supercapacitor applications: A review. Mater Today Commun 25:101691. https://doi.org/10.1016/j.mtcomm.2020.101691

[28]   S.K. Tarik Aziz, M. Umekar, I. Karajagi, S.K. Riyajuddin, K.V.R. Siddhartha, A. Saini, A.K. Potbhare, R.G. Chaudhary, V. Vishal, P.C. Ghosh, A. Dutta. A Janus cerium-doped bismuth oxide electrocatalyst for complete water splitting, Cell Reports: Physical Science, 2022, 3(11) 101106.

[29]   Shinde P V., Shinde NM, Shaikh SF, Lee D, Yun JM, Woo LJ, Al-Enizi AM, Mane RS, Kim KH (2020) Room-temperature synthesis and CO2-gas sensitivity of bismuth oxide nanosensors. RSC Adv 10:17217–17227. https://doi.org/10.1039/D0RA00801J

[30]   Deng P, Wang H, Qi R, Zhu J, Chen S, Yang F, Zhou L, Qi K, Liu H, Xia BY (2020) Bismuth Oxides with Enhanced Bismuth-Oxygen Structure for Efficient Electrochemical Reduction of Carbon Dioxide to Formate. ACS Catal 10:743–750. https://doi.org/10.1021/ACSCATAL.9B04043/SUPPL_FILE/CS9B04043_SI_001.PDF

[31]   Köhler R, Siebert D, Kochanneck L, Ohms G, Viöl W (2019) Bismuth Oxide Faceted Structures as a Photocatalyst Produced Using an Atmospheric Pressure Plasma Jet. Catalysts 9:533. https://doi.org/10.3390/CATAL9060533

[32]   Chouke, Prashant B., Kanhaiya M. Dadure, Ajay K. Potbhare, Ganesh Bhusari, A. Mondal, Karan Chaudhary, V. Singh, Martin F. Desimone, R. G. Chaudhary, and Dhanraj T. Masram, Biosynthesized δ-Bi2O3 Nanoparticles from Crinum viviparum Flower Extract for Photocatalytic Dye Degradation and Molecular Docking, ACS Omega, 7 (2022), 20983-20993.

[33]   Qiu Y, Fan H, Chang X, Dang H, Luo Q, Cheng Z (2018) Novel ultrathin Bi2O3 nanowires for supercapacitor electrode materials with high performance. Appl Surf Sci 434:16–20. https://doi.org/10.1016/J.APSUSC.2017.10.171

[34]   Leontie L, Caraman M, Alexe M, Harnagea C (2002) Structural and optical characteristics of bismuth oxide thin films. Surf Sci 507–510:480–485. https://doi.org/10.1016/S0039-6028(02)01289-X

[35]   Trivedi M, Tallapragada R, Branton L, Trivedi D, Nayak G, Latiyal O, Jana S (2015) Evaluation of Atomic, Physical, and Thermal Properties of Bismuth Oxide Powder: An Impact of Biofield Energy Treatment. Am J Nano Res Appl 3:94–98. https://doi.org/10.11648/J.NANO.20150306.11

[36]   Astuti Y, Fauziyah A, Nurhayati S, Wulansari A, Andianingrum R, Hakim A, Bhaduri G (2016) Synthesis of α-Bismuth oxide using solution combustion method and its photocatalytic properties. IOP Conf Ser Mater Sci Eng 107:1–8. https://doi.org/10.1088/1757-899X/107/1/012006

[37]   Depablos-rivera O, Martínez A, Rodil SE (2021) Interpretation of the Raman spectra of bismuth oxide thin fi lms presenting different crystallographic phases. J Alloys Compd 853:157245. https://doi.org/10.1016/j.jallcom.2020.157245

[38]   Zhang L, Hashimoto Y, Taishi T, Nakamura I, Ni QQ (2011) Fabrication of flower-shaped Bi2O3 superstructure by a facile template-free process. Appl Surf Sci 257:6577–6582. https://doi.org/10.1016/J.APSUSC.2011.02.081

[39]   Abu-Dief AM, Mohamed WS (2017) α-Bi2O3 nanorods: synthesis, characterization and UV-photocatalytic activity. Mater Res Express 4:035039. https://doi.org/10.1088/2053-1591/AA6712

[40]   Li C, He P, Dong F, Liu H, Jia L, Liu D, Du L, Liu H, Wang S, Zhang Y (2019) An efficient and facile one-step synthesis strategy: Bismuth oxide with controllable size and

shape for high-performance supercapacitors. Mater Lett 245:29–32.
https://doi.org/10.1016/j.matlet.2019.02.098

[41] Ambare R, Shinde P, Nakate U, Lokhande B, Mane R (2018) Sprayed Bismuth Oxide Interconnected Nanoplates Supercapacitor Electrode Materials. Appl Surf Sci 453:215–219. https://doi.org/10.1016/j.apsusc.2018.05.090

[42] Chu D, Wu Y, Wang L (2022) Synthesis and characterization of novel coral spherical bismuth oxide. Results Chem 4:100448. https://doi.org/10.1016/j.rechem.2022.100448

[43] Sun J, Wang J, Li Z, Yang Z, Yang S (2015) Controllable synthesis of 3D hierarchical bismuth compounds with good electrochemical performance for advanced energy storage devices. RSC Adv 5:51773–51778. https://doi.org/10.1039/C5RA09760F

[44] Huang X, Yan J, Zeng F, Yuan X, Zou W, Yuan D (2013) Facile preparation of orange-like Bi2O2.33 microspheres for high performance supercapacitor application. Mater Lett 90:90–92. https://doi.org/10.1016/J.MATLET.2012.09.019

[45] Zheng FL, Li GR, Ou YN, Wang ZL, Su CY, Tong YX (2010) Synthesis of hierarchical rippled Bi2O3 nanobelts for supercapacitor applications. Chem Commun 46:5021–5023. https://doi.org/10.1039/C002126A

[46] V S, R. G. B, B. J. L, R. C. A (2023) Electro-Synthesized Bismuth Oxide Nanomaterials on Flexible Substrate Electrode for Supercapacitor Application. ES Energy Environ. https://doi.org/10.30919/esee944

[47] Han Y, Li L, Liu Y, Li X, Qi X, Song L (2018) Fabrication of Strontium Bismuth Oxides as Novel Battery-Type Electrode Materials for High-Performance Supercapacitors. 2018.

[48] Teli AM, Bhat TS, Beknalkar SA, Mane SM, Chaudhary LS, Patil DS, Pawar SA, Efstathiadis H, Cheol Shin J (2022) Bismuth manganese oxide based electrodes for asymmetric coin cell supercapacitor. Chem Eng J 430:133138. https://doi.org/10.1016/j.cej.2021.133138

[49] Danamah HM, Raut SD, Shaikh ZA, Mane RS (2022) Chemical Synthesis of Bismuth Oxide and Its Ionic Conversion to Bismuth Sulphide for Enhanced Electrochemical Supercapacitor Energy Storage Performance. J Electrochem Soc 169:120537. https://doi.org/10.1149/1945-7111/acaac9

[50] Jo S, Pak S, Lee Y-W, Cha S, Hong J, Sohn JI (2023) Enhancing the Electrochemical Energy Storage Performance of Bismuth Ferrite Supercapacitor Electrodes via Simply Induced Anion Vacancies. Int J Energy Res 2023:1–9. https://doi.org/10.1155/2023/2496447

[51] Prasath A, Athika M, Duraisamy E, Selva Sharma A, Sankar Devi V, Elumalai P (2019) Carbon Quantum Dot-Anchored Bismuth Oxide Composites as Potential Electrode for Lithium-Ion Battery and Supercapacitor Applications. ACS Omega 4:4943–4954. https://doi.org/10.1021/acsomega.8b03490

[52] Singh S, Sahoo RK, Shinde NM, Yun JM, Mane RS, Kim KH (2019) Synthesis of Bi2O3-MnO2 Nanocomposite Electrode for Wide-Potential Window High Performance Supercapacitor. Energies 12:3320. https://doi.org/10.3390/en12173320

[53]   Mane SA, Kashale AA, Kamble GP, Kolekar SS, Dhas SD, Patil MD, Moholkar A V., Sathe BR, Ghule A V. (2022) Facile synthesis of flower-like Bi_2O_3 as an efficient electrode for high performance asymmetric supercapacitor. J Alloys Compd 926:166722. https://doi.org/10.1016/j.jallcom.2022.166722

[54]   Kumar V, Soam A, Sahoo PK, Panda HS (2021) Enhancement of electrochemical properties of carbon solution doped bismuth ferrite for supercapacitor application. Mater Today Proc 41:165–171. https://doi.org/10.1016/j.matpr.2020.08.515

[55]   Ji Z, Dai W, Zhang S, Wang G, Shen X, Liu K, Zhu G, Kong L, Zhu J (2020) Bismuth oxide/nitrogen-doped carbon dots hollow and porous hierarchitectures for high-performance asymmetric supercapacitors. Adv Powder Technol 31:632–638. https://doi.org/10.1016/j.apt.2019.11.018

[56]   Wang L, He Q, Xiao F, Yang L, Jiang Y, Su R, He P, Lei H, Jia B, Tang B (2024) Three-dimensional hierarchical nanosheets based spherical bismuth metal-organic frameworks: Controllable synthesis and high performance for supercapacitor. Electrochim Acta 484:144082. https://doi.org/10.1016/j.electacta.2024.144082.

[57]   Jalalah M, Sasmal A, Nayak AK, Harraz FA (2023) Rapid, external acid-free synthesis of Bi_2WO_6 nanocomposite for efficient supercapacitor application. J Taiwan Inst Chem Eng 143:104697. https://doi.org/10.1016/j.jtice.2023.104697.

[58]   Shinde NM, Ghule BG, Raut SD, Narwade SH, Pak JJ, Mane RS (2021) Hopping Electrochemical Supercapacitor Performance of Ultrathin BiOCl Petals Grown by a Room-Temperature Soft-Chemical Process. Energy & Fuels 35:6892–6897. https://doi.org/10.1021/acs.energyfuels.1c00308.

[59]   Cui Y, Cheng Q-Y, Wu H, Wei Z, Han B-H Graphene oxide-based benzimidazole-crosslinked networks for high-performance supercapacitors. https://doi.org/10.1039/c3nr01480k

Green Synthesis and Emerging Applications of Frontier Nanomaterials   Materials Research Forum LLC
Materials Research Foundations 169 (2024) 331-364   https://doi.org/10.21741/9781644903278-13

Chapter 13

Green synthesis and applications of ceria-based nanomaterials

Usman Lawal Usman[1], Nakshatra Bahadur Singh[2,3] and Narendra P. Singh[4*]

[1] Department of Biology, Umaru Musa Yar'adua University, Katsina- Nigeria

[2]Department of Chemistry and Biochemistry, Sharda University, Greater Noida, India

[3]Research & Development Cell, Sharda University, Greater Noida, India

[4]Department of Chemistry, Udai Pratap Autonomous College, Varanasi, India

* usman.usman@umyu.edu.ng; napratap.singh@gmail.com

Abstract

In recent years, there has been a growing interest in developing environmentally and sustainable methods for the synthesis of nanomaterials (NMs). Among various NMs, ceria-based NMs have gained significant attention due to their unique properties and diverse applications. This chapter provides a comprehensive overview of the green synthesis approaches employed for the fabrication of ceria-based NMs and their subsequent applications. The chapter begins with an introduction to ceria-based NMs, highlighting their exceptional physicochemical properties such as high surface area, redox capabilities, and oxygen storage capacity. Subsequently, it delves into the concept of green synthesis, emphasizing the significance of using environmentally benign routes for NMs fabrication. Various green synthesis methods, including biological, template-mediated, and microwave-assisted techniques, are discussed in detail, highlighting their advantages, limitations, and applicability to ceria-based NMs. Furthermore, the chapter explores the wide-ranging applications of ceria-based NMs in different fields, such as catalysis, energy storage and conversion, environmental remediation, and biomedical applications. Specific examples and case studies are presented to illustrate the effectiveness of ceria-based NMs in these applications. Additionally, the chapter discusses the potential challenges and future perspectives associated with the green synthesis and applications of ceria-based NMs, including scalability, stability, and toxicity considerations. By adopting sustainable and environmentally friendly approaches, the synthesis and applications of ceria-based NMs can contribute to the development of cleaner and more efficient technologies for a sustainable future.

Keywords

Cerium-based NMs, Green Synthesis, Adsorbents, Biomedical, Environmental Mitigation

Contents

1.   Introduction

Nanomaterials, with their unique properties and functionalities at the nanoscale, have revolutionized various fields of science and technology. The ability to control the size, shape, and composition of NMs has opened up new possibilities for developing advanced materials with enhanced properties [1]. Nanomaterials exhibit distinct physical, chemical, and optical properties compared to their bulk counterparts, making them highly attractive for a wide range of applications, including electronics, catalysis, energy storage, and medicine. The conventional synthesis methods of NMs often involve the use of hazardous chemicals and energy-intensive processes, leading to environmental pollution and sustainability concerns [2]. Green synthesis, also known as sustainable or eco-friendly synthesis, offers a promising alternative by utilizing environmentally benign and renewable resources, minimizing waste generation, and reducing energy consumption [3]. Green synthesis approaches involve the use of biological, plant-based, or environmentally friendly chemical methods to fabricate NMs. Ceria, or cerium oxide (CeO_2) has emerged as a remarkable NMs due to its unique properties and wide-ranging applications. Ceria-based NMs possess excellent catalytic, redox, and oxygen storage/release properties, which can be attributed to the high oxygen vacancy concentration and mixed valence states of cerium ions [4]. The size, shape, and surface properties of ceria NMs significantly influence their performance in various applications [5]. By adopting green synthesis approaches, the properties of ceria-based NMs can be further enhanced, making them more suitable for specific applications.

This chapter aims to explore the green synthesis methods for ceria-based NMs and their applications. It provides an overview of various green synthesis methods employed for the fabrication of ceria-based NMs, including sol-gel synthesis, hydrothermal and solvothermal methods, and plant-mediated synthesis. Each method was discussed in terms of its process conditions, and potential for tailoring the properties of ceria NMs. Furthermore, the characterization techniques used to evaluate the properties and structure of green-synthesized ceria-based NMs were highlighted. The influence of green synthesis approaches on the properties of ceria-based NMs. The effects of dopants, surface modifications, and particle size control on the properties and applications of ceria NMs were elucidated. Finally, the chapter highlighted various applications of green-synthesized ceria-based NMs. Specifically, their utilization in environmental applications, such as catalysis, energy storage, and environmental sensing. Additionally, their potential in biomedical applications, including drug delivery, imaging, and biosensing was discussed.

2.   Natural extracts and bioresources in green synthesis

The utilization of natural extracts and bioresources has gained significant attention in the green synthesis of ceria-based NMs. This approach capitalizes on the rich reserves of bioactive compounds present in various plants, microorganisms, and agricultural waste materials. Green synthesis using natural extracts offers several advantages, including sustainability, biocompatibility, low toxicity, and cost-effectiveness [6]. Below are some highlighted utilizations of natural extracts and bioresources and their applications in the synthesis of ceria-based NMs.

2.1   Plant extracts

Plant-mediated synthesis is one of the most common green synthesis methods, where plant extracts serve as surfactant, reducing, shielding, capping and stabilizing agents. Different parts of plants, such as leaves, stems, roots, and fruits contain a diverse array of phytochemicals, including phenols, flavonoids, terpenoids, alkaloids, and proteins [7]. These bioactive compounds possess inherent reducing properties, which facilitate the reduction of metal precursors into ceria-based nanoparticles (NPs). Moreover, the capping action of these compounds stabilizes the formed NPs, preventing agglomeration. Plant extracts offer a sustainable and eco-friendly alternative to conventional chemical-reducing agents, contributing to green chemistry principles [8].

2.2   Microorganisms

Various microorganisms, such as bacteria, fungi, and algae, can be employed in the green synthesis of ceria-based NMs. These microorganisms can generate extracellular enzymes or metabolites with reducing capabilities, enabling the formation of NPs. Microbial cells or culture filtrates can be utilized as bioreactors in the synthesis process. Green synthesis using microorganisms is advantageous as it involves mild reaction conditions, uses low-cost substrates, and can be readily scaled up for large-scale production. Additionally, the use of microorganisms allows for the biosynthesis of complex nanostructures and the potential for intracellular nanoparticle synthesis [9].

2.3   Agricultural waste materials

Agricultural waste materials, such as fruit peels, tea waste, rice husks, and corn cobs, are abundant and renewable resources that can be harnessed for green synthesis. These waste materials contain bioactive compounds that can act as reducing and capping agents for nanoparticle synthesis. Utilizing agricultural waste not only reduces waste disposal problems but also adds value to otherwise discarded resources [10]. The synthesis of ceria-based NMs using agricultural waste is an eco-friendly and economically viable approach.

2.4   Marine bioresources

Marine organisms, including seaweeds, algae, and marine bacteria, offer a vast pool of bioactive compounds that can be utilized in green synthesis. The unique biochemical composition of marine bioresources makes them promising candidates for the synthesis of ceria-based NMs. Additionally, marine ecosystems are less explored in terms of their green synthesis potential, providing opportunities for novel discoveries and applications [11].

3.   Green synthesis methods for ceria-based NMs

Green synthesis methods offer sustainable and environmentally friendly approaches to fabricating ceria-based NMs. These methods utilize eco-friendly resources, benign solvents, and energy-efficient processes, minimizing the environmental impact associated with traditional synthesis techniques [12]. Table 1, explores various green synthesis methods employed for the production of ceria NPs, highlighting their advantages, process conditions, and potential for tailoring NMs properties.

Table 1: Green Synthesis Methods for Ceria-Based NMs

Method	Environmental Advantages	Key Parameters	Particle Size and Morphology Control
Sol-Gel Synthesis	Use of non-toxic precursors	Temperature, pH, concentration of reactants	Addition of surfactants, aging time and temperature, use of
Hydrothermal/ Solvothermal	Water as a solvent, mild reaction conditions	Temperature, pressure, pH, seeding agents	Reaction time and temperature, use of surfactants or templates
Plant-Mediated Synthesis	Use of natural extracts/bioresources, environmentally friendly	Extraction conditions, reaction temperature, pH	Optimization of extraction conditions, use of different plant extracts
Microwave-Assisted	Rapid and efficient synthesis, enhanced control over reaction conditions	Microwave power, solvent, reaction time	Reaction time and temperature, use of additives or templates
Green Electrochemical	Use of non-toxic electrolytes, minimal waste generation	Electrolyte composition, voltage, current	Electrode material, reaction time and temperature, electrode surface modification
Green Sonochemical	Energy-efficient synthesis, enhanced reaction kinetics	Ultrasonic power, temperature, reaction time	Intensity and duration of ultrasonic irradiation, solvent composition, presence of additives

3.1   Sol-gel synthesis

Sol-gel synthesis is a widely used green method for fabricating ceria NPs. It involves the hydrolysis and condensation of precursor compounds under mild conditions, typically using water or alcohol as solvents. The sol-gel approach allows precise control over particle size, composition, and morphology by adjusting the reaction parameters such as precursor concentration, pH, and temperature. Green sol-gel synthesis methods often utilize bio-based precursors or environmentally friendly solvents, reducing the reliance on hazardous chemicals and minimizing waste generation [13].

Green Synthesis and Emerging Applications of Frontier Nanomaterials   Materials Research Forum LLC
Materials Research Foundations 169 (2024) 331-364   https://doi.org/10.21741/9781644903278-13

3.2   Hydrothermal and solvothermal methods

Hydrothermal and solvothermal methods involve the synthesis of ceria NPs under high-pressure and high-temperature conditions in aqueous or organic solvents, respectively. These methods provide excellent control over particle size, shape, and crystallinity. Green hydrothermal and solvothermal synthesis strategies focus on using non-toxic solvents, optimizing reaction conditions, and reducing energy consumption [14]. Additionally, the utilization of natural extracts or bioresources as reducing or stabilizing agents in these methods offers sustainable alternatives to traditional chemical reagents.

3.3   Plant-mediated synthesis

Plant-mediated synthesis has emerged as an eco-friendly and sustainable approach for fabricating ceria NPs. This method utilizes plant extracts or plant-derived compounds as reducing agents and stabilizers. The reducing and capping agents present in plant extracts facilitate the reduction of metal precursors and the stabilization of resulting NPs. Plant-mediated synthesis offers advantages such as low cost, ease of scalability, and the use of renewable resources [15]. It also provides an opportunity to explore the synergistic effects of phytochemicals on the synthesis and properties of ceria-based NMs.

3.4   Microwave-assisted synthesis

Microwave-assisted synthesis is a rapid and efficient green synthesis method for ceria-based NMs. In this approach, microwave irradiation is utilized to heat the reaction mixture, resulting in accelerated reaction kinetics and reduced reaction times. Microwave-assisted synthesis offers advantages such as improved reaction control, enhanced crystallinity, and narrower size distribution of NPs [16]. It also allows for energy and time savings compared to conventional heating methods.

3.5   Ultrasound-assisted synthesis

Ultrasound-assisted synthesis involves the application of high-frequency ultrasound waves to facilitate the synthesis of ceria-based NMs. The cavitation effect induced by ultrasound creates localized hotspots and intense mixing, leading to enhanced mass transfer and reaction rates. Ultrasound-assisted synthesis offers benefits such as reduced reaction times, improved particle size control, and enhanced crystallinity of the NPs [17]. It is a green and energy-efficient method that eliminates the need for high temperatures and toxic solvents.

3.6   Green electrochemical synthesis

Green electrochemical synthesis represents an eco-friendly approach to the production of ceria-based NMs. It involves the use of electrochemical cells and electrodes to facilitate the reduction of metal ions and subsequent NPs formation [18]. Electrochemical synthesis offers precise control over particle size, morphology, and composition. It also enables the synthesis of NMs with high purity and avoids the use of hazardous reagents. The technique is scalable, energy-efficient, and allows for the recycling of reaction components [19].

3.7 Green templating approaches

Green templating approaches involve the use of natural templates, such as proteins, peptides, or natural fibers, to guide the formation of ceria-based NMs. These templates provide a structured framework for the nucleation and growth of NPs, resulting in controlled size, shape, and surface properties. Green templating methods offer advantages such as improved particle uniformity, enhanced stability, and the possibility of creating complex nanostructures (NSs). The templates are biodegradable and can be easily removed after nanoparticle synthesis, leaving behind well-defined NMs [20].

3.9 Bio-inspired synthesis

Bioinspired synthesis draws inspiration from biological systems and processes to create ceria-based NMs. This approach utilizes biomolecules, such as enzymes, proteins, or peptides, to mediate the synthesis and assembly of NPs [21]. By mimicking biological processes, bio-inspired synthesis offers advantages such as precise control over particle size, shape, and functionality. It also allows for the synthesis of hierarchical structures and the incorporation of bioactive compounds into the NMs. Bioinspired synthesis methods align with the principles of green chemistry, emphasizing sustainability and the utilization of renewable resources [22]. These green synthesis approaches expand the repertoire of environmentally friendly methods for the production of ceria-based NMs. They offer advantages such as enhanced control over particle properties, reduced reaction times, energy efficiency, and the utilization of eco-friendly materials and techniques. These approaches contribute to the development of sustainable nanotechnology and promote the use of green and clean synthesis practices.

4. Principles of green synthesis of nanomaterial

Green synthesis is guided by a set of principles that prioritize sustainability, environmental friendliness, and the reduction of hazardous materials and waste. These principles are rooted in the field of green chemistry and serve as a framework for developing environmentally conscious synthesis methods for ceria-based NMs. The following principles outline the key considerations in green synthesis as shown in Figure 1:

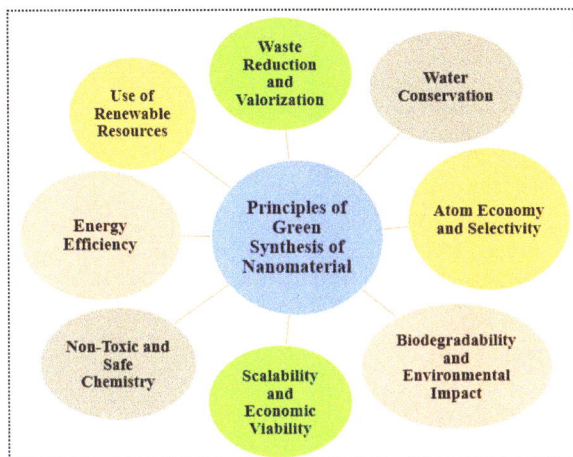

Figure 1: Principles of green synthesis of NMs

4.1 Use of renewable resources

Green synthesis approaches prioritize the use of renewable resources as raw materials. This includes utilizing bio-based precursors, natural extracts, or plant-derived compounds as reducing agents, stabilizers, or capping agents. By relying on renewable resources, the dependency on non-renewable resources can be reduced, contributing to a more sustainable synthesis process [23].

4.2 Non-toxic and safe chemistry

Green synthesis methods emphasize the use of non-toxic chemicals and solvents. This helps ensure the safety of researchers and minimizes the environmental impact associated with hazardous substances. By replacing harmful reagents with benign alternatives, the risk of exposure to toxic materials is reduced, making the synthesis process safer and more sustainable [23].

4.3 Energy efficiency

Green synthesis approaches strive to optimize energy consumption. This can be achieved by employing energy-efficient techniques such as microwave-assisted synthesis or Sonochemical methods, which reduce the overall energy requirements compared to conventional heating methods. Additionally, process parameters, such as reaction time and temperature, are optimized to minimize energy usage while maintaining desired product quality [24].

4.4 Waste reduction and valorization

Green synthesis methods aim to minimize waste generation by optimizing reaction conditions and resource utilization. This includes reducing the use of excess reactants, maximizing the conversion of starting materials into desired products, and minimizing the generation of unwanted by-products or waste streams. Furthermore, waste materials generated during the synthesis process can be explored for potential valorization or recycling, contributing to a circular economy approach [25].

4.5   Water conservation

Green synthesis approaches consider water conservation by minimizing water usage and implementing water recycling strategies. This can be achieved by utilizing water-efficient reaction setups, optimizing reaction stoichiometry, and employing closed-loop systems that allow for the reuse of water or solvent media [26].

4.6   Scalability and economic viability

Green synthesis methods aim to be scalable and economically viable for large-scale production. This involves considering the availability and cost of raw materials, energy requirements, and process scalability. By incorporating these factors into the synthesis design, green synthesis methods facilitate the practical application and commercialization of ceria-based NMs [27].

4.7   Atom economy and selectivity

Green synthesis methods aim to maximize atom economy, which refers to the efficient utilization of atoms in the synthesis process [23]. This involves minimizing the use of excess reagents and optimizing reaction conditions to promote selective reactions, thereby reducing the formation of undesired by-products. By focusing on high atom economy and selectivity, green synthesis approaches maximize resource efficiency and minimize waste generation.

4.8   Biodegradability and environmental impact

Green synthesis methods prioritize the use of biodegradable materials and the reduction of environmental impact. This includes selecting solvents, stabilizers, and reducing agents that are biodegradable and have minimal adverse effects on ecosystems [28]. By minimizing the persistence of synthesis by-products and ensuring the biodegradability of materials, green synthesis methods contribute to environmental sustainability. By incorporating these principles into the design and implementation of green synthesis methods for ceria-based NMs, researchers can contribute to a more sustainable and environmentally friendly approach to nanotechnology. These principles guide the development of synthesis routes that are efficient, safe, cost-effective, and in harmony with environmental preservation. Green synthesis plays a crucial role in advancing the field of ceria-based NMs while addressing the challenges of sustainability and responsible manufacturing.

5.   Environmental advantages of green synthesis for ceria-based NMs

The utilization of green synthesis approaches for the production of ceria-based NMs offers significant environmental advantages over traditional synthesis methods. These advantages stem from the incorporation of sustainable practices and the use of environmentally friendly materials throughout the synthesis process. Below are some of the environmental benefits of green synthesis for ceria-based NMs.

5.1   Reduced energy consumption

Green synthesis methods prioritize energy efficiency, aiming to minimize energy consumption during the synthesis process. By optimizing reaction conditions, such as temperature and time, researchers can reduce the energy requirements for heating, cooling, and overall reaction kinetics.

This results in lower energy-related greenhouse gas emissions, contributing to the mitigation of climate change and overall energy conservation [29].

5.2 Minimization of hazardous chemicals

Green synthesis approaches emphasize the use of non-toxic or low-toxicity chemicals, reducing the reliance on hazardous materials [30]. Traditional synthesis methods often employ toxic solvents, catalysts, or stabilizers that pose risks to human health and the environment. In contrast, green synthesis methods focus on the selection of environmentally friendly solvents, catalysts, and stabilizers, minimizing the release of harmful substances during the synthesis process and subsequent application of ceria-based NMs.

5.3 Water-based reactions

Many green synthesis approaches for ceria-based NMs utilize water as a solvent, taking advantage of its abundance, low cost, and non-toxic nature. Water-based reactions reduce the dependence on organic solvents, which are often derived from petrochemicals and contribute to environmental pollution. Water can be readily recycled and treated, minimizing its environmental impact and enabling more sustainable production processes [31].

5.4 Waste minimization

Green synthesis methods aim to minimize waste generation by optimizing reactant ratios, enhancing precursor utilization efficiency, and reducing the formation of undesired by-products. This leads to higher product yields and less waste disposal, decreasing the environmental burden associated with synthesis. Furthermore, the utilization of green solvents and catalysts ensures that the waste generated during the synthesis process is less hazardous and easier to manage [32].

5.5 Sustainable resource utilization

Green synthesis approaches prioritize the utilization of sustainable resources, such as renewable feedstocks or agricultural waste, in the synthesis of ceria-based NMs. By reducing the reliance on non-renewable resources, green synthesis methods contribute to the conservation of natural resources and promote a circular economy [33].

5.6 Biocompatibility and biodegradability

Green synthesis methods prioritize the use of biocompatible materials, making the resulting ceria-based NMs more suitable for biomedical and environmental applications. Biodegradable components, such as bio-based solvents or stabilizers, facilitate the degradation and disposal of NMs after use, reducing their potential long-term environmental impact [34]. The environmental advantages of green synthesis for ceria-based NMs are significant. By implementing sustainable practices, reducing the use of hazardous chemicals, minimizing waste generation, and utilizing renewable resources, researchers can contribute to a more environmentally friendly and sustainable approach to nanomaterial synthesis. These environmental benefits align with the broader goals of achieving a greener and more sustainable future.

6.   Reaction parameters for green synthesis of ceria-based NMs

In the green synthesis of ceria-based NMs, careful consideration of reaction parameters is crucial to ensure the sustainability and efficiency of the synthesis process. These parameters include reaction temperature, reaction time, pH, precursor concentration, and the use of catalysts or additives. By optimizing these parameters, researchers can achieve desirable properties of ceria NPs while minimizing environmental impact [12]. Figure 2 depicted the key reaction parameters for green synthesis of ceria-based NMs.

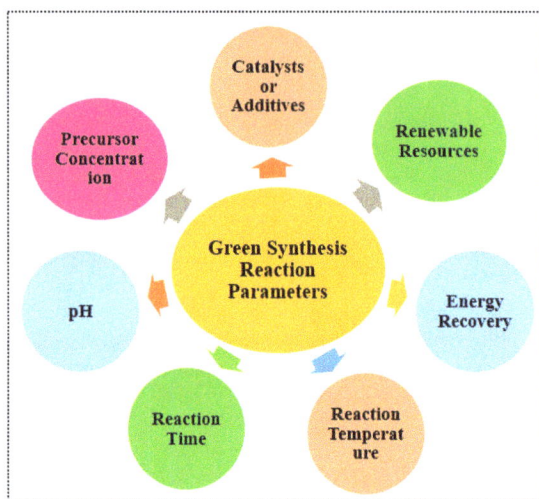

Figure 2: Some reaction parameters for green synthesis of NMs

6.1   Reaction temperature

The reaction temperature plays a vital role in determining the nucleation, growth, and crystallinity of ceria NPs. Green synthesis methods aim to minimize the energy input by optimizing the reaction temperature. Lower reaction temperatures not only reduce energy consumption but also minimize the potential for side reactions or undesired phase transformations. However, it is essential to maintain a temperature that is sufficient for the desired reaction kinetics and nanoparticle formation [35].

6.2   Reaction time

The reaction time influences the extent of nucleation and growth of ceria NPs. Longer reaction times often lead to increased particle size and crystallinity. Green synthesis approaches focus on minimizing the reaction time to reduce energy consumption and improve process efficiency [36]. By carefully controlling the reaction time, researchers can achieve the desired particle size and crystallinity while minimizing unnecessary energy consumption.

6.3 *pH*

The pH of the reaction medium significantly affects the formation and stability of ceria NPs. It can influence the hydrolysis and condensation reactions involved in nanoparticle synthesis. Green synthesis methods aim to use mild and environmentally friendly pH conditions to avoid the use of strong acids or bases that can be hazardous or require extensive neutralization processes. Maintaining a near-neutral or slightly basic pH range is often preferred for green synthesis, ensuring safer and more sustainable reaction conditions [37].

6.4 Precursor concentration

The concentration of precursors, such as cerium salts, in the reaction mixture, affects the particle size, morphology, and composition of ceria NPs [38]. Optimizing the precursor concentration is essential to control these parameters while minimizing waste generation. Green synthesis methods strive to use lower precursor concentrations to reduce material usage and waste production, without compromising the desired nanoparticle characteristics.

6.5 Catalysts or additives

Catalysts or additives can play a crucial role in enhancing the synthesis efficiency, controlling particle size, or facilitating specific reactions during green synthesis. Green catalysts, derived from renewable resources or natural compounds, are preferred to minimize the environmental impact [39]. The judicious selection and optimization of catalysts or additives can promote desired reactions, improve reaction kinetics, and enable the synthesis of high-quality ceria NPs.

6.6 Renewable resources

Green synthesis methods prioritize the use of renewable resources as starting materials or reaction media. This includes the utilization of bio-derived solvents, renewable feedstocks, or agricultural waste products. By replacing petroleum-based chemicals with renewable alternatives, the environmental impact of the synthesis process can be significantly reduced [40].

6.7 Energy recovery

Green synthesis approaches emphasize energy recovery and recycling. The heat generated during the reaction can be recovered and utilized for subsequent reactions or other energy requirements. Integration of energy recovery systems contributes to the overall sustainability and energy efficiency of the synthesis process [41].

By carefully optimizing these reaction parameters, green synthesis methods for ceria-based NMs can achieve controlled particle size, morphology, composition, and crystallinity while minimizing environmental impact. The judicious selection and fine-tuning of these parameters are essential for sustainable and efficient synthesis processes, aligning with the goals of green chemistry and sustainable nanomaterial synthesis.

7. Plant-mediated synthesis of ceria-based NMs

Plant-mediated synthesis, also known as photosynthesis or green synthesis using plant extracts, has emerged as a sustainable and environmentally friendly approach for the synthesis of ceria-based NMs. This method harnesses the inherent reducing and stabilizing properties of plant extracts to facilitate the formation of NPs. Plant-mediated synthesis offers numerous advantages, including the use of renewable resources, low toxicity, and the potential for large-scale production [15]. Below are some of the principles and applications of plant-mediated synthesis for ceria-based NMs.

7.1   Principles of plant-mediated synthesis

Plant-mediated synthesis relies on the unique biochemical composition of plant extracts, which contain various bioactive compounds, such as phenolics, flavonoids, alkaloids, terpenoids, and proteins. These compounds act as reducing and stabilizing agents, enabling the conversion of metal precursors into NPs. The phytochemicals present in plant extracts interact with the metal ions, leading to their reduction and subsequent nucleation and growth into nanoscale particles. The use of plant extracts as reducing agents offers a sustainable and green alternative to traditional chemical reducing agents [42].

7.2   Plant selection and extract preparation

Different plant species possess varying bioactive compounds, making plant selection a critical factor in plant-mediated synthesis. Plants with high concentrations of reducing and stabilizing agents are preferred. Commonly used plants include medicinal herbs, fruits, vegetables, and agricultural waste materials. After plant selection, the extract is prepared by grinding or crushing plant material and extracting the bioactive compounds using solvents, such as water, ethanol, or other environmentally friendly solvents. The resulting extract serves as the reducing and stabilizing agent in the synthesis process [43].

7.3   Nanoparticle formation and stabilization

Plant extracts facilitate the reduction of metal ions and subsequent nucleation and growth of ceria-based NPs. The bioactive compounds in the plant extract act as reducing agents, donating electrons to the metal ions, leading to their reduction and subsequent formation of NPs. The phytochemicals also serve as stabilizing agents, preventing particle aggregation and providing stability to the NPs. The composition of the plant extract and its concentration play a crucial role in controlling the size, shape, and stability of the synthesized NPs [44].

7.4   Advantages of plant-mediated synthesis

Plant-mediated synthesis offers several advantages over traditional synthesis methods for ceria-based NMs [45,15]:

Eco-Friendly and Sustainable: Plant-mediated synthesis utilizes renewable plant resources and eliminates the need for hazardous reducing agents, making it an environmentally friendly approach. It reduces the environmental impact associated with traditional chemical synthesis methods.

Low Toxicity and Biocompatibility: Plant extracts are generally non-toxic and biocompatible, making the resulting ceria-based NMs suitable for biomedical and environmental applications.

These NMs have potential in areas such as drug delivery, sensing, catalysis, and environmental remediation.

Cost-Effective and Scalable: Plant-mediated synthesis is cost-effective and scalable due to the abundance of plant resources and the ease of extract preparation. It offers a viable alternative to expensive and complex chemical synthesis methods.

Controlled Particle Size and Shape: Plant-mediated synthesis allows for controlled synthesis of ceria-based NPs with specific sizes and shapes by modulating the concentration and composition of the plant extract. This control over particle properties is crucial for tailoring their desired applications.

Synergistic Effects: Plant extracts often contain multiple bioactive compounds, which can exhibit synergistic effects on the synthesis process. These synergistic effects enhance the reduction efficiency, stability, and functionality of the synthesized ceria-based NPs.

8.   Particle size and morphology control in green synthesis

One of the key objectives in the green synthesis of ceria-based NMs is precise control over particle size and morphology. The size and morphology of NPs have a significant impact on their physical, chemical, and catalytic properties, making them crucial factors for various applications. Green synthesis approaches employ different strategies to achieve the desired particle size and morphology while minimizing the environmental impact [46]. Below are some methods and techniques used to control particle size and morphology in the green synthesis of ceria-based NMs.

8.1   Selection of precursor and stabilizers

The choice of precursor and stabilizers in the synthesis process can influence the nucleation and growth of NPs, thereby affecting the particle size and morphology [36]. Green synthesis methods carefully select precursors and stabilizers that promote controlled particle growth and hinder excessive agglomeration. By understanding the interactions between precursors and stabilizers, researchers can tailor the synthesis conditions to achieve desired particle sizes and shapes.

8.2   Reaction parameters

Reaction parameters, such as temperature, reaction time, and pH, play a critical role in controlling particle size and morphology. Green synthesis methods optimize these parameters to achieve the desired outcomes. Lower reaction temperatures and shorter reaction times are often favored to limit particle growth and agglomeration. Maintaining a specific pH range can also influence the formation of specific crystal facets or control the growth kinetics of ceria NPs [47]. By fine-tuning these parameters, researchers can achieve precise control over particle size and morphology.

8.3   Use of templates and surfactants

Templates and surfactants are commonly employed in green synthesis methods to direct particle growth and shape. Templates can act as sacrificial structures around which NPs form, determining the final particle morphology. Surfactants, on the other hand, can control the surface energy and interactions, influencing the nucleation and growth processes. Green synthesis methods employ biocompatible or bio-based templates and surfactants to ensure the sustainability and biocompatibility of the resulting ceria-based NMs [48].

8.4   Solvent engineering

The choice of solvent and its properties can influence particle size and morphology in green synthesis. Solvent engineering involves the selection of environmentally friendly solvents with specific characteristics that control particle growth. For example, the use of polar solvents can promote slow and controlled nucleation, leading to smaller particle sizes. Non-polar solvents, on the other hand, can facilitate anisotropic growth and the formation of specific crystal facets. Green solvent choices contribute to the sustainability and eco-friendliness of the synthesis process [23].

8.5   Use of additives and dopants

Incorporating additives or dopants in the synthesis process can modify the growth kinetics and crystal structure, leading to controlled particle size and morphology. Green synthesis methods explore the use of environmentally friendly additives or dopants to tailor the particle growth and morphology. These additives can influence surface energies, crystallographic orientation, or lattice defects, enabling precise control over particle characteristics [49].

8.6   Time-Dependent Processes

Some green synthesis approaches employ time-dependent processes, such as sequential addition of precursors or controlled aging periods, to achieve specific particle sizes or morphologies [50]. By carefully designing the synthesis process and incorporating time-dependent steps, researchers can manipulate the nucleation and growth kinetics, resulting in desired particle size and morphology.

8.7   Characterization

Continuous monitoring and characterization of the synthesis process can provide valuable feedback for controlling particle size and morphology. By analyzing the intermediate stages or using in situ monitoring techniques, researchers can make adjustments to the synthesis parameters, ensuring the desired particle characteristics are achieved. Controlling particle size and morphology is crucial for tailoring the properties of ceria-based NMs. Green synthesis approaches employ a combination of precursor selection, reaction parameter optimization, template and surfactant utilization, solvent engineering, and the incorporation of additives to achieve precise control. By implementing these strategies, researchers can produce ceria-based NMs with desired properties while minimizing the environmental impact and promoting sustainability in nanomaterial synthesis.

9.   Green synthesis approaches for tailoring ceria-based NMs

Ceria-based NMs synthesized using green methods offer tremendous opportunities for tailoring their properties to meet specific application requirements. Various green synthesis approaches that enable the control and customization of ceria-based nanomaterial properties are highlighted below.

9.1   Doping and alloying

Doping and alloying refer to the introduction of foreign elements into the ceria lattice to modify its properties. Green synthesis approaches facilitate the incorporation of dopants or alloying elements during nanomaterial synthesis, resulting in controlled changes in the crystal structure, electronic structure, and surface properties. Doping with transition metals, rare earth elements, or other elements can enhance the NMs' catalytic activity, redox properties, or optical properties [51]. Green synthesis methods allow for the precise control of dopant concentration, ensuring the desired functionality while minimizing the use of toxic or hazardous substances. These approaches utilize green reducing agents, bioresources, or natural extracts as dopant sources, thereby reducing the reliance on toxic or hazardous chemicals. Some commonly employed green synthesis approaches for doped ceria NPs include [52]:

Co-precipitation: In this method, cerium and dopant precursors are simultaneously precipitated from aqueous solutions under controlled conditions. Green reducing agents or plant extracts can be used to reduce the precursors and promote the formation of doped ceria NPs. The co-precipitation approach allows for homogeneous distribution of dopant ions within the ceria lattice.

Hydrothermal/Solvothermal Synthesis: Hydrothermal or solvothermal methods involve the reaction of precursors in a closed vessel under high temperature and pressure conditions. Green solvents and natural extracts can be used as reaction media, and dopant precursors can be introduced to achieve controlled doping. The hydrothermal/solvothermal approach enables the synthesis of doped ceria NPs with controlled particle size, morphology, and dopant distribution.

Plant-mediated Synthesis: Plant-mediated synthesis utilizes plant extracts as reducing agents and stabilizers for the synthesis of doped ceria NPs. The plant extracts contain bioactive compounds that can reduce and stabilize the dopant ions, promoting their incorporation into the ceria lattice during nanoparticle formation. This approach offers a sustainable and eco-friendly route for doping ceria NPs.

Microwave-assisted Synthesis: Microwave-assisted synthesis involves the use of microwave irradiation to accelerate the synthesis process. Green reducing agents and dopant precursors can be mixed in a suitable solvent and exposed to microwave irradiation, resulting in rapid and efficient doping of ceria NPs. This method offers advantages such as shorter reaction times and improved control over particle size and dopant distribution. Green synthesis approaches for doped ceria NPs offer several benefits, including reduced environmental impact, enhanced sustainability, and the potential utilization of natural resources. These methods enable precise control over dopant concentration, size, morphology, and crystalline structure, resulting in tailor-made ceria NPs with enhanced properties and functionalities [53]. Doped ceria NPs synthesized through green approaches find applications in catalysis, energy conversion, environmental remediation, and other fields where customized materials with specific properties are required.

9.2   Surface functionalization

Green synthesis approaches enable the surface functionalization of ceria-based NMs with various organic and inorganic species. Surface functionalization can enhance the NMs ' stability, dispersibility, selectivity, or compatibility with specific environments. Green methods utilize environmentally friendly reagents, such as natural extracts, bioresources, or plant-based reducing agents, to modify the surface chemistry of NMs [54]. Functionalization techniques include grafting organic molecules, forming self-assembled monolayers, or attaching NPs to the surface. Surface functionalization allows for the customization of NMs ' interactions with other materials, biological systems, or targeted applications.

9.3 Size and morphology control

Green synthesis approaches offer strategies to control the size and morphology of ceria-based NMs. By carefully adjusting the synthesis conditions, such as reaction temperature, precursor concentration, or reaction time, the size and shape of the NMs can be precisely tailored. Green methods often utilize natural extracts, biomolecules, or green solvents as reaction media, promoting controlled nucleation and growth of nanocrystals [55]. Size control is crucial for achieving specific properties related to quantum confinement effects, catalytic activity, or optical properties. Morphology control enables the synthesis of NMs with desired shapes, such as NPs, nanorods, nanowires, or nanocubes, which exhibit unique properties and functionalities.

9.4 Hierarchical and porous structures

Green synthesis approaches allow for the creation of hierarchical and porous structures in ceria-based NMs. Hierarchical structures combine multiple length scales, such as microscale structures with nanoscale features, leading to enhanced functionality and performance. Green methods employ templates, sacrificial templates, or self-assembly processes to fabricate hierarchical architectures sustainably. Porous structures offer increased surface area, improved mass transport, and higher catalytic activity. Green synthesis techniques facilitate the formation of porous structures by utilizing natural or bio-based templates, such as proteins, cellulose, or plant extracts, which can be selectively removed or transformed into porous networks [56].

9.5 Tailored nanostructures

Green synthesis approaches enable the synthesis of ceria-based NMs with tailored nanostructures, such as core-shell structures, hollow structures, or composite structures. Core-shell structures involve the encapsulation of ceria cores within a shell made of another material, offering enhanced stability, controlled release, or synergistic effects. Hollow structures possess empty interiors, providing space for guest materials, drug delivery, or encapsulation applications. Composite structures combine ceria with other functional materials, such as metals, metal oxides, or carbon-based materials, to achieve enhanced properties, such as improved conductivity, enhanced catalytic performance, or efficient energy storage [57].

10. Green ligands for surface functionalization

Surface functionalization of ceria NPs involves the attachment of ligands onto the nanoparticle surface to modify its chemical and physical properties. Green ligands, derived from environmentally friendly and sustainable sources, have gained significant attention due to their low toxicity, abundance, and potential for replacing conventional ligands derived from hazardous or non-renewable materials [58]. Some of the commonly used green ligands for the surface functionalization of ceria NPs are highlighted below.

a. Biomolecules

Biomolecules such as proteins, peptides, amino acids, and carbohydrates are widely used as green ligands for surface functionalization. They offer several advantages, including high specificity, biocompatibility, and diverse functional groups for conjugation. Biomolecules can be attached to the nanoparticle surface through covalent bonding or affinity-based interactions, providing tailored functionalities and enabling applications in biomedicine, biosensing, and biocatalysis [59].

b. Natural Extracts and Polyphenols

Natural extracts, such as plant extracts, fruits, or tea extracts, contain polyphenols that can act as green ligands for surface functionalization. Polyphenols possess multiple functional groups, including hydroxyl and carboxyl groups, which can interact with the nanoparticle surface through coordination or covalent bonding. Polyphenols offer antioxidant properties, stability, and the potential for tailoring the surface chemistry of ceria NPs for various applications [60].

c. Bio-Based Polymers

Bio-based polymers, such as cellulose, chitosan, or starch, can serve as green ligands for surface functionalization. These polymers are derived from renewable sources and offer a wide range of functional groups for nanoparticle surface attachment. Bio-based polymers provide stability, dispersibility, and biocompatibility to ceria NPs. They can be modified or chemically functionalized to introduce specific functionalities, making them suitable for applications in drug delivery, tissue engineering, and environmental remediation [61].

d. Organic Acids and Organic Solvents

Organic acids, including citric acid, tartaric acid, or malic acid, can act as green ligands for surface functionalization. These ligands offer carboxyl groups that can interact with the nanoparticle surface through coordination or electrostatic interactions. Organic solvents, such as alcohols or esters derived from renewable sources, can also be used as green ligands for surface modification. They provide solubility, stability, and tailored surface chemistry to ceria NPs.

e. Green Surfactants

Green surfactants, such as biosurfactants or surfactants derived from renewable sources, are gaining attention as green ligands for surface functionalization. These surfactants can stabilize and disperse ceria NPs while providing specific functionalities through their hydrophilic and hydrophobic regions [62]. Green surfactants offer advantages such as low toxicity, biodegradability, and reduced environmental impact compared to traditional surfactants.

The use of green ligands for surface functionalization of ceria NPs offers a sustainable and environmentally friendly approach. These ligands provide tailored functionalities, stability, and compatibility for various applications. Furthermore, the utilization of green ligands reduces the dependence on hazardous chemicals, promotes sustainability, and aligns with the principles of green synthesis and nanomaterial fabrication.

11. Characterization techniques for green-synthesized ceria-based NMs

Characterizing the properties and structure of green-synthesized ceria-based NMs is crucial for understanding their behavior and optimizing their performance in various applications. In this section, we discuss the key characterization techniques commonly employed to evaluate the properties of ceria-based NMs synthesized using green methods [53]. These techniques provide valuable insights into the size, morphology, crystal structure, composition, surface properties, and other relevant characteristics of the NMs as depicted in Table 2.

Table 2: Characterization techniques for green-synthesized ceria-based NMs

Technique	Principle and Application	Material Obtained
X-ray Diffraction (XRD)	Crystal structure determination and phase identification	The phase composition, crystal structure, lattice parameters
Transmission Electron Microscopy (TEM)	Imaging and structural analysis of NPs	Particle size, shape, morphology, lattice fringes
Scanning Electron Microscopy (SEM)	Surface morphology and elemental analysis	Surface morphology, particle size distribution, elemental composition
Energy-Dispersive X-ray Spectroscopy (EDX)	Elemental composition analysis and mapping	Elemental composition, distribution of elements
Fourier-Transform Infrared Spectroscopy (FTIR)	Chemical bonding and functional group analysis	Functional groups present, chemical bonding information
UV-Vis Spectroscopy	Optical properties and bandgap determination	Absorption and bandgap energy of the NMs
Raman Spectroscopy	Vibrational modes and structural information	Crystal structure, defects, chemical composition
Surface Area Analysis	Determination of specific surface area	Surface area, pore size distribution, pore volume
Dynamic Light Scattering (DLS)	Particle size and size distribution analysis	Hydrodynamic diameter, polydispersity index
Zeta Potential Analysis	Surface charge and stability analysis	Electrophoretic mobility, colloidal stability

11.1   Structural characterization techniques

X-ray Diffraction (XRD): XRD is used to analyze the crystal structure and phase composition of ceria-based NMs. By measuring the diffraction pattern of X-rays scattered by the NMs, XRD enables the identification of crystalline phases and the determination of their crystallographic properties, such as lattice parameters and crystal size. Transmission Electron Microscopy (TEM): TEM allows direct observation of the size, shape, and morphology of individual ceria-based NPs. It provides high-resolution images that reveal detailed structural features, such as lattice fringes and defects. Additionally, electron diffraction in TEM can be used to confirm the crystalline nature and orientation of the NMs. Scanning Electron Microscopy (SEM): SEM provides information on the surface morphology and size distribution of ceria-based NMs. It uses a focused electron beam to scan the sample surface, producing high-resolution images. SEM is particularly useful for evaluating the overall shape, surface roughness, and agglomeration behavior of the NMs.

11.2   Chemical and compositional characterization techniques

Energy-Dispersive X-ray Spectroscopy (EDS): EDS is employed to determine the elemental composition and chemical distribution within ceria-based NMs. By detecting characteristic X-ray emissions from the sample, EDS provides qualitative and quantitative information about the elements present, their relative abundances, and their spatial distribution. X-ray Photoelectron Spectroscopy (XPS): XPS is utilized to investigate the surface chemistry and oxidation states of ceria-based NMs. It involves bombarding the sample surface with X-rays and measuring the kinetic energy and intensity of emitted photoelectrons. XPS can identify the elements present, determine their chemical states, and assess surface contamination or functionalization.

11.3   Optical and spectroscopic characterization techniques

UV-Visible Spectroscopy: UV-Visible spectroscopy is employed to study the optical properties of ceria-based NMs. It measures the absorption and reflection of light in the UV and visible range, providing information about the bandgap energy, electronic transitions, and optical properties of the NMs. Raman Spectroscopy: Raman spectroscopy is used to investigate the vibrational modes and molecular structure of ceria-based NMs. By measuring the inelastic scattering of monochromatic light, Raman spectroscopy provides information about the bonding, crystal symmetry, and lattice vibrations of the NMs.

11.4   Surface characterization techniques

Brunauer-Emmett-Teller (BET) Analysis: BET analysis is employed to determine the specific surface area and porosity of ceria-based NMs. It utilizes gas adsorption isotherms to assess the NMs ' surface area, pore size distribution, and pore volume. BET analysis provides important information related to the NMs ' surface reactivity and adsorption capacities. Fourier Transform Infrared Spectroscopy (FTIR): FTIR spectroscopy is used to investigate the surface functional groups and chemical bonding of ceria-based NMs. It measures the absorption of infrared radiation by the NMs, providing information about the presence and nature of chemical groups on the surface.

These characterization techniques, among others, play a crucial role in assessing the structural, chemical, optical, and surface properties of green-synthesized ceria-based NMs. By combining multiple techniques, researchers can obtain comprehensive insights into the NMs composition, morphology, size distribution, crystal structure, and surface characteristics [63]. This knowledge is vital for tailoring the synthesis parameters, optimizing the performance, and advancing the applications of ceria-based NMs in various fields.

12.   Applications of green synthesized ceria-based NMs

Ceria-based NMs synthesized through green approaches have gained significant attention in various environmental applications due to their unique properties, high surface area, and catalytic activity [5]. Table 3, depicts the utilization of green synthesized ceria-based NMs in environmental and biomedical applications [19].

12.1   Catalytic applications

Green synthesized ceria-based NMs exhibit excellent catalytic activity and have been widely employed in environmental remediation processes. They can effectively remove pollutants, such as volatile organic compounds (VOCs), nitrogen oxides (NOx), carbon monoxide (CO), and various organic pollutants. The high surface area and redox properties of ceria NPs enable the activation of oxygen molecules and facilitate pollutant oxidation or reduction reactions. Green synthesis methods ensure the use of environmentally friendly materials and minimize the generation of hazardous by-products [64].

Table 3: Environmental and biomedical applications of green-synthesized ceria-based NMs

Category	Environmental Applications	Biomedical Applications
Catalytic Applications	*Pollution control:* Green catalysts for air and water pollution remediation, degradation of organic pollutants	*Drug delivery systems:* Green-synthesized NPs for targeted drug delivery, controlled release of therapeutics
	Wastewater treatment: Removal of heavy metals, organic contaminants, and dyes	*Targeted therapy:* Ceria-based NMs for targeted cancer therapy, disease treatment
	Air purification: Catalytic removal of pollutants, such as NOx, CO, and VOCs	*Biomedical imaging:* Contrast agents for enhanced imaging and diagnostics
Energy Storage and Conversion	*Energy storage:* Green-synthesized NMs for batteries, supercapacitors	*Tissue engineering:* Ceria-based scaffolds for tissue regeneration and repair
	Energy conversion: Use of ceria-based NMs in fuel cells, solar cells	*Regenerative medicine:* Green-synthesized NMs for tissue regeneration and organ repair
Environmental Sensing and Monitoring	*Pollution detection:* Green nanosensors for the detection of pollutants in air, water, and soil	*Diagnostics:* Green-synthesized NMs for disease diagnostics and biomarker detection
	Water quality monitoring: Ceria-based NMs for the detection of heavy metals, contaminants, and pathogens	*Point-of-care testing:* Biosensors based on green-synthesized NMs for rapid and portable diagnostics

12.2 Water treatment and purification

Ceria-based NMs have shown great potential for water treatment and purification applications. Green synthesized ceria NPs can be utilized for the removal of heavy metals, such as arsenic, lead, and cadmium, from contaminated water sources. The NPs can adsorb heavy metal ions onto their surfaces or undergo redox reactions to convert them into less toxic forms. Additionally, ceria NPs can act as catalysts in advanced oxidation processes (AOPs) for the degradation of organic pollutants and the removal of harmful microorganisms from water [65].

12.3 Energy storage and conversion

Green synthesized ceria-based NMs are also promising for energy storage and conversion applications. They can be utilized as catalysts or additives in fuel cells, hydrogen production, and energy storage devices. Ceria NPs exhibit excellent oxygen storage capacity and redox properties, which are crucial for enhancing the performance of energy conversion systems [66]. By employing green synthesis methods, the production of ceria-based NMs for energy applications can be made more sustainable and environmentally friendly.

12.4   Photocatalysis

Green synthesized ceria NPs have demonstrated significant potential in photocatalytic applications. They can be utilized as photocatalysts for the degradation of organic pollutants and the generation of renewable energy through water splitting or CO_2 reduction. The high surface area and unique electronic properties of ceria NPs facilitate the absorption of light and promote charge separation, leading to efficient photocatalytic reactions [67]. Green synthesis approaches ensure the use of eco-friendly materials and promote sustainable photocatalysis.

12.5   Environmental sensing

Green synthesized ceria-based NMs have also been explored for environmental sensing applications. Their surface properties can be modified to detect specific environmental pollutants or analytes. Ceria NPs can be functionalized with appropriate ligands or receptors to selectively interact with target molecules, enabling sensitive and selective detection [68]. The utilization of green synthesis methods ensures the development of environmentally friendly and biocompatible sensors.

12.6   Drug delivery systems

Green synthesized ceria-based NMs have shown promise as drug delivery systems due to their high surface area, excellent stability, and ability to encapsulate and release therapeutic agents. These NMs can be functionalized with biocompatible coatings and loaded with drugs, proteins, or nucleic acids, allowing targeted and controlled release of therapeutic payloads. The controlled drug release provided by ceria-based NMs helps improve treatment efficacy, minimize side effects, and enhance patient compliance [69].

12.7   Imaging agents

Ceria-based NMs have unique optical and magnetic properties that make them suitable for various imaging modalities. By modifying the surface of these NMs with fluorescent dyes or contrast agents, they can be used as imaging probes for techniques such as fluorescence imaging, magnetic resonance imaging (MRI), and computed tomography (CT) imaging [70]. The use of green synthesized ceria-based NMs as imaging agents enables high-resolution imaging, early disease detection, and monitoring of therapeutic responses.

12.8   Diagnostic assays

Green synthesized ceria-based NMs can be employed in diagnostic assays for the detection and quantification of biomarkers associated with diseases. These NMs can be functionalized with specific recognition elements, such as antibodies or aptamers, to selectively capture and detect target biomolecules. The interactions between the biomarkers and the functionalized NMs can generate measurable signals, enabling sensitive and rapid diagnostic assays for various diseases, including cancer, infectious diseases, and cardiovascular disorders [71].

12.9   Therapeutics and tissue engineering

Green synthesized ceria-based NMs possess unique antioxidant and anti-inflammatory properties, making them potential candidates for therapeutic applications. These NMs can scavenge reactive oxygen species (ROS) and reduce oxidative stress, which is associated with many diseases.

Additionally, their surface can be modified to enhance biocompatibility and promote cellular interactions, making them suitable for tissue engineering applications [72]. Ceria-based NMs hold promise in regenerative medicine, wound healing, and the development of biomaterials with enhanced healing properties.

12.10 Biosensors and point-of-care devices

The exceptional properties of green synthesized ceria-based NMs, such as high surface area, catalytic activity, and electrical conductivity, make them ideal for biosensing applications. These NMs can be integrated into biosensors and point-of-care devices for the detection of biomarkers, pathogens, and toxins. The rapid response, sensitivity, and portability of ceria-based nanosensors facilitate on-site and in-field diagnostics, enabling early disease detection and personalized medicine approaches [73].

13. Challenges in green synthesis approaches

Green synthesis approaches for ceria-based NMs offer numerous advantages, such as sustainability, biocompatibility, and reduced environmental impact. However, several challenges still exist in the field of green synthesis. Figure 3 shows some of the current challenges associated with green synthesis approaches for ceria-based NMs [32]:

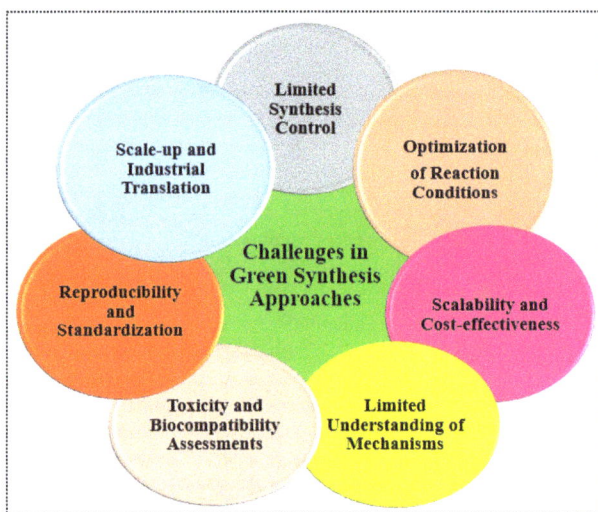

Figure 3: Some challenges associated with green synthesis approaches

Limited Synthesis Control

Green synthesis methods often involve complex reaction mechanisms and multiple parameters, making it challenging to achieve precise control over the size, shape, and composition of the synthesized NMs. This lack of control can result in variations in the properties and performance of the NMs, affecting their desired applications. Developing strategies to enhance synthesis control and achieve consistent and reproducible results is a key challenge.

Optimization of Reaction Conditions

Green synthesis approaches often require specific reaction conditions, such as temperature, pressure, pH, and reaction time, to achieve the desired nanomaterial properties. Optimizing these reaction conditions for green synthesis methods can be time-consuming and resource-intensive. Identifying the optimal reaction parameters that promote efficient and eco-friendly synthesis while ensuring the desired product quality remains a challenge.

Scalability and Cost-effectiveness

While green synthesis methods are known for their environmental benefits, they may face challenges when it comes to scalability and cost-effectiveness. Some green synthesis approaches may require expensive raw materials or energy-intensive processes, limiting their practical application on a larger scale. Developing scalable and cost-effective green synthesis approaches for ceria-based NMs is crucial for their commercialization and widespread adoption.

Limited Understanding of Mechanisms

Green synthesis approaches often involve complex biological, chemical, or physical mechanisms that are not yet fully understood. The lack of comprehensive understanding hinders the optimization and control of green synthesis processes. Further research is needed to unravel the underlying mechanisms and establish fundamental principles that guide the green synthesis of ceria-based NMs.

Toxicity and Biocompatibility Assessments

While green synthesis approaches are generally considered more environmentally friendly and biocompatible, it is crucial to conduct thorough toxicity and biocompatibility assessments. Understanding the potential adverse effects of green-synthesized NMs on human health and the environment is essential for their safe application. Standardized protocols for toxicity testing and comprehensive evaluation of NMs long-term effects are needed to ensure their safe use.

Reproducibility and Standardization

Reproducibility and standardization are key challenges in green synthesis approaches. Variations in experimental conditions, precursor quality, and synthesis protocols can lead to inconsistent results, hindering the comparability and reliability of research findings. Establishing standardized protocols, quality control measures, and guidelines for green synthesis methods is necessary to promote reproducibility and facilitate collaboration among researchers.

Scale-up and Industrial Translation

Moving from laboratory-scale synthesis to industrial-scale production remains a challenge for green synthesis approaches. Ensuring that the environmentally friendly synthesis methods developed in the lab can be successfully scaled up while maintaining the desired product quality and performance requires careful optimization and process engineering. Collaborations between academia, industry, and regulatory bodies are essential to overcome the barriers and facilitate the industrial translation of green synthesis approaches.

Addressing these current challenges in green synthesis approaches for ceria-based NMs will contribute to the development of more efficient, cost-effective, and environmentally friendly synthesis methods. Continued research, technological advancements, and interdisciplinary collaborations are crucial to overcome these challenges and promote the widespread adoption of green synthesis approaches in various industries.

14.   Future perspectives

The field of green synthesis approaches for ceria-based NMs is continually evolving, and several potential future developments and improvements can enhance their efficacy and applicability. Below are some areas with promising prospects:

Advanced Catalyst Design: Future developments can focus on the design and synthesis of advanced catalysts by tailoring the composition, structure, and morphology of ceria-based NMs. This can involve incorporating dopants, controlling defects, and optimizing the surface properties to enhance catalytic activity and selectivity. Understanding the structure-function relationships at the nanoscale can guide the design of highly efficient and specific catalysts for various applications.

Multifunctional NMs*:* Future research can explore the integration of ceria-based NMs with other functional materials to create multifunctional hybrids. These hybrids can exhibit combined properties, such as catalysis, energy storage, and sensing, enabling versatile applications. By carefully designing the interfaces and interactions between different components, synergistic effects can be achieved, leading to enhanced performance and novel functionalities.

Nanoscale Characterization Techniques: Advances in nanoscale characterization techniques will provide valuable insights into the structure, morphology, and properties of green-synthesized ceria-based NMs. Techniques such as in situ characterization, high-resolution microscopy, and spectroscopy can help understand the dynamic behavior of NMs during synthesis and operating conditions. These techniques will contribute to a deeper understanding of the synthesis mechanisms and enable precise control over nanomaterial properties.

Computational Modeling and Simulation: Computational modeling and simulation can play a significant role in predicting and optimizing the properties of green-synthesized ceria-based NMs. By combining theoretical calculations with experimental data, it is possible to gain insights into the nanoscale processes, predict material behavior, and accelerate the discovery of novel green synthesis approaches. Computational methods can also assist in understanding the mechanisms of catalytic reactions and guide the design of more efficient catalysts.

Green Solvents and Reaction Conditions: Research can focus on the development of novel green solvents and reaction conditions for green synthesis approaches. Green solvents with low toxicity and environmental impact can replace traditional organic solvents, promoting greener and more sustainable synthesis methods. Moreover, optimizing reaction parameters, such as temperature, pressure, and pH, can enhance the efficiency, selectivity, and reproducibility of green synthesis approaches.

Scalable Manufacturing Processes: Innovations in scalable manufacturing processes are essential to bridge the gap between laboratory-scale synthesis and industrial production of green-synthesized ceria-based NMs. Process optimization, flow chemistry, continuous synthesis methods, and automation can contribute to cost-effective and large-scale production. Developing sustainable and efficient manufacturing processes will facilitate the widespread adoption of green-synthesized NMs in commercial applications.

Conclusions

The development and application of green-synthesized ceria-based NMs have emerged as a promising and sustainable approach in the field of nanoscience and nanotechnology. The utilization of environmentally friendly synthesis methods, coupled with the unique properties of ceria-based NMs, opens up a wide range of applications in diverse fields, including catalysis, energy storage, environmental remediation, and biomedicine. In this chapter, we have explored the overview of ceria-based NMs, the importance of green synthesis approaches, various green synthesis methods such as sol-gel synthesis, hydrothermal and solvothermal synthesis, plant-mediated synthesis, microwave-assisted synthesis, electrochemical methods, and sonochemical methods. We have also discussed the characterization techniques used for analyzing green-synthesized ceria-based NMs, including X-ray diffraction (XRD), transmission electron microscopy (TEM), scanning electron microscopy (SEM), energy-dispersive X-ray spectroscopy (EDX), Fourier-transform infrared spectroscopy (FTIR), UV-Visible spectroscopy, Raman spectroscopy, and surface area analysis techniques. Furthermore, we have examined how green synthesis approaches can be employed to tailor the properties of ceria-based NMs, including doping, surface modification, and control of particle size and morphology. These strategies enable the optimization of NMs for specific applications, enhancing their catalytic activity, stability, biocompatibility, and sensing capabilities. We have also explored the significant applications of green-synthesized ceria-based NMs in environmental remediation, energy storage and conversion, as well as biomedical applications such as therapeutics, imaging, and biosensing. These applications demonstrate the potential of green-synthesized ceria-based NMs to address critical challenges in areas such as pollution control, renewable energy, and healthcare. Thus, green-synthesized ceria-based NMs hold immense promise for sustainable technological advancements and have the potential to revolutionize various industries. By employing environmentally friendly synthesis approaches, tailoring their properties, and exploring diverse applications, these NMs can contribute to a greener and more sustainable future. Continued research, innovation, and collaboration among scientists, engineers, and policymakers are crucial to further unlock the potential of green-synthesized ceria-based NMs and translate them into practical solutions for the benefit of society and the environment.

References

[1] F.D. Guerra, M.F.Attia, D.CWhitehead, F. Alexis, Nanotechnology for Environmental Remediation: Materials and Applications. Molecules 23(2018)1760. https://doi.org/10.3390/molecules23071760

[2] S.F.Ahmed, M. Mofijur, N. Rafa, A.T. Chowdhury, S. Chowdhury, M. Nahrin, A.B.M.S. Islam, H.C.Ong, Green approaches in synthesising nanomaterials for environmental nanobioremediation: Technological advancements, applications,

benefits and challenges. Environ Res. 204(2022)111967. doi: 10.1016/j.envres.2021.111967. Epub 2021 Aug 25. PMID: 34450159.

[3] U.L. Usman, N. B. Singh, B.K. Allam, S. Banerjee, Biogenic Synthesis of Zinc Oxide/Chitosan Nanocomposite using Callistemon citrinus (Bottle Brush) for Photocatalytic Degradation of Methylene Blue. Macromol Symp. Macromol. Symp. 407(2023) 2100357: https://doi.org/10.1002/masy.202100357

[4] K.R.B. Singh, V.Nayak, T.Sarkar, R.P.Singh. Cerium oxide nanoparticles: Properties, biosynthesis and biomedical application. RSC Advances. RSC Adv., 10(2020) 27194-27214. https://doi.org/10.1039/D0RA04736H.

[5] R.G. Chaudhary, P.B. Chouke, R.Bagade, A.K. Potbhare, Molecular docking and antioxidant activity of *Cleome simplicifolia* assisted synthesis of cerium oxide nanoparticles, Mater. Today: Procs, 2020, 29 (4), 1085-1090: doi.org/10.1016/j.matpr.2020.05.062.

[6] Usman Lawal Usman, Nakshatra Bahadur Singh, Bharat Kumar Allam, Sushmita Banerjee, Plant extract mediated synthesis of Fe3O4-chitosan composite for the removal of lead ions from aqueous solution. Mater. Today: Proc 60 (2022) 1140-1149; https://doi.org/10.1016/j.matpr.2022.02.311

[7] P.B. Chouke, T. Shrirame, A.K. Potbhare, A. Mondal, A.R. Chaudhary, S. Mondal, S.R. Thakare, E. Nepovimova, M. Valis, K. Kuca, R. Sharma, R.G. Chaudhary, Bioinspired metal/metal oxide nanoparticles: A road map to potential applications, Mater. Today Adv. 16(2022) 100314. https://doi.org/10.1016/j.mtadv.2022.100314.

[8] F. Khan, M. Shariq, M. Asif, M. A. Siddiqui, P. Malan, F. Ahmad. Green Nanotechnology: Plant-Mediated Nanoparticle Synthesis and Application. Nanomaterials (Basel). 12(4)(2022)673. doi: 10.3390/nano12040673. PMID: 35215000; PMCID: PMC8878231.

[9] S. Ghosh , R. Ahmad, K. Banerjee, M. F. AlAjmi, S. Rahman, Mechanistic Aspects of Microbe-Mediated Nanoparticle Synthesis. Frontiers in Microbiology. Front. Microbiol. 12(2021)638068. https://doi.org/10.3389/fmicb.2021.638068

[10] Francisco Rodríguez-Félix, Abril Zoraida Graciano-Verdugo, María Jesús Moreno-Vásquez, Irlanda Lagarda-Díaz, Carlos Gregorio Barreras-Urbina, Lorena Armenta-Villegas, Alberto Olguín-Moreno, José Agustín Tapia-Hernández, "Trends in Sustainable Green Synthesis of Silver Nanoparticles Using Agri-Food Waste Extracts and Their Applications in Health", Journal of Nanomaterials, vol. 2022, Article ID 8874003, 37 pages, 2022. https://doi.org/10.1155/2022/8874003

[11] A. Jain, Algae-mediated synthesis of biogenic nanoparticles. Advances in Natural Sciences: Nanoscience and Nanotechnology. Adv. Nat. Sci: Nanosci. Nanotechnol. 13(4)(2022) 043001. DOI 10.1088/2043-6262/ac996a.

[12] M. S. Samuel, M. Ravikumar, J. A. John, E. Selvarajan, H. Patel, P. S..Chander, J. Soundarya, ,S.Vuppala, R.Balaji, N. Chandrasekar, A Review on Green Synthesis of Nanoparticles and Their Diverse Biomedical and Environmental Applications. Catalysts 12(2022), 459. https://doi.org/10.3390/catal12050459

[13]   Dmitry Bokov, Abduladheem Turki Jalil, Supat Chupradit, Wanich Suksatan, Mohammad Javed Ansari, Iman H. Shewael, Gabdrakhman H. Valiev, Ehsan Kianfar, "Nanomaterial by Sol-Gel Method: Synthesis and Application", Advances in Materials Science and Engineering, (2021), 5102014, 21 pages https://doi.org/10.1155/2021/5102014

[14]   M. Nyoka, Y. E. Choonara, P. Kumar, P.P.D. Kondiah, V. Pillay, Synthesis of Cerium Oxide Nanoparticles Using Various Methods: Implications for Biomedical Applications. Nanomaterials (Basel). 10(2) (2020) 242. doi: 10.3390/nano10020242. PMID: 32013189; PMCID: PMC7075153.

[15]   A.K. Potbhare, R.G. Chaudhary, V. Sonkusare, A. Mondal, A.R. Rai, H.D. Juneja. Phytosynthesis of nearly monodisperse CuO nanospheres using *Phyllanthus Reticulatus/Conyza Bonariensis* and its antioxidant/antibacterial assays, Materials Science & Engineering C, 2019, 99, 783-793. https://doi.org/10.1016/j.msec.2019.02.010.

[16]   Nayak J, Devi C, Vidyapeeth L. Microwave assisted synthesis: a green chemistry approach. Int. Res. J. Pharm. Appl. Sci. 2016;3(5):278-285.

[17]   Christos Vaitsis, Maria Mechili, Nikolaos Argirusis, Eirini Kanellou, Pavlos K. Pandis, Georgia Sourkouni, Antonis Zorpas and Christos Argirusis, Ultrasound-Assisted Preparation Methods of Nanoparticles for Energy-Related Applications. In: Nanotechnology and the Environment. (2020). DOI: 10.5772/intechopen.92802

[18]   S. Arndt , D. Weis, K. Donsbach, S. R. Waldvogel, The "Green" Electrochemical Synthesis of Periodate. Angew Chemie - Int Ed. 59, (2020) 7969-8300 https://doi.org/10.1002/anie.202002717

[19]   Mohamed Madani , Shimaa Hosny , Dalal Mohamed Alshangiti , Norhan Nady , Sheikha A. Alkhursani , Huda Alkhaldi , Samera Ali Al-Gahtany , Mohamed Mohamady Ghobashy, Ghalia A. Gabe, "Green synthesis of nanoparticles for varied applications: Green renewable resources and energy-efficient synthetic routes" Nanotechnology Reviews, 11(1) (2022) 731-759. https://doi.org/10.1515/ntrev-2022-0034

[20]   F. S. Irwansyah, A. R. Noviyanti, D. R. Eddy, R. Risdiana, Green Template-Mediated Synthesis of Biowaste Nano-Hydroxyapatite: A Systematic Literature Review. Molecules 27(2022)5586. https://doi.org/10.3390/molecules27175586.

[21]   . Mondal, M.S. Umekar, G.S. Bhusari, P.B. Chouke, T. Lambat, S. Mondal, R.G. Chaudhary, S.H. Mahmood, Biogenic synthesis of metal/metal oxide nanostructured materials, Curr. Pharm. Biotechnol. 22 (13) 2021, 1782-1793. https://doi.org/10.2174/1389201022666210111122911.

[22]   Palaniyandi Velusamy, Govindarajan Venkat Kumar, Venkadapathi Jeyanthi Jayabrata Das, and Raman Pachaiappan, Bio-Inspired Green Nanoparticles: Synthesis, Mechanism, and Antibacterial Application. Toxicol Res. 32(2016) 95–102, https://doi.org/10.5487/TR.2016.32.2.095

[23.] R. G. Chaudhary, N. B. Singh, A. R. Daddemal-Chaudhary and Rohit Sharma, Review on Agrobiowaste-mediated nanohybrids for removal of toxic heavy metals from wastewater, ChemistrySelect, 2024, 9(4) e202304230. https://doi.org/10.1002/slct.202304230.

[24]   I. V. Machado, J. R. N. dos Santos, M. A. P. Januario, A.G. Corrêa, Greener organic synthetic methods: Sonochemistry and heterogeneous catalysis promoted multicomponent reactions. Ultrasonics Sonochemistry 78 (2021) 105704. https://doi.org/10.1016/j.ultsonch.2021.105704.

[25]   P. Lisbona, S. Pascual, V. Pérez. Waste to energy: Trends and perspectives, Chemical Engineering Journal Advances 14 (2023) 100494. https://doi.org/10.1016/j.ceja.2023.100494.

[26]   S. Bauer, M. Wagner, Possibilities and Challenges of Wastewater Reuse— Planning Aspects and Realized Examples. Water (Switzerland). Water 14(2022) 1619. https://doi.org/10.3390/w14101619

[27]   Caroline Visentin Adan William da Silva Trentin, Adeli Beatriz Braun, Antônio Thomé, Lifecycle assessment of environmental and economic impacts of nano-iron synthesis process for application in contaminated site remediation. Journal of Cleaner Production 231 (2019) 307e319. https://doi.org/10.1016/j.jclepro.2019.05.236

[28]   L. Soltys, O. Olkhovyy, T.Tatarchuk, M. Naushad, Green Synthesis of Metal and Metal Oxide Nanoparticles: Principles of Green Chemistry and Raw Materials. Magnetochemistry.7(11)(2021)145.https://doi.org/10.3390/magnetochemistry711014 5.

[29]   Neil Osterwalder, Christian Capello, Konrad Hungerbu"hler, Wendelin J. Stark, Energy Consumption During Nanoparticle Production: How Economic is Dry Synthesis?. J Nanopart Res 8(2006) 1–9, https://doi.org/10.1007/s11051-005-8384-7.

[30]   M. S. Umekar, G. S. Bhusari, T. Bhoyar, V. Devthade, B. P. Kapgate, A. P. Potbhare, R. G. Chaudhary and A. A. Abdala, Graphitic carbon nitride-based photocatalysts for environmental remediation of organic pollutants, Current Nanoscience, 19 (2) (2023), 148-169. https://doi.org/10.2174/1573413718666220127123935.

[31]   M. Branca, M.Ibrahim, D. Ciuculescu, K. Philippot, C. Amiens, Water Transfer of Hydrophobic Nanoparticles: Principles and Methods. In Handbook of Nanoparticles; Aliofkhazraei M., Ed., 1st; Springer: New York, (2015)1279–1311. 10.1007/978-3-319-15338-4_29

[32]   Shuaixuan Ying, Zhenru Guan, Polycarp C. Ofoegbu , Preston Clubb , Cyren Rico, Feng He, Jie Hong, Green synthesis of nanoparticles: Current developments and limitations. Environmental Technology and Innovation. 26 (2022) 102336. https://doi.org/10.1016/j.eti.2022.102336

[33]   Mohammadreza Khalaj· Mohammadreza Kamali, Tejraj M Aminabhavi M Elisabete V Costa, Raf Dewil, Lise Appels, Isabel Capel, Sustainability insights into the synthesis of engineered nanomaterials - Problem formulation and considerations.

Environ Res. 220(2023)115249. doi: 10.1016/j.envres.2023.115249. Epub 2023 Jan 9. PMID: 36632884.

[34]   E. R. Sadiku, O. Agboola , I. D. Ibrahim A. Babu Reddy, M. Bandla , P. N. Mabalane, et al. Synthesis of Bio-Based and Eco-Friendly Nanomaterials for Medical and BioMedical Applications. In: Materials Horizons: From Nature to Nanomaterials Green Biopolymers and their Nanocomposites, (2019) 283-312. https://doi.org/10.1007/978-981-13-8063-1_13.

[35]   X.C.Jiang, W. M. Chen, C. Y. Chen, S. X. Xiong, A. B. Yu, Role of Temperature in the Growth of Silver Nanoparticles Through a Synergetic Reduction Approach. Nanoscale Res Lett. 6(1)(2011)32. doi: 10.1007/s11671-010-9780-1. Epub 2010 Sep 23. PMID: 27502655; PMCID: PMC3211407.

[36]   N.T.K. Thanh, N. Maclean S. Mahiddine, Mechanisms of nucleation and growth of nanoparticles in solution. Chemical Reviews. 114(2014)7610−7630

[37]   Nan-Chun Wu, Er-Wei Shi, Yan-Qing Zheng, Wen-Jun Li, Effect of pH of medium on hydrothermal synthesis of nanocrystalline cerium(IV) oxide powders. J Am Ceram Soc. 85(10)(2005)2462 – 2468. DOI: 10.1111/j.1151-2916.2002.tb00481.x

[38]   Al-Hada NM, Md. Kasmani R, Kasim H, Al-Ghaili AM, Saleh MA, Banoqitah EM,

[Alhawsawi AM, Baqer AA, Liu J, Xu S, et al. The Effect of Precursor Concentration on the Particle Size, Crystal Size, and Optical Energy Gap of CexSn1−xO2 Nanofabrication.Nanomaterials. (8)(2021)2143.https://doi.org/10.3390/nano11082143

[39]   D. Rodríguez-Padrón, A. M. Puente-Santiago, A.M. Balu, M. J.Muñoz-Batista , R. Luque, Environmental Catalysis: Present and Future. ChemCatChem. 2019 https://doi.org/10.1002/cctc.201801248.

[40.] Phebe Asantewaa Owusu and Samuel Asumadu-Sarkodie, A review of renewable energy sources, sustainability issues and climate change mitigation, Cogent Engineering 3(2016)1167990DOI: 10.1080/23311916.2016.1167990.

[41]   W.] Chen, Z. Huang, K. J. Chua, Sustainable energy recovery from thermal processes: a review. Energ Sustain Soc 12(2022)46; https://doi.org/10.1186/s13705-022-00372-2.

[42]   L. Berta, N.A. Coman, A. Rusu, C. Tanase, A Review on Plant-Mediated Synthesis of Bimetallic Nanoparticles, Characterisation and Their Biological Applications. Materials (Basel). 14(24)(2021)7677; doi: 10.3390/ma14247677. PMID: 34947271; PMCID: PMC8705710.

[43]   A. Altemimi , N. Lakhssassi, A. Baharlouei, D.G. Watson, D.A. Lightfoot, Phytochemicals: Extraction, Isolation, and Identification of Bioactive Compounds from Plant Extracts. Plants (Basel). 6(4)(2017)42,. doi: 10.3390/plants6040042. PMID: 28937585; PMCID: PMC5750618.

[44]   J.O. Adeyemi, A.O. Oriola, D. C. Onwudiwe, A. O. Oyedeji, Plant Extracts Mediated Metal-Based Nanoparticles: Synthesis and Biological Applications. Biomolecules 12(2022) 627. https://doi.org/10.3390/biom12050627

[45]   P. Bhilkar, A. Bodhne, S. Yerpude, R. Madankar, S. Somkuwar, A. Chaudhary, A. Lambat, M. Desimone, R. Sharma, R. Chaudhary, Phyto-derived Metal Nanoparticles: Prominent Tool for Biomedical Applications, OpenNano, 14 (2023) 100192. https://doi.org/10.1016/j.onano.2023.100192.

[46]   P. K. Dikshit, J. Kumar, A. K. Das, S. Sadhu, S. Sharma, S. Singh, P. K. Gupta, B. S. Kim, Green Synthesis of Metallic Nanoparticles: Applications and Limitations. Catalysts 11(2021) 902. https://doi.org/10.3390/catal11080902

[47]   M. Shah, D. Fawcett, S. Sharma, S. K. Tripathy, G.E.J. Poinern , Green Synthesis of Metallic Nanoparticles via Biological Entities. Materials (Basel). 8(11)⊗2015)7278-7308. doi: 10.3390/ma8115377. PMID: 28793638; PMCID: PMC5458933.

[48]   M. Pérez-Page, E. Yu, J. Li, M. Rahman, D. M. Dryden, R. Vidu, P. Stroeve, Template-based syntheses for shape controlled nanostructures. Adv Colloid Interface Sci. 234(2016)51-79. doi: 10.1016/j.cis.2016.04.001. Epub 2016 Apr 20. PMID: 27154387.

[49]   S. S. Mohtar, F. Aziz, A. F. Ismail, N. S. Sambudi, H. Abdullah , A. N. Rosli , B. Ohtani . Impact of Doping and Additive Applications on Photocatalyst Textural Properties in Removing Organic Pollutants: A Review. Catalysts. 11(10)(2021)1160. https://doi.org/10.3390/catal11101160

[50]   J. Singh, T. Dutta, K. H. Kim, M. Rawat, P. Samaddar, P. Kumar 'Green' synthesis of metals and their oxide nanoparticles: applications for environmental remediation. J Nanobiotechnol 16(2018) 84 ; https://doi.org/10.1186/s12951-018-0408-4

[51]   A. Bandyopadhyay , B. J. Sarkar, S. Sutradhar, J. Mandal, P. K. Chakrabarti , Synthesis, structural characterization, and studies of magnetic and dielectric properties of Gd3+ doped cerium oxide (Ce0.90Gd0.10O2−δ). J Alloys Compd. 865(2021) 158838, https://doi.org/10.1016/j.jallcom.2021.158838.

[52]   N. Joudeh, D. Linke, Nanoparticle classification, physicochemical properties, characterization, and applications: a comprehensive review for biologists. J Nanobiotechnol 20,(2022)262 ; https://doi.org/10.1186/s12951-022-01477-8

[53]   Nitin Kumar Sharma, Jyotsna Vishwakarma, Summi Rai, Taghrid S Alomar, Najla AlMasoud, Ajaya Bhattarai, Green Route Synthesis and Characterization Techniques of Silver Nanoparticles and Their Biological Adeptness. ACS Omega. 7(31)(2022)27004-27020. doi: 10.1021/acsomega.2c01400. PMID: 35967040; PMCID: PMC9366950.

[54]   Karolina Wieszczycka, Katarzyna Staszak, Marta J. Woz´niak-Budych, Jagoda Litowczenko, Barbara M. Maciejewska , Stefan Jurga., Surface functionalization –

The way for advanced applications of smart materials. Coordination Chemistry Reviews. 436(2021)213846 https://doi.org/10.1016/j.ccr.2021.213846.

[55]   F. Charbgoo, M. B. Ahmad, M. Darroudi Cerium oxide nanoparticles: green synthesis and biological applications. Int J Nanomedicine.12(2017)1401-1413; doi: 10.2147/IJN.S124855. PMID: 28260887; PMCID: PMC5325136.

[56]   X. Y. Yang, L. H. Chen, Y. Li, J. C. Rooke, C. Sanchez, B. L. Su, Hierarchically porous materials: Synthesis strategies and structure design. Chemical Society Reviews. Chem. Soc. Rev., 46(2017) 481-558. https://doi.org/10.1039/C6CS00829A.

[57]   N.B. Singh, R.G. Chaudhary, M.F. Desimone, A. Agrawal, S.K. Shukla, Green synthesized nanomaterials for safe technology in sustainable agriculture, Current Pharmaceutical Biotechnology, 24 (2023) 61-85. https://doi.org/10.2174/1389201023666220608113924.

[58]   M.A Neouze, U Schubert. Surface Modification and Functionalization of Metal and Metal Oxide Nanoparticles by Organic Ligands. Monatsh Chem 139(2008) 183–195; https://doi.org/10.1007/s00706-007-0775-2

[59]   D Sharma, S Kanchi, K Bisetty. Biogenic synthesis of nanoparticles: A review. Arabian journal of Chemistry 12(2019) 3576–3600. http://dx.doi.org/10.1016/j.arabjc.2015.11.002.

[60]   S.M. Amini, A Akbari. Metal nanoparticles synthesis through natural phenolic acids. IET Nanobiotechnol. 13(8)(2019) 771-777. doi: 10.1049/iet-nbt.2018.5386. PMID: 31625516; PMCID: PMC8676617.

[61]   J.B. Ricardo, Pinto, Carlos D Luís, Marques A.A.P. Paula, A.J. Silvestre, C.S. Freire. An overview of luminescent bio-based composites. Journal of Applied Polymer Science. Appl. Polym. Sci. 131(2014)41169; DOI: 10.1002/app.41169.

[62]   T Benvegnu, D Plusquellec, L Lemiègre. Surfactants from renewable sources: Synthesis and applications. In: Monomers, Polymers and Composites from Renewable Resources. 2008. DOI: 10.1016/B978-0-08-045316-3.00007-7.

[63]   U.V. Gaikwad, AR. Golhar, NK Choudhari, AR. Chaudhari. Structural characterization techniques of materials. In: AIP Conference Proceedings. 2104, 020027 (2019); https://doi.org/10.1063/1.5100395 2104.

[64]   Umekar, M. S., Bhusari, G. S., Potbhare, A. K., Mondal, A., Kapgate, B. P., Desimone, M. F., & Chaudhary, R. G. (2021). Bioinspired reduced graphene oxide based nanohybrids for photocatalysis and antibacterial applications. *Curr. Pharm. Biotechnol.*, *22*(13),1759-1781. http://dx.doi.org/10.2174/1389201022666201231115826

[65]   J. Yang, B. Hou, J. Wang, B. Tian, J. Bi, N. Wang, X. Li, X. Huang. Nanomaterials for the Removal of Heavy Metals from Wastewater. Nanomaterials (Basel).9(3)(2019)424. doi: 10.3390/nano9030424. PMID: 30871096; PMCID: PMC6473982.

[66]   K. Kowsuki, R . Nirmala, Y.H. Ra, R. Navamathavan, Recent advances in cerium oxide-based nanocomposites in synthesis, characterization, and energy storage

applications: A comprehensive review. Results Chem. 5(1)(2023)100877. DOI: 10.1016/j.rechem.2023.100877

[67] M. Pavel, C. Anastasescu, R.N. State, A .Vasile, F. Papa, I. Balint. Photocatalytic Degradation of Organic and Inorganic Pollutants to Harmless End Products: Assessment of Practical Application Potential for Water and Air Cleaning. Catalysts 13(2023,) 1380. https://doi.org/10.3390/catal13020380

[68] P.C. Ray, H. Yu, P. P. Fu, Toxicity and environmental risks of nanomaterials: challenges and future] needs. J Environ Sci Health C Environ Carcinog Ecotoxicol Rev. 1(2009)1-35. doi:]   10.1080/10590500802708267. PMID: 19204862; PMCID: PMC2844666.

[69] Jayanta Kumar Patra, Gitishree Das, Leonardo Fernande Fraceto, Estefania Vangelie Ramos Campos, Maria del Pilar Rodriguez-Torres, Laura Susana Acosta-Torres, Luis Armando Diaz-Torres, Renato]   Grillo, Mallappa Kumara Swamy, Shivesh Sharma Solomon Habtemariam, Han-Seung Shin, Nano] based drug delivery systems: recent developments and future prospects. J Nanobiotechnol 16, 71 (2018). https://doi.org/10.1186/s12951-018-0392-8.

[70.] A. Farooq, S. Sabah, S. Dhou, N. Alsawaftah, G. Husseini. Exogenous Contrast Agents in Photoacoustic Imaging: An In Vivo Review for Tumor Imaging. Nanomaterials (Basel). 12(3)(2022)393. doi: 10.3390/nano12030393. PMID: 35159738; PMCID: PMC8840344.

[71] Saman Sargazi, Iqra Fatima, Maria Hassan Kiani, Vahideh Mohammadzadeh, Rabia Arshad, Muhammad Bilal , Abbas Rahdar, Ana M. Díez-Pascual, Razieh Behzadmehr, Fluorescent-based nanosensors for selective detection of a wide range of biological macromolecules: A comprehensive review. International Journal of Biological Macromolecules. 206(2022)115-147 https://doi.org/10.1016/j.ijbiomac.2022.02.137.

[72] X. Zheng, P. Zhang, Z. Fu, S. Meng, L. Dai, H. Yang, Applications of nanomaterials in tissue engineering. RSC Advances. RSC Adv. 11(2021)19041-19058. https://doi.org/10.1039/D1RA01849C.

[73] S. Pandey,Advance Nanomaterials for Biosensors. Biosensors 12(2022) 219. https://doi.org/10.3390/bios12040219.

Green Synthesis and Emerging Applications of Frontier Nanomaterials   Materials Research Forum LLC
Materials Research Foundations 169 (2024) 365-397   https://doi.org/10.21741/9781644903278-14

Chapter 14

Green synthesis of gold nanoparticles

Matthew Ogoe[1], Komal Janiyani[1], Harjeet Singh[1, 2], Shivani R Pandya [1, 2*]

[1] Research and Development Cell, Parul University, Vadodara, Gujarat, India

[2] Narnarayan Shahtri Institute of Technology, Institute of Forensic Science and Cyber Security, Jetalpur, Ahmedabad, Gujarat, India

* shivpan02@gmail.com

Abstract

The green synthesis of gold nanoparticles (AuNPs) represents a revolutionary shift in nanotechnology, aligning with sustainability and environmental conservation principles. Traditional methods for synthesizing AuNPs often involve toxic chemicals and harsh conditions, raising concerns about their environmental and health impacts. Green synthesis leverages natural resources such as plant extracts, microorganisms, and benign chemical agents to produce AuNPs in an eco-friendly manner. This chapter covers the basic ideas and historical development of green chemistry concerning the synthesis of nanoparticles. Critical analysis is given to several biological, chemical, and physical approaches to green synthesis, emphasizing the mechanisms of stabilization and reduction made possible by phytochemicals and biomolecules. This chapter further discusses important analytical methods for determining the physicochemical characteristics of green synthesized AuNPs and assuring their appropriateness for various applications. The benefits of green synthesis are highlighted, including improved biocompatibility, cost-effectiveness, and environmental benefits. The chapter also covers the extensive applications of AuNPs in biomedicine, environmental remediation, and industrial catalysis, demonstrating their versatile utility. Notwithstanding the promising perspectives, issues with scalability, repeatability, and regulatory barriers still exist. Future directions on how cutting-edge technology like machine learning and artificial intelligence can be used to optimize green synthesis processes are discussed. The potential of green synthesis to transform the production of nanoparticles and support technological innovation and sustainable development is highlighted in this chapter, to facilitate the adoption and application of green synthesis in nanotechnology by tackling existing constraints and investigating potential avenues for advancement.

Keywords

Green Synthesis, Gold Nanoparticles, Nanotechnology, Phytochemicals, Biocompatibility, Biomedical Applications, Green Chemistry

Contents

1.   Introduction

The term "nanotechnology" describes the synthesis and utilization of materials with a nanoscale presence, or sizes between 1- 100 nm (1). Nowadays, physics, chemistry, engineering, and biology are all included in the study of nanotechnology, which is regarded as a multidisciplinary field (2). The recent decades have seen a surge in this scope of technology due to the reported unique physiochemical characteristics, ratio of surface-to-volume, biocompatibility, as well as the electrical characteristics of nanomaterials over the bulk counterparts that have warranted their patronage and numerous applications in diverse fields such as cosmetics (3), oil and gas (4), biomedical (5), catalysis (6), cancer therapies (7–9) , diagnosing diseases, and delivering of drug (10–12) among others. Chemical and physical synthesis are the main approaches in conventional/traditional nanoparticle synthesis. While the physical synthesis approach comprises pulsed laser deposition, laser evaporation, and microwave-assisted combustion, the chemical method of nanoparticle synthesis involves procedures like sol-gel method, chemical reduction, polyol synthesis, and precipitation (13). However, these techniques come with several restrictions and disadvantages, including the use of hazardous materials, expensive production costs, risky practices, harnessing heat, and high pressure to create dangerous byproducts that are bad for the environment and people's health (14)The possible risks surpass the advantages of conventional approaches for synthesizing nanomaterials, considering the numerous associated risks to both researchers and the environment. Due to these identified shortcomings, traditional methods of nanoparticle synthesis have fallen out of favor, making way for a more sustainable, eco-friendly, less/non-toxic means of producing nanomaterials, "the Green Synthesis approach."

Green synthesis is a safe, economical, ecologically harmless, and hygienic means of producing nanomaterials (15). By utilizing different bio-active molecules/compounds found in microbes, plants, and plant extracts like proteins, microorganisms like specific enzymes (16), different amino acid groups (13), and different chemical compositions (17) bacteria, fungi, yeast, algae, and some plants serve as substratum for environmentally friendly production of nanomaterials (18),. Figure 1 illustrates the various techniques for nanoparticle synthesis. Several bioactive compounds were employed as reducing, capping, or stabilizing agents that have a significant influence on the morphology, sizes, form, and overall applications of green-synthesized NPs (19).

Gold nanoparticles (AuNPs), with their distinct physiochemical properties, have significantly advanced the field of nanotechnology and continue to maintain their high level of research interest (20). The unique characteristics of AuNPs, such as high coefficient of X-ray absorption, radioactiveness, and localized surface plasmon resonance (LSPR), make them effective for the diagnosis and therapy of tumors (20,21) . AuNPs can form chemical bonds with chemical groups/compounds that comprise Nitrogen and Sulphur, which permits extensive functionalization, and surface modification with exceptional biocompatibility, targeting, and drug delivery capabilities(22).

This chapter thoroughly analyses cutting-edge studies in the "green synthesis" routes of AuNPs employing various natural materials. It specifically highlights biosynthesis methods involving different parts of plants and intracellular and extracellular synthesis methods using microbes while also discussing their broad applications. Green synthesis's advantages over conventional chemical techniques are thoroughly examined, alongside their recent applications, drawbacks, and potential opportunities in this burgeoning field.

2.   Green synthesis of gold nanoparticles

Anastas and Warner pioneered Green Synthesis in the release of their book "Green Chemistry" in the early 2000s (23). The 12 green chemistry principles were presented to decrease or do away with the utilization of hazardous materials during chemical reactions while encouraging sustainability, safety, and environmental responsibility (24). Over time, these guidelines have provided direction for green synthesis; their implementation limited the application of hazardous solvents and reagents, minimized waste, enhanced process safety, increased energy efficiency, and lessened environmental effects (25). To synthesize AuNPs sustainably, natural substrates derived from bacteria, fungi, and algae, together with extracts of plants, are employed as reducing, capping, and stabilizing materials. This section also examines the features and attributes of these sustainably produced nanoparticles.

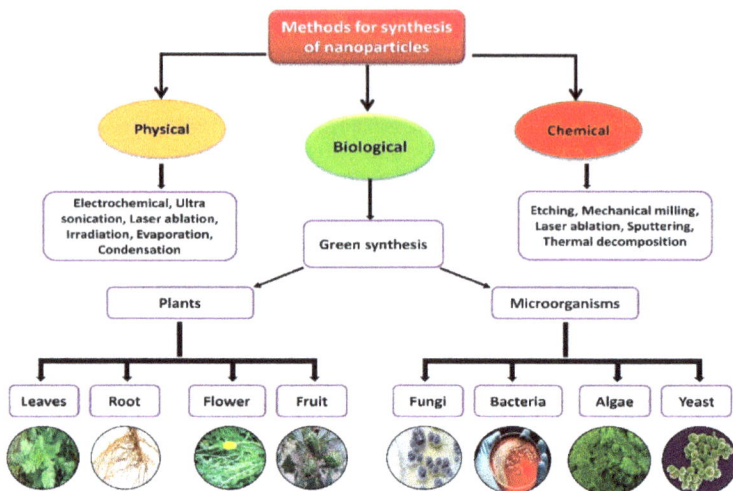

Figure 1. Different approaches of nanoparticle production. Reproduced from open-source article https://creativecommons.org/licenses/by/4.0/. (26)

1.1 Plant-mediated method of AuNPs synthesis.

Phytonanotechnology has attracted interest over time thanks to its quick, inexpensive, and environmentally friendly nanoparticle synthesis method (27). Plants are composed of various bio-components, including flavonoids, phytosterols, terpenoids, quinones, and others, whose functional groups accelerate AuNPs stabilization, reduction, and capping (28). Leaves are the most typically employed component of plants, although it has been found that practically every part of the plant can mediate AuNPs formation (27). The suggested plant-mediated process for AuNP production is depicted in Figure 2. Production of AuNPs has been feasible with a variety of extracts of plant species. *Garcinia mangostana* aqueous peel extract produced spherical AuNPs, of 47.92 ± 4.92 nm size (29). Highly stable, spherical AuNPs of size 50-80 nm were synthesized utilizing a leaf extraction from *Salix alba L.* (white willow). This plant has high phenolic contents, antipyretic, and anti-inflammatory qualities (30). FTIR spectra verified that aromatic groups, amines, and amides were reducing and capping the gold NPs (30). Phenolic and alcoholic hydroxyl groups in *Curcumae kwangsiensis* aqueous leaf extract reduced Au ions to Au metals, resulting in spherical 8 to 25 nm sized AuNPs that exhibited cytotoxicity to SK-OV-3, PA-1, and SW-626 ovarian carcinoma cell lines (31). Apart from leaves, plant seeds, barks, different fruits, flowers, and peels obtained from fruit are also utilized in synthesizing AuNPs from plants. The work of (32) used cinnamon bark extract as a stabilizing and reducing material to produce AuNP with a dispersion of particle sizes of around 35 nm. A list of various plants employed in synthesization AuNPs is displayed in Table 1.

Figure 2. Proposed mechanism for plant-mediated AuNP synthesis. Reproduced from open-source article https://creativecommons.org/licenses/by/4.0/. (26)

Table 1. list of various plants that used to synthesize AuNPs

Plant species	Plant part	Bio-molecule involved in reduction and/or stabilization	Size(nm) of AuNP	Morphology of AuNPs	Ref.
Fragaria ananassa	Leaves	Phenols, flavonoids, terpenoids	5-31	Spherical	(33)
Ribes nigrum	Leaves	Phenols, flavonoids, terpenoids	6-44	Spherical	(33)
Ribes uva-crispa	Leaves	Phenols, flavonoids, terpenoids	8-47	Spherical	(33)
Garcinia mangostana	Leaves, fruit	Flavonoids, Phenols	43-52	spherical	(29)
Justicia glauca	Leaves	Alkaloids, flavonoids, steroids, terpenoids	32	spherical and hexagonal	(34)
Cassia auriculata	Leaves	Polysaccharides and flavonoids	15-25	Spherical and triangular crystalline	(35)
Olive plant	Leaves	Luteolin-7-glucoside, apigenin-7-glucoside, oleuropein, and proteins	50-100	Spherical and anisotropic	(36)
Mangifera indica	leaves	Thiamine, terpenoids, and flavonoids.	17-20	Spherical	(37)

Citrus maxima	fruit	Proteins, terpenes, and ascorbic acid	15-35	Spherical	(38)
Lonicera Japonica	flowers	Amino acids	8	Triangular and tetrahedral	(39)
Moringa oleifera	flowers	Flavonoids, carotenoids, phenols, sterols, and amino acids	3-5	Spherical or near-spherical	(9)
Mango peel	Fruit peels	Polyphenols, flavonoids, carotenoids, and vitamins	6-18	Quasi-spherical	(40)
Musa paradisiaca (banana)	Peels	Glycosides, alkaloids flavonoids, and tannins	50	Spherical	(8)
Pogostemon benghalensis	leaves	Proteins	10-50	Triangular and spherical	(41)
salvia officinalis	seed	Polyphenols	6-20	Spherical	(42)

1.2   Microbe-mediated green synthesis of AuNPs

1.2.1   Bacteria-mediated synthesis of AuNPs

Green production of gold NPs utilizing various microorganisms (bacteria, algae, fungi, and others), has several benefits, including a low-cost growing medium, simplicity of handling and processing, and the capability of adsorbing and converting different ions of metals into NPs; it has gained much attention in the scope of industrial microbiology (28,43). *Bacillus marisflavi* YCIS MN 5 has been shown by (44) to create AuNPs extracellularly at room temperature. The generated AuNPs demonstrated exceptional catalytic dye degradation in methylene blue and congo red. The reduction reaction demonstrated to adhere to pseudo-first-order kinetics, with reaction amount constants of methylene blue dye and that of Congo red being 0.2484 min^{-1} and 0.2192 min^{-1}, respectively (44). Gram-negative bacteria belonging to the Betaproteobacteria class, Leptothrix, secreted extracellular RNA that could reduce Au^{3+} to create spherical AuNPs measuring 5 nm in an ambient condition (45). (46), used a cell-free product of *Bacillus subtilis* in the green synthesization of AuNPs. Gold-thiol bonds were observed in the reduction process, and the AuNPs were then stabilized by the denatured bacterial protein that acted as a capping agent. Table 2 mentions a list of other bacteria species used in AuNP synthesis and the morphology of the nanoparticles.

1.2.2   Fungi-mediated synthesis of AuNPs

Compared to other microbes, fungal species, including yeast, toadstools, molds, and mushrooms, have lately been integrated with the fabrication of nanoparticles owing to their ability to release copious volumes of protein (47). When it comes to industry and lab culture, fungi are more superficial than other microorganisms and are known to generate AuNPs more quickly and in more quantity (48). *Penicillium aculeatum* can generate 60 nm-sized spherical AuNPs that were shown to be stable at roughly -30 mV. This is probably because the protein coatings on AuNPs' surfaces offer strong electrostatic repulsive forces (49). Bio-reduction of ions of gold to create AuNPs is facilitated by polyphenols and specific proteins found in the marine endophytic fungus *Penicillium citrinum.* The produced AuNPs ranged from sizes 60 to 80 nm and had an irregular shape (47). Some fungal species that were effectively engaged in producing gold NPs are listed in Table 2.

1.2.3 Algal-mediated synthesis of AuNPs

Algae represent significant bio-manufacturers due to their efficient synthesis of proteins, pigments, phytonutrients, and other nutrients, thereby becoming pivotal in nanoparticle production over time (50). Utilizing biomolecules like enzymes and pigments for both capping and reducing agents involves efficient energy utilization, their hyperaccumulation of heavy metals, their far greater rate of CO_2 sequestration, and the absence of harmful byproducts are all reasons why cyanobacteria and microalgae are becoming more and more popular in the field of green NP synthesis (51). Their developed nanoparticles may find application as preservatives, antioxidants, stimulants, antibacterial, antiviral, and anticancer agents due to the diverse biological effects of their secondary metabolites including phenolics, alkaloids, flavonoids, glycosides, and plant acids (50). Table 2 lists the algae species and the corresponding biomolecules that are utilized as capping, stabilizing, and reducing substances in green AuNP synthesis.

Table 2. List of various microorganisms utilized in AuNPs synthesis

Species	Reducing and/or stabilizing biomolecules	Size(nm) of AuNP	Morphology of AuNPs	Ref
Bacteria				
Bacillus cereus	Bacterial proteins	20–50	Spherical	(52)
Bacillus subtilis 168	Ketose and aldose	5–25	Octahedral	(53)
Lactobacillus Kimchicus	NADPH-dependent reductase, sugars	5–15	Spherical	(54)
Fusarium oxysporum	Bacterial proteins	20 – 50	Spherical, Hexagonal	(52)
Halomonas salina	Nitrate reductase enzymes	5 -25	Spherical	(28)
Fungi				
Aspergillus sp. WL-Au	OH, COH, NH, CH, COC and CO functional groups	4 - 29	Spherical	(55)
Inonotus obliquus (Chaga mushroom)	Polyphenols	11.0–37.7	Spherical, Hexagonal, Triangle	(56)
Sclerotium rolfsii	NADPH-dependent enzyme	10–15	Spherical	(57)
Botrytis cinerea	Reductase enzymes	1 - 100	Hexagonal triangular, , spherical, pyramidal and decahedral,	(58)
Aspergillus foetidus	Proteins	30 - 50	Spherical	(59)

Algae				
Galaxaura elongata	Stearic acid, gallic acid oleic acid, 11-eicosenoic acid	3.85 –77.13	Spherical	(60)
Tetraselmis kochinensis (live)	Cytoplasmic and cell wall enzymes	5 – 35	Spherical, triangular	(61)
Ecklonia cava	Amine and hydroxyl group	30	Spherical /triangular	(62)
Cystoseira baccata	Phenol, vitamins, terpenoids and sterols	8	Spherical	(7)
Padina gymnospora	Algal cell wall polysaccharide hydroxyl groups	53–67	Spherical	(63)
Chlorella vulgaris	Carboxyl, amino and hydroxyl functional groups	2–10	Anisotropic	(64)

1.3   Advantages of green synthesis

As previously mentioned, green synthesis has several benefits over traditional chemical and physical techniques concerning environmental sustainability. Economic, energy economy, harmless, lower waste generation, and enhanced suitability for pharmaceutical and biomedical applications are among the numerous benefits it provides (65,66). Furthermore, these NPs exhibit enhanced pharmacological activity compared to those produced through chemical or physical processes. Various biological agents provide distinct comparative advantages in synthesizing nanomaterials, as summarized in Table 3.

Table 3. Advantages of NPs synthesized via green methods

Synthesis Components	Advantages	Ref.
Bacteria	Synthesized nanomaterials display minimal crystal flaws, uniform size distribution, high purity, monodispersity, and other advantageous properties. Thermophilic bacteria are particularly beneficial for extracellular manufacturing processes.	(67)
Fungi	Generates substantial quantities of enzymes with diverse practical applications. Shows robust metal tolerance and bioaccumulation capabilities. Supports scalability, efficient downstream processing, biomass management, economic viability, and large-scale manufacturing. Generates uniform nanoparticles with distinct morphologies.	(68)
Yeast	Capable of producing a wide range of nanomaterials (NMs). Efficiently produces bulk quantities of metal nanoparticles (NPs).	(69)

Viruses	Efficient bio-templates for synthesizing two-dimentional and three-dimensional nanomaterials. Synthesized particles exhibit unique features including enhanced biodegradability, biocompatibility, monodispersity, stability, and potential toxicity reduction through manipulation. Certain viruses enhance nanoporous material binding for specific applications. Viral nanoparticles from plant viruses are stable, nontoxic, biocompatible, harmless, and amenable to chemical and genetic modifications for interacting with diverse biological molecules.	(70)
Micro and Macro-algae	Abundantly available. Capable of accumulating heavy metals. Rapid synthesis capability.	(71)
Actinomycetes	Genetically modifiable to enhance the production of nanoparticles (NMs) with improved size and polydispersity. Produces valuable secondary metabolites such as antibiotics. Facilitates easier scalability for large-scale production.	(72)
Phyto synthesis	Rapid, environmentally friendly, and non-pathogenic synthesis method. Capable of single-step approach. Exhibits biocompatibility and biodegradability. Potential for scalable production. Capable of heavy metal accumulation and detoxification. Enables rapid and stable synthesis, surpassing microbial methods. Offers morphological versatility via diverse plant sources.	(73)(74)

3.   Green synthesized aunps: An analysis of their characteristics

Particle size distribution, aggregation, wettability, adsorption potential, zeta potential, and interaction surface features are some of the methods utilized for characterization to understand better the properties and uses of synthesized nanoparticles (75–77). Precise and thorough characterization of these nanoparticles is necessary to ensure their biological functionality, safety profile, and manufacturing repeatability (78). To achieve this, a broad array of physicochemical methods is used to describe the artificially produced NPs precisely. These methods include ultraviolet-visible (UV-Vis) spectroscopy Fourier transform infrared spectroscopy (FTIR), attenuated total reflection (ATR), Raman spectroscopy, photoluminescence analysis (PL), dynamic light scattering (DLS), UV-visible diffuse reflectance spectroscopy (UV-DRS), transmission electron microscopy (TEM), scanning electron microscopy (SEM), atomic force microscopy (AFM), field emission scanning electron microscopy (FE-SEM), X-ray diffraction (XRD), X-ray photoelectron spectroscopy (XPS), energy dispersive X-ray analysis (EDAX), thermogravimetric differential thermal analysis (TG-DTA), and nuclear magnetic resonance (NMR) (75,77,79–87). In this section, a few of the characterization techniques have been discussed.

3.1   Optoelectronic properties and spectroscopy

3.1.1   UV-Visible spectroscopy (UV-Vis)

UV-vis, a pivotal analysis method, is employed to gain a comprehensive understanding of the optical and electrical properties of nanomaterials. This method involves the absorption of specific wavelengths within the UV-visible region. The UV-Vis spectrum for the stability study of gold nanoparticles produced from green tea, Green Coconut water and Zimbro tea solutions, as shown in Figure 3, is a testament to its role in understanding nanomaterial properties. The recorded Lambda max of all solutions between 530 – 540 nm indicates the successful synthesis of AuNPs. (88)UV-vis was instrumental in confirming the successful synthesis of AuNPs from endophytic fungi *Fusarium solani*. The absorption spectra, as shown in Figure 4 a, revealed a maximum peak at 563 nm, a characteristic of AuNPs. The UV-vis spectrum of AuNPs synthesized using the algae extract of *Turbinaria conoides*, also shown in Figure 4 b, further confirmed the synthesis. The spectrum displayed three bands, including an observable SPR band at 525 nm in the visible region, characteristic of anisotropic nanoparticle formation. (89)

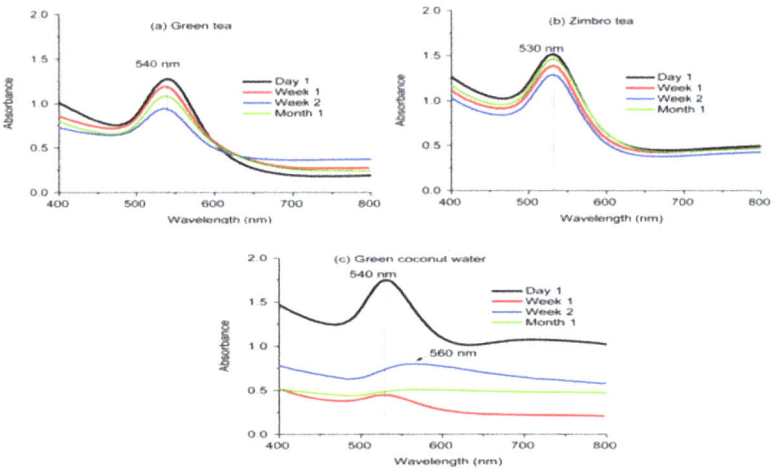

Figure 3. UV-Vis spectra of distinct gold nanoparticle solutions were reduced and stabilized utilizing (a) Camellia sinensis (green tea), (b) Cistus ladanifer (Zimbro tea), and (c) Cocos nucifera (green coconut) water at 25°C. Reproduced from open-source article https://creativecommons.org/licenses/by/4.0/_ (88).

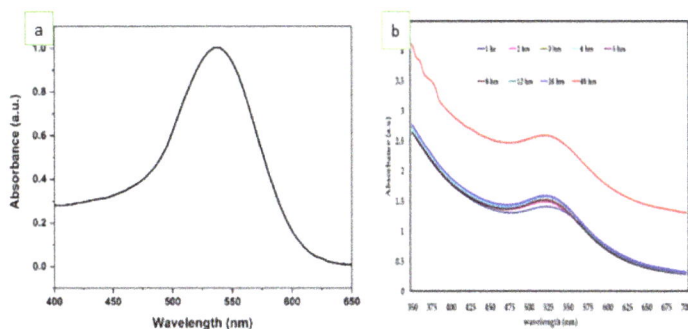

Figure 4. a. UV– vis absorption spectra of AuNPs synthesized endophytic fungi Fusarium solani. b. UV–vis absorption spectra of gold NPs synthesized utilizing the marine alga Turbinaria conoides. Reproduced from open-source articles https://creativecommons.org/licenses/by/4.0/ (89–91)

3.2 Functional group identification

3.2.1 Fourier transform infrared spectroscopy (FTIR)

FTIR measurements are conducted during nanoparticle characterization to determine the potential biomolecules in charge of the reduction, capping subsequent stabilization of the AuNPs (92). Following a thorough drying process, a refined mixture containing the nanoparticles is pulverized using potassium bromide (KBr) pellets before analysis (93). By displaying the locations of bands linked to certain functional groups and bond types and strengths, a recorded spectrum can provide details about molecular structures and interactions (94). The comparison of the FT-IR spectra of AuNP produced with a combination of husk extract from *Acacia nilotica* (AN) and fruit extract from *Olea europaea* (OE) (OEAN extract) showed distinct peaks at 2031.16 cm^{-1}, 2336.36 cm^{-1}, 2159.87 cm^{-1}, 1635.36 cm^{-1} and 3276.13 cm^{-1} for the nanoparticles as shown in Figure 5A. These peaks correspond to functional groups like OH, O-H bending, C=C stretching, N-H and C=O, indicating the involvement of residual plant extract compounds in the reduction and stabilization of the AuNPs (95). Similarly, the FT-IR analysis of AuNPs synthesized using the *Fusarium solani* ATLOY-8 isolated from *Chonemorpha fragrans* exhibited distinct peaks corresponding to different functional groups. The observed peak at 1413 cm^{-1} was attributed to the amide II bands of proteins present in the sample. A band at 1041 cm^{-1} indicated C-N stretching vibrations, while the identified band at 690 cm^{-1} indicated C-H stretching vibrations confirming the presence of proteins or polypeptides illustrated in Figure 5B (91). AuNPs were synthesized using *Mentha spicata* essential oil, it was initially analyzed via FTIR spectroscopy to identify the reducing agents in the oil. The essential oil displayed absorption bands at 2922, 1674, 1436, 1369, 1246, 1144, 1110, 1056, 892, 802, and 704 cm^{-1}, which changed after AuNP synthesis, exhibiting bands at 2955, 2360, 1436, 1369, 1248, 1111, and 897 cm^{-1}. The identified peak at 2955 cm^{-1} indicated the presence of C–H groups in monoterpenes such as carvone and limonene, suggesting their role in reducing and capping gold ions. The band at 2360 cm^{-1} is attributed to C≡C triple bonds in AuNPs, while a peak at 1676 cm^{-1} signifies a stretch C=O bond. The lack of a peak at 1144 cm^{-1} indicates the involvement of stretch C–O groups during the synthesis process. Shifts in peaks from 1246 to 1248 cm^{-1}, 1110 to 1111 cm^{-1}, and 892 to 897 cm^{-1} after AuNP formation reflect changes due to nanoparticle synthesis (90).

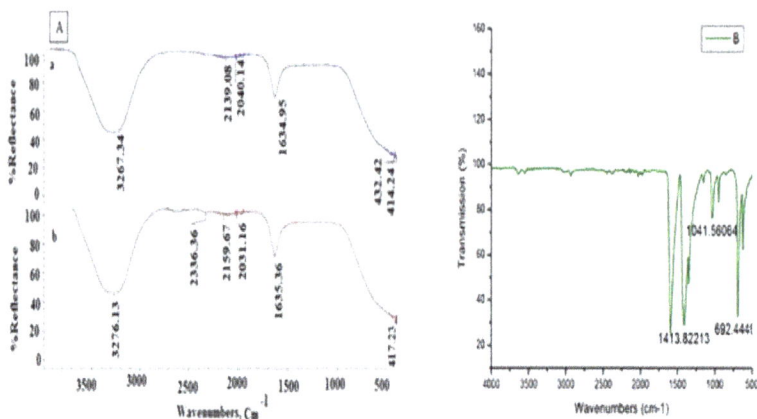

Figure 5. FT-IR spectra of (A) AuNPs from OEAN extract, (B) Fusarium solani ATLOY-8, highlighting functional groups that are involved in nanoparticle synthesis and their stabilization. Reprinted from open sources https://creativecommons.org/licenses/by/4.0/. (91,95)

3.3 Microstructure and surface analysis

3.3.1 Dynamic light scattering (DLS)

DLS studies are primarily utilized to determine particle size dispersion and to acquire the mean hydrodynamic diameter of the sample. Zeta potential (ZP) values are derived from measuring the surface electric charge of particles, offering insights into nanoparticle stability (96). Figure 6 A is DLS examination of camptothecin-stabilized AuNPs obtained from the endophytic strain *Fusarium solani* ATLOY –8 (91). The DLS histogram and zeta potential for AuNPs in solutions of Zimbro tea, green coconut water and green tea, are displayed in Figure 6 B. Distinct particle size distributions for each stabilizer were found by the DLS analysis, showing that they were coated with phytochemicals with distinct molecular weights (88).

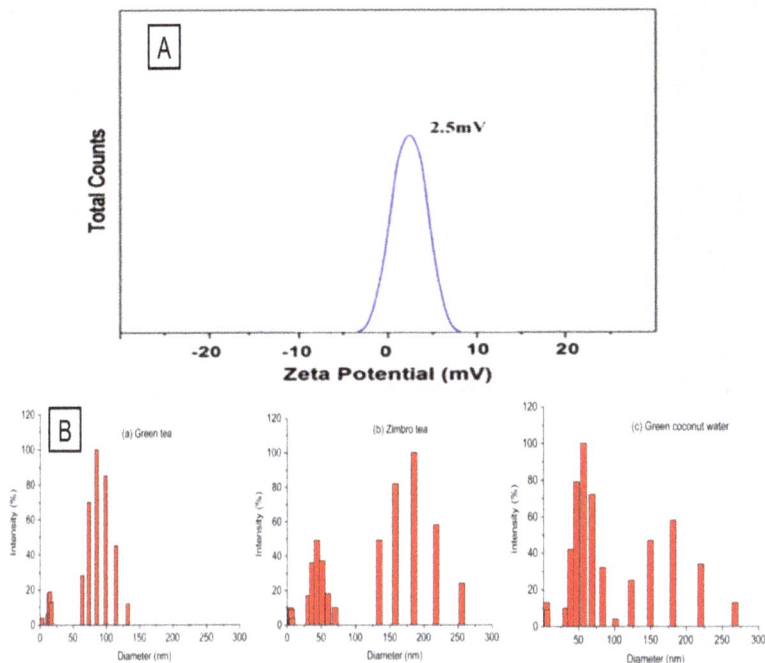

Figure 6. (A) Zeta potential and stability analysis of camptothecin-stabilized AuNPs. (B) DLS and zeta potential measurements for AuNPs stabilized with (a) green tea, (b) Zimbro tea, and (c) green coconut water, showing particle size distributions and surface charges. Reprinted from an open-source https://creativecommons.org/licenses/by/4.0/_ (91) (88).

3.4 Electron microscopy (TEM/SEM)

Over many years, the TEM has been a highly developed instrument widely used in several scientific domains because of its exceptional capacity to deliver structural and chemical data greater than a wide variety of length scales, including atomic dimensions (97). Through a thin specimen and an electron beam of constant current, TEM creates an image incorporating information obtained from the electrons that are transported through the sample. TEM is often the most used technique to examine the morphology, shape, and size of nanoparticles. It gives clear pictures of the sample and the most precise estimation of the uniformity of the particles (94). High-resolution surface imaging can be achieved with SEM, a popular technique that can also be used to study materials at the nanoscale (94). SEM and TEM analysis have been employed in the characterizing several green synthesized AuNPs. The majority of the Au nanoparticles synthesized using Galaxssaura elongate were spherical, according to TEM analysis, as shown in Figure 7. Additionally, trace quantities of hexagonal, rod, triangular, and truncated triangular nanoparticles were discovered in addition to spherical nanoparticles (98). Figure 8 is the Transmission Electron Microscope of AuNPs produced with *Acinetobacter* sp. GWRVA25 reveals spherical, monodispersed crystalline AuNPs of the size range of 20 - 100 nm under different conditions of pH and temperature modifications (99). The SEM analysis of green-produced AuNPs mediated by *Carduus edelbergii* is displayed in Figure 9. The produced particles had uniform distribution and spherical shape (100)

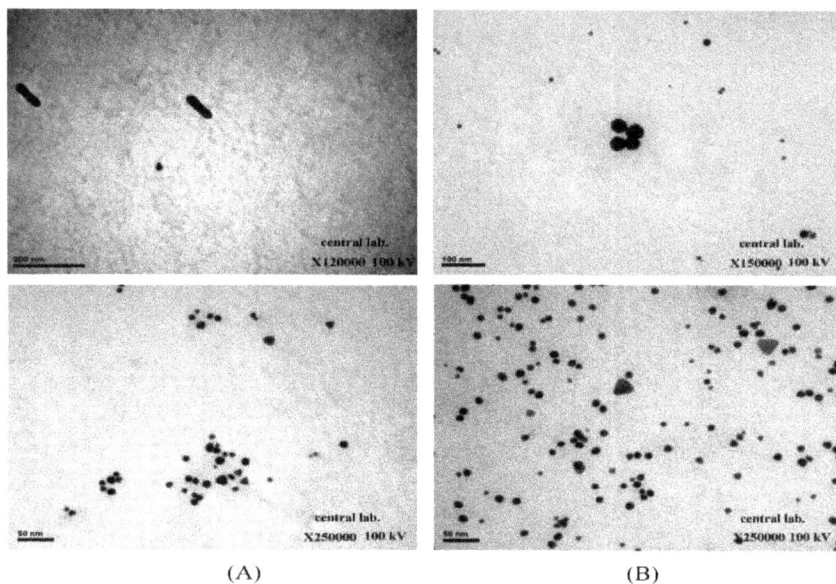

(A)   (B)

Figure 7. TEM micrograph of AuNPs produced in an algal powder (A) and algal ethanolic extract by reducing HAuCl4 ions (B) of Galaxaura elongate. Reproduced from open source article https://creativecommons.org/licenses/by/4.0/ (98).

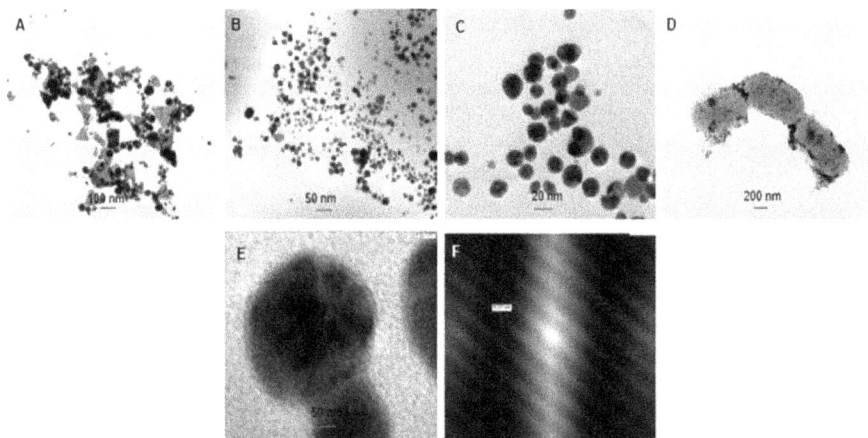

Figure 8. TEM images of AuNPs produced with Acinetobacter sp. GWRVA25. Employing (A, B) HAuCl4 at 1 and 0.5 mM, (C, D) at 37 and 50°C, (E) a magnified image of ideal AuNPs, and (F) lattice and fringes on AuNP. Reproduced from open source article https://creativecommons.org/licenses/by/4.0/ (99)

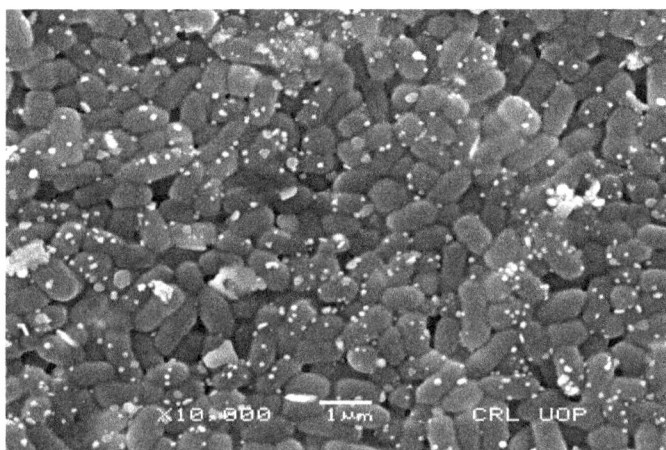

Figure 9. SEM image of AuNPs mediated by Carduus edelbergii. Reproduced from open-source article https://creativecommons.org/licenses/by/4.0/_(100)

3.5   Atomic force microscopy (AFM)

AFM provides high-resolution, three-dimensional imaging and quantitative measurements of mechanical properties, which makes it an essential tool for surface profiling and nanomechanical investigation of nanomaterials. The manufactured nanoparticle's size and surface shape are examined using AFM imaging. One benefit of AFM imaging is that it doesn't require surface modification or coating beforehand. AFM has been used to undertake topological characterization of tiny NPs ($\leq$6 nm), like ion-doped Y_2O_3, without any extra treatment (94). Figure 10 shows the surface morphology of AuNPs obtained using *Ribes nigrum* fruit extract using AFM. The average particle diameter was found to be 12.7 ± 3.2 nm, consistent with SEM microscopy findings (101).

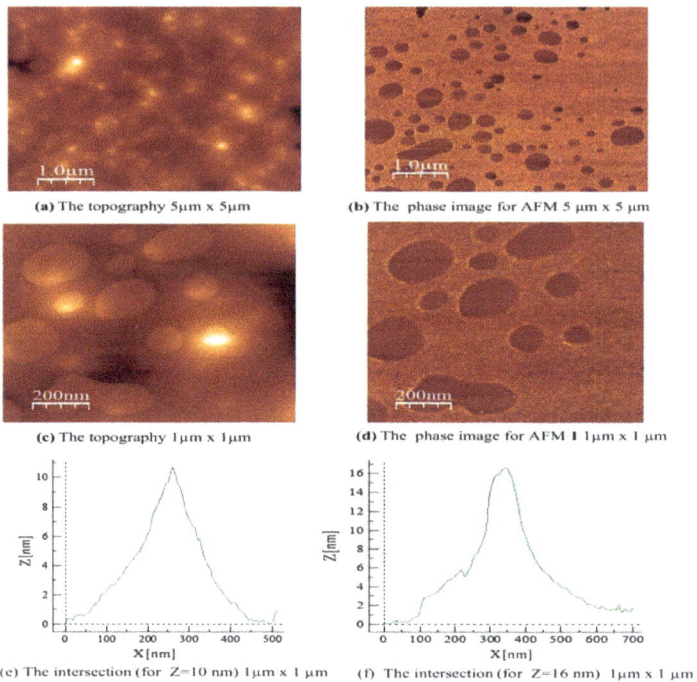

(a) The topography 5μm x 5μm

(b) The phase image for AFM 5 μm x 5 μm

(c) The topography 1μm x 1μm

(d) The phase image for AFM I 1μm x 1 μm

(e) The intersection (for Z=10 nm) 1μm x 1 μm

(f) The intersection (for Z=16 nm) 1μm x 1 μm

Figure 10. AFM pictures of AuNPs produced using Ribes nigrum fruit extract. Reproduced from open-source article https://creativecommons.org/licenses/by/4.0/_(101)

3.5 Crystallography and structural analysis

3.6.1 X-ray diffraction (XRD)

XRD stands out as the most widely adopted technique for NP characterization. Typically, XRD delivers info about phase composition, crystal lattice structure, lattice constants, and the size of the crystalline grains within the NPs (94).

Crystal structure and phase consistency of camptothecin-stabilized AuNPs produced by *Fusarium solani* was determined using X-ray diffraction (XRD) examination covering 2θ between 5° and 80°. Figure 11 A is the XRD pattern showing distinct peaks at 2θ angles of 32.19°, 38.32°, 46.16°, 57.50°, and 76.81°, corresponding to the crystalline structure of camptothecin (CPT) with AuNPs (91). Figure 11 B is the XRD diffractogram of AuNPs synthesized from green tea, stabilized with green tea showing five distinct peaks at approximately 2θ = 38°, 45°, 65°, 78°, and 82° (88). In Figure 11 C, the XRD diffractogram of gold NP synthesized with algal extract of *T. conoides* shows diffraction peaks occurring at *2θ* values of 38.36° (111), 44.13° (200), 64.78° (220), and 77.98° (311) (89).

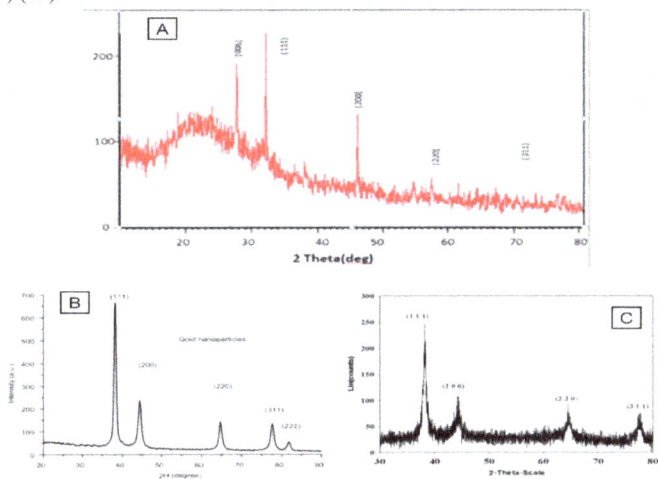

Figure 11. XRD patterns of green-synthesized AuNPs: (A) Camptothecin-stabilized AuNPs by Fusarium solani. (B) AuNPs synthesized from green tea. (C) AuNPs synthesized using T. conoides algal extract. Reproduced from open-source article https://creativecommons.org/licenses/by/4.0/_ (88,89,91)

4.   Applications of green synthesized AuNPs

Nanoparticles of metals like gold and silver are frequently utilized in targeted drug delivery, imaging, diagnostics, and therapeutic interventions. As a result, they have had a substantial impact on medicine. Their enormous surface area, small size, lack of cytotoxicity, and modifiable optical, physical, and chemical characteristics are the reasons behind this. These nanoparticles typically exhibit resonance under a magnetic field that changes over time, enabling them to generate sufficient heat to target tumor cells as heat-inducing agents selectively. AuNPs manufactured via green methods have demonstrated anticancer activity against various types of cancer cells. Notably, MCF-7 breast cancer cells (102), A 549 human lung cancer cells (59,103,104), and human keratinocyte cell lines (103), HL-60 leukemia cells (105) They have all been found to undergo apoptosis. Figure 12 shows the multifunctional applications of green synthesized AuNPs.

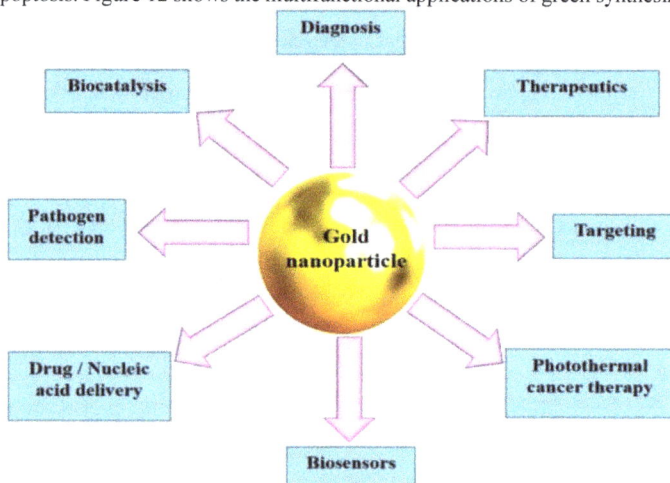

Figure 12. Multifunctional applications of AuNPs span medical purposes including drug administration and diagnostic techniques, catalytic processes, sensing and detection in biosensors, antimicrobial activity, and environmental remediation. Reprinted from an open-source https://creativecommons.org/licenses/by/4.0/_ (106)

4.1   Antibacterial potential

Antibacterials, commonly known as antibiotics, are antimicrobial drugs that prevent and treat bacterial contamination by killing or inhibiting bacterial growth. Green-synthesized AuNPs demonstrated antibacterial properties across multiple sectors (107). *Acorus calamus* rhizome synthesized AuNP exhibited excellent antibacterial activity without using a binder. Applying chosen NPs to the cloth specimens that were mobilized demonstrated effective antibacterial action against Gram-positive *Staphylococcs aureus* and Gram-negative *E. coli* (108). With LIVE/DEAD BacLight, a novel method tested against skin infections and *Brevi* bacteria on linens for the bio-functionalization and coloring of various materials using green technology, demonstrated significant antibacterial potential (109). In a scientific investigation, *E. coli* strains were subjected to a 10:1 blend of eco-friendly synthesized gold nanoparticles (AuNPs) utilizing Macadamia nut shell extract. The evaluation of antibacterial showed that the inhibitory zone was doubled compared to the crude extract (110,111). **Figure 13** Green-synthesized nanoparticles show promise in decreasing the building of fresh biofilms and altering present ones, thereby increasing the sensitivity of bacteria to antibacterial treatments. For example, AuNPs produced from

Trachyspermum ammi extract inhibited biofilm production by *L. monocytogenes* and *S. marcescens* (84) **Fig. 14**.

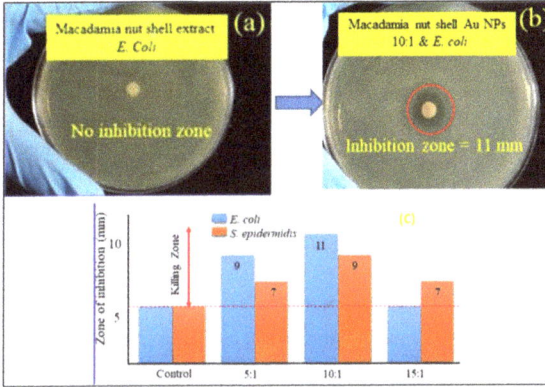

Figure 13. Exemplary petri dish samples include (a) a sterile disc with plant extract only; (b) E. coli treated with a 10:1 reaction mixture of plant extract and green-synthesized AuNPs; and (c) Antibacterial assessment results showing a 2X increase in the zone of inhibition. Reproduced from open-source article https://creativecommons.org/licenses/by/4.0/_ (110).

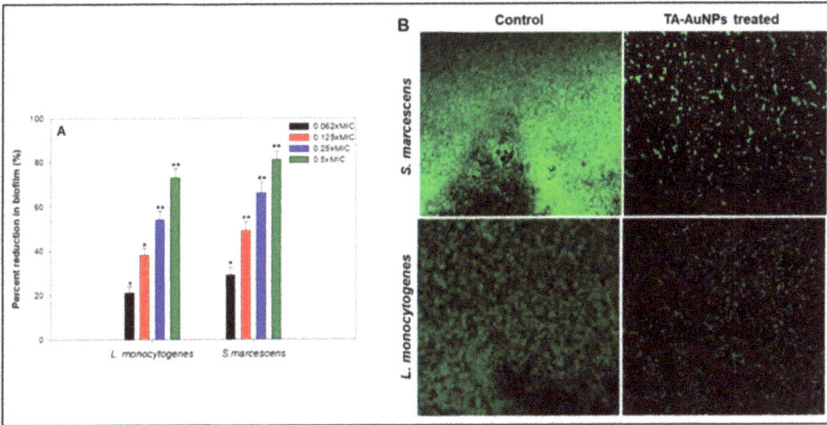

Figure 14. (A) The effect of TA-AuNP sub-MICs on the formation of biofilms. (B) Confocal laser scanning microscopy images showing the biofilms of L. Monocytogenes and S. marcescens with and without 0.5× MICs of TA-AuNPs (AuNPs generated from Trachyspermum ammi e extract). Reproduced from open-source article https://creativecommons.org/licenses/by/4.0/_(84).

4.2   Anticancer activity

Regarding cancer treatment, NPs synthesized via environmentally friendly processes exhibit the potential to tackle cancer-related challenges. These NPs can be tailored to selectively attach to tumor markers or cancer cells through antibodies or ligands, thereby enabling battered and precise delivery of therapeutic agents (112). Nanoparticles synthesized sustainably also enhance drug stability, bioavailability, and solubility by encapsulating anticancer agents, thereby overcoming challenges associated with conventional chemotherapy (113). AuNPs synthesized from the hydroethanolic extract of Brazilian Red Propolis and its constituents demonstrated a cytotoxic effect against PC-3 cells. (114). Integration of green synthesized AuNPs with carboxymethylcellulose to form a nanocomposite (ACMC-AuNPs) enhances their potency, as evidenced by the IC50 values obtained (2.56 ± 0.19 µg/mL) observed in MCF-7 breast cancer cells. Moreover, CMC-AuNPs were reported to halt the G1/G0 phases of the cell cycle, highlighting their potential as potent cancer therapy agents (115).

4.3   Bioimaging

The non-intrusive technique for monitoring biological processes throughout a specific period frame is called bioimaging. It helps describe the 3D structure of specimens without causing physical disturbance, and it does not impede any of the many life activities, such as breathing, movement, or other physiological functions. Additionally, it helps to make the connection between the observation of all the tissues in multicellular animals and the subcellular structures (116,117). Green synthesized AuNPs exhibit distinct optical properties, particularly surface plasmon resonance, which enhances imaging contrast and resolution (118–120). They are also appropriate for both *in vitro* and *in vivo* bioimaging applications due to their low toxicity and biocompatibility (121,122). Kotcheriakota et al. conducted research wherein they synthesized highly biocompatible gold nanoparticles intended for bioimaging applications within the near-infrared (NIR) spectrum, specifically at 820 nm, using extracts from the *Zinnia elegans* plant. The produced gold nanoparticles, termed AuZE, demonstrated green fluorescence upon excitation at 350 nm, alongside red fluorescence observed at 710 nm within the NIR spectrum. Remarkably, following intraperitoneal administration in C57BL6 mice, AuZE dispersed extensively throughout the mice's brains without necessitating specific ligands. This resulted in vivid red fluorescence detectable within the NIR range (excitation at 710 nm, emission at 820 nm), as confirmed by non-invasive imaging techniques and ICP-OES analysis. (123)

4.4   Bio-sensor

AuNPs have emerged as versatile components in sensor technology, pivotal for detecting and interpreting environmental changes or events. These nanomaterials effectively convert detected signals into digital data processed by computers to generate specific outputs. Integrating AuNPs in sensor applications represents a significant advancement in green technology, extending beyond traditional uses such as temperature, pressure, and flow measurements. (116,117). AuNPs synthesized using patuletin isolated from *Tagetes patula* were evaluated for their sensing capabilities using fourteen different drugs, demonstrating reliable detection abilities across most compounds except for piroxicam, which exhibited luminescence quenching effects (124).

AuNPs made with *Cerasus serrulata* leaf extract effectively detected hydrazine, showing a wide linear detection range (5-272 nM) and a low detection limit (0.05 µM) (125).

4.5 Catalytic activity

With the aid of a catalyst in a process, the transition state can be reached with lower activation energy, enabling the reaction to reach completion while maintaining the overall free energy of the reactants and products unchanged (126). The biosynthesis of AuNPs has sparked interest owing to the growing demand for catalysts in numerous applications, specifically in non-toxic processes and fuel cell technology, since their surface can be used to selectively oxidize reactions (127) Mung bean starch (MBS) is a bio-agent for synthesizing AuNPs without requiring additional stabilizing or capping agents. The ensuing AuNPs were applied as heterogenous catalysts to 4-nitrophenol reduction (128).

5. Challenges and future perspective

The green synthesis technique has been shown to achieve success by generating a range of nanoparticles, including AuNPs, with different shapes and characteristics in recent times. Even with these noteworthy developments, the green synthesis approach is not without its difficulties and restrictions. Here is an overview of the current problems in Figure 15. While there has been progress in controlling the morphologies of nanomaterials (NMs), precise management of their dimensions, particularly their size, remains a significant challenge. Because of batch-to-batch variability, significant quantities of homogenous nanomaterials (NMs) cannot yet be produced via laboratory-controlled biosynthesis. As such, there is a pressing demand for ecologically benign, stable, and scalable continuous flow-based synthesis methods.

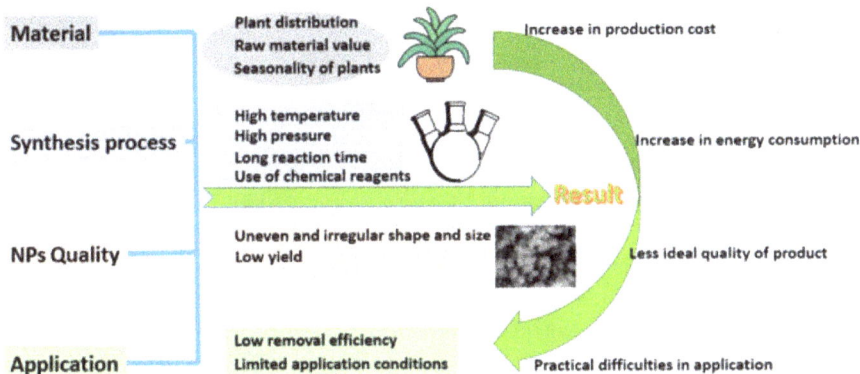

Figure 15. Challenges and limitations in green synthesis technology: areas for future research. Reproduced from open-source article https://creativecommons.org/licenses/by/4.0/_(129)

Issues with reactant locations, solvent selection, mixing times, surface tension, flow velocity, viscosity, and product contamination prevention currently hamper the application of continuous synthesis procedures. It is imperative to overcome these obstacles to facilitate the low-cost, high-throughput commercial production of NMs. As green nanomaterial synthesis is still in its infancy, there is still much that various natural resources can do to produce biogenic NMs. This includes creating NMs using various biopolymers, microbes, and waste materials. Because green synthesis offers remarkable benefits that chemical or physical synthesis methods cannot, research into this field must be expanded.

Conclusion

Because of their unique qualities, AuNPs have a vast spectrum of uses in electronics, biology, sensors, optics, and catalysis industries. While there are many physicochemical techniques for creating nanoparticles (NPs), green synthesis is a viable method that can make a variety of nanoparticles, encompassing gold nanomaterials, in an easy, economical, and ecologically responsible way. Various plant components and microorganisms, like algae, yeast, bacteria, and fungi, are employed as natural feedstock throughout the biosynthesis process of green synthesis. Compared to traditional chemical and physical approaches, this method has several benefits, including improved safety because it doesn't use dangerous chemicals or involve any externally harmful compounds when synthesizing NP. Biosynthesized nanoparticles with strong anticancer, antibacterial, anti-inflammatory, antioxidant, and catalytic activity are just a few of their advantageous characteristics. The field of green synthesis is continuously evolving, focusing on advancing our understanding of AuNP biogenesis, characterization, and functionalization. Comprehensive knowledge of green chemistry principles and further research are pivotal to advancing this discipline. By tackling current challenges through standardized protocols and innovative techniques, there is significant potential to revolutionize the production of gold NPs, benefiting both scientific experimentation and economic applications.

References

[1] Anu Mary Ealia S, Saravanakumar MP. A review on the classification, characterisation, synthesis of nanoparticles and their application. IOP Conf Ser Mater Sci Eng. 2017 Nov;263:032019. https://doi.org/10.1088/1757-899X/263/3/032019

[2] Cai W. Applications of gold nanoparticles in cancer nanotechnology. Nanotechnol Sci Appl. 2008 Sep;Volume 1:17-32. https://doi.org/10.2147/NSA.S3788

[3] Mu L, Sprando RL. Application of Nanotechnology in Cosmetics. Pharm Res. 2010 Aug 21;27(8):1746-9. https://doi.org/10.1007/s11095-010-0139-1

[4] Mokhatab S, Fresky MA, Islam MR. Applications of Nanotechnology in Oil and Gas E&P. J Pet Technol. 2006 Apr 1;58(04):48-51. https://doi.org/10.2118/0406-0048-JPT

[5] Wagner AM, Knipe JM, Orive G, Peppas NA. Quantum dots in biomedical applications. Acta Biomater. 2019 Aug;94:44-63. https://doi.org/10.1016/j.actbio.2019.05.022

[6] EL-Sheshtawy HS, El-Hosainy HM, Shoueir KR, El-Mehasseb IM, El-Kemary M. Facile immobilization of Ag nanoparticles on g-C3N4/V2O5 surface for enhancement of post-illumination, catalytic, and photocatalytic activity removal of organic and inorganic pollutants. Appl Surf Sci. 2019 Feb;467-468:268-76. https://doi.org/10.1016/j.apsusc.2018.10.109

[7] González-Ballesteros N, Prado-López S, Rodríguez-González JB, Lastra M, Rodríguez-Argüelles MC. Green synthesis of gold nanoparticles using brown algae Cystoseira baccata:

Its activity in colon cancer cells. Colloids Surfaces B Biointerfaces. 2017 May;153:190-8. https://doi.org/10.1016/j.colsurfb.2017.02.020

[8] Vijayakumar S, Vaseeharan B, Malaikozhundan B, Gopi N, Ekambaram P, Pachaiappan R, et al. Therapeutic effects of gold nanoparticles synthesized using Musa paradisiaca peel extract against multiple antibiotic resistant Enterococcus faecalis biofilms and human lung cancer cells (A549). Microb Pathog. 2017 Jan;102:173-83. https://doi.org/10.1016/j.micpath.2016.11.029

[9] Anand K, Gengan RM, Phulukdaree A, Chuturgoon A. Agroforestry waste Moringa oleifera petals mediated green synthesis of gold nanoparticles and their anti-cancer and catalytic activity. J Ind Eng Chem. 2015 Jan;21:1105-11. https://doi.org/10.1016/j.jiec.2014.05.021

[10] Tahir N, Madni A, Balasubramanian V, Rehman M, Correia A, Kashif PM, et al. Development and optimization of methotrexate-loaded lipid-polymer hybrid nanoparticles for controlled drug delivery applications. Int J Pharm. 2017 Nov;533(1):156-68. https://doi.org/10.1016/j.ijpharm.2017.09.061

[11] Zhao X, Li F, Li Y, Wang H, Ren H, Chen J, et al. Co-delivery of HIF1α siRNA and gemcitabine via biocompatible lipid-polymer hybrid nanoparticles for effective treatment of pancreatic cancer. Biomaterials. 2015 Apr;46:13-25. https://doi.org/10.1016/j.biomaterials.2014.12.028

[12] Mandal B, Bhattacharjee H, Mittal N, Sah H, Balabathula P, Thoma LA, et al. Core-shell-type lipid-polymer hybrid nanoparticles as a drug delivery platform. Nanomedicine Nanotechnology, Biol Med. 2013 May;9(4):474-91. https://doi.org/10.1016/j.nano.2012.11.010

[13] Yaduvanshi N, Jaiswal S, Tewari S, Shukla S, Wabaidur SM, Dwivedi J, et al. Palladium nanoparticles and their composites: Green synthesis and applications with special emphasis to organic transformations. Inorg Chem Commun. 2023 May;151:110600. https://doi.org/10.1016/j.inoche.2023.110600

[14] Pourzahedi L, Eckelman MJ. Comparative life cycle assessment of silver nanoparticle synthesis routes. Environ Sci Nano. 2015;2(4):361-9. https://doi.org/10.1039/C5EN00075K

[15] Santo-Orihuela PL, Desimone MF, Catalano PN. Green Synthesis: A Land of Complex Nanostructures. Curr Pharm Biotechnol. 2023 Jan;24(1):3-22. https://doi.org/10.2174/1389201023666220512094533

[16] Birla SS, Tiwari VV, Gade AK, Ingle AP, Yadav AP, Rai MK. Fabrication of silver nanoparticles by Phoma glomerata and its combined effect against Escherichia coli , Pseudomonas aeruginosa and Staphylococcus aureus. Lett Appl Microbiol. 2009 Feb;48(2):173-9. https://doi.org/10.1111/j.1472-765X.2008.02510.x

[17] Ganesh Babu MM, Gunasekaran P. Production and structural characterization of crystalline silver nanoparticles from Bacillus cereus isolate. Colloids Surfaces B Biointerfaces. 2009 Nov;74(1):191-5. https://doi.org/10.1016/j.colsurfb.2009.07.016

[18] Bansal V, Rautaray D, Ahmad A, Sastry M. Biosynthesis of zirconia nanoparticles using the fungus Fusarium oxysporum. J Mater Chem. 2004;14(22):3303. https://doi.org/10.1039/b407904c

[19] Shamaila S, Zafar N, Riaz S, Sharif R, Nazir J, Naseem S. Gold Nanoparticles: An Efficient Antimicrobial Agent against Enteric Bacterial Human Pathogen. Nanomaterials. 2016 Apr 14;6(4):71. https://doi.org/10.3390/nano6040071

[20] Bai X, Wang Y, Song Z, Feng Y, Chen Y, Zhang D, et al. The Basic Properties of Gold Nanoparticles and their Applications in Tumor Diagnosis and Treatment. Int J Mol Sci. 2020 Apr 3;21(7):2480. https://doi.org/10.3390/ijms21072480

[21] Khan I, Saeed K, Khan I. Nanoparticles: Properties, applications and toxicities. Arab J Chem. 2019 Nov;12(7):908-31. https://doi.org/10.1016/j.arabjc.2017.05.011

[22] Hammami I, Alabdallah NM, jomaa A Al, kamoun M. Gold nanoparticles: Synthesis properties and applications. J King Saud Univ - Sci. 2021 Oct 1;33(7):101560. https://doi.org/10.1016/j.jksus.2021.101560

[23] Hassan AI, Saleh HM. Principles of Green Chemistry. Mater Horizons From Nat to Nanomater [Internet]. 2021 [cited 2024 Jun 22];15-32. Available from: https://link.springer.com/chapter/10.1007/978-981-33-6897-2_2 https://doi.org/10.1007/978-981-33-6897-2_2

[24] de Marco BA, Rechelo BS, Tótoli EG, Kogawa AC, Salgado HRN. Evolution of green chemistry and its multidimensional impacts: A review. Saudi Pharm J. 2019 Jan 1;27(1):1-8. https://doi.org/10.1016/j.jsps.2018.07.011

[25] Logeswari P, Silambarasan S, Abraham J. Ecofriendly synthesis of silver nanoparticles from commercially available plant powders and their antibacterial properties. Sci Iran. 2013 Jun 1;20(3):1049-54.

[26] Khan F, Shariq M, Asif M, Siddiqui MA, Malan P, Ahmad F. Green Nanotechnology: Plant-Mediated Nanoparticle Synthesis and Application. Nanomaterials. 2022. https://doi.org/10.3390/nano12040673

[27] Amina SJ, Guo B. <p>A Review on the Synthesis and Functionalization of Gold Nanoparticles as a Drug Delivery Vehicle</p>. Int J Nanomedicine. 2020 Dec;Volume 15:9823-57. https://doi.org/10.2147/IJN.S279094

[28] Lee KX, Shameli K, Yew YP, Teow SY, Jahangirian H, Rafiee-Moghaddam R, et al. Recent developments in the facile bio-synthesis of gold nanoparticles (AuNPs) and their biomedical applications. Int J Nanomedicine. 2020;15:275-300. https://doi.org/10.2147/IJN.S233789

[29] Lee KX, Shameli K, Miyake M, Ahmad Khairudin NBB, Mohamad SEB, Hara H, et al. Gold Nanoparticles Biosynthesis: A Simple Route for Control Size Using Waste Peel Extract. IEEE Trans Nanotechnol. 2017 Nov;16(6):954-7. https://doi.org/10.1109/TNANO.2017.2728600

[30] Islam NU, Jalil K, Shahid M, Rauf A, Muhammad N, Khan A, et al. Green synthesis and biological activities of gold nanoparticles functionalized with Salix alba. Arab J Chem. 2019 Dec;12(8):2914-25. https://doi.org/10.1016/j.arabjc.2015.06.025

[31] Chen J, Li Y, Fang G, Cao Z, Shang Y, Alfarraj S, et al. Green synthesis, characterization, cytotoxicity, antioxidant, and anti-human ovarian cancer activities of Curcumae kwangsiensis

leaf aqueous extract green-synthesized gold nanoparticles. Arab J Chem. 2021 Mar;14(3):103000. https://doi.org/10.1016/j.arabjc.2021.103000

[32] ElMitwalli OS, Barakat OA, Daoud RM, Akhtar S, Henari FZ. Green synthesis of gold nanoparticles using cinnamon bark extract, characterization, and fluorescence activity in Au/eosin Y assemblies. J Nanoparticle Res [Internet]. 2020 Oct 1 [cited 2024 Jun 24];22(10):1-9. Available from: https://link.springer.com/article/10.1007/s11051-020-04983-8 https://doi.org/10.1007/s11051-020-04983-8

[33] Stozhko NY, Bukharinova MA, Khamzina EI, Tarasov A V., Vidrevich MB, Brainina KZ. The Effect of the Antioxidant Activity of Plant Extracts on the Properties of Gold Nanoparticles. Nanomaterials. 2019 Nov 21;9(12):1655. https://doi.org/10.3390/nano9121655

[34] Karuppiah C, Palanisamy S, Chen SM, Emmanuel R, Muthupandi K, Prakash P. Green synthesis of gold nanoparticles and its application for the trace level determination of painter's colic. RSC Adv. 2015;5(21):16284-91. https://doi.org/10.1039/C4RA14988B

[35] Ganesh Kumar V, Dinesh Gokavarapu S, Rajeswari A, Stalin Dhas T, Karthick V, Kapadia Z, et al. Facile green synthesis of gold nanoparticles using leaf extract of antidiabetic potent Cassia auriculata. Colloids Surfaces B Biointerfaces. 2011 Oct;87(1):159-63. https://doi.org/10.1016/j.colsurfb.2011.05.016

[36] Khalil MMH, Ismail EH, El-Magdoub F. Biosynthesis of Au nanoparticles using olive leaf extract: 1st Nano Updates. Arab J Chem. 2012 Oct 1;5(4):431-7. https://doi.org/10.1016/j.arabjc.2010.11.011

[37] Philip D. Rapid green synthesis of spherical gold nanoparticles using Mangifera indica leaf. Spectrochim Acta Part A Mol Biomol Spectrosc. 2010 Nov;77(4):807-10. https://doi.org/10.1016/j.saa.2010.08.008

[38] Yu J, Xu D, Guan HN, Wang C, Huang LK, Chi DF. Facile one-step green synthesis of gold nanoparticles using Citrus maxima aqueous extracts and its catalytic activity. Mater Lett. 2016 Mar;166:110-2. https://doi.org/10.1016/j.matlet.2015.12.031

[39] Nagajyothi PC, Lee SE, An M, Lee KD. Green Synthesis of Silver and Gold Nanoparticles Using Lonicera Japonica Flower Extract. Bull Korean Chem Soc. 2012 Aug 20;33(8):2609-12. https://doi.org/10.5012/bkcs.2012.33.8.2609

[40] Yang N, WeiHong L, Hao L. Biosynthesis of Au nanoparticles using agricultural waste mango peel extract and its in vitro cytotoxic effect on two normal cells. Mater Lett. 2014 Nov;134:67-70. https://doi.org/10.1016/j.matlet.2014.07.025

[41] Paul B, Bhuyan B, Dhar Purkayastha D, Dey M, Dhar SS. Green synthesis of gold nanoparticles using Pogestemon benghalensis (B) O. Ktz. leaf extract and studies of their photocatalytic activity in degradation of methylene blue. Mater Lett. 2015 Jun;148:37-40. https://doi.org/10.1016/j.matlet.2015.02.054

[42] Oueslati MH, Ben Tahar L, Harrath AH. Synthesis of ultra-small gold nanoparticles by polyphenol extracted from Salvia officinalis and efficiency for catalytic reduction of p-nitrophenol and methylene blue. Green Chem Lett Rev. 2020 Jan 2;13(1):18-26. https://doi.org/10.1080/17518253.2019.1711202

[43] Akintelu SA, Yao B, Folorunso AS. Bioremediation and pharmacological applications of gold nanoparticles synthesized from plant materials. Heliyon. 2021 Mar;7(3):e06591. https://doi.org/10.1016/j.heliyon.2021.e06591

[44] Nadaf NY, Kanase SS. Biosynthesis of gold nanoparticles by Bacillus marisflavi and its potential in catalytic dye degradation. Arab J Chem. 2019 Dec;12(8):4806-14. https://doi.org/10.1016/j.arabjc.2016.09.020

[45] Kunoh T, Takeda M, Matsumoto S, Suzuki I, Takano M, Kunoh H, et al. Green Synthesis of Gold Nanoparticles Coupled with Nucleic Acid Oxidation. ACS Sustain Chem Eng. 2018 Jan 2;6(1):364-73. https://doi.org/10.1021/acssuschemeng.7b02610

[46] Lim K, Macazo FC, Scholes C, Chen H, Sumampong K, Minteer SD. Elucidating the Mechanism behind the Bionanomanufacturing of Gold Nanoparticles Using Bacillus subtilis. ACS Appl Bio Mater. 2020 Jun 15;3(6):3859-67. https://doi.org/10.1021/acsabm.0c00420

[47] Manjunath HM, Joshi CG, Raju NG. Biofabrication of gold nanoparticles using marine endophytic fungus - Penicillium citrinum. IET Nanobiotechnology. 2017 Feb 14;11(1):40-4. https://doi.org/10.1049/iet-nbt.2016.0065

[48] Qu Y, Pei X, Shen W, Zhang X, Wang J, Zhang Z, et al. Biosynthesis of gold nanoparticles by Aspergillum sp. WL-Au for degradation of aromatic pollutants. Phys E Low-dimensional Syst Nanostructures. 2017 Apr;88:133-41. https://doi.org/10.1016/j.physe.2017.01.010

[49] Barabadi H, Honary S, Ali Mohammadi M, Ahmadpour E, Rahimi MT, Alizadeh A, et al. Green chemical synthesis of gold nanoparticles by using Penicillium aculeatum and their scolicidal activity against hydatid cyst protoscolices of Echinococcus granulosus. Environ Sci Pollut Res. 2017 Feb 4;24(6):5800-10. https://doi.org/10.1007/s11356-016-8291-8

[50] Sampath S, Madhavan Y, Muralidharan M, Sunderam V, Lawrance AV, Muthupandian S. A review on algal mediated synthesis of metal and metal oxide nanoparticles and their emerging biomedical potential. J Biotechnol. 2022 Dec;360:92-109. https://doi.org/10.1016/j.jbiotec.2022.10.009

[51] Radulescu DM, Surdu VA, Ficai A, Ficai D, Grumezescu AM, Andronescu E. Green Synthesis of Metal and Metal Oxide Nanoparticles: A Review of the Principles and Biomedical Applications. Int J Mol Sci. 2023 Oct 20;24(20):15397. https://doi.org/10.3390/ijms242015397

[52] Pourali P, Badiee SH, Manafi S, Noorani T, Rezaei A, Yahyaei B. Biosynthesis of gold nanoparticles by two bacterial and fungal strains, Bacillus cereus and Fusarium oxysporum, and assessment and comparison of their nanotoxicity in vitro by direct and indirect assays. Electron J Biotechnol. 2017 Sep;29:86-93. https://doi.org/10.1016/j.ejbt.2017.07.005

[53] Narayanan KB, Sakthivel N. Biological synthesis of metal nanoparticles by microbes. Vol. 156, Advances in Colloid and Interface Science. 2010. p. 1-13. https://doi.org/10.1016/j.cis.2010.02.001

[54] Markus J, Mathiyalagan R, Kim YJ, Abbai R, Singh P, Ahn S, et al. Intracellular synthesis of gold nanoparticles with antioxidant activity by probiotic Lactobacillus kimchicus DCY51 T isolated from Korean kimchi. Enzyme Microb Technol. 2016 Dec;95:85-93. https://doi.org/10.1016/j.enzmictec.2016.08.018

[55] Shen W, Qu Y, Pei X, Li S, You S, Wang J, et al. Catalytic reduction of 4-nitrophenol using gold nanoparticles biosynthesized by cell-free extracts of Aspergillus sp. WL-Au. J Hazard Mater. 2017 Jan;321:299-306. https://doi.org/10.1016/j.jhazmat.2016.07.051

[56] Lee KD, Nagajyothi PC, Sreekanth TVM, Park S. Eco-friendly synthesis of gold nanoparticles (AuNPs) using Inonotus obliquus and their antibacterial, antioxidant and cytotoxic activities. J Ind Eng Chem. 2015 Jun;26:67-72. https://doi.org/10.1016/j.jiec.2014.11.016

[57] Narayanan KB, Sakthivel N. Facile green synthesis of gold nanostructures by NADPH-dependent enzyme from the extract of Sclerotium rolfsii. Colloids Surfaces A Physicochem Eng Asp. 2011 May;380(1-3):156-61. https://doi.org/10.1016/j.colsurfa.2011.02.042

[58] Barabadi H, Honary S, Ebrahimi P, Mohammadi MA, Alizadeh A, Naghibi F. Microbial mediated preparation, characterization and optimization of gold nanoparticles. Brazilian J Microbiol. 2014 Dec;45(4):1493-501. https://doi.org/10.1590/S1517-83822014000400046

[59] Roy S, Das TK, Maiti GP, Basu U. Microbial biosynthesis of nontoxic gold nanoparticles. Mater Sci Eng B. 2016 Jan;203:41-51. https://doi.org/10.1016/j.mseb.2015.10.008

[60] Abdel-Raouf N, Al-Enazi NM, Ibraheem IBM. Green biosynthesis of gold nanoparticles using Galaxaura elongata and characterization of their antibacterial activity. Arab J Chem. 2017 May;10:S3029-39. https://doi.org/10.1016/j.arabjc.2013.11.044

[61] Senapati S, Syed A, Moeez S, Kumar A, Ahmad A. Intracellular synthesis of gold nanoparticles using alga Tetraselmis kochinensis. Mater Lett. 2012 Jul;79:116-8. https://doi.org/10.1016/j.matlet.2012.04.009

[62] Venkatesan J, Manivasagan P, Kim SK, Kirthi AV, Marimuthu S, Rahuman AA. Marine algae-mediated synthesis of gold nanoparticles using a novel Ecklonia cava. Bioprocess Biosyst Eng. 2014 Aug 14;37(8):1591-7. https://doi.org/10.1007/s00449-014-1131-7

[63] Singaravelu G, Arockiamary JS, Kumar VG, Govindaraju K. A novel extracellular synthesis of monodisperse gold nanoparticles using marine alga, Sargassum wightii Greville. Colloids Surfaces B Biointerfaces. 2007 May;57(1):97-101. https://doi.org/10.1016/j.colsurfb.2007.01.010

[64] Annamalai J, Nallamuthu T. Characterization of biosynthesized gold nanoparticles from aqueous extract of Chlorella vulgaris and their anti-pathogenic properties. Appl Nanosci. 2015 Jun 12;5(5):603-7. https://doi.org/10.1007/s13204-014-0353-y

[65] Hassaan MA. Green Synthesis of Ag and Au Nanoparticles from Micro and Macro Algae - Review. Int J Atmos Ocean Sci [Internet]. 2018;2(1):10. Available from: https://www.researchgate.net/profile/Mohamed-Hassaan-2/publication/330317044_Green_Synthesis_of_Ag_and_Au_Nanoparticles_from_Micro_and__Macro_Algae_-_Review/links/5c5ec90292851c48a9c4eae7/Green-Synthesis-of-Ag-and-Au-Nanoparticles-from-Micro-and-Macro-Alga https://doi.org/10.11648/j.ijaos.20180201.12

[66] Ijaz I, Gilani E, Nazir A, Bukhari A. Detail review on chemical, physical and green synthesis, classification, characterizations and applications of nanoparticles [Internet]. Vol. 13, Green Chemistry Letters and Reviews. Taylor and Francis Ltd.; 2020. p. 59-81. Available

from: https://www.tandfonline.com/doi/abs/10.1080/17518253.2020.1802517
https://doi.org/10.1080/17518253.2020.1802517

[67] Fang X, Wang Y, Wang Z, Jiang Z, Dong M. Microorganism assisted synthesized nanoparticles for catalytic applications. Energies. 2019;12(1).
https://doi.org/10.3390/en12010190

[68] Abdelghany TM, Al-Rajhi AMH, Abboud MA Al, Alawlaqi MM, Magdah AG, Helmy EAM, et al. Recent Advances in Green Synthesis of Silver Nanoparticles and Their Applications: About Future Directions. A Review [Internet]. Vol. 8, BioNanoScience. Springer New York LLC; 2018. p. 5-16. Available from: https://link.springer.com/article/10.1007/s12668-017-0413-3 https://doi.org/10.1007/s12668-017-0413-3

[69] Olobayotan I, Akin-Osanaiye B. Biosynthesis of silver nanoparticles using baker's yeast, Saccharomyces cerevisiae and its antibacterial activities. Access Microbiol. 2019;1(1A).
https://doi.org/10.1099/acmi.ac2019.po0316

[70] Thangavelu RM, Ganapathy R, Ramasamy P, Krishnan K. Fabrication of virus metal hybrid nanomaterials: An ideal reference for bio semiconductor. Arab J Chem. 2020;13(1):2750-65.
https://doi.org/10.1016/j.arabjc.2018.07.006

[71] Bhuyar P, Rahim MHA, Sundararaju S, Ramaraj R, Maniam GP, Govindan N. Synthesis of silver nanoparticles using marine macroalgae Padina sp. and its antibacterial activity towards pathogenic bacteria. Beni-Suef Univ J Basic Appl Sci. 2020;9(1).
https://doi.org/10.1186/s43088-019-0031-y

[72] Eid AM, Fouda A, Niedbała G, Hassan SED, Salem SS, Abdo AM, et al. Endophytic streptomyces laurentii mediated green synthesis of Ag-NPs with antibacterial and anticancer properties for developing functional textile fabric properties. Antibiotics [Internet]. 2020;9(10):1-18. Available from: https://www.mdpi.com/2079-6382/9/10/641/htm https://doi.org/10.3390/antibiotics9100641

[73] Hamedi S, Shojaosadati SA. Rapid and green synthesis of silver nanoparticles using Diospyros lotus extract: Evaluation of their biological and catalytic activities. Polyhedron. 2019;171:172-80. https://doi.org/10.1016/j.poly.2019.07.010

[74] Jain N, Jain P, Rajput D, Patil UK. Green synthesized plant-based silver nanoparticles: therapeutic prospective for anticancer and antiviral activity. Vol. 9, Micro and Nano Systems Letters. Society of Micro and Nano Systems; 2021. https://doi.org/10.1186/s40486-021-00131-6

[75] Khan SA, Shahid S, Lee CS. Green synthesis of gold and silver nanoparticles using leaf extract of clerodendrum inerme; characterization, antimicrobial, and antioxidant activities. Biomolecules [Internet]. 2020;10(6). Available from: https://www.mdpi.com/2218-273X/10/6/835 https://doi.org/10.3390/biom10060835

[76] Zaeem A, Drouet S, Anjum S, Khurshid R, Younas M, Blondeau JP, et al. Effects of biogenic zinc oxide nanoparticles on growth and oxidative stress response in flax seedlings vs. In vitro cultures: A comparative analysis. Biomolecules [Internet]. 2020;10(6):1-16. Available from: https://www.mdpi.com/2218-273X/10/6/918 https://doi.org/10.3390/biom10060918

[77] Srihasam S, Thyagarajan K, Korivi M, Lebaka VR, Mallem SPR. Phytogenic generation of NiO nanoparticles using stevia leaf extract and evaluation of their in-vitro antioxidant and antimicrobial properties. Biomolecules [Internet]. 2020;10(1). Available from: https://www.mdpi.com/2218-273X/10/1/89 https://doi.org/10.3390/biom10010089

[78] Noah N. Green synthesis: Characterization and application of silver and gold nanoparticles. In: Green Synthesis, Characterization and Applications of Nanoparticles. Elsevier; 2018. p. 111-35. https://doi.org/10.1016/B978-0-08-102579-6.00006-X

[79] Mickymaray S. One-step synthesis of silver nanoparticles using saudi arabian desert seasonal plant Sisymbrium irio and antibacterial activity against multidrug-resistant bacterial strains. Biomolecules [Internet]. 2019;9(11). Available from: https://www.mdpi.com/2218-273X/9/11/662 https://doi.org/10.3390/biom9110662

[80] Khan SA, Shahid S, Shahid B, Fatima U, Abbasi SA. Green synthesis of MNO nanoparticles using abutilon indicum leaf extract for biological, photocatalytic, and adsorption activities. Biomolecules [Internet]. 2020;10(5). Available from: https://www.mdpi.com/2218-273X/10/5/785 https://doi.org/10.3390/biom10050785

[81] Alshehri AA, Malik MA. Phytomediated photo-induced green synthesis of silver nanoparticles using Matricaria chamomilla L. and its catalytic activity against rhodamine B. Biomolecules [Internet]. 2020;10(12):1-24. Available from: https://www.mdpi.com/2218-273X/10/12/1604 https://doi.org/10.3390/biom10121604

[82] Singh R, Hano C, Nath G, Sharma B. Green biosynthesis of silver nanoparticles using leaf extract of carissa carandas l. And their antioxidant and antimicrobial activity against human pathogenic bacteria. Biomolecules [Internet]. 2021;11(2):1-11. Available from: https://www.mdpi.com/2218-273X/11/2/299 https://doi.org/10.3390/biom11020299

[83] Ahmad H, Venugopal K, Rajagopal K, Britto S De, Nandini B, Pushpalatha HG, et al. Green synthesis and characterization of zinc oxide nanoparticles using eucalyptus globules and their fungicidal ability against pathogenic fungi of apple orchards. Biomolecules [Internet]. 2020;10(3). Available from: https://www.mdpi.com/2218-273X/10/3/425 https://doi.org/10.3390/biom10030425

[84] Perveen K, Husain FM, Qais FA, Khan A, Razak S, Afsar T, et al. Microwave-assisted rapid green synthesis of gold nanoparticles using seed extract of trachyspermum ammi: Ros mediated biofilm inhibition and anticancer activity. Biomolecules. 2021;11(2):1-16. https://doi.org/10.3390/biom11020197

[85] Ansari MA, Murali M, Prasad D, Alzohairy MA, Almatroudi A, Alomary MN, et al. Cinnamomum verum bark extract mediated green synthesis of ZnO nanoparticles and their antibacterial potentiality. Biomolecules [Internet]. 2020;10(2). Available from: https://www.mdpi.com/2218-273X/10/2/336 https://doi.org/10.3390/biom10020336

[86] Viana RLS, Fidelis GP, Medeiros MJC, Morgano MA, Alves MGCF, Passero LFD, et al. Green synthesis of antileishmanial and antifungal silver nanoparticles using corn cob xylan as a reducing and stabilizing agent. Biomolecules [Internet]. 2020;10(9):1-21. Available from: https://www.mdpi.com/2218-273X/10/9/1235 https://doi.org/10.3390/biom10091235

[87] Prasad KS, Prasad SK, Ansari MA, Alzohairy MA, Alomary MN, Alyahya S, et al. Tumoricidal and bactericidal properties of znonps synthesized using cassia auriculata leaf

extract. Biomolecules [Internet]. 2020;10(7):1-14. Available from: https://www.mdpi.com/2218-273X/10/7/982 https://doi.org/10.3390/biom10070982

[88] Geraldes AN, da Silva AA, Leal J, Estrada-Villegas GM, Lincopan N, Katti K V, et al. Green Nanotechnology from Plant Extracts: Synthesis and Characterization of Gold Nanoparticles. Adv Nanoparticles [Internet]. 2016;05(03):176-85. Available from: http://info.submit4journal.com/id/eprint/1905/ https://doi.org/10.4236/anp.2016.53019

[89] Rajeshkumar S, Malarkodi C, Gnanajobitha G, Paulkumar K, Vanaja M, Kannan C, et al. Seaweed-mediated synthesis of gold nanoparticles using Turbinaria conoides and its characterization. J Nanostructure Chem. 2013;3(1). https://doi.org/10.1186/2193-8865-3-44

[90] Moosavy MH, de la Guardia M, Mokhtarzadeh A, Khatibi SA, Hosseinzadeh N, Hajipour N. Green synthesis, characterization, and biological evaluation of gold and silver nanoparticles using Mentha spicata essential oil. Sci Rep [Internet]. 2023;13(1). Available from: https://www.nature.com/articles/s41598-023-33632-y https://doi.org/10.1038/s41598-023-33632-y

[91] Clarance P, Luvankar B, Sales J, Khusro A, Agastian P, Tack JC, et al. Green synthesis and characterization of gold nanoparticles using endophytic fungi Fusarium solani and its in-vitro anticancer and biomedical applications. Saudi J Biol Sci. 2020;27(2):706-12. https://doi.org/10.1016/j.sjbs.2019.12.026

[92] Elavazhagan T, Arunachalam KD. Memecylon edule leaf extract mediated green synthesis of silver and gold nanoparticles. Int J Nanomedicine [Internet]. 2011;6:1265-78. Available from: https://www.tandfonline.com/doi/abs/10.2147/IJN.S18347 https://doi.org/10.2147/IJN.S18347

[93] Jayaseelan C, Ramkumar R, Rahuman AA, Perumal P. Green synthesis of gold nanoparticles using seed aqueous extract of Abelmoschus esculentus and its antifungal activity. Ind Crops Prod [Internet]. 2013;45:423-9. Available from: https://www.sciencedirect.com/science/article/pii/S092666901200653X https://doi.org/10.1016/j.indcrop.2012.12.019

[94] Mourdikoudis S, Pallares RM, Thanh NTK. Characterization techniques for nanoparticles: comparison and complementarity upon studying nanoparticle properties. Nanoscale. 2018;10(27):12871-934. https://doi.org/10.1039/C8NR02278J

[95] Awad MA, Eisa NE, Virk P, Hendi AA, Ortashi KMOO, Mahgoub AASA, et al. Green synthesis of gold nanoparticles: Preparation, characterization, cytotoxicity, and anti-bacterial activities. Mater Lett. 2019;256. https://doi.org/10.1016/j.matlet.2019.126608

[96] Rajeshkumar S, Bharath L V. Mechanism of plant-mediated synthesis of silver nanoparticles - A review on biomolecules involved, characterisation and antibacterial activity [Internet]. Vol. 273, Chemico-Biological Interactions. 2017. p. 219-27. Available from: https://www.sciencedirect.com/science/article/pii/S0009279717304799 https://doi.org/10.1016/j.cbi.2017.06.019

[97] Smith DJ. Characterization of Nanomaterials Using Transmission Electron Microscopy. In: Nanocharacterisation. The Royal Society of Chemistry; 2015. p. 1-29. https://doi.org/10.1039/9781782621867-00001

[98] Abdel-Raouf N, Al-Enazi NM, Ibraheem IBM. Green biosynthesis of gold nanoparticles using Galaxaura elongata and characterization of their antibacterial activity. Arab J Chem. 2017; https://doi.org/10.1016/j.arabjc.2013.11.044

[99] Nadhe SB, Wadhwani SA, Singh R, Chopade BA. Green Synthesis of AuNPs by Acinetobacter sp. GWRVA25: Optimization, Characterization, and Its Antioxidant Activity. Front Chem. 2020; https://doi.org/10.3389/fchem.2020.00474

[100] Jamil S, Dastagir G, Foudah AI, Alqarni MH, Yusufoglu HS, Alkreathy HM, et al. Carduus edelbergii Rech. f. Mediated Fabrication of Gold Nanoparticles; Characterization and Evaluation of Antimicrobial, Antioxidant and Antidiabetic Potency of the Synthesized AuNPs. Molecules. 2022 Oct 7;27(19). https://doi.org/10.3390/molecules27196669

[101] Dobrucka R, Dlugaszewska J, Kaczmarek M. Antimicrobial and cytostatic activity of biosynthesized nanogold prepared using fruit extract of Ribes nigrum. Arab J Chem. 2019;12(8):3902-10. https://doi.org/10.1016/j.arabjc.2016.02.009

[102] MR KP. Applications of the Green Synthesized Gold Nanoparticles-Antimicrobial Activity, Water Purification System and Drug Delivery System. Nanosci Technol Open Access [Internet]. 2015;2(2):1-4. Available from: https://pdfs.semanticscholar.org/538a/37615f28939d41021402b53c95a4e27095ad.pdf https://doi.org/10.15226/2374-8141/2/2/00126

[103] Wang C, Mathiyalagan R, Kim YJ, Castro-Aceituno V, Singh P, Ahn S, et al. Rapid green synthesis of silver and gold nanoparticles using Dendropanax morbifera leaf extract and their anticancer activities. Int J Nanomedicine [Internet]. 2016;11:3691-701. Available from: https://www.tandfonline.com/doi/abs/10.2147/IJN.S97181 https://doi.org/10.2147/IJN.S97181

[104] Patra S, Mukherjee S, Barui AK, Ganguly A, Sreedhar B, Patra CR. Green synthesis, characterization of gold and silver nanoparticles and their potential application for cancer therapeutics. Mater Sci Eng C [Internet]. 2015;53:298-309. Available from: https://www.sciencedirect.com/science/article/pii/S0928493115300497 https://doi.org/10.1016/j.msec.2015.04.048

[105] Geetha R, Ashokkumar T, Tamilselvan S, Govindaraju K, Sadiq M, Singaravelu G. Green synthesis of gold nanoparticles and their anticancer activity. Cancer Nanotechnol. 2013;4(4-5):91-8. https://doi.org/10.1007/s12645-013-0040-9

[106] Santhosh PB, Genova J, Chamati H. Review Green Synthesis of Gold Nanoparticles: An Eco-Friendly Approach [Internet]. Vol. 4, Chemistry (Switzerland). Multidisciplinary Digital Publishing Institute; 2022. p. 345-69. Available from: https://www.mdpi.com/2624-8549/4/2/26/htm https://doi.org/10.3390/chemistry4020026

[107] Ibrahim NA, Eid BM, Abdel-Aziz MS. Green synthesis of AuNPs for eco-friendly functionalization of cellulosic substrates. Appl Surf Sci [Internet]. 2016;389:118-25. Available from: https://www.sciencedirect.com/science/article/pii/S0169433216315136 https://doi.org/10.1016/j.apsusc.2016.07.077

[108] Ganesan RM, Prabu HG. Synthesis of gold nanoparticles using herbal Acorus calamus rhizome extract and coating on cotton fabric for antibacterial and UV blocking applications. Arab J Chem [Internet]. 2019;12(8):2166-74. Available from:

https://www.sciencedirect.com/science/article/pii/S1878535214003682
https://doi.org/10.1016/j.arabjc.2014.12.017

[109] Velmurugan P, Shim J, Bang KS, Oh BT. Gold nanoparticles mediated coloring of fabrics and leather for antibacterial activity. J Photochem Photobiol B Biol [Internet]. 2016;160:102-9. Available from: https://www.sciencedirect.com/science/article/pii/S1011134416300823 https://doi.org/10.1016/j.jphotobiol.2016.03.051

[110] Dang H, Fawcett D, Poinern GEJ. Green synthesis of gold nanoparticles from waste macadamia nut shells and their antimicrobial activity against Escherichia coli and Staphylococcus epidermis. Int J Res Med Sci [Internet]. 2019;7(4):1171. Available from: https://researchportal.murdoch.edu.au/view/pdfCoverPage?instCode=61MUN_INST&filePid =1313701803000789 1&download=true https://doi.org/10.18203/2320-6012.ijrms20191320

[111] Jiang M, Li S, Ming P, Guo Y, Yuan L, Jiang X, et al. Rational design of porous structure-based sodium alginate/chitosan sponges loaded with green synthesized hybrid antibacterial agents for infected wound healing. Int J Biol Macromol. 2023;237. https://doi.org/10.1016/j.ijbiomac.2023.123944

[112] Awad NS, Salkho NM, Abuwatfa WH, Paul V, AlSawaftah NM, Husseini GA. Tumor vasculature vs tumor cell targeting: Understanding the latest trends in using functional nanoparticles for cancer treatment. OpenNano. 2023;11. https://doi.org/10.1016/j.onano.2023.100136

[113] Ma X, Lee C, Zhang T, Cai J, Wang H, Jiang F, et al. Correction to: Image-guided selection of Gd@C-dots as sensitizers to improve radiotherapy of non-small cell lung cancer (Journal of Nanobiotechnology, (2021), 19, 1, (284), 10.1186/s12951-021-01018-9). Vol. 20, Journal of Nanobiotechnology. BioMed Central Ltd; 2022. https://doi.org/10.1186/s12951-021-01184-w

[114] Botteon CEA, Silva LB, Ccana-Ccapatinta G V, Silva TS, Ambrosio SR, Veneziani RCS, et al. Biosynthesis and characterization of gold nanoparticles using Brazilian red propolis and evaluation of its antimicrobial and anticancer activities. Sci Rep. 2021;11(1). https://doi.org/10.1038/s41598-021-81281-w

[115] Doghish AS, Hashem AH, Shehabeldine AM, Sallam AAM, El-Sayyad GS, Salem SS. Nanocomposite based on gold nanoparticles and carboxymethyl cellulose: Synthesis, characterization, antimicrobial, and anticancer activities. J Drug Deliv Sci Technol. 2022;77. https://doi.org/10.1016/j.jddst.2022.103874

[116] Pang YX, Li X, Zhang X, Yeoh JX, Wong C, Manickam S, et al. The synthesis of carbon-based quantum dots: A supercritical fluid approach and perspective. Vol. 27, Materials Today Physics. Elsevier Ltd; 2022. https://doi.org/10.1016/j.mtphys.2022.100752

[117] Malik N, Arfin T, Khan AU. Graphene nanomaterials: Chemistry and pharmaceutical perspectives. In: Nanomaterials for Drug Delivery and Therapy. Elsevier; 2019. p. 373-402. https://doi.org/10.1016/B978-0-12-816505-8.00002-3

[118] Si P, Razmi N, Nur O, Solanki S, Pandey CM, Gupta RK, et al. Gold nanomaterials for optical biosensing and bioimaging. Nanoscale Adv. 2021 May;3(10):2679-98. https://doi.org/10.1039/D0NA00961J

[119] Sarfraz N, Khan I. Plasmonic Gold Nanoparticles (AuNPs): Properties, Synthesis and their Advanced Energy, Environmental and Biomedical Applications. Chem - An Asian J. 2021 Apr;16(7):720-42. https://doi.org/10.1002/asia.202001202

[120] Wu Y, Ali MRK, Chen K, Fang N, El-Sayed MA. Gold nanoparticles in biological optical imaging. Nano Today. 2019 Feb;24:120-40. https://doi.org/10.1016/j.nantod.2018.12.006

[121] Sargazi S, Laraib U, Er S, Rahdar A, Hassanisaadi M, Zafar MN, et al. Application of Green Gold Nanoparticles in Cancer Therapy and Diagnosis. Vol. 12, Nanomaterials. MDPI; 2022. https://doi.org/10.3390/nano12071102

[122] Bharadwaj KK, Rabha B, Pati S, Sarkar T, Choudhury BK, Barman A, et al. Green synthesis of gold nanoparticles using plant extracts as beneficial prospect for cancer theranostics. Vol. 26, Molecules. MDPI; 2021. https://doi.org/10.3390/molecules26216389

[123] Kotcherlakota R, Nimushakavi S, Roy A, Yadavalli HC, Mukherjee S, Haque S, et al. Biosynthesized Gold Nanoparticles: In Vivo Study of Near-Infrared Fluorescence (NIR)-Based Bio-imaging and Cell Labeling Applications. ACS Biomater Sci Eng. 2019 Oct;5(10):5439-52. https://doi.org/10.1021/acsbiomaterials.9b00721

[124] Ateeq M, Shah MR, ul Ain N, Bano S, Anis I, Lubna, et al. Green synthesis and molecular recognition ability of patuletin coated gold nanoparticles. Biosens Bioelectron [Internet]. 2015;63:499-505. Available from: https://www.sciencedirect.com/science/article/pii/S0956566314005776 https://doi.org/10.1016/j.bios.2014.07.076

[125] Karthik R, Chen SM, Elangovan A, Muthukrishnan P, Shanmugam R, Lou BS. Phyto mediated biogenic synthesis of gold nanoparticles using Cerasus serrulata and its utility in detecting hydrazine, microbial activity and DFT studies. J Colloid Interface Sci [Internet]. 2016;468:163-75. Available from: https://www.sciencedirect.com/science/article/pii/S0021979716300467 https://doi.org/10.1016/j.jcis.2016.01.046

[126] Hegedus LL, Mccabe RW. Catalyst Poisoning. Catal Rev [Internet]. 1981;23(3):377-476. Available from: https://books.google.com/books/about/Catalyst_Poisoning.html?id=RbVTAAAAMAAJ https://doi.org/10.1080/03602458108079641

[127] Qu Y, Shen W, Pei X, Ma F, You S, Li S, et al. Biosynthesis of gold nanoparticles by Trichoderma sp. WL-Go for azo dyes decolorization. J Environ Sci (China) [Internet]. 2017;56:79-86. Available from: https://www.sciencedirect.com/science/article/pii/S1001074216307628 https://doi.org/10.1016/j.jes.2016.09.007

[128] Chairam S, Konkamdee W, Parakhun R. Starch-supported gold nanoparticles and their use in 4-nitrophenol reduction Starch-supported gold nanoparticles in 4-nitrophenol reduction. J Saudi Chem Soc. 2017;21(6):656-63. https://doi.org/10.1016/j.jscs.2015.11.001

[129] Ying S, Guan Z, Ofoegbu PC, Clubb P, Rico C, He F, Hong J. Green synthesis of nanoparticles: Current developments and limitations. Environ. Technol. Innov. 2022 May;26:102336. https://doi.org/10.1016/j.eti.2022.102336

About the Editors

Martin F. Desimone

Dr. Martín F. Desimone is Professor in the Department of Chemical Sciences in the Faculty of Pharmacy and Biochemistry of the University of Buenos Aires and holds a Principal Researcher position at CONICET, Argentina. Additionally, he holds a Visiting Professor position at the Universidade Federal do Rio Grande in Brazil. He studied Pharmacy, Biochemistry and received his Ph. D. from the University of Buenos Aires, Argentina. He was a postdoctoral visiting scientist at the University of Basque Country in Spain and held a Maître de conferences position in the Collège de France working in the Laboratoire de Chimie de la Matière Condensée de Paris (UMR 7574 Sorbonne Université, CNRS, Collège de France). He leads a research team working on nanocomposites and hybrid materials combining biomaterials with nanomaterials and biological active molecules for biomedical applications such as wound healing, tissue engineering, drug delivery, 3D printing and stimuli-responsive materials. He has established fruitful collaborations with national and international teams contributing to the work of eleven Ph D and three International Master in Biomedical Sciences (IMBS, Germany-Argentina) students who concluded their work with his direction. He is the recipient of several awards including awards from the National Academy of Pharmacy and Biochemistry, the Award "Innovar" in the category applied research received in two consecutive years (2016 and 2017) from the Ministry of Science, Technology and Productive Innovation (MINCYT, Argentina), the distinction "Dr. José A. Balseiro" XV edition, received in The Argentine Senate, which is the upper house of the Argentine National Congress, International Association of Advanced Materials medal of the year 2018 (Sweden), and the "Academic Excellence" from the University of Buenos Aires in four consecutive years, among others. He is cofounder of the start-up enterprise Hybridon which is producing nanotechnological applications. He has a record of 40 invited talks at national and international conferences. He has co-authored over 135 international scientific publications, 20 book chapters, and 4 patents.

Dr Martín F. DESIMONE
Departamento de Ciencias Químicas.
IQUIMEFA-CONICET
Facultad de Farmacia y Bioquímica, Universidad de Buenos Aires
(1113) Junin 956 Piso 3. Buenos Aires. Argentina.
E-mail: desimone@ffyb.uba.ar / martinfdesimone@gmail.com
Tel: +54-11-52874332

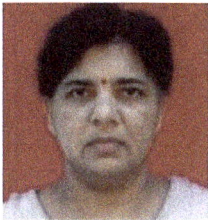

Rajshree B. Jotania

Dr. Rajshree B. Jotania is a professor of Physics, Department of Physics, Electronics and Space science, at Gujarat University, Ahmedabad, India. She was Junior Research Fellow (DAE-BRNS project) during 1987 to 1989 at Physics Department, Saurashtra University, Rajkot, India. She obtained a few regional and national awards for contribution toward scientific research. She worked at National Chemical Laboratory, Pune, India for few months as a Summer Visiting Teacher Fellow in 2005 and as a Visiting Scientist Fellow in 2011. She possesses 34 years of teaching experience at UG and PG level. She is a member of board of studies at few Universities of Gujarat, India and a Mentor of DST-INSPIRE (Department of Science and Technology-Innovation in Science Pursuit for Inspired Research) program. She has published more than 160 papers in various research journals and conference proceedings. She has delivered more than 25 invited talks at various DST-INSPIRE Internship science camp in India. She has edited six books entitled 'Ferrites and ceramic composites' (Vol. I & II, Trans Tech Publisher (TTP), Switzerland), Magnetic Oxides and Ceramic Composites (Vol. I & II, Materials Research Forum, LLC, USA), Magnetic nanoparticles for biomedical applications (Materials Research Forum, LLC, USA) and Green Nanomaterials for Clean and Sustainable Environment (Current Nanoscience, Bentham Science, Netherlands) She visited Singapore (three times), Malaysia, USA (New York), North Africa (Tunisia twice) and Canada for research work. She has attended more than 55 international, national conferences/symposiums/seminars/ academy meeting and worked as a chair person as well as delivered invited talks in a few international and national conferences. She possesses a life membership of eight professional bodies and she has guided seven Ph.D, twelve M. Phil students. At present five research students are working under her guidance for Ph. D. To date she has completed five research projects of various agencies. She has worked as a deputy coordinator, DRS (SAP-I) program of University Grant Commission, India. She is ex-secretary of an Indian Association of Physics Teacher (IAPT), Gujarat chapter (RC-07). Recently she is associated with a few universities for scientific co-operation in nano-magnetic and nanotechnology field. She is a member of board of management at Gujarat university, Ahmedabad, India.

Prof. Rajshree B. Jotania
Department of Physics, Electronics and Space science,
University school of sciences,
Gujarat University,
Ahmedabad, Gujarat, India,
Email: rajshree_jotania@yahoo.co.in

Dr. Ratiram Gomaji Chaudhary

Dr. Ratiram Gomaji Chaudhary is presently working as Associate Professor and Head, Post Graduate Department of Chemistry, Seth Kesarimal Porwal College Kamptee. Recently, he has received a prestigious award 'Best Researcher Award-2023' from RTM Nagpur University, Nagpur, and Shikshan Prasarak Mandal, Kamptee. He is SENATE member of RTM Nagpur University, Nagpur. He is a BoS Member of Chemistry and also worked as Task force Member. Worked as chairperson and member of several committees of RTM Nagpur University. His research areas of interest are biogenic synthesis, phytosynthesis, metal oxide/graphene-based nanohybrids, antimicrobial assay, docking, and photocatalysis. He has completed one Major Research Project funded by SERB/DST. He is a recognized supervisor of Nagpur University, and under his guidance awarded: 06 Ph.D. students, working-06, and MSc Projects completed-75. He has been awarded two times with 'Rajiv Gandhi National Fellowship Award' funded by University Grant Commission (UGC) as JRF for pursuing M.Phil and Ph.D degrees. He has published books-06, book chapters-25, and regular articles-110, and review articles-25 in the peer-reviewed SCI/Scopus indexing journals. He is a recipient of several awards. He is a reviewer of more than 75 peer-reviewed journals of reputed publishers. He has reviewed more 300 articles. He worked as a Guest Editor for Material Today: Proceeding, Current Pharmaceutical Biotechnology, Current Pharmaceutical Design, Current Nanoscience and Jordon Journal of Physics. He worked as a Review Editor for Frontiers in Bioengineering and Biotechnology; Frontiers in Molecular Biosciences; and Frontiers in Materials: Polymeric and Composite Materials etc. Moreover, he has organized several National/State level conferences/International webinars, delivered several Invited/Resource talks/chairs the session in various international scientific events. Also, he is a member of various scientific organizations like ACS, ISCA, ISCAS, ITAS, ITS, ACT etc.

Dr. Ratiram G. Chaudhary
Associate Professor & Head
Post Graduate Department of Chemistry
Seth Kesarimal Porwal College of Arts, Science and Commerce, Kamptee,
RTM Nagpur University, Nagpur, Maharashtra-441001, India
Email: chaudhary_rati@yahoo.com
Personal website: https://bionanomat.org/